T0406730

Neuroimmunology of the Skin

Richard D. Granstein • Thomas A. Luger
Editors

Neuroimmunology of the Skin

Basic Science to Clinical Practice

Richard D. Granstein, MD
Cornell University
Weill Cornell Medical College
Department of Dermatology
1300 York Ave.
New York NY 10021
USA
rdgranst@med.cornell.edu

Thomas A. Luger, MD
Universitätsklinikum Münster
Medizinische Einrichtungen
Klinik und Poliklinik für
Von-Esmarch-Str. 56
48149 Münster
Germany
Luger@uni-muenster.de

ISBN 978-3-540-35986-9 e-ISBN 978-3-540-35989-0

DOI: 10.1007/978-3-540-35989-0

Library of Congress Control Number: 2008928261

Cover design: eStudio Calamar, Spain

Printed on acid-free paper

9 8 7 6 5 4 3 2 1

springer.com

Preface

It has long been noted anecdotally that affect, psychological state and neurologic state have influences on inflammatory skin diseases. Disorders such as psoriasis, atopic dermatitis, acne and rosacea, among many others, are reported to become exacerbated by stress. Furthermore, it is widely believed that stress alters cutaneous immunity. However, mechanisms responsible for these effects have remained incompletely understood. Scientific evidence for an influence of the nervous system on immune and inflammatory processes in the skin has been developed only relatively recently. This area of research has now become intensely active and fruitful. Although neurocutaneous immunology is a young field, it is now accepted that the nervous system plays a major role in regulating immune and inflammatory events within the skin. Data has been obtained demonstrating the influences of neuroendocrine hormones as well as neuropeptides, neurotransmitters, nucleotides and other products of nerves on immune cells and immune processes. Much of the data obtained over the past few years suggests that neurologic influences have implications for immunity and inflammation, not just in the skin, but also in many other organ systems. These findings have important implications for understanding pathology and pathophysiology. Most importantly, they suggest novel new approaches to prevention and treatment of many disorders.

As scientific activities in neurocutaneous immunology have expanded, the need for a comprehensive, up-to-date textbook summarizing the current state of the field became apparent. This book includes sections dealing with the major areas of research ongoing in neuroimmunology. These include basic neuroimmunology of the skin, stress effects in cutaneous immunity, neurobiology of skin appendages and the role of the nervous system in the pathophysiology of skin disorders.

We believe that this book will be useful to investigators studying the effects of the nervous system and psychologic state on the physiology and pathophysiology of the skin. Also, clinicians with an interest in inflammatory skin diseases will find this book to be quite useful.

In addition to finding this book to be a useful scientific and clinical resource, we hope that the reader finds it to be both fascinating and enjoyable.

New York and Münster — R.D. Granstein
2008 — T.A. Luger

Contents

List of Contributors

John C. Ansel
Department of Dermatology
University of Colorado at Denver and Health Sciences Center
Aurora, CO, USA

Cheryl A. Armstrong
Department of Dermatology
University of Colorado at Denver and Health Sciences Center
Aurora, CO, USA

Anna M. Bender
Department of Dermatology
Weill Cornell Medical College
New York, USA
amb2044@columbia.edu

S.S. Biel
Beiersdorf AG
P.O. Box 550, Unnastraße 48
20245 Hamburg, Germany
stefan.biel@beiersdorf.com

Frank Birklein
Department of Neurology
Johannes Gutenberg University
Langenbeckstr. 1
55101 Mainz, Germany
birklein@neurologie.klinik.uni-mainz.de

Markus Böhm
Department of Dermatology
University of Münster
Von-Esmarch-Str. 58
48149 Münster, Germany
bohmm@uni-muenster.de

Thomas Brzoska
Department of Dermatology
University of Münster
Von-Esmarch-Str. 58
48149 Münster, Germany
brzoska@uni-muenster.de

Hongqing Chen
Department of Dermatology
University of Colorado at Denver and Health Sciences Center
Aurora, CO, USA

Firdaus S. Dhabhar
Department of Psychiatry & Behavioral Sciences
Immunology Program,
Neuroscience Institute, & Cancer Center
Stanford University
300 Pasteur Drive, MC 5135
Stanford, CA 94305-5135
fdhabhar@stanford.edu

Christopher G. Engeland
University of Illinois at Chicago
College of Dentistry
Dept. of Periodontics
801 S Paulina St., M/C 859, Room 458
Chicago IL 60612, USA
engeland@uic.edu

John J. Finlay-Jones
Telethon Institute for Child Health Research
Centre for Child Health Research
University of Western Australia
P.O. Box 855
West Perth 6872, Australia
johnfj@ichr.uwa.edu.au

Shelley Gorman
Telethon Institute for Child Health Research
Centre for Child Health Research
University of Western Australia, P.O. Box 855
West Perth 6872, Australia
shelleyg@ichr.uwa.edu.au

Richard D. Granstein
Cornell University
Weill Cornell Medical College
Department of Dermatology
1300 York Ave.
New York NY 10021, USA
rdgranst@med.cornell.edu

Madhulika A. Gupta
Department of Psychiatry
Schulich School of Medicine and Dentistry
University of Western Ontario
London, ON, Canada
magupta@uwo.ca

Prue H. Hart
Telethon Institute for Child Health Research
Centre for Child Health Research
University of Western Australia
PO Box 855
West Perth 6872, Australia
prueh@ichr.uwa.edu.au

Yozo Ishiuji
Beiersdorf AG
P.O. Box 550, Unnastraße 48
20245 Hamburg, Germany

Karin Loser
Department of Dermatology
University of Münster
Von-Esmarch-Str. 58
48149 Münster, Germany
loserk@uni-muenster.de

Yi Lu
Department of Dermatology
University of Colorado at Denver and Health Sciences Center
Aurora, CO, USA

Thomas A. Luger
Universitätsklinikum Münster
Medizinische Einrichtungen
Klinik und Poliklinik für
Von-Esmarch-Str. 56
48149 Münster, Germany
Luger@uni-muenster.de

Georges Maestroni
Istituto Cantonale di Patologia
Center for Experimental Pathology
P.O. Box 6601
Locarno, Switzerland
georges.maestroni@ti.ch

A. Martin
Beiersdorf AG
P.O. Box 550, Unnastraße 48
20245 Hamburg, Germany

Phillip T. Marucha
University of Illinois at Chicago
College of Dentistry, Department of Periodontics
801 S Paulina St., M/C 859, Room 458
Chicago IL 60612, USA
marucha@uic.edu

Dieter Metze
Department of Dermatology
University of Münster
Von-Esmarchstrasse 58
48149 Münster, Germany
metzed@uni-muenster.de

Eva M.J. Peters
Psycho-Neuro-Immunology, Charité Center 12
Internal Medicine and Dermatology
Department of Psychosomatic Medicine and Psychotherapy, Neuroscience Research Center
Campus Mitte, Charité Platz 1
10117 Berlin, Germany
eva.peters@charite.de; frl_peters@yahoo.com

Siba P. Raychaudhuri
1911, Geneva Place
Davis, CA 95618, USA
raysiba@aol.com

Smriti K. Raychaudhuri
Department of Genetics, Stanford
University School of Medicine
Palo Alto, CA 94305, USA

Brian Rothschild
Department of Dermatology
University of Colorado at Denver and Health
Sciences Center
Aurora, CO, USA
brian.rothschild@gmail.com

Tanja Schlereth
Department of Neurology
Johannes Gutenberg University
Langenbeckstr. 1
55101 Mainz, Germany

Thomas E. Scholzen
Ludwig-Boltzmann Institute of Cell Biology
and Immunobiology of the Skin
Department of Dermatology
University of Münster
Münster, Germany
thomas.scholzen@spirig.ch

Kristina Seiffert
Division of Dermatology and Cutaneous Sciences
Michigan State University
4120 Biomedical and Physical Science Building
East Lansing, MI 48824, USA
krs@msu.edu

Peter I. Song
Department of Dermatology
University of Colorado at Denver and Health
Sciences Center
Aurora, CO, USA

Sonja Ständer
Clinical Neurodermatology
Department of Dermatology
University of Münster
Von-Esmarch-Strasse 58
48149 Münster, Germany
sonja.staender@uni-muenster.de

Antje Steinhoff
Department of Dermatology
University of Münster
Münster, Germany

Martin Steinhoff
Department of Dermatology and Boltzmann
Institute for Immunobiology of the Skin
University Hospital Münster
Von-Esmarch-Str. 58
48149 Münster, Germany
msteinho@uni-muenster.de

L. Terstegen
Beiersdorf AG
P.O. Box 550, Unnastraße 48
20245 Hamburg, Germany

Desmond J. Tobin
Medical Biosciences Research, School of Life Sciences
University of Bradford, Bradford
West Yorkshire BD7 1DP, UK
d.tobin@bradford.ac.uk

Elhav Weinstein
Columbia University School of Physicians and
Surgeons (EW)
New York, NY, USA
elhav@mac.com

Katrin Wilke
Beiersdorf AG
P.O. Box 550, Unnastraße 48
20245 Hamburg, Germany
katrin.wilke@beiersdorf.com

Gil Yosipovitch
Department of Dermatology
Wake Forest University Medical Center
Winston-Salem, NC 27157
gyosipov@wfubmc.edu

Christos C. Zouboulis
Departments of Dermatology
Venereology, Allergology and Immunology
Dessau Medical Center, Auenweg 38
06847 Dessau, Germany
christos.zouboulis@klinikum-dessau.de

Section I

Basic Neuroimmunology of the Skin

1 Neuroanatomy of the Skin

D. Metze

Contents

Key Features

- The skin is equipped with afferent sensory and efferent autonomic nerves.
- The sensory system contains receptors for touch, temperature, pain, itch, and various other physical and chemical stimuli.
- Antidromic propagation of the impulses may directly elicit an inflammatory reaction.
- The autonomic nervous system maintains cutaneous homeostasis by regulating vasomotor functions, pilomotor activities, and glandular secretion.
- Contact of neural structures with various immune cells allows for a strong interaction between the nervous and the immune systems.

Synonyms Box: Axons, cytoplasmic extensions of neurons located in the central nervous system and ganglia; Schwann cell–axon complex, primary neural functional unit; myelinated nerve fiber, axon enwrapped by concentric layers of Schwann cell membranes; nonmyelinated nerves, polyaxonal units in the cytoplasm of Schwann cells; "free" nerve endings, axons covered by extensions of Schwann cells and a basal lamina; corpuscular nerve endings, composed of neural and nonneural components

1.1 Structure of Cutaneous Nerves

The body must be equipped with an effective communication and control system for protection in a constantly changing environment. For this purpose, a dense network of highly specialized afferent sensory and efferent autonomic nerve branches occurs in all cutaneous layers. The sensory system contains receptors for touch, temperature, pain, itch, and various other physical and chemical stimuli. The autonomous system plays a crucial role in maintaining cutaneous homeostasis by regulating vasomotor functions, pilomotor activities, and glandular secretion.

The integument is innervated by large, cutaneous branches of musculocutaneous nerves that arise segmentally from the spinal nerves. In the face, branches of the trigeminal nerve are responsible for cutaneous innervation. Nerve trunks enter the subcutaneous fat tissue, divide, and form a branching network at the dermal subcutaneous junction. This deep nervous plexus supplies the deep vasculature, adnexal structures, and sensory receptors. In the dermis small nerve bundles ascend along with the blood vessels and lymphatic vessels and form a network of interlacing nerves beneath the

R.D. Granstein and T.A. Luger (eds.), *Neuroimmunology of the Skin*,

epidermis, that is, the superficial nerve plexus of the papillary dermis [56,63].

The cutaneous nerves contain both sensory and autonomic nerve fibers. The sensory nerves conduct afferent impulses along their cytoplasmic processes to the cell body in the dorsal root ganglia or, as in the face, in the trigeminal ganglion. Cutaneous sensory neurons are unipolar. One branch of a single axon extends towards the periphery and the other one toward the central nervous system. It has been calculated that 1,000 afferent nerve fibers innervate one square centimeter of the skin. The sensory innervation is organized in well defined segments or dermatomes, and still, an overlapping innervation may occur. Since postganglionic fibers originate in sympathetic chain ganglia where preganglionic fibers of several different spinal nerves synapse, the autonomic nerves supply the integument in a different pattern. In the skin, autonomic postganglionic fibers are codistributed with the sensory nerves until they arborize into the terminal autonomic plexus that supplies skin glands, blood vessels, and arrector pili muscles [56].

The larger nerve trunks are surround by epineurial connective-tissue sheaths that disintegrate in the dermis where perineurial layers and the endoneurium envelope the primary neural functional unit, that is, the Schwann cell–axon complex. The multilayered perineurium consists of flattened cells and collagen fibers and serves mechanical as well as barrier functions (Fig. 1.1). The perineurial cells are surrounded by a basement membrane, possess intercellular tight junctions of the zonula occludens type, and show high pinocytotic activity. The endoneurium is composed of fine connective tissue fibers, fibroblasts, capillaries, and, occasionally, a few macrophages and mast cells. The endoneural tissue is separated from the Schwann cells by a basement membrane and serves as nutritive functions for the Schwann cells [8].

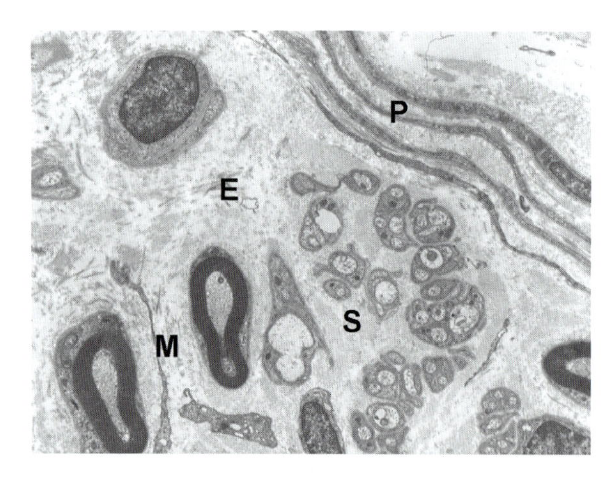

Fig. 1.1 The multilayered perineurium (P) and the endoneurium (E) envelope the primary neural functional unit, that is, the Schwann cell–axon complex. In nonmyelinated nerves, several axons are found in the cytoplasm of a Schwann cell forming the polyaxonal unit (S). In myelinated nerves the Schwann cell forms concentric myelin layers (M). Electron microscopy

The Schwann cell–axon complex consists of the cytoplasmic processes of the neurons that propulse the action potentials and the enveloping Schwann cell. The peripheral axon may be myelinated or unmyelinated. In myelinated nerve fibers, the Schwann cell membranes wrap themselves around the axon repeatedly; thus forming the regular concentric layers of the myelin sheath. In nonmyelinated nerves, several axons are found in the cytoplasm of Schwann cells forming characteristic polyaxonal units. However, these axons are invested with only a single or a few layers of Schwannian plasma membranes, without formation of thick lipoprotein sheaths [8]. This arrangement suggests a crucial role of Schwann cells for development, mechanical protection, and function of the nerves. In addition, the Schwann cells serve as a tube to guide regenerating nerve fibers. The axons are long and thin cytoplasmic extensions of neurons located in the central nervous system and ganglia that may reach a length of more than 100 cm. Ultrastructurally, the cytoplasm of the axons contains neurofilaments belonging to the intermediate filament family, mitochondria, longitudinally orientated endoplasmic reticulum, neurotubuli, and small vesicles that represent packets of neurotransmitter substances en route to the nerve terminal [68].

Cutaneous nerves contain both myelinated and unmyelinated nerve fibers and the number of myelinated fibers is decreased in the upper dermis. In general, myelinated type A-fibers correspond to motor neurons of striated muscles and a subgroup of sensory neurons, whereas unmyelinated type C-fibers constitute autonomic and sensory fibers. The myelinization of the axons allow for a high conduction velocity of 4–70 m s^{-1} as compared to a lower speed of 0.5–2 m s^{-1} in the unmyelinated fibers. The sensory myelinated fibers are further divided on the basis of their diameter and conduction speed into rapidly conducting Aβ- and slowly conducting Aδ- subcategories. Since the conduction velocity of action potentials of individual axons remains constant and myelinated and unmyelinated fibers show no overlap, this feature is a useful tool in the classification of sensory nerve fibers. Several neurophysiological experiments have shown that the Aβ-fibers conduct tactile sensitivity, whereas Aδ- and C-fibers transmit temperature, noxious sensations, and itch [40].

In the upper dermis, small myelinated nerve fibers are surrounded only by a monolayer of perineurial cells and a small endoneurium, while in thin peripheral branches of unmyelinated nerve fibers perineural sheaths are absent [7]. After losing their myelin sheaths, cutaneous nerves terminate either as free nerve endings or in association with receptors, such as Merkel cells or special nerve end-organs. The existence of intraepidermal nerves was a matter of debate for a long time. By means of silver impregnation techniques, histochemistry, and immunohistochemistry, nerve fibers could be identified in all layers of the epidermis [26]. Intraepidermal nerves run a straight or tortuous course and even may branch with a density of 2–10 fibers per 1,000 keratinocytes or 114 fibers per epidermal area of one square millimeter. However, there is a large variation on different body sites [31,32]. Measurement of intraepidermal nerve density can be used for discrimination of neuropathic diseases [34]. Intraepidermal free nerve endings mediate sensory modalities, but additional neurotrophic functions on epidermal cells have been proposed. Beyond that, a close contact between calcitonin gene-related peptide (CGRP) containing nerves and Langerhans cells have been demonstrated [27]. Neuroimmunologic functions have been supported by the finding that neuropeptides such as CGRP are able to modulate the antigen presenting function of Langerhans cells [2].

By routine light microscopy, only larger nerve bundles and some of the corpuscular nerve endings can be detected. Silver impregnation with silver salts, vital and in vitro methylene blue-staining, and histochemical reactivity for acetylcholinesterase will highlight fine nerve fibers. Peripheral nerves can be immunostained for a variety of proteins, such as myelin basic protein (a component of the myelin sheath), leu 7 (CD57, a marker for a subset of natural killer lymphocytes that cross-reacts with an epitope associated with myelin proteins), CD56 (N-CAM, an adhesion molecule), protein-gene-product 9.5 (PGP 9.5), nerve growth receptor, clathrin, synaptophysin (membrane protein of neural vesicles), neurofilament proteins (intermediate filaments of neurons), neuron specific enolase, and calcium-binding S-100 (expressed in neurons and Schwann cells) [43].

1.2 Sensory Receptors

The sensory receptors of the skin are built either by free or corpuscular nerve endings. Corpuscular endings contain both neural and non-neural components and are of two main types: non-encapsulated Merkel's "touch spots" and encapsulated receptors [48,20]. In the past, many of the free and corpuscular nerve endings in man and animals have been associated with specific sensory functions according to their distribution and complex architecture. However, since identification of specific sensory modalities within individual terminal axons is not always possible by means of neurophysiological techniques, many of the assumptions remain speculative.

1.2.1 Free Nerve Endings

In humans, the "free" nerve endings do not represent naked axons but remain covered by small cytoplasmic extensions of Schwann cells and a basal lamina; the latter may show continuity with that of the epidermis. The terminal endings are positioned intraepidermally, in the papillary dermis, and around skin appendages. By confocal laser scanning microscopy, the bulk of free nerve endings could be demonstrated just below the dermoepidermal junction [63]. Only recently, a subpopulation of nonpeptidergic, nociceptive neurons could be identified that terminate in the upper layers of the epidermis distinct from CGRP positive intraepidermal nerves with a different central projection [70]. In hairy skin, a single Schwann cell may enclose multiple ramifying nerve endings from one or more myelinated stem axons, leading to overlapping perceptions of low discrimination. On the contrary, the fine, punctate discrimination in the skin of palms and soles can be attributed to the fact that one or more axonal branches of a single nerve fiber terminate within the area of one dermal papilla. Since these brushlike, "penicillate" nerve fibers have only a few cell organelles, they are assumed to represent rapidly adapting receptors [11]. Multiple sensory modalities such as touch, temperature, pain, and itch may be attributed to the free nerve endings of "polymodal" C-fibers. In addition, some of the myelinated Aδ-fibers may account for particular subqualities of pain and itch [60].

1.2.2 Merkel Cells and Merkel's Touch Spot

Free nerve endings may be associated with individual Merkel cells of the epidermis. Single Merkel cells can be found in low numbers among the basal keratinocytes at the tips of the rete ridges in glabrous skin of fingertips, lip, gingiva, and nail bed. In hair follicles, abundant Merkel cells are enriched in two belt-like clusters, one in the deep infundibulum and one in the isthmus region [44]. No Merkel cells are present in the deep follicular portions, including the bulb, or in the dermis. Merkel cells possess a cytokeratin skeleton of characteristic low-molecular-weight and form desmosomal junctions with the neighboring keratinocytes. At the ultrastructural level, they are easily identified by membrane-limited granules with a central dense core. The structure of these cytoplasmic granules closely resembles neurosecretory granules in neurons and neuroendocrine cells. Likewise, the Merkel cells contain a battery of neuropeptides and neurotransmitter-like substances, such as vasoactive intestinal peptide, calcitonin gene-related peptide, substance P, neuron-specific enolase, synaptophysin, met-enkephalin, and chromogranin A [15].

A single arborizing myelinated nerve may supply as many as 50 Merkel cells. The dermal surface of the unmyelinated nerve terminal is enclosed in the Schwann cell membrane whose basement membrane is laterally continuous with the basement membrane of the epidermis. The upper surface of the flattened axon is in direct contact with the Merkel cell and contains many vesicles and mitochondria [23]. A cluster of Merkel cell–axon complexes at the base of a thickened plaque of the epidermis near a hair follicle in conjunction with a highly vascular underlying dermis constitutes the hair disc (Haarscheibe of Pinkus). The non-encapsulated Merkel's "touch spots" have been only recently shown to be innervated by C- and A- fibers, indicating multimodal sensory functions [50]. However, Merkel cell–axon complexes also have been demonstrated in the external root sheath of hairs and even in ridged palmar and plantar skin close to the site where the eccrine duct enters the epidermis. The presence of neurotransmitter-like substances in the dense-core granules suggests the Merkel cell to act as a receptor that transmits a stimulus to the adherent dermal nerve in a synaptic mode.

Far beyond their sensory functions, Merkel cells are speculated to have neurotrophic functions and to participate in the paracrine and autocrine regulation of inflammatory diseases [45]. Only recently, the intriguing questions as to the role of Merkel cells in hair biology have been raised [46].

1.2.3 Pacinian Corpuscle

The encapsulated receptors of the skin possess a complex structure and function as a rapidly adapting mechanoreceptor, the Pacinian corpuscle being the archetype. The Pacinian corpuscles are distributed throughout the dermis and subcutis, with greatest concentration on the soles and palms, and with less frequency on the nipples and extragenital areas [67]. These receptors are large structures of 0.5–4 mm in length and 0.3–0.7 mm in diameter. The characteristic multilaminar structure resembles an onion and contains an unmyelinated axon in the center. The capsule consists of an outer zone of multilayered perineurial cells and fibrous connective tissue, a middle zone composed of collagen fibers, elastic fibers, and fibroblasts, and an inner zone made up of Schwann cells that are closely packed around the nerve fiber [12,47]. The Pacinian corpuscles are innervated by a single myelinated sensory axon, which loses its sheaths as it passes the core of the corpuscle. Fluid filled spaces in the outer zones account for the loose arrangement of the lamellae and the spaces as seen upon routine histology. The lamellated structure may function as a mechanical filter that, on the one hand, amplifies any applied compressing or distorting force, and on the other hand, restricts the range of response. The Pacinian corpuscles are the only cutaneous receptors where, after isolation, direct evidence for mechanical perception and transmission could be demonstrated [37]. Of further interest is the close association of this type of mechanoreceptor with adjacent glomerular arteriovenous anastomoses, implying a function in the regulation of blood flow [12].

1.2.4 Meissner's Corpuscles and Mucocutaneous End-Organs

The Meissner's corpuscles are located beneath the epidermal–dermal junction between the rete ridges, with the highest density on the palmar and plantar skin. The sites of their greatest concentration are the fingertips, where approximately every fourth papilla contains a Meissner corpuscle. These end organs are elongated structures, orientated perpendicularly to the skin surface, and, by averaging 20–40 × 150 μm in size, occupy a major part of the papilla. This neural end-organ consists of modified Schwann cells stacked transversely [24]. After losing their myelin sheath one or more axons enter the bottom of the corpuscle, ramify, and pursue an upward spiral in-between the laminar Schwann cells. The axons end in bulbous terminals that contain mitochondria and vesicles. The Meissner corpuscles do not possess a true capsule but collagen fibers and elastic tissue components have an intimate relationship with the neural structures [10].

Mucocutaneous end-organs and genital corpuscles closely resemble Meissner corpuscles and are found at junctions of hairy skin and mucous membranes, such as the vermillion border of the lips, eyelids, clitoris, labia minores, prepuce, glans, and the perianal region. Although these end-organs are not recognized in routinely stained sections, silver impregnation methods, acetylcholinesterase stainings, immunhistochemistry, and electron microscopy reveal irregular loops of nerve terminals surrounded by concentric lamellar processes of modified Schwann cells [42]. Mucocutaneous end-organs are mainly distributed in the glabrous skin, but they can be also found throughout the skin of the face where they have been recognized as Krause's end-bulbs [47].

Since the Meissner corpuscle and its variants do not possess a true capsule derived from the perineurium, they may be alternatively regarded as highly specialized free nerve endings that are mechanoreceptors sensitive to touch [21].

1.2.5 Sensory Receptors of Hair Follicles

Although man is not equipped with sinus hairs, for example, the vibrissae of cats and rats, hair follicles of all human body sites have a complex nerve supply well fulfilling important tactile functions. Hair follicles are innervated by fibers that arise from myelinated nerves in the deep dermal plexus, ramify, and run parallel to and encircle the lower hair follicles. Consequently, some of the nerve fibers terminate at the upper part of the hair stem in lanceolate endings enfolded in Schwann cells lying parallel to the long axis of the hair follicle in a palisaded array [10]. In addition, other nerve fibers form the pilo-Ruffini corpuscle that encircles the hair follicle just below the sebaceous duct. This sensory organ consists of branching nerve terminals enclosed in a unique connective tissue compartment [22]. The perifollicular nerve endings are believed to be slow-adapting mechanoreceptors that respond to the bending of hairs [5]. A further subtle network of nerves can be found around the hair infundibula that may form synapses with Merkel cells of the interfollicular epidermis. However, hair follicles themselves possess Merkel cell–axon complexes among their epithelia. Vellus and terminal hairs may differ in the complexity of innervation.

1.3 Autonomic Innervation

The effector component of the cutaneous nervous system is of autonomic nature and serves manifold sociosexual and vital functions by regulating sweat gland secretion, pilomotor activities, and blood flow. The autonomic innervation of the skin mostly belongs to the sympathetic division of the autonomic nervous system. The postganglionic nerve fibers run in peripheral nerves to the skin, where they are codistributed with the sensory nerves until they arborize into a terminal autonomic plexus that surrounds the effector structures. On histologic grounds alone, it is not possible to distinguish nerve fibers of the autonomic system from those of the sensory system. Interestingly, in congenital sensory neuropathy where only autonomic nerves are preserved, sweat glands, arrector pili muscles, and blood vessels are the only innervated structures [7]. Although the cutaneous nerves comprise both sensory and sympathetic fibers, the autonomic dermatome is not precisely congruent with the sensory dermatome since the postganglionic nerves from a single ramus originate from preganglionic fibers of several different spinal cord segments [9].

Histochemically, three classes of postganglionic nerve fibers can be differentiated. Adrenergic fibers synthesize and store catecholamines that can be visualized in the nerve terminals by fluorescence microscopy. In some terminals, norepinephrine may be stored in dense core vesicles. Cholinergic fibers contain acetylcholine, which is stored in synaptic vesicles of the nerve endings. Cholinergic fibers are cholinesterase-positive throughout their entire length and thus must be considered, at least physiologically, to be parasympathetic. The non-adrenergic, non-cholinergic fibers contain adenosine triphosphate (ATP) or related purines (purinergic fibers). The terminal endings of all of the sympathetic nerve fibers show axonal beading. At the ultrastructural level, the varicosities of the different classes of autonomic nerves variably contain mitochondria, agranular vesicles, small and large granular vesicles, and large opaque vesicles [9].

1.3.1 Sweat Glands

The sweat glands are enclosed by a basketlike network of nerves, the density of innervation being much greater around the eccrine glands than the apocrine glands. The glands are innervated by autonomic fibers, some of which have been shown to contain catecholamines.

Accordingly, the periglandular nerve terminals revealed ultrastructural features of adrenergic fibers. Occasional cholinergic nerve endings were found in the vicinity of the secretory ducts [64]. Because many nerve endings have been found in closer proximity to the capillaries than to the glandular epithelia, the concept of a neurohumoral mode of transmission was supported [28].

Apocrine secretion is thought to result primarily from adrenergic activity. Thus, the glands can be stimulated by local and systemic administration of adrenergic agents. Likewise, myoepithelia of isolated axillary sweat glands have been shown to contract in response to phenylephrine or adrenalin but not acetylcholine [51]. Since denervation does not prevent a response to emotional stimulation, apocrine glands may be further stimulated humorally by circulating catecholamins to secrete fluid and pheromones.

In contrast to the ordinary sympathetic innervation, the major neurotransmitter released from the periglandular nerve endings is acetylcholine. Cholinergic stimulation is the most potent factor in the widespread eccrine sweating for regulation of temperature. In addition to acetylcholine, catecholamins, vasoactive intestinal peptide (VIP), and atrial natriuretic (ANP) have been detected in the periglandular nerves. Norepinephrine and VIP can not be regarded as effective as acetylcholine but they synergistically amplify acetylcholine-induced cAMP accumulation, which is an important second messenger in the metabolism of secretory cells [52,53]. Myoepithelial cells contract in response to cholinergic but not adrenergic stimulation [51,54]. In view of the fact that ANP functions as a diuretic and causes vasodilation, it may assist the sweat glands in regulating water and electrolyte balance. The functional significance of other periglandular neuropeptides such as calcitonin gene-related peptide (CGRP) and galanin for the regulation of sweating is still obscure [62].

The assumption that periglandular catecholamines directly induce sweating during periods of emotional stress seems unlikely because both emotional and thermal sweating can be inhibited by atropine. Emotional sweating, which is usually confined to the palms, soles, axilla, and, more variably, to the forehead, may be controlled by particular parts of the hypothalamic sweat centers under the influence of the cortex without input from thermosensitive neurons [55].

Regulation of body temperature is the most important function of eccrine sweat glands. The preoptic hypothalamic areas contain thermosensitive neurons that detect changes in the internal body temperature. Local heating of this temperature control center induces sweating, vasodilation, and panting that enhance heat loss. Conversely, experimental cooling causes vasoconstriction and shivering. In addition to thermoregulatory sweating due to an increased body temperature, skin temperature also has an influence on the sweating rate. The warm-sensitive-neurons in the hypothalamus can be activated by afferent impulses from the cutaneous thermoreceptors [6]. Efferent nerve fibers from the hypothalamic sweat center descend and, after synapsing, reach the periglandular sympathetic nerves.

1.3.2 Sebaceous Glands

Sebaceous gland secretion presumably is not under direct neural control but depends upon circulating hormones. The dense network of nonmyelinated nerve fibers that have been found to be wrapped around Meibomian glands in the eyelids may also function as sensory organs [47]. However, there is evidence that neuropeptides and proopiomelanocortin derivatives produced by peripheral nerves and cellular constituents of the epidermis participate in the regulation of sebum secretion [58,4].

1.3.3 Arrector Pili Muscle

The nerves of the arrector pili muscles arise from the perifollicular nerve network. Adrenergic nerve terminals lie within 20–100 nm of adjacent smooth muscle cells. By activating alpha-receptors on the smooth muscle cells of the hair erectors, the hairs are pulled in an upright position producing a "goose-flush" upon emotional and cold-induced stimulation.

1.3.4 Blood Vessels

Depending on their location in the body, blood vessels are variably innervated. The autonomic system mediates the constriction and dilation of the vessel walls and of the arteriovenous anastomoses and, thus, contributes to the regulation of the cutaneous circulation. Blood flow is essential for tissue nutrition but also is involved in many other functions such as control of the body temperature and tumescence of the genitalia.

The vast majority of vessels in the dermis are surrounded by nerves, which run along with them but do not innervate them. Studies of cutaneous vessels have not focused specifically on their innervation. Unmyelinated nerve fibers ensheathed by Schwann cells were found to be disposed in the adventitia of arterial vessels but neither nerves nor nerve endings have been observed between the muscle cells of the media [68]. In the arteriovenous anastomoses of the glomus organ that bypass the capillary circulation at the acral body sites, numerous nonmyelinated nerves ensheathed by Schwann cells are present peripheral to the glomus cells [18].

Ultrastructural and histochemical studies showed that the microcirculation is innervated by adrenergic, cholinergic, and prurinergic fibers. While adrenergic fibers mediate strong vasoconstriction, acetylcholine acts as a vasodilator [13]. Additionally, various neuropeptides are involved in the regulation of the microvascular system of the skin, such as VIP and peptide histidine isoleucine that directly relax smooth muscle cells. It is also assumed that neuropeptides, synergistically with mast cell and other endogenous factors, are involved in the induction of edema by increasing the permeability of post-capillary venules [3]. Beyond that, neurohormones such as melanocyte stimulating hormone (MSH) directly modulate cytokine production and adhesion molecule expression of endothelial cells, which were found to express receptors specific for this peptide [39].

In the central nervous system, the primary centers that regulate and integrate blood flow are the hypothalamus, medulla oblongata, and spinal cord. The vasodilator and vasoconstrictor areas in the medulla oblongata integrate messages from higher cortical centers, the hypothalamus, the baroreceptors, chemoreceptors, and somatic afferent fibers. These major vasomotor centers on the brain stem regulate blood flow and blood pressure via the sympathetic ganglia. Episodic flushing may be associated with a variety of emotional disturbances and environmental influences. Beyond that, there exist spinal vasomotor reflexes that are segmentally or regionally arranged in the spinal cord.

Activation of the sympathetic nervous system by the heat production center in the preoptic region of the hypothalamus reduces the blood flow in the skin and, consequently, decreases the transfer of heat to the body surface. Conversely, in response to heat, blood warmer than normal passes the hypothalamus and inhibits the heat-promotion mechanisms. The blood vessels will dilate upon inhibition of the sympathetic stimulation, allowing for rapid loss of heat. Vasodilation also occurs reflexively through direct warming of the skin surface upon release of the vasoconstrictor tone. This reflex may either originate in cutaneous receptors or by central nervous system stimulation [35].

1.4 Nerves and the Immune System

The function of sensory nerves not only comprises conduction of nociceptive information to the central nervous system for further processing, but sensory fibers also have the capacity to respond directly to noxious stimuli by initiating a local inflammatory reaction. Noxious stimulation of polymodal C-fibers produces action potentials that travel centrally to the spinal cord and, in a retrograde fashion, along the ramifying network of axonal processes. The antidromic impulses that start from the branching points cause the secretion of neuropeptides stored along the peripheral nerves. As a consequence of their effects over vessels, glands, and resident inflammatory cells in the close proximity, a neurogenic inflammation is induced. This "axon-reflex" model is partly responsible for the triple response of Lewis. A firm, blunt injury evokes a primary local erythema, followed by a wave of arteriolar vasodilation that extends beyond the stimulated area (flare reaction). Subsequently, increased permeability of the postcapillary venoles leads to plasma extravasation and edema, that is, a wheal reaction in the area of the initial erythema [9].

The triple response can be elicited by the administration of histamine, various neuropeptides, and antidromic electric stimulation of sensory nerves and can be abolished by denervation and local anesthetics. Among others, substance P, neurokinin A, somatostatin, and calcitonin gene-related peptide play a major role in the axon-flare reaction. The inhibition of the axon-reflex vasodilation by topical pretreatment with capsaicin, a substance P depleting substance, provides direct evidence for a neurogenic component of inflammation [65].

However, the nature of the flare and wheal reaction is far more complex than previously thought. Beyond direct initiation of vasodilation, leakage of plasma and inflammatory cells, neuropeptides may exert their effects via the activation of mast cells [16]. Some morphological findings suggest an interaction of sensory nerves with mast cells as they have been observed in close proximity to myelinated, unmyelinated, and substance P-containing nerves [66, 57]. Electric stimulation of rat nerves was associated with an increase in degranulating mast cells [33]. As a result, neuropeptides seem to have the capacity to degranulate mast cells. However, even potent mast cell activating neuropeptides induce histamine release in vitro only when added in relatively high concentrations [14]. Other experiments and stimulation of nerves in mast cell-deficient mice support the notion that mast cells were not essential for neurogenic inflammation [3]. The recent observation of histamine-immunoreactive nerves in the skin of Sprague–Dawley rats even suggest a more direct route of cutaneous histamine effects, mediated exclusively by the peripheral nervous system [30].

There is increasing evidence for a synergistic function between neuropeptides and inflammatory mediators. Moreover, the polymodal C-fibers have proinflammatory actions, but their excitability is itself increased in the presence of inflammatory mediators. In view of this positive feedback, it can be speculated that the nervous system may be involved in augmenting and self-sustaining an inflammatory response [41]

Recent studies strongly suggest an interaction between the nervous system and immune system far beyond than that described for the classical model of axon-flare. The close anatomical association of the cutaneous nerves with inflammatory or immuno-competent cells and the well recognized immunomodulatory effect of many neuropeptides indicate the existence of a neuroimmunological network (Fig. 1.2). Nerves have been described in the Peyer's patches and the spleen and, after release of substance P, may influence T cell proliferation and homing [61,69]. Likewise, neuropeptides have been discussed to play a role in the lymph node response to injected antigens [25] and to stimulate B-cell immunoglobulin production [36]. By release of calcitonin gene-related peptide (CGRP) and substance P, some cutaneous nerve fibers may activate polymorphonuclear cells [49] and stimulate macrophages [38]. Secretory neuropeptides further stimulate endothelial cells to transport preformed adhesion molecules, such as P- and E-selectin from intracellular Weibel-Palade bodies to the endothelial surface and, thereby, enhances chemotactic functions [59]. Moreover, on the one hand, substance P stimulates the production of proinflammatory as well as immunomodulating cytokines and, on the other hand, cytokines such as interleukin-1 enhance the production of substance P in sympathetic neurons [39,1,17].

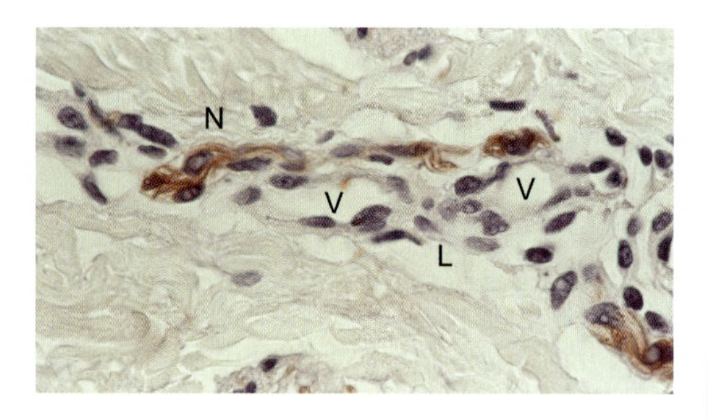

Fig. 1.2 Close anatomical association of the nerve fibers (N) with postcapillary venoles (V) and inflammatory cells (L) is the precondition for many neuro-immunologic functions. Immunostaining for S100

In human epidermis, nerve fibers are intimately associated with Langerhans cells. Immunohistochemically, these intraepidermal nerve fibers contained CGRP and seemed to be capable of depositing CGRP at or near Langerhans cells [2]. In addition, another neuropeptide, that is, melanocyte stimulating hormone (α-MSH), was recently detected in nerves as well as several cells in the skin [39,29]. Like CGRP, α-MSH was also demonstrated to inhibit the function of immunocompetent cells and to induce tolerance to potent contact allergens [27,19]. These findings strongly support the concept of an interaction between the immune- and neuroendocrine system in the skin.

In conclusion, the complex innervation of the skin with sensory nerve fibers that potentially release a variety of neuropeptides implies a participation of neuroimmunological mechanisms in the pathophysiology of skin diseases and substantiates the old notion that stress and emotional state can affect the development and course of many dermatoses.

Summary for the Clinician

The skin possesses a complex communication and control system for protection of the organism in a constantly changing environment. The cutaneous nerves form a dense network of afferent sensory and efferent autonomic that branches in all cutaneous layers. The sensory system is composed of receptors for touch, temperature, pain, itch, and various other physical and chemical stimuli. Stimuli are either processed in the central nervous system or may directly elicit an inflammatory reaction by antidromic propagation of the impulses. The autonomic nervous system maintains cutaneous homeostasis by regulating vasomotor functions, pilomotor activities, and glandular secretion. Skin biopsies allow for diagnosis and differentiation of various forms of neuropathies. Beyond that, a close contact of neural structures with various immune cells implicates a strong interaction between the nervous and the immune systems. Examination of the neuroanatomy of the skin is the first step to understanding the sensory, autonomic, and immunological functions of the skin.

References

1. Ansel JC, Brown JR, Payan DG, et al (1993) Substance P selectively activates TNF-alpha gene expression in murine mast cells. J Immunol 150:4478–4485
2. Asahina A, Hosoi J, Grabbe S, et al (1995) Modulation of Langerhans cell function by epidermal nerves. J Allergy Clin Immunol 96:1178–1182
3. Baraniuk JN (1997) Neuropeptides in the skin. In: Bos J (ed) The Skin Immune System, 2nd edn. CRC Press, Boca Raton, pp 311–326
4. Bhardwaj RS, Luger TA (1994) Proopiomelanocortin production by epidermal cells: Evidence for an immune neuroendocrine network in the epidermis. Arch Dermatol Res 287:85–90
5. Biemesderfer D, Munger BL, Binck J, et al (1978) The pilo–Ruffini complex: A non-sinus hair and associated slowly-adapting mechanoreceptor in primate facial skin. Brain Res 142:197–222
6. Boulant JA (1981) Hypothalamic mechanisms in thermoregulation. Fed Proc 40:2843–2850
7. Bourlond A, Winkelmann RK (1966) Nervous pathways in papillary layer of human skin: An electron microscopic study. J Invest Dermatol 47:193–204
8. Breathnach AS (1977) Electron microscopy of cutaneous nerves and receptors. J Invest Dermatol 69:8–26
9. Burnstock G (1977) Autonomic neuroeffector junctions-reflex vasodilatation in the skin. J Invest Dermatol 69:47–57
10. Cauna N (1966) Fine structure of the receptor organs and its probable functional significance. In: DeReuck AVS, Knight J (eds) CIBA Foundation Symposium: Touch, Heat and Pain. Churchill, London, pp 117–127
11. Cauna N (1980) Fine morphological characteristics and microtopography of the free nerve endings of the human digital skin. Anat Rec 198:643–656
12. Cauna N, Mannan G (1958) The structure of human digital Pacinian corpuscles (Corpuscula Lamellosa) and its functional significance. J Anat 92:1–20
13. Coffman JD, Cohen RA (1987) A cholinergic vasodilator mechanism in human finger. Am J Physiol 252:H594–H599
14. Ebertz JM, Hirschman CA, Kettelkamp NS, et al (1987) Substance P-induced histamine release in human cutaneous mast cells. J Invest Dermatol 88:682–685
15. Fantini F, Johansson O (1995) Neurochemical markers in human cutaneous Merkel cells. An immunohistochemical investigation. Exp Dermatol 4:365–371
16. Foreman J, Jordan C (1983) Histamine release and vascular changes induced by neuropeptides. Agents Actions 13:105–116
17. Freidin M, Kessler JA (1991) Cytokine regulation of substance P expression in sympathetic neurons. Proc Natl Acad Sci U S A 88:3200–3203
18. Goodmann TF (1972) Fine structure of the cells of the Suquet-Hoyer canal. J Invest Dermatol 59:363–369
19. Grabbe S, Bhardwaj RS, Steinert M, et al (1996) Alpha-melanocyte stimulating hormone induces hapten-specific tolerance in mice. J Immunol 156:473–478
20. Halata Z (1975) The mechanoreceptors of the mamammalian skin: Ultrastructure and morphological classification. Adv Anat Embryol Cell biol 50: 3–77
21. Halata Z, Munger BL (1980) The sensory innervation of primate eyelid. Anat Rec 198:657–670
22. Halata Z, Munger BL (1981) Identification of the Ruffini Corpuscle in human hair skin. Cell Tissue Res 219:437–440
23. Hashimoto K (1972) Fine structure of Merkel cell in human oral mucosa. J Invest Dermatol 58:381–387
24. Hashimoto K (1973) Fine structure of the Meissner corpuscle of human palmar skin. J Invest Dermatol 60:20–28
25. Helme RD, Eglezos A, Dandie GW, et al (1987) The effect of substance P on the regional lymph node antibody response to antigenic stimulation in capsaicin-pretreated rats. J Immunol 139:3470–3473
26. Hilliges M, Wang L, Johansson O (1995) Ultrastructural evidence for nerve fibers within all vital layers of the human epidermis. J Invest Dermatol 104:134–137
27. Hosoi J, Murphy GF, Egan CL, et al (1993) Regulation of Langerhans cell function by nerves containing calcitonin gene-related peptide. Nature 363: 159–163
28. Jenkinson DM, Montgomery I, Elder HY (1978) Studies on the nature of the peripheral sudomotor control mechanism. J Anat 125:625–639
29. Johansson O, Ljungberg A, Han SW, et al (1991) Evidence for gamma melanocyte stimulating hormone containing nerves and neutrophilic granulocytes in the human skin by immunofluorescence. J Invest Dermatol 96:852–856
30. Johansson O, Virtanen M, Hilliges M (1995) Histaminergic nerves demonstrated in the skin. A new direct mode of neurogenic inflammation? Exp Dermatol 4:93–96
31. Johansson O, Wang L, Hilliges M, et al (1999) Intraepidermal nerves in human skin: PGP 9.5 immunohistochemistry with special reference to the nerve density in skin from different body regions. J Peripher Nerv Syst 4:43–52
32. Kawakami T, Ishihara M, Mihara M (2001) Distribution density of intraepidermal nerve fibers in normal skin. J Dermatol 28:63–70
33. Kiernan JA (1972) The involvement of mast cells in vasodilatation due to axon reflexes in injured skin. Q J Exp Physiol Cogn Med Sci 57:311–317
34. Koskinen M, Hietaharju A, Kylaniemi M, et al (2005) A quantitative method for assessment of intraepidermal nerve fibers in small fiber neuropathy. J Neurol 252:789–794
35. Kranink KK (1991) Temperature regulation and the skin. In: Goldsmith LA (ed) Physiology, Biochemistry, and Molecular Biology of the Skin. Oxford University Press, New York, Oxford, pp 1085–1095
36. Laurenzi MA, Persson MA, Dalsgaard CJ, et al (1989) Stimulation of human B lymphocyte differentiation by the neuropeptides substance P and neurokinin A. Scand J Immunol 30:695–701
37. Loewenstein WR, Skalak R (1966) Mechanical transmission in a Pacinian corpuscle. An analysis and a theory. J Physiol 182:346–378
38. Lotz M, Vaughan JH, Carson DA (1988) Effect of neuropeptides on production of inflammatory cytokines by human monocytes. Science 241:1218–1221
39. Luger TA, Bhardwaj RS, Grabbe S, et al (1996) Regulation of the immune response by epidermal cytokines and neurohormones. J Dermatol Sci 13:5–10
40. Lynn B (1991) Cutaneous sensation. In: Goldsmith LA (ed) Physiology, Biochemistry, and Molecular Biology of the Skin. Oxford University Press, New York, Oxford, pp 779–815

41. Lynn B, Shakhanbeh J (1988) Neurogenic inflammation in the skin of the rabbit. Agents Actions 25:228–230
42. MacDonald DM, Schmitt D (1979) Ultrastructure of the human mucocutaneous end organ. J Invest Dermatol 72:181–186
43. Metze D (2004) Skin nerve anatomy: Neuropeptide distribution and its relationship to itch. In: Yosipovitch G, Greaves M, Fleischer A, McGlone F (eds) Itch: Basic Mechanisms and Therapy. Marcel Dekker, New York, pp 71–86
44. Moll I (1994) Merkel cell distribution in human hair follicles of the fetal and adult scalp. Cell Tissue Res 277:131–138
45. Moll I, Bladt U, Jung EG (1992) Distribution of Merkel cells in acute UVB erythema. Arch Dermatol Res 284:271–274
46. Moll I, Paus R, Moll R (1996) Merkel cells in mouse skin: Intermediate filament pattern, localization, and hair cycle-dependent density. J Invest Dermatol 106:281–286
47. Montagna W, Kligman AM, Carlisle KS (1992) Atlas of normal human skin. Springer, New York Berlin Heidelberg
48. Munger BL, Ide C (1988) The structure and function of cutaneous sensory receptors. Arch Histol Cytol 51:1–34
49. Payan DG, Levine JD, Goetzl EJ (1984) Modulation of immunity and hypersensitivity by sensory neuropeptides. J Immunol 132:1601–1604
50. Reinisch CM, Tschachler E (2005) The touch dome in human skin is supplied by different types of nerve fibers. Ann Neurol 58:88–95
51. Sato K (1980) Pharmacological responsiveness of the myoepithelium of the isolated human axillary apocrine sweat gland. Br J Dermatol 103:235–243
52. Sato K, Sato F (1981) Pharmacologic response of isolated single eccrine sweat glands. Am J Physiol 240:R44–R52
53. Sato K, Sato F (1987) Effect of VIP on sweat secretion and cAMP accumulation in isolated simian eccrine glands. Am J Physiol 253:R935–R941
54. Sato K, Nishiyama A, Kobayashi M (1979) Mechanical properties and functions of the myoepithelium in the eccrine sweat gland. Am J Physiol 237:C177–C184
55. Sato K, Kang WH, Saga K, et al (1989) Biology of sweat glands and their disorders. II. Disorders of sweat gland function. J Am Acad Dermatol 20:713–726
56. Sinclair DC (1973) Normal anatomy of sensory nerves and receptors. In: Jarrett A (ed) The Physiology and Pathophysiology of the Skin, Vol 2, The Nerves and Blood Vessels. Academic Press, London, pp 347–402
57. Skofitsch G, Savitt JM, Jacobovitz DM (1985) Suggestive evidence for a functional unit between mast cells and substance P fibers in the rat diaphragm and mesentery. Histochemistry 82:5–8
58. Slominski A, Paus R, Wortsman J (1993) On the potential role of proopiomelanocortin in skin physiology and pathology. Mol Cell Endocrinol 93:C1–C6
59. Smith CH, Barker JN, Morris RW, et al (1993) Neuropeptides induce rapid expression of endothelial cell adhesion molecules and elicit granulocytic infiltration in human skin. J Immunol 151:3274–3282
60. Staender S, Steinhoff M, Schmelz M, et al (2003) Neurophysiology of pruritus. Cutaneous elicitation of itch. Arch Dermatol 139:1463–1470
61. Stanisz AM, Scicchitano R, Dazin P, et al (1987) Distribution of substance P receptors on murine spleen and Peyer's patch T and B cells. J Immunol 139:749–754
62. Tainio H, Vaalasti A, Rechardt L (1987) The distribution of substance P-, CGRP-, galanin- and ANP-like immunoreactive nerves in human sweat glands. Histochem J 19:375–380
63. Tschachler E, Reinisch CM, Mayer C, et al (2004) Sheet preparations expose the dermal nerve plexus of human skin and render the dermal nerve end organ accessible to extensive analysis. J Invest Dermatol 122:177–182
64. Uno H (1977) Sympathetic innervation of the sweat glands and piloerector muscle of macaques and human beings. J Invest Dermatol 69:112–130
65. Wallengren J, Möller H (1986) The effect of capsaicin on some experimental inflammations in human skin. Acta Derm Venereol 66:375–380
66. Wiesner-Menzel L, Schulz B, Vakilzadeh F, et al (1981) Electron microscopical evidence for a direct contact between nerve fibres and mast cells. Acta Derm Venereol 61:465–469
67. Winkelmann RK (1965) Nerve changes in aging skin. In: Montagna W (ed) Advances in the Biology of Skin, vol 6, Aging. Pergamon, Oxford, p 51
68. Winkelmann RK (1967) Cutaneous nerves. In: Zelickson AV (ed) Ultrastructure of normal and abnormal skin. Lea&Febiger, Philadelphia, pp 202–227
69. Zhu LP, Chen D, Zhang SZ, et al (1984) Observation of the effect of substance P on human T and B lymphocyte proliferation. Immunol Commun 13:457–464
70. Zylka MJ, Rice FL, Anderson DJ (2005) Topographically distinct epidermal nociceptive circuits revealed by axonal tracers targeted to Mrgprd. Neuron 45:17–25

Neuroreceptors and Mediators 2

S. Ständer and T.A. Luger

Content

Key Features

- Cutaneous unmyelinated, polymodal sensory C-fibers have afferent functions to mediate cold, warmth, touch, pain, and itch to the CNS.
- Polymodal sensory C-fibers mediate also efferent functions by the release of neuropeptides.
- CGRP released from sensory nerves has an impact on keratinocyte differentiation, cytokine expression, and apoptosis.
- SP from sensory fibers trigger skin mast cell degranulation upon acute immobilization stress in animals.
- Histamine released from mast cells may act on keratinocytes to enhance production and release of nerve growth factor.
- NGF sensitizes different neuroreceptors, including transient receptor potential V1 (TrpV1).
- Cannabinoid agonist exhibit peripheral antinociceptive effects possibly by stimulation of β-endorphin release from keratinocytes.

Synonyms Box: Itch, puritus

Abbreviations: *AD* Atopic dermatitis, *CB* Cannabinoid receptor, *CGRP* Calcitonin gene-related peptide, *CNS* Central nervous system, *DRG* Dorsal root ganglia, *GDNF* Glial cell line-derived neurotrophic factor, *ETA* Endothelin receptor A, *ETB* Endothelin receptor B, *LC* Langerhans cells, *Mrgprs* Mas-related G-protein coupled receptors, *NGF* Nerve growth factor, *PAR-2* Proteinase-activated receptor-2, *PEA* Palmitoylethanolamine, *PKR* Prokineticin receptor, *SP* Substance P, *THC* Tetrahydrocannabinol, *TrkA* Tyrosine kinase A, *Trp* Transient receptor potential, *VIP* Vasoactive intestinal peptide

2.1 Introduction

Acting as border to the environment, the skin reacts to external stimuli such as cold, warmth, touch, destruction (pain), and tickling [e.g., by parasites (itch)]. The modality-specific communication is transmitted to the

R.D. Granstein and T.A. Luger (eds.), *Neuroimmunology of the Skin*,

central nervous system (CNS) by specialized nerve fibers and sensory receptors. In the skin, dermal myelinated nerve fibers such as Aβ- and Aδ-fibers transmit touch and other mechanical stimuli (e.g., stretching the skin) and fast-conducting pain [46]. Unmyelinated C-fibers in the papillary dermis and epidermis are specialized to stimuli such as cold, warmth, burning, or slow conducting pain and itch [41,84]. In the epidermis, two major classes of sensory nerve fibers can be distinguished (Table 2.1) by their conduction velocity, reaction to trophic stimuli (e.g., nerve growth factor, glial cell-line derived neurotrophic factor), and expression of neuropeptides and neuroreceptors [3,115,116]. This complex system enables the CNS to clearly distinguish between incoming signals from different neurons in quality and localization. Moreover, C-fibers have contacts and maintain cross-talk with other skin cells such as keratinocytes, Langerhans cells, mast cells, and inflammatory cells. This enables sensory nerves to function not only as an afferent system that conducts stimuli from the skin to the CNS, but also as an efferent system that stimulates cutaneous cells by secreting several kinds of neuropeptides. In addition, sensory sensations can be modified in intensity and quality by this interaction (Table 2.2). In this chapter, an overview is given on the neuroreceptors and mediators of C-fibers involved in the sensory system of the skin and their communication with other skin cells.

Table 2.1 The two major epidermal C-fiber classes[a]

	Peptidergic	Non-peptidergic
Conducting velocity	0.5 m s^{-1}	1.0 m s^{-1}
Diameter	0.3–1.0 μm	0.3–1.0 μm
Localization in epidermis	Up to stratum spinosum	Up to granular layer
Receptors (receptor for growth factors, other receptors)	trkA, p75, e.g. Histamine receptor, Trp-group	c-RET, binding of Isolectin B4, Mrgprd, TrpV1
Neurotransmitters	Peptidergic, e.g., SP, CGRP	Non-peptidergic
Trophic factor (both present in keratinocytes)	Nerve growth factor (NGF)	Glial cell-line derived neurotrophic factor (GDNF)
Function	Itch, cold, warmth, burning pain, noxious heat	Mechanical stimuli, warmth, pain

[a] 98% of all epidermal nerve fibers

Table 2.2 Function of neuroreceptors on C-fibers

Receptor	Ligand	Function
Histamine receptors: H1–H4	Histamine	Pruritus (H1 and H4 receptor), neurogenic inflammation; sensitized by bradykinin, prostaglandins
Endothelin receptors: A, B	Endothelin 1, 2, 3	ETA: Pruritus, mast cell degranulation, inflammation, increase of TNF-alpha, IL-6, VEGF, TGF-beta1 ETB: suppression of pruritus
TrpV1	Noxious heat (>42°C), protons, capsaicin, anandamide	Cold, heat, burning pain, burning pruritus, noxious heat sensitized by NGF, galanin, bradykinin
TrpV2	Noxious heat (>52°C)	Pain induced by heat
TrpV3	Warmth (>33°C)	Warmth
TrpV4	Warmth (~25°C)	Warmth
TrpM8 (on Aδ-fibers)	Cold (8–28°C), menthol, icilin	Cold
TrpA1 (AnkTM1)	Noxious cold (<17°C), wasabi, horseradish, mustard	Pain induced by cold, burning

(continued)

Table 2.2 (continued)

Receptor	Ligand	Function
PAR-2	Tryptase, trypsin	Pruritus, neurogenic inflammation
Opioid receptors: Mu-, delta-receptor	Endorphins, enkephalins	Suppression of pain, pruritus, and neurogenic inflammation
Cannabinoid receptors CB1, CB2	Cannabinoids CB1: anandamide CB2: PEA	Suppression of itch, pain and neurogenic inflammation, release of opioids

2.2 Neurojunctions with Cutaneous Cells and Efferent Functions of the Skin Nervous System

Unmyelinated C-fibers are found in the papillary dermis as well as in the epidermis up to the granular layer. Electron microscopic and confocal scanning microscopy investigations demonstrated C-fibers having contacts to keratinocytes by slightly invaginating into keratinocyte cytoplasm [12,33,36]. These neuro-epidermal junctions are discussed as representing synapses [12] since the adjacent plasma membranes of keratinocytes were slightly thickened, closely resembling post-synaptic membrane specializations in nervous tissues. The nerve fibers cross-talk with the connected cells and exert, in addition to sensory function, trophic and paracrine functions. These efferent functions are mediated by neuropeptides [e.g., substance P (SP), calcitonin gene-relate peptide (CGRP), vasoactive intestinal polypeptide (VIP)] released upon antidromic activation of the peripheral terminals of unmyelinated C-fibers [77]. For example, nerve fibers were reported to influence epidermal growth and keratinocyte proliferation [38]. CGRP released from sensory nerves was demonstrated to have an impact on keratinocyte differentiation, cytokine expression, and apoptosis through intracellular nitric oxide (NO) modulation and stimulation of nitric oxide synthase (NOS) activity [24]. This connection also has an influence on several diseases; for example, wound healing is disturbed in diabetic patients due to small fiber neuropathy and decreased release of SP from nerve fibers [32].

Neuronal connections to Langerhans cells [31,37], melanocytes [34], and Merkel cells [58] have also been demonstrated. It was observed that CGRP-containing C-nerve fibers were associated with epidermal Langerhans cells (LC), and CGRP was found to be present at the surface of some cells. Further, CGRP was shown to inhibit LC antigen presentation [37]. In a confocal microscopic analysis, intraepidermal nerve ending contacts with melanocytes were found [34]. Thickening of apposing plasma membranes between melanocytes and nerve fibers, similar to contacts observed in keratinocytes, were confirmed. Stimulation of cultured human melanocytes with CGRP, SP, or vasoactive intestinal peptide (VIP) led to increased DNA synthesis rate of melanocytes by the cAMP pathway in a concentration- and time-dependent manner mediated [34].

In the papillary dermis, direct connections between unmyelinated nerve fibers and mast cells were found [53,109]. It is debated whether this connection has relevance in healthy human skin [105]. However, experimental studies showed that intradermally injected SP induces release of histamine via binding to NKR on mast cells and thereby acts as a pruritogen [15]. Other investigations demonstrated SP-induced release of pruritogenic mediators from mast cells under pathologic conditions [70,99]. Furthermore, a connection between neuropeptides, mast cells, and stress could be shown in animal studies [82]. Acute immobilization stress triggered skin mast cell degranulation via SP from unmyelinated nerve fibers. Pruritus, whealing, and axon-reflex erythema due to histamine release appear in human skin after intradermal injection of VIP, neurotensin, and secretin. Also somatostatin was reported to stimulate histamine release from human skin mast cells [15].

Neuropeptides such as SP and CGRP act on blood vessels inducing dilatation and plasma extravasation, resulting in neurogenic inflammation with erythema and edema [94]. SP upregulates adhesion molecules such as intercellular adhesion molecule 1 (ICAM-1) [73], is chemotactic for neutrophils [5], and induces release of cytokines such as interleukin (IL)-2 or IL-6 from them [18]. In sum, release of neuropeptides from nerve fibers enables dermal inflammation by acting on vessels and on inflammatory cells. Interestingly, increased SP-immunoreactive nerve fibers have been observed in certain inflammatory skin diseases such as psoriasis, atopic dermatitis, and prurigo nodularis [1,42,43].

2.3 Histamine Receptors

Histamine and the receptors H1 to H4 have been the most thoroughly studied mediator and neuroreceptors for decades. Lewis reported 70 years ago that intradermal injection of histamine provokes redness, wheal, and flare (so called triple response of neurogenic inflammation) accompanied by pruritus [52]. Accordingly, histamine is used for most experimental studies investigating neurogenic inflammation and itching [78]. Histamine is stored in mast cells and keratinocytes while H1 to H4 receptors are present on sensory nerve fibers and inflammatory cells [35,100]. Thus, histamine-induced itch may be evoked by release from mast cells or keratinocytes. Only recently it was reported that, in addition to histamine receptor 1 (H1), H3 and H4 receptors on sensory nerve fibers are also involved in pruritus induction in mice [6,96]. Interestingly, histamine released from mast cells may act on keratinocytes to enhance production and release of nerve growth factor (NGF) [47]. In turn, NGF induces histamine release from mast cells and sensitizes different neuroreceptors, including transient receptor potential V1 (TrpV1) [113]. Current studies suggest that histamine also regulates SP release via prejunctional histamine H3 receptors that are located on peripheral endings of sensory nerves [67]. This may have an impact on SP-dependent diseases such as ulcerations. Accordingly, a current study demonstrated that mast cell activation and histamine are required for normal cutaneous wound healing [106].

2.4 Endothelin Receptors

Endothelin (ET) -1, -2, and -3 produced by endothelial cells and mast cells induce neurogenic inflammation associated with burning pruritus [48,108]. Endothelin binds to two different receptors, endothelin receptor A (ETA) and ETB, which are present on mast cells [57]. Injected into the skin, ET-1 induces mast cell degranulation and mast cell-dependent inflammation [59]. Furthermore, ET-1 induces TNF-α and IL-6 production, enhanced VEGF production, and TGF-β1 expression by mast cells [57]. ET-1 was therefore identified to participate in pathological conditions of various disorders via its multi-functional effects on mast cells under certain conditions. For example, ET-1 contributes to ultraviolet radiation (UVR)-induced skin responses such as tanning or inflammation by involvement of mast cells [59]. Interestingly, ET-1 was also identified to display potent pruritic actions in the mouse, mediated to a substantial extent via ETA while ETB exerted an antipruritic role [101].

2.5 Trp-Family

The transient receptor potential (TRP) family of ion channels is constantly growing and to date comprises more than 30 cation channels, most of which are permeable for Ca^{2+}. On the basis of sequence homology, the Trp family can be divided into seven main subfamilies: the TrpC ("Canonical") family, the TrpV ("Vanilloid") family, the TrpM ("Melastatin") family, the TrpP ("Polycystin") family, the TrpML ("Mucolipin") family, the TrpA ("Ankyrin") family, and the TrpN ("NOMPC") family. Concerning a role in cutaneous nociception, the TrpV and the TrpM groups are both expressed on sensory nerve fibers with different functions [68].

2.5.1 TrpV1: The Capsaicin Receptor

The TrpV1 receptor (vanilloid receptor, VR1) is expressed on central and peripheral neurons [68]. In the skin, the TrpV1 receptor is present on sensory C- and Aδ-fibers [87]. Different types of stimuli activate the receptor such as low pH (<5.9), noxious heat (>42°C), the cannabinoid/endovanilloid anandamide, leukotrien B4, and exogenous capsaicin. Trp receptors act as nonselective cation-channels, which open after stimulation and enable ions inward into the nerve fiber, resulting in a depolarization. As a result, for example, after capsaicin application, TrpV1 is stimulated to either transmit burning pain or a burning pruritus. Because of antidromic activation, C-fibers release neuropeptides, which mediate neurogenic inflammation. Upon chronic stimulation, TrpV1 receptor signaling exhibits desensitization in a

Ca2+-dependent manner, such as upon repeated activation by capsaicin or protons [111]. The desensitized receptor is permanently opened with a following steady-state of cations intra- and extracellular. This hinders depolarization of nerve fiber and the transmission of either itch or burning pain. Moreover, neuropeptides such as SP are depleted from the sensory nerve fibers; the axonal transport of both neuropeptides and NGF in the periphery is slowed. This mechanism is used therapeutically upon long-term administration of capsaicin for relief of both localized pain and localized pruritus. Clinically, the first days of the therapy are accompanied by burning, erythema, or flare induced by the neurogenic inflammation. After this initial phase, pain and itch sensations are depressed as was demonstrated in many studies and case reports [83]. Like the histamine receptor, the TrpV1 receptor may be sensitized by bradykinin and prostaglandins, as well as by NGF [39,81,113], with lowering of the activation threshold and facilitated induction of pain and itch. For example, instead of noxious heat, moderate warmth may activate a sensitized receptor.

The topical calcineurin inhibitors pimecrolimus and tacrolimus have been introduced during the past years as new topical anti-inflammatory therapies. The only clinically relevant side-effect is initial burning and stinging itch with consequent rapid amelioration of pruritus. This resembles neurogenic inflammation induced by activation of the TrpV1 receptor. Recent animal studies provide evidence that both calcineurin inhibitors bind to the TrpV1 [80,90]. It was demonstrated that topical application of pimecrolimus and tacrolimus is followed by an initial release of SP and CGRP from primary afferent nerve fibers in mouse skin [90]. Animal studies proved that the Ca2+-dependent desensitization of TrpV1 receptor might be, in part, regulated through channel dephosphorylation by calcineurin [61,111].

2.5.2 Thermoreceptors

2.5.2.1 *Heat Receptors: TrpV2, TrpV3, TrpV4*

Three transient receptor potential (Trp) receptors are activated by different ranges of warmth or heat. TrpV2 is activated by noxious heat above 52°C; TrpV3 mediates warm temperature above 33°C, and TrpV4 also is activated by temperature around 25°C [13,14,50,71,98,110]. TrpV4 may also act as a cold receptor as shown by the binding of camphor, which induces a cold-feeling [71]. All three thermoreceptors are also present on keratinocytes. Recent animal studies suggest that skin surface temperature has an influence on epidermal permeability barrier. At temperatures 36–40°C, barrier recovery was accelerated. Temperatures of 34 or 42°C led to a delayed barrier recovery [19]. This suggested that TrpV is involved in epidermal barrier homeostasis. However, all of these receptors were defined quite recently and their expression patterns in the skin as well as detailed non-neuronal function await further exploration.

2.5.2.2 *Cold Receptors: TrpM8, TrpA1*

TrpM8 (CMR1) is a cold receptor expressed on myelinated Aδ-fibers that is stimulated by 8–28°C. Also menthol and icilin activate the TrpM8 and thereby may act as a therapeutic tool in the cold-mediated suppression of itch [68]. Another cold receptor, TrpA1 (ANKTM1), has a lower activation temperature (<17°C) compared to the TrpM8 receptor and is also activated by wasabi, horseradish, mustard, bradykinine, as well as tetrahydrocannabinol (THC) [45,71,95]. TrpA1 is found in a subset of nociceptive sensory neurons where it is co-expressed with-TrpV1 but not TrpM8. It was shown that lowering the skin temperature by cooling reduced the intensity of experimentally induced itch [11]. A similar effect was achieved with menthol, although the skin temperature was not decreased [11]. It was concluded that these findings suggest a central inhibitory effect of cold sensitive Aδ-fiber activation on itch. A role in cold hyperalgesia in inflammatory and neuropathic pain is assumed; however, the underlying mechanisms of this enhanced sensitivity to cold are poorly understood [65]. It has been speculated that cold hyperalgesia occurs by NGF mediating an increase in TrpA1 receptors on nerve fibers.

2.6 Proteinase-Activated Receptor 2

The proteinase-activated receptor-2 (PAR-2) was demonstrated on sensory nerve fibers and is activated by mast cell mediators such as tryptase [92]. Activation leads to induction of pruritus and neurogenic inflammation comparable to effects induced upon histamine release from mast cells [66,102]. In atopic dermatitis (AD), PAR-2 expression was enhanced on primary afferent nerve fibers in the lesional skin, suggesting that this receptor is involved in pathophysiology of pruritus in AD [93]. This may also explain the inefficacy of antihistamines in AD as they do not block the tryptase-mast cell axis. Cutaneous mast cells also express PAR-2, suggesting an additional autocrine mechanism [62]. PAR-2 was recently suggested to also be involved in pain mechanisms. Activation of PAR-2 is reported to induce sensitization of primary nociceptors along with hyperalgesia [21]. Together, these results suggest PAR-2 to be involved in cutaneous nociception mainly during inflammation.

2.7 Opioid Receptors

Two opioid receptors, the μ- and δ-receptor, have been demonstrated on sensory nerve fibers [85,86]. Opioid peptides such as β-endorphin, enkephalins, and endomorphins act on capsaicin-sensitive nerve fibers to inhibit the release of inflammatory neuropeptides such as SP, neurokinin A, and CGRP [51,74]. In previous studies, it was shown that peripheral opioid receptors mediate antinociceptive effects preferentially by activation of the μ-receptor and less by δ-receptor [91]. Application of peripheral morphine inhibited responses to both mechanical and thermal stimuli in inflamed skin, suggesting that peripheral opioids might modulate pain responses [107]. These findings suggest that peripheral opioid receptors act as inhibitory receptors in the skin.

In contrast, in the central nervous system, clinical and experimental observations suggest that pruritus can be evoked or intensified by endogenous or exogenous opioids [7,29,49,76]. For example, systemically administrated morphine suppresses pain but induce pruritus [97]. This phenomenon can be explained by activation of spinal opioid receptors, mainly mu- and to a lesser extent kappa- and delta-opioid receptors, on pain transmitting neurons, which induce analgesia, often combined with induction of pruritus [78]. Reversing this effect by mu-opioid antagonists results thereby in inhibition of pruritus. Accordingly, several experimental studies have demonstrated that different mu-opioid receptor antagonists may significantly diminish pruritus [8,60].

2.8 Cannabinoid Receptors

Up to now, two cannabinoid receptors, CB1 and CB2, have been defined precisely by their wide expression in the CNS and on immune cells [20,56,63]. CB1 was described as being densely localized in the CNS; recent studies revealed an additional expression of CB1 in peripheral tissue, that is, primary afferent neurons [2,16,72]. CB2 receptors were mainly found in the periphery, for example, on T-lymphocytes, mast cells [26,30], and also on rat spinal cord [112]. Both receptors were recently found to be expressed also on cutaneous sensory nerve fibers, mast cells, and keratinocytes [88].

Topical application of cannabinoid agonists leads to inhibition of pain, pruritus, and neurogenic inflammation [23,75,89]. During inflammation, CB1 expression in primary afferent neurons and transport to peripheral axons is increased and contributes thereby to enhanced antihyperalgesic efficacy of locally administered CB1 agonist [4]. In addition, it was demonstrated that injections of the CB2 agonist palmitoylethanolamine (PEA) may inhibit experimental NGF-induced thermal hyperalgesia [27].

However, the antinociceptive effects are believed to be mediated in part by opioid and vanilloid mechanisms and not directly by activation of cannabinoid receptors. For example, it was shown that the CB1 agonist anandamide binds to the TrpV1 receptor [114] and that topical cannabinoids directly inhibit TrpV1 functional activities via a calcineurin pathway [69]. Moreover, it was demonstrated that the antinociceptive effects of CB2 agonists can be prevented by the μ-opioid receptor-antagonist naloxone [28,104]. Interestingly, the cannabinoid agonist AM1241 stimulates β-endorphin release from rat skin tissue and from cultured human keratinocytes [40]. In sum, cannabinoid receptors seem to exert a central role in cutaneous nociception mediating direct and indirect effects and therefore represent interesting targets for the development of antinociceptive therapies.

2.9 Trophic Factors

2.9.1 Nerve Growth Factor

Neurotrophins have been found in recent years to play a major role in skin homeostasis and inflammatory diseases. One member of this family, NGF, has several regulatory functions in cutaneous nociception, cutaneous nerve development, and reconstruction after injury through action on peptidergic C-fibers [9,64]. In epidermal keratinocytes, NGF production underlies neuropeptide release. After release of neuropeptides by a nociceptive stimulus, an upregulation of the expression of NGF and NGF secretion from the keratinocytes is induced [17]. Released NGF acts on skin nerves to sensitize neuroreceptors towards noxious thermal, mechanical, and chemical stimuli (see above) and is transported along the axon towards the dorsal root ganglia (DRG) to induce upregulation of a variety of proteins involved in neuronal growth and sensitivity. These mechanisms lead to altered peripheral nociception, for example, facilitated induction of pruritus and pain. For example, prolonged treatment of rats with moderate doses of NGF is sufficient to stimulate neuropeptide synthesis in primary afferent neurons without causing long-lasting changes in thermal nociceptive threshold [79]. Moreover, application of NGF also enhances capsaicin-evoked thermal hyperalgesia [10]. In cutaneous inflammatory diseases, NGF was demonstrated to be over-expressed in prurigo nodularis, and in AD where it is speculated to contribute to the neurohyperplasia of the disease [22,44], as well as in allergic diseases [64].

2.9.2 Glial Cell Line-Derived Neurotrophic Factor (GDNF)

During embryonic development, nociceptors are dependent on NGF, but a large subpopulation lose this dependence during embryonic and postnatal life and become responsive to the transforming growth factor beta family member, glial cell line-derived growth factor (GDNF). The family comprises members such as artemin, neurturin and glial cell line-derived growth factor, which are involved in the induction and maintenance of pain and hyperalgesia [3]. These factors act on non-peptidergic C-fibers [115,116] and expression of GDNF in the skin can change mechanical sensitivity [3]. More importantly, GDNF sensitizes thermal nociceptors towards cold or heat hyperalgesia by potentiation of TrpV1 signaling or increased expression of TrpA1 [25,54,55]. In the DRG, exposure to GDNF, neurturin, or artemin potentate TrpV1 function at doses 10–100 times lower than NGF. Moreover, GDNF family members induced capsaicin responses in a subset of neurons that were previously insensitive to capsaicin [55]. Exposure of nerve fibers to GDNF induces, in addition, expression of prokineticin receptors (PKR) in the nonpeptidergic population of neurons. These receptors cause heat hyperalgesia by sensitizing TRPV1 [103].

Summary for the Clinician

The skin is equipped with a dense network of specialized nerve fibers for the sensation of external stimuli such as cold, warmth, tough, pain, and itch. During the past years, many neuroreceptors were identified on sensory nerve fibers, which mediate these sensations and respond to external stimuli. The chronification of sensations such as pain and itch underlie complex mechanisms such as sensitization of neuroreceptors. Several modern therapies are yet identified to interact with these mechanism and to achieve the clinical relief of peripheral pain and pruritus.

References

1. Abadia Molina F, Burrows NP, Jones RR, et al (1992) Increased sensory neuropeptides in nodular prurigo: a quantitative immunohistochemical analysis. Br J Dermatol 127:344–351
2. Ahluwalia J, Urban L, Capogna M, et al (2000) Cannabinoid 1 receptors are expressed in nociceptive primary sensory neurons. Neuroscience 100:685–688
3. Albers KM, Woodbury CJ, Ritter AM, et al (2006) Glial cell-line-derived neurotrophic factor expression in skin alters the mechanical sensitivity of cutaneous nociceptors. J Neurosci 26:2981–2990
4. Amaya F, Shimosato G, Kawasaki Y, et al (2006) Induction of CB1 cannabinoid receptor by inflammation in primary afferent neurons facilitates antihyperalgesic effect of peripheral CB1 agonist. Pain 124:175–183
5. Ansel JC, Armstrong CA, Song I, et al (1997). Interactions of the skin and nervous system. J Investig Dermatol Symp Proc 2:23–26.
6. Bell JK, McQueen DS, Rees JL (2004) Involvement of histamine H4 and H1 receptors in scratching induced by histamine receptor agonists in Balb C mice. Br J Pharmacol 142:374–380
7. Bernstein JE, Grinzi RA (1981) Butorphanol-induced pruritus antagonized by naloxone. J Am Acad Dermatol 5:227–228
8. Bernstein JE, Swift RM (1979) Relief of intractable pruritus with naloxone. Arch Dermatol 115:1366–1367

9. Botchkarev VA, Yaar M, Peters EM, et al (2006) Neurotrophins in skin biology and pathology. J Invest Dermatol 126:1719–1727
10. Bowles WR, Sabino M, Harding-Rose C, et al (2006) Chronic nerve growth factor administration increases the peripheral exocytotic activity of capsaicin-sensitive cutaneous neurons. Neurosci Lett 403:305–308
11. Bromm B, Scharein E, Darsow U, et al (1995) Effects of menthol and cold on histamine-induced itch and skin reactions in man. Neurosci Lett 187:157–160
12. Chateau Y, Misery L (2004) Connections between nerve endings and epidermal cells: are they synapses? Exp Dermatol 13:2–4
13. Chung MK, Lee H, Caterina MJ (2003) Warm temperatures activate TRPV4 in mouse 308 keratinocytes. J Biol Chem 278:32037–32046
14. Chung MK, Lee H, Mizuno A, et al (2004) TRPV3 and TRPV4 mediate warmth-evoked currents in primary mouse keratinocytes. J Biol Chem 279:21569–1575
15. Church MK, el-Lati S, Caulfield JP (1991) Neuropeptide-induced secretion from human skin mast cells. Int Arch Allergy Appl Immunol 94:310–318
16. Coutts AA, Irving AJ, Mackie K, et al (2002) Localisation of cannabinoid CB(1) receptor immunoreactivity in the guinea pig and rat myenteric plexus. J Comp Neurol 448:410–422
17. Dallos A, Kiss M, Polyanka H, et al (2006) Effects of the neuropeptides substance P, calcitonin gene-related peptide, vasoactive intestinal polypeptide and galanin on the production of nerve growth factor and inflammatory cytokines in cultured human keratinocytes. Neuropeptides 40:251–263
18. Delgado AV, McManus AT, Chambers JP, et al (2003) Production of tumor necrosis factor-alpha, interleukin 1-beta, interleukin 2, and interleukin 6 by rat leukocyte subpopulations after exposure to substance P. Neuropeptides 37:355–361
19. Denda M, Sokabe T, Fukumi-Tominaga T, et al (2007) Effects of skin surface temperature on epidermal permeability barrier homeostasis. J Invest Dermatol 127:654–659
20. Devane WA, Dysarz FA IIIrd, Johnson MR, et al (1988) Determination and characterization of a cannabinoid receptor in rat brain. Mol Pharmacol 34:605–613
21. Ding-Pfennigdorff D, Averbeck B, Michaelis M (2004) Stimulation of PAR-2 excites and sensitizes rat cutaneous C-nociceptors to heat. Neuroreport 15:2071–2075
22. Dou YC, Hagstromer L, Emtestam L, et al (2006) Increased nerve growth factor and its receptors in atopic dermatitis: an immunohistochemical study. Arch Dermatol Res 298:31–37
23. Dvorak M, Watkinson A, McGlone F, et al (2003) Histamine induced responses are attenuated by a cannabinoid receptor agonist in human skin. Inflamm Res 52:238–245
24. E Y, Golden SC, Shalita AR, et al (2006) Neuropeptide (calcitonin gene-related peptide) induction of nitric oxide in human keratinocytes in vitro. J Invest Dermatol 126:1994–2001
25. Elitt CM, McIlwrath SL, Lawson JJ, et al (2006) Artemin overexpression in skin enhances expression of TRPV1 and TRPA1 in cutaneous sensory neurons and leads to behavioral sensitivity to heat and cold. J Neurosci 26:8578–8587
26. Facci L, Dal Toso R, Romanello S, et al (1995) Mast cells express a peripheral cannabinoid receptor with differential sensitivity to anandamide and palmitoylethanolamide. Proc Natl Acad Sci U S A 92:3376–3380
27. Farquhar-Smith WP, Rice AS (2003) A novel neuroimmune mechanism in cannabinoid-mediated attenuation of nerve growth factor-induced hyperalgesia. Anesthesiology 99:1391–1401
28. Fattore L, Cossu G, Spano MS, et al (2004) Cannabinoids and reward: interactions with the opioid system. Crit Rev Neurobiol 16:147–158
29. Fjellner B, Hägermark O (1982) Potentiation of histamine-induced itch and flare response in human skin by the enkephalin analogue FK 33–824, beta-endorphin and morphine. Arch Dermatol Res 274:29–37
30. Galiegue S, Mary S, Marchand J, et al (1995) Expression of central and peripheral cannabinoid receptors in human immune tissues and leukocyte subpopulations. Eur J Biochem 232:54–61
31. Gaudillere A, Misery L, Souchier C, et al (1996) Intimate associations between PGP9.5-positive nerve fibres and Langerhans cells. Br J Dermatol 135:343–344
32. Gibran NS, Jang YC, Isik FF, et al (2002) Diminished neuropeptide levels contribute to the impaired cutaneous healing response associated with diabetes mellitus. J Surg Res 108:122–128
33. Gray EG (1963) Electron microscopy of presynaptic organelles of the spinal cord. J Anat 97:101–106
34. Hara M, Toyoda M, Yaar M, et al (1996) Innervation of melanocytes in human skin. J Exp Med 184:1385–1395
35. Hill SJ (1990) Distribution, properties, and functional characteristics of three classes of histamine receptor. Pharmacol Rev 42:45–83
36. Hilliges M, Wang L, Johansson O (1995) Ultrastructural evidence for nerve fibers within all vital layers of the human epidermis. J Invest Dermatol 104:134–137
37. Hosoi J, Murphy GF, Egan CL, et al (1993) Regulation of Langerhans cell function by nerves containing calcitonin gene-related peptide. Nature 363:159–163
38. Hsieh ST, Lin WM (1999) Modulation of keratinocyte proliferation by skin innervation. J Invest Dermatol 113:579–586
39. Hu HJ, Bhave G, Gereau RW IVth (2002) Prostaglandin and protein kinase A-dependent modulation of vanilloid receptor function by metabotropic glutamate receptor 5: potential mechanism for thermal hyperalgesia. J Neurosci 22:7444–7452
40. Ibrahim MM, Porreca F, Lai J, et al (2005) CB2 cannabinoid receptor activation produces antinociception by stimulating peripheral release of endogenous opioids. Proc Natl Acad Sci U S A 102(8):3093–3098
41. Ikoma A, Steinhoff M, Ständer S, et al (2006) Neurobiology of pruritus. Nat Rev Neurosci 7:535–547
42. Jarvikallio A, Harvima IT, Naukkarinen A (2003) Mast cells, nerves and neuropeptides in atopic dermatitis and nummular eczema. Arch Dermatol Res 295:2–7
43. Jiang WY, Raychaudhuri SP, Farber EM (1998) Double-labeled immunofluorescence study of cutaneous nerves in psoriasis. Int J Dermatol 37:572–574
44. Johansson O, Liang Y, Emtestam L (2002) Increased nerve growth factor- and tyrosine kinase A-like immunoreactivities

in prurigo nodularis skin – an exploration of the cause of neurohyperplasia. Arch Dermatol Res 293:614–619
45. Jordt SE, Bautista DM, Chuang HH, et al (2004) Mustard oils and cannabinoids excite sensory nerve fibres through the TRP channel ANKTM1. Nature 427:260–265
46. Julius D, Basbaum AI (2001) Molecular mechanisms of nociception. 413:203–210
47. Kanda N, Watanabe S (2003) Histamine enhances the production of nerve growth factor in human keratinocytes. J Invest Dermatol 121:570–577
48. Katugampola R, Church MK, Clough GF (2000) The neurogenic vasodilator response to endothelin-1: a study in human skin in vivo. Exp Physiol 85:839–846
49. Ko MC, Naughton NN (2000) An experimental itch model in monkeys: characterization of intrathecal morphine-induced scratching and antinociception. Anesthesiology 92:795–805
50. Lee H, Caterina MJ (2005) TRPV channels as thermosensory receptors in epithelial cells. Pflugers Arch 451:160–167
51. Lembeck F, Donnerer J, Bartho L (1982) Inhibition of neurogenic vasodilation and plasma extravasation by substance P antagonists, somatostatin and [D-Met^2, Pro^5]-enkephalinamide. Eur J Pharmacol 85:171–176
52. Lewis T, Grant RT, Marvin HM (1929) Vascular reactions of the skin to injury. Heart 14:139–160
53. Liang Y, Marcusson JA, Jacobi HH, et al (1998) Histamine-containing mast cells and their relationship to NGFr-immunoreactive nerves in prurigo nodularis: a reappraisal. J Cutan Pathol 25:189–198
54. Lindfors PH, Voikar V, Rossi J, et al (2006) Deficient nonpeptidergic epidermis innervation and reduced inflammatory pain in glial cell line-derived neurotrophic factor family receptor alpha2 knock-out mice. J Neurosci 26:1953–1960
55. Malin SA, Molliver DC, Koerber HR, et al (2006) Glial cell line-derived neurotrophic factor familiy members sensitize nociceptors in vitro and produce thermal hyperalgesia in vivo. J Neurosci 26:8588–8599
56. Matsuda LA, Lolait SJ, Brownstein MJ, et al (1990) Structure of a cannabinoid receptor and functional expression of the cloned cDNA. Nature 346:561–564
57. Matsushima H, Yamada N, Matsue H, et al (2004) The effects of endothelin-1 on degranulation, cytokine, and growth factor production by skin-derived mast cells. Eur J Immunol 34:1910–1919
58. Merkel F (1875) Tastzellen und Tastkorperchen bei den Haustieren und beim Menschen. Arch Mikrosk Anat 11:636–652
59. Metz M, Lammel V, Gibbs BF, et al (2006) Inflammatory murine skin responses to UV-B light are partially dependent on endothelin-1 and mast cells. Am J Pathol 169:815–822
60. Metze D, Reimann S, Beissert S, et al (1999) Efficacy and safety of naltrexone, an oral opiate receptor antagonist, in the treatment of pruritus in internal and dermatological diseases. J Am Acad Dermatol 41:533–539
61. Mohapatra DP, Nau C (2005) Regulation of Ca2+-dependent desensitization in the vanilloid receptor TRPV1 by calcineurin and cAMP-dependent protein kinase. J Biol Chem 280:13424–13432.
62. Moormann C, Artuc M, Pohl E, et al (2006) Functional characterization and expression analysis of the proteinase-activated receptor-2 in human cutaneous mast cells. J Invest Dermatol 126:746–755
63. Munro S, Thomas KL, Abu-Shaar M (1993) Molecular characterization of a peripheral receptor for cannabinoids. Nature 365:61–65
64. Nockher WA, Renz H (2006) Neurotrophins in allergic diseases: from neuronal growth factors to intercellular signaling molecules. J Allergy Clin Immunol 117:583–589
65. Obata K, Katsura H, Mizushima T, et al (2005) TRPA1 induced in sensory neurons contributes to cold hyperalgesia after inflammation and nerve injury. J Clin Invest 115:2393–2401
66. Obreja O, Rukwied R, Steinhoff M, et al (2006) Neurogenic components of trypsin- and thrombin-induced inflammation in rat skin, in vivo. Exp Dermatol 15:58–65
67. Ohkubo T, Shibata M, Inoue M, et al (1995) Regulation of substance P release mediated via prejunctional histamine H3 receptors. Eur J Pharmacol 273:83–88
68. Patapoutian A (2005) TRP channels and thermosensation. Chem Senses 30(Suppl 1):i193–i194
69. Patwardhan AM, Jeske NA, Price TJ, et al (2006) The cannabinoid WIN 55,212–2 inhibits transient receptor potential vanilloid 1 (TRPV1) and evokes peripheral antihyperalgesia via calcineurin. Proc Natl Acad Sci U S A 103:11393–11398
70. Payan DG, Goetzl EJ (1987) Dual roles of substance P: modulator of immune and neuroendocrine functions. Ann N Y Acad Sci 512:465–475
71. Peier AM, Reeve AJ, Andersson DA, et al (2002) A heat-sensitive TRP channel expressed in keratinocytes. Science 296:2046–2049
72. Pertwee RG (1999) Evidence for the presence of CB1 cannabinoid receptors on peripheral neurones and for the existence of neuronal non-CB1 cannabinoid receptors. Life Sci 65:597–605
73. Quinlan KL, Song IS, Naik SM, et al (1999) VCAM-1 expression on human dermal microvascular endothelial cells is directly and specifically up-regulated by substance P. J Immunol 162:1656–1661
74. Ray NJ, Jones AJ, Keen P (1991) Morphine, but not sodium cromoglycate, modulates the release of substance P from capsaicin-sensitive neurones in the rat trachea in vitro. Br J Pharmacol 102:797–800
75. Rukwied R, Watkinson A, McGlone F, et al (2003) Cannabinoid agonists attenuate capsaicin-induced responses in human skin. Pain 102:283–288
76. Sakurada T, Sakurada S, Katsuyama S, et al (2000) Evidence that N-terminal fragments of nociceptin modulate nociceptin-induced scratching, biting and licking in mice. Neurosci Lett 279:61–64
77. Schmelz M, Schmidt R, Bickel A, et al (1998) Innervation territories of single sympathetic C fibers in human skin. J Neurophysiol 79:1653–1660
78. Schmelz M, Schmidt R, Weidner C, et al (2003) Chemical response pattern of different classes of C-nociceptors to pruritogens and algogens. J Neurophysiol 89:2441–2448
79. Schuligoi R, Amann R (1998) Differential effects of treatment with nerve growth factor on thermal nociception and on calcitonin gene-related peptide content of primary afferent neurons in the rat. Neurosci Lett 252:147–149

80. Senba E, Katanosaka K, Yajima H, et al (2004) The immunosuppressant FK506 activates capsaicin- and bradykinin-sensitive DRG neurons and cutaneous C-fibers. Neurosci Res 50:257–262
81. Shu X, Medell LM (1998) Nerve growth factor acutely sensitizes the response of adult rat sensory neurons to capsaicin. J Neurosci 18:8947–8959
82. Singh LK, Pang X, Alexacos N, et al (1999) Acute immobilization stress triggers skin mast cell degranulation via corticotropin releasing hormone, neurotensin, and substance P: a link to neurogenic skin disorders. Brain Behav Immun 13:225–239
83. Ständer S, Metze D (2004) Treatment of pruritic skin diseases with topical capsaicin. In: Yosipovitch G, Greaves MW, Fleischer AB, McGlone F (eds) Itch: Basic Mechanisms and Therapy. Marcel Dekker, New York, pp 287–304
84. Ständer S, Schmelz M (2006) Chronic itch and pain – similarities and differences. Eur J Pain10:473–478
85. Ständer S, Gunzer M, Metze D, et al (2002) Localization of μ-opioid receptor 1A on sensory nerve fibers in human skin. Regul Pept 110:75–83
86. Ständer S, Steinhoff M, Schmelz M, et al (2003) Neurophysiology of pruritus. Cutaneous elicitation of itch. Arch Dermatol 139:1463–1470
87. Ständer S, Moormann C, Schumacher M, et al. (2004). Expression of vanilloid receptor subtype 1 (VR1) in cutaneous sensory nerve fibers, mast cells and epithelial cells of appendage structures. Exp Dermatol 13:129–139
88. Ständer S, Schmelz M, Metze D, et al (2005) Distribution of cannabinoid receptor 1 (CB1) and 2 (CB2) on sensory nerve fibers and adnexal structures in human skin. J Dermatol Sci 38:177–188
89. Ständer S, Reinhardt HW, Luger TA (2006) Topische Cannabinoid-Agonisten: Eine effektive, neue Möglichkeit zur Behandlung von chronischem Pruritus. Hautarzt 57:801–807
90. Ständer S, Ständer H, Seeliger S, et al (2007) Topical pimecrolimus (SDZ ASM 981) and tacrolimus (FK 506) transiently induces neuropeptide release and mast cell degranulation in murine skin. Br J Dermatol 156:1020–1026
91. Stein C, Millan MJ, Shippenberg TS, et al (1989) Peripheral opioid receptors mediating antinociception in inflammation. Evidence for involvement of mu, delta and kappa receptors. J Pharmacol Exp Ther 248:1269–1275
92. Steinhoff M, Vergnolle N, Young SH, et al (2000) Agonists of proteinase-activated receptor 2 induce inflammation by a neurogenic mechanism. Nat Med 6:151–158
93. Steinhoff M, Neisius U, Ikoma A, et al (2003) Proteinase-activated receptor-2 mediates itch: a novel pathway for pruritus in human skin. J Neurosci 23:6176–6180
94. Steinhoff M, Stander S, Seeliger S, et al (2003) Modern aspects of cutaneous neurogenic inflammation. Arch Dermatol 139:1479–1488
95. Story GM, Peier AM, Reeve AJ, et al (2003) ANKTM1, a TRP-like channel expressed in nociceptive neurons, is activated by cold temperatures. Cell 112:819–829
96. Sugimoto Y, Iba Y, Nakamura Y, et al (2004) Pruritus-associated response mediated by cutaneous histamine H3 receptors. Clin Exp Allergy 34:456–459
97. Summerfield JA (1981) Pain, itch and endorphins. Br J Dermatol 105:725–726
98. Tamura S, Morikawa Y, Senba E (2005) TRPV2, a capsaicin receptor homologue, is expressed predominantly in the neurotrophin-3-dependent subpopulation of primary sensory neurons. Neuroscience 130:223–228
99. Thomsen JS, Sonne M, Benfeldt E, et al (2002) Experimental itch in sodium lauryl sulphate-inflamed and normal skin in humans: a randomized, double-blind, placebo-controlled study of histamine and other inducers of itch. Br J Dermatol 146:792–800
100. Togias A (2003) H1-receptors: localization and role in airway physiology and in immune functions. J Allergy Clin Immunol 112(Suppl):S60–S68
101. Trentin PG, Fernandes MB, D'Orleans-Juste P, et al (2006) Endothelin-1 causes pruritus in mice. Exp Biol Med (Maywood) 231:1146–1151
102. Ui H, Andoh T, Lee JB, et al (2006) Potent pruritogenic action of tryptase mediated by PAR-2 receptor and its involvement in anti-pruritic effect of nafamostat mesilate in mice. Eur J Pharmacol 530:172–178
103. Vellani V, Colucci M, Lattanzi R, et al (2006) Sensitization of transient receptor potential vanilloid 1 by the prokineticin receptor agonist Bv8. J Neurosci 26:5109–5116
104. Vigano D, Rubino T, Parolaro D (2005) Molecular and cellular basis of cannabinoid and opioid interactions. Pharmacol Biochem Behav 81:360–368
105. Weidner C, Klede M, Rukwied R, et al (2000) Acute effects of substance P and calcitonin gene-related peptide in human skin – a microdialysis study. J Invest Dermatol 115:1015–1020
106. Weller K, Foitzik K, Paus R, et al (2006) Mast cells are required for normal healing of skin wounds in mice. FASEB J 20:2366–2368
107. Wenk HN, Brederson JD, Honda CN (2006) Morphine directly inhibits nociceptors in inflamed skin. J Neurophysiol 95:2083–2097
108. Wenzel RR, Zbinden S, Noll G, et al (1998) Endothelin-1 induces vasodilation in human skin by nociceptor fibres and release of nitric oxide. Br J Clin Pharmacol 45:441–446
109. Wiesner-Menzel L, Schulz B, Vakilzadeh F, et al (1981) Electron microscopal evidence for a direct contact between nerve fibres and mast cells. Acta Derm Venereol 61:465–469
110. Woodbury CJ, Zwick M, Wang S, et al (2004) Nociceptors lacking TRPV1 and TRPV2 have normal heat responses. J Neurosci 24:6410–6415
111. Wu ZZ, Chen SR, Pan HL (2005) Transient receptor potential vanilloid type 1 activation down-regulates voltage-gated calcium channels through calcium-dependent calcineurin in sensory neurons. J Biol Chem 280:18142–18151
112. Zhang J, Hoffert C, Vu HK, et al (2003) Induction of CB2 receptor expression in the rat spinal cord of neuropathic but not inflammatory chronic pain models. Eur J Neurosci 17:2750–2754
113. Zhang X, Huang J, McNaughton PA (2005) NGF rapidly increases membrane expression of TRPV1 heat-gated ion channels. Embo J 24:4211–4223
114. Zygmunt PM, Petersson J, Andersson DA, et al (1999) Vanilloid receptors on sensory nerves mediate the vasodilator action of anandamide. Nature 400:452–457
115. Zylka MJ (2005) Nonpeptidergic circuits feel your pain. Neuron 47:771–772 (Comment on: Neuron 2005 47:787–793)
116. Zylka MJ, Rice FL, Anderson DJ (2005) Topographically distinct epidermal nociceptive circuits revealed by axonal tracers targeted to Mrgprd. Neuron 45:17–25

Autonomic Effects on the Skin

3

F. Birklein and T. Schlereth

Contents

Key Features

- This chapter describes anatomy and function of the sympathetic nervous system in the skin.
- Sympathetic activity mainly serves maintenance of body temperature (thermoregulation).
- Sympathetic vasoconstrictor activity can be evaluated by quantifying vasoconstrictor reflexes.
- Examples of disorders leading to vasoconstrictor abnormalities are Raynaud's syndrome or Complex Regional Pain Syndrome.
- Almost unique in humans sweating is also controlled by the sympathetic activity.
- Sweating can be quantified using sudometry, which helps to diagnose sweating disorders like hyperhidrosis or autonomic neuropathies.
- More recently the role of the sympathetic nervous system for cellular and neurogenic inflammation was detected.

Synonyms Box: Noradrenaline, norepinephrine; Vasodilation, vasodilation; Anhidrosis, lack of sweating

3.1 Anatomy of the Autonomic Nervous System

The autonomic nervous system has two major functional parts – the sympathetic and parasympathetic systems. The parasympathetic system mainly regulates inner organs and glands, it participates in blood pressure and heart rate regulation and it holds an important role in uro-genital function. The impact of the parasympathetic nervous system on the skin, however, is of negligible importance in humans. The skin is mainly innervated by the sympathetic nervous system. Sympathetic preganglionic neurons that regulate effector cells in the skin originate in the lateral cell columns of the thoracic spinal cord. They project to the paravertebral ganglia of the sympathetic trunk and synapse with postganglionic neurons that travel with peripheral nerves and innervate the effector cells. Preganglionic sympathetic neurons themselves are under direct central control involving centers in the brain stem and hypothalamus [23]. Sympathetic efferent activity to the skin mainly subserves thermoregulation. Heat loss or gain mainly depends on activity of hypothalamic thermoregulatory neurons, but not exclusively. In a clinical study investigating stroke patients, we have been able to show that the cerebral cortex exerts an

R.D. Granstein and T.A. Luger (eds.), *Neuroimmunology of the Skin*,

inhibitory function on the tonically active sympathetic nervous system [33]. Cortico-hypothalamic pathways project to the contralateral hypothalamus. Pathways then descend via the brainstem to the ipsilateral preganglionic neurons in the intermediolateral cell columns of the spinal cord [40]. Central disinhibition after stroke acutely leads to increased activity of preganglionic sudo - and vasomotor neurons [44]. In the skin itself the main structures under sympathetic control are blood vessels and sweat glands, and – less important for humans – hair follicles, in particular the arrector pili muscles. In addition to these classical effector organs, the sympathetic nervous system may have an impact on immune cells and primary afferent neurons, in particular in chronic inflammatory and neuropathic conditions. While blood vessels are innervated by postganglionic noradrenergic sympathetic nerve fibers, sweat glands are innervated by cholinergic postganglionic sympathetic fibers. This duality of either noradrenergic or cholinergic innervation is necessary to achieve thermoregulation. In a cold environment, noradrenergic nerves become activated, leading to vasoconstriction while cholinergic neurons are inhibited and sweat production stops. In a hot environment, activation patterns are just the opposite.

3.2 Sympathetic Vasomotor Control in the Skin

Microcirculation in the skin is anatomically organized in two horizontal plexus: The upper one just below the epidermis and the lower one between dermis and subcutaneous tissue. Larger arteries, which are densely innervated by sympathetic axons, arise from the deeper tissue and feed both plexus. Smaller arterioles interconnect the deep and the superficial plexus and shunt blood from arterioles to venules [25]. They are innervated by only a few sympathetic axons on the surface of the vessels. All these blood vessels belong to the thermoregulatory skin blood flow system. Capillaries that originate from the superficial plexus and extend to the dermal papillae are usually not innervated by sympathetic neurons, or if they are, only sparsely. Capillary blood flow has only minor thermoregulatory, but major nutritional function, since capillaries deliver oxygen to the tissue [27].

Thermoregulatory skin blood flow ranges from 10–20 ml min^{-1} per 100 g in a thermo-neutral environment [12]. During severe heat stress, thermoregulatory skin blood flow can increase to 150–200 ml min^{-1} per 100 g generating heat loss of about 4,000 kJ h^{-1}. During cold exposure, skin flow can be reduced to <1 ml min^{-1} per 100 g. This variation is mainly achieved by activity changes of vasoconstrictor sympathetic neurons. High sympathetic activity, regardless of thermoregulatory, baroreceptor, or stress origin, leads to high noradrenaline signaling and subsequently to arteriolar vasoconstriction and cold pale skin. Low sympathetic activity reduces noradrenaline signaling and arterioles passively dilate, causing the skin to warm up. An active sympathetic vasodilation is discussed controversially: It is unlikely to occur in the finger tips but it may occur on hairy skin. The most probable mechanisms of active vasodilation are related to sudomotor activation (see below) or primary afferent activation [26].

Maintenance of vasomotor tone, on the one hand, and timely responses to homeostatic challenges, on the other, are typical features of skin vasomotion. Therefore, activation of sympathetic vasoconstrictors is organized in different reflex arcs involving different levels of the nervous system. The veno-arteriolar reflex depends solely on very peripheral sympathetic nerve function [54]. It can be elicited during nerve trunk blockade, thus it works like a local axon reflex [19]. Distension of veins by lowering the limb induces arteriolar constriction that reduces skin blood flow. The veno-arteriolar reflex disappears, corresponding to the degeneration of postganglionic sympathetic nerve fibers, after surgical sympathectomy. Skin vasoconstriction is also under baroreceptor control. A baroreceptor reflex can be elicited by, for example, standing up or deep inspiratory gasps, the latter causing a phasic vasoconstriction [7], which is activated by chest excursion. The afferent part of this reflex uses spinal afferents. Another vasoconstrictor reflex employs thermoreception. Cold exposure of the skin activates cold nociceptors and cold fibers [15] and induces a more tonic vasoconstrictor activation. Finally, the brain itself does not only exert inhibitory functions on the sympathetic nervous system (see above). In particular, limbic brain regions, involved in generating emotional stress responses, induce vasoconstriction [32]. For example, mental arithmetic under a certain time limit induces a moderate but long-lasting peripheral vasoconstriction.

The major transmitter mediating vasoconstriction in human skin is noradrenaline. Neuropeptide Y coexists in sympathetic fibers and may play a role in cutaneous vasoconstriction [27]. In a microdialysis study, we were able to show that human skin blood flow is under the control of both alpha1- and alpha2- receptors on smooth muscles in the arteriolar walls. We applied different concentrations of noradrenaline (alpha1- and alpha2-agonist), clonidine (alpha2-agonist), and phenylephrine (alpha1-agonist) via intradermal microdialysis fibers into the skin of healthy volunteers. Skin blood flow was visualized by laser-Doppler imaging scans and quantified in a skin area close to the membranes (Fig. 3.1). Alpha1- and alpha2-agonists cause equipotent and dose-dependent vasoconstriction [59]. In contrast to thermoregulatory blood flow, catecholamines exert a more differentiated action on nutritive blood flow in capillaries. Nutritive blood flow, however, constitutes only 10% of total skin blood flow. Nutritive blood flow is under alpha-receptor vasoconstrictive and also under beta-receptor vasodilatory control [27]. One might speculate that under physiological conditions both mechanisms neutralize each other. In human subcutaneous resistance vessels, it has been shown that one important alpha1-receptor subtype in the contractile responses to noradrenaline is the alpha1A-receptor [24]. In skin blood flow the alpha2C-receptor has been found to be important for vasoconstriction [57].

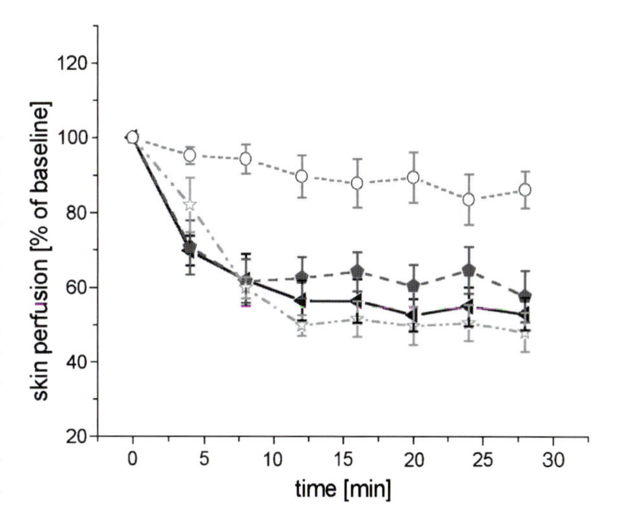

Fig. 3.1 As compared to saline (*circles*), 10^{-4} M solution of noradrenaline (*triangles*), clonidine (*pentagons*), and phenylephrine (*stars*) caused significant skin vasoconstriction when perfused through dermal microdialysis membranes. There was no difference between alpha1-R (phenylephrine) and alpha2 –R (clonidine) stimulation

As explained above, noradrenaline is released from sympathetic nerves on the outer surface of blood vessels. Noradrenaline, therefore, first activates adrenoreceptors on smooth muscles causing vasoconstriction. There are, however, further alpha-receptors located on endothelial cells inside blood vessels. In rodents it has been shown that activation of these alpha2-receptors induces release of nitric oxide (NO) [8] and prostacyclin [35]; both are strong vasodilators. To explore whether this mechanism could also be important in human skin we performed another microdialysis study (Fig. 3.2) [20]. The alpha2-agonist clonidine was delivered to the skin in pharmacological concentrations. As expected, the first reaction was vasoconstriction. After a few minutes, however, continuous delivery of clonidine then leads to local vasodilation. The switch from vasoconstriction to vasodilation strongly depends on clonidine concentration. Blocking the endothelial nitric oxide synthase (eNOS) by N^G^-monomethyl-L-arginine (NMMA) or cyclooxygenase (COX) by acetylsalicylic acid prevented clonidine-induced vasodilation. This means the most important second messengers mediating catecholamine-induced vasodilation in humans are nitric oxide and prostaglandins. The endothelial alpha2-receptor is a G-protein coupled receptor, which activates eNOS and NO is produced from L-arginine [14]. When NO is released from endothelium cells, it diffuses to the

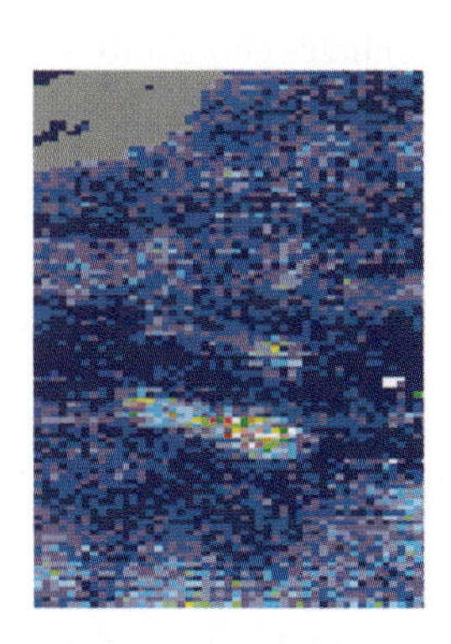

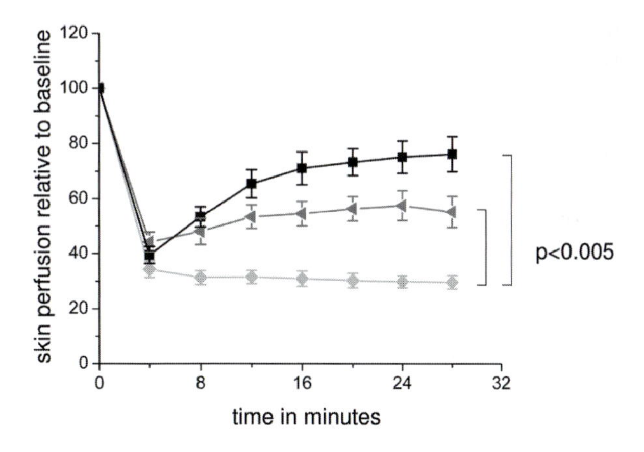

Fig. 3.2 While clonidine 10–4 M (*diamonds*) caused long lasting vasoconstriction, higher concentrations (5×10^{-4} M, *triangles* and 10^{-3} M, *squares*) induced significant revasodilation after initial vasoconstriction. This vasodilation is shown in a specimen skin perfusion picture on the left as a bright stripe

smooth muscles and mediates vasodilation via cGMP [11]. In consequence, intracellular Ca2+ and the sensitivity of the contractile system to Ca2+ decrease, leading to relaxation of the smooth muscle cells [10]. The exact mechanism of alpha2-receptor coupling to prostaglandin synthesis is not clear. However, it is known that prostaglandins are produced in endothelial cells by activation of phospholipase A2, cyclooxygenase, and prostacyclin synthase. After diffusion, prostaglandins mediate relaxation of smooth muscle cells and thereby induce vasodilation, which again is Ca2+-dependent [56]. The most important alpha2-receptor subtype might be the alpha2D-receptors [16]. To achieve vasodilation by clonidine we had to apply very high concentrations into the interstitial space. This means that under physiological conditions catecholamines, which are released from sympathetic nerves on the outer surfaces of blood vessels, probably will not induce vasodilation. However, catecholamines that occur in the circulation could easily reach endothelial alpha2-receptors. The origin of these catecholamines is the adrenal glands and some spill-over from sympathetic nerve terminals during sustained activation. Clinical examples that might be related to catecholamine endothelial dependent vasodilation might be the cyclic re-vasodilation, which occurs during sustained cold exposure to avoid tissue damage, purple fingers in some labile subjects with high sympathetic tone, or some forms of complex regional pain syndromes [2].

3.2.1 Measurement of Skin Vasoconstrictor Activity

Skin blood flow can be quantitatively measured by plethysmography, laser-Doppler flowmetry, or more indirectly by measuring skin temperature (e.g., by thermography) [4]. Under stable thermoregulatory conditions (acclimatized subjects, temperature-controlled lab) different vasconstrictor reflexes can be elicited by a number of autonomic maneuvers. The decrease of skin blood is quantified. The most often used autonomic maneuvers are the veno-arteriolar reflex, inspiratory gasps, the Valsalva maneuver, and the cold pressor test. In general, all of these reflexes are highly variable and there is usually a broad overlap between health and disease. Skin temperature recording is more stable, but in healthy subjects absolute skin temperature has a broad range. Left–right differences are more stable. Physiologically, skin temperature difference between corresponding sites is less than 1°C [55]. Greater differences might be of diagnostic value.

3.2.2 Examples of Sympathetic Disorders Leading to Vasoconstrictor Abnormalities

Peripheral denervation, for example, by neuropathies or any other nerve lesion first leads to increased skin temperature in the denervated skin areas, which is due to the loss of vasoconstrictor activity. Later on, increased temperature turns into cold skin [47]. The reason for this change of symptoms is that vasoconstrictor alpha-receptors develop so-called denervation supersensitivity. This means low concentrations of catecholamines suffice to maximally induce vasoconstriction due to "sensitized" alpha receptors. Another disorder of altered sympathetic vasoconstrictor activity is Complex Regional Pain Syndrome (CRPS) [4]. In CRPS, skin discoloration and skin temperature changes are associated with chronic pain, for which sympathetic blocking is sometimes mandatory (Fig. 3.3) [1].

3.3 Regulation of Eccrine Sweating

Humans and apes use eccrine sweat production for heat dissipation [46]. Sweating is initiated by cholinergic sympathetic signaling to sweat glands. Thermoregulatory sweating can be mainly found on hairy skin; emotionally induced sweating is more or less restricted to the glabrous skin on palms and soles.

Thermoregulation is controlled centrally and peripherally. An increase of core or skin temperature activates thermo-receptor afferent fibers and simultaneously or subsequently thermoregulatory brain centers in the reticular formation, nucleus raphe magnus [22] and particularly in the hypothalamus (preoptical nucleus) [6]. The increase of activity of thermoregulatory neurons in the hypothalamus initiates sudomotor drive. The efferent pathways are the same as for vasoconstriction. The transmitter of the pre- as well as the postganglionic fibers is acetylcholine. This is in contrast to vasoconstrictor neurons, in which only the preganglionic synapses are cholinergic. Like all other postganglionic sympathetic fibers, sudomotor efferents belong to the class of unmyelinated C-fibres, which branch into different nerve terminals [43]. Acetylcholine, which is released after activation, finally binds to muscarinic (M3) receptors at the eccrine sweat gland [29]. The pathway for emotional sweating is slightly different. Emotional sweating starts by activation of the limbic system in the brain (insula, amygdalae, hippocampus, cingulate gyrus). Efferents then travel through the brainstem and activate preganglionic

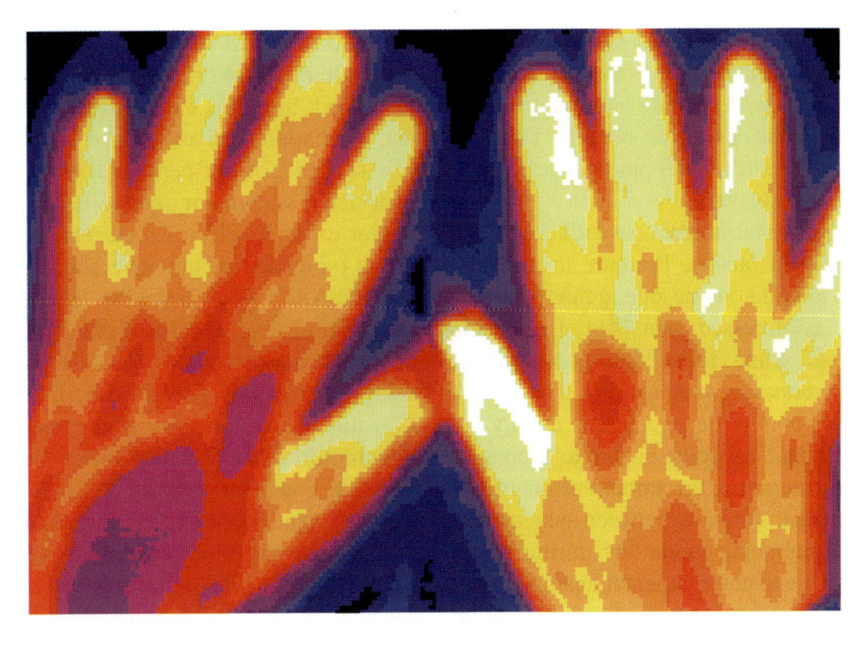

Fig. 3.3 Thermography allows to examine chronic symptoms of altered sympathetic innervation. A pair of hands from a chronic CRPS patient is shown. The left affected side was 3°C colder than the right side. This is indicated by darker colors

sympathetic neurons. The remaining pathway is the same as for thermoregulatory sweating.

In the peripheral nervous system, sudomotor efferent activity is unique because it can be self-amplifying. If a sudomotor axon reflex is initiated, for example, by acetylcholine, which binds to nicotinergic acetylcholine receptors on peripheral sudomotor axons, continuous spreading of the sudomotor axon reflex area can be observed [49]. This is in sharp contrast to, for example, nociceptive axon-reflexes. The mechanism of spreading might be that acetylcholine, released from sudomotor endings, activates nearby sudomotor fibers and thereby initiates a new axon reflex (Fig. 3.4). Thus, sweating spreads into the surrounding skin. This mechanism might contribute to sweat gland overactivity in hyperhidrosis. Another peripheral influence on sweating comes from primary afferent fibers. Upon activation, C-fibers release neuropeptides with induction of neurogenic inflammation. Neuropeptides dilate skin blood vessels and increase skin temperature. Both factors could indirectly increase local sweat rate [37], but calcitonin-gene related peptide (CGRP), the most abundant neuropeptide in human skin, also has some direct neuronal effects. In a recent study, we found that cholinergic sweating was significantly increased by local CGRP in physiological concentrations but not further amplified if higher concentrations of CGRP were applied. Other neuropeptides, such as

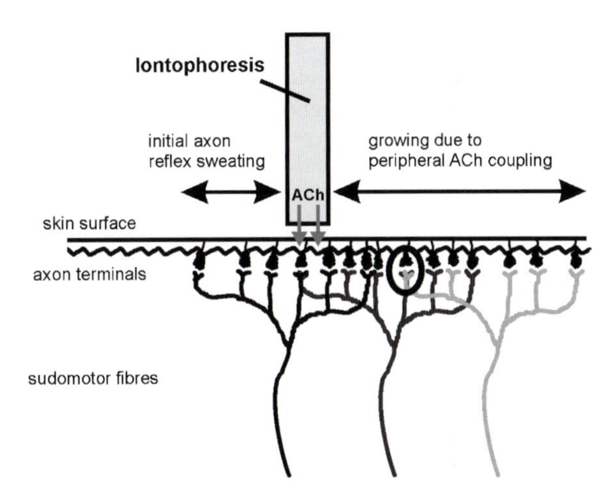

Fig. 3.4 Acetylcholine iontophoresis activates sudomotor neurons. This activation is conducted centrally (antidromically). At peripheral branching points, some of the action potentials travel to the periphery again, inducing axon reflex sweating. Where nerve endings meet in the periphery (*marked by circle*), new sudomotor units are activated, inducing a new axon reflex and thereby a spreading of the sweat response

vasoactive intestinal peptide or substance P, had no effects [51]. Previous patch clamp studies revealed that CGRP itself has an inhibitory effect on nicotinic ACh-receptors. However, in human skin CGRP is rapidly degraded by peptidases, and smaller fragments of CGRP ($CGRP_{1-6}$, $CGRP_{1-5}$, or $CGRP_{1-4}$) enhance the activity of nicotinic ACh-receptors [13]. Thus, depending on its metabolism, CGRP could either inhibit or enhance sweating. From our results we could assume that under physiological conditions CGRP enhances sweating.

In addition to neuronal control, sweating is influenced by several local factors. Pressure to the skin, for example, due to lying on one side, quickly inhibits sweating on the compressed side [42]. Local skin warming increases while local skin cooling (i.e., vasoconstriction) reduces sweating [36]. The reason for the strong temperature dependence is mainly the temperature dependent capacity of single glands to produce sweat. Furthermore, the amount of functional sweat glands varies substantially between subjects and depends on environmental conditions during childhood [41]. Exposition to environmental heat in hot climates and regular exercise increase the size of sweat glands and their function – but not their number.

3.3.1 Measurement of Sweating

In contrast to vasoconstriction, sweating can be reliably measured for clinical diagnosis. It can be visualized with iodine starch staining [34]. Thereby, dyshidrotic areas after central or peripheral nerve lesions can be identified. Sweating can be provoked by drinking hot tea, irradiation with an infrared lamp, or by exercises [3]. The integrity of peripheral sudomotor axons can be tested if sudomotor axon reflex sweating is elicited by, for example, electrical current [38] or cholinergic drugs (so-called quantitative sudomotor axon reflex test, QSART) (Fig. 3.5) [28]. The amount of sweat evaporation over a defined skin area is then measured with sweat capsules, the spatial extension of this axon reflex can be visualized by iodine staining. Wet skin turns violet and dry skin remains whitish [48]. Sudomotor axon reflexes differ between different body sites, with largest volumes on the lower legs, and between genders, with men sweating more intensely than women [5]. Since sudomotor axon reflex cannot be elicited soon after peripheral nerve lesions, QSART can be used for the evaluation of sudomotor function in peripheral nerve diseases such as neuropathies [28].

3.3.2 Some Sweating Disorders

Disturbances of sweat gland function can be very unpleasant for the patient, in the case of hyperhidrosis,

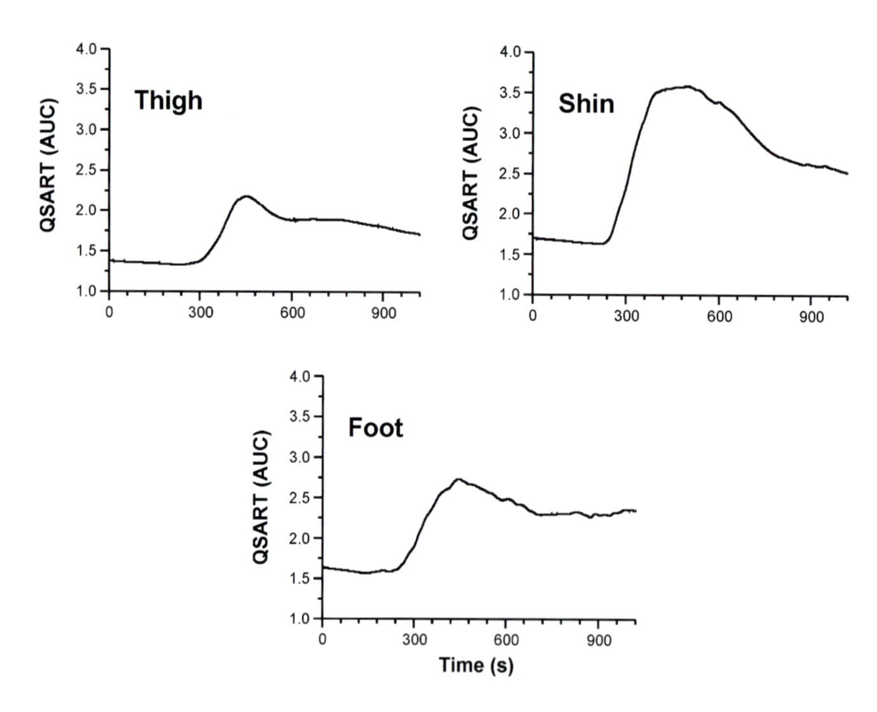

Fig. 3.5 Quantitative sudomotor axon reflex test (QSART). Example from one subject shows somatotopic arrangement of sweating on the leg

or even dangerous, in the case of anhidrosis. Lack of functional nerve growth factor receptor (TRK-A) causes patients to suffer from congenital insensitivity to pain with anhidrosis (CIPA). These patients have multiple infections and mutilations since trophic function of both sudomotor and primary afferents is missing [39]. Impaired sweating can also be found in polyneuropathies like diabetes mellitus. Hyperhidrosis occurs in hyperthyroidism or phaeochromocytoma [53], and many people suffer from localized hyperhidrosis on the sole, palms, and axillae [17]. This syndrome usually does not reflect a pathological condition, but suffering in daily activities can provide the need for treatment. Treatment of hyperhidrosis could be managed by tab water iontophoresis, aluminum hydroxide ointment, or botulinum toxin injections. In a clinical study, we found that BoNT (botulinum neurotoxin) type B seems to affect autonomic functions more than BoNT type A [50]. In severe cases of localized hyperhidrosis endoscopic sympathectomy has been proven to be successful [47].

3.4 Sympathetic Nervous System and Inflammation

There are some clinical symptoms that suggest a link between sympathetic nervous system and immune reaction in human skin. In CRPS patients, who are characterized by sympathetic dysfunction, Langerhans cells were found extensively in skin biopsies [9]. Furthermore, in these patients sympatholytic procedures ameliorate pain but also inflammation and edema. As another example, after hemispherical stroke, which amplifies sympathetic outflow to the paretic side by disinhibition, T-cell responses in the delayed-type hypersensitivity reaction were significantly attenuated [52]. All these findings triggered in-depth studies on the mechanisms of sympathetic–immune interaction.

The sympathetic nervous system influences dendritic cells, which play a crucial role in the innate immune response. Dendritic cells express a variety of adrenergic receptors, with alpha1-Rs playing a stimulatory and beta2-Rs an inhibitory effect on dendritic cell migration. It has been shown that beta2-Rs in skin and bone marrow-derived dendritic cells when stimulated respond to noradrenaline (NE) by decreased interleukin-12 (IL-12) and increased the anti-inflammatory IL-10 production, which in turn downregulates inflammatory cytokine production [31]. The situation can dramatically change in inflammation, where immune cells downregulate their expression of beta2-Rs and upregulate their expression of alpha1-RS [18]. Alpha1-adrenoceptors stimulate the production and release of proinflammatory cytokines. If alpha1-Rs were to become expressed by the resident or recruited immune cells, then sympathetic activation would be predicted to cause pain and other inflammatory signs via cytokine release. Another phenomenon suggesting the involvement of the sympathetic nervous system in allergic responses is the fact that contact sensitizers under certain circumstance inhibit the local noradrenaline turnover in the skin. Thereby the immune reaction could be amplified [30]. A more indirect contribution of the sympathetic nervous system to inflammation symptoms is the fact that lymph nodes and lymph vessels are innervated by catecholaminergic sympathetic nerves. If the sympathetic nervous system is overactive due to, for example, chronic pain, lymphatic vessel constriction could amplify inflammation-related edema [21].

3.5 Summary

The sympathetic nervous system has many physiological influences on human skin. It regulates vasoconstriction and sometimes even vasodilation by adrenergic signaling and it controls sweating by cholinergic fibers. This means that it controls thermoregulation and stress reactions in human skin. Furthermore, the adrenergic part of the sympathetic nervous system could also affect the inflammation response, either suppressing it in intact skin or enhancing it if inflammation is already present. Furthermore, it is clinically very important that the sympathetic nervous system might contribute to chronic pain. In intact skin noradrenaline is not able to excite nociceptors and cause pain. After peripheral nerve lesions, and probably under the guidance of nerve growth factor, the number of functional adrenoreceptors on primary afferents increases so that sympathetic activation and noradrenaline release could excite nociceptors [45]. This phenomenon is called sympathetically maintained pain. Very recently one interesting explanation for sympathetically maintained pain was proposed [58]. Soon after an experimental nerve lesion in rats, sympathetic nerve fibers – similar to peptidergic primary afferent fibers – sprout into the upper dermis. These sympathetic nerve fibers are then in close association with afferent fibers. This never occurs in intact skin and could be the anatomic basis for a sympathetic–nociceptive interaction in neuropathic pain. Whether it really does so must be addressed by functional studies. Until then, the role of the autonomic nervous system in the skin will remain incompletely understood.

Acknowledgements

This work was supported by the German research foundation (grants Bi 579/1 and Bi 579/4), the German Research Network on Neuropathic Pain and the Rhineland-Palatinate Foundation for Innovation. We thank Mr. Stuart Turner for help with the manuscript preparation.

Summary for the Clinician

Autonomic innervation of human skin is of efferent sympathetic origin subserving mainly thermoregulation. Sympathetic fibers reach the skin by traveling with major nerves to the periphery. In cold environment sympathetic vasoconstrictor drive is enhanced, leading to pale cold skin. Simultaneously, sudomotor drive is reduced and sweat production is stopped. Thereby heat loss is avoided. In hot environment, activation pattern is just the opposite: vasoconstrictor drive is reduced, skin vessels dilate, and the skin becomes warm and red. Sudomotor drive is enhanced and sweating starts. Thereby heat dissipation is achieved.

Sympathetic vasoconstriciton is mediated by noradrenaline release from sympathetic nerves terminating on the outer surface of skin blood vessels. Noradrenaline equipotentially binds to alpha1 and alpha2-receptors on smooth muscle cells in the arteriolar wall causing vasoconstriction. Circulating catecholamines (e.g., adrenaline) in the blood, however, also reach alpha2-receptors on endothelial cells inside arteriols. Then via second messengers vasodilation is induced. This duality of catecholamine effects helps to avoid skin ischemia, for example, during prolonged cold exposure.

Sudomotor fibers activate sweat glands via the release of acetylcholine. Acetylcholine binds to muscarinic acetylcholine receptors at the sweat glands and nicotinergic receptors at the peripheral sudomotor terminals. The latter mechanism reinforces sudomotor activity. As a side effect acetylcholine also induces active vasodilation and thereby reinforces passive vasodilation by reduction of sympathetic vasoconstrictor activity. The resulting warm skin then further leads to increased sweat output, while, for example, pressure on the skin reduces sweating. In contrast to vasoconstriction, sweating can be quantitatively measured and the result might help to diagnose autonomic failure in neuropathies.

Beside this "classical" sympathetic function, ongoing research also suggested an augmentative effect of catecholamines on inflammatory cells in the skin, and in particular after peripheral nerve lesions, catecholamines could directly excite primary afferent nociceptive fibers. If this is confirmed in future studies one can assume that the sympathetic nervous system might also participate in the generation or maintenance of chronic inflammatory or neuropathic pain.

References

1. Baron R, Schattschneider J, Binder A, et al (2002) Relation between sympathetic vasoconstrictor activity and pain and hyperalgesia in complex regional pain syndromes: a case-control study. Lancet 359:1655–1660
2. Birklein F (2005) Complex regional pain syndrome (CRPS). In: Cervero F, Jensen TS (eds) Handbook of Clinical Neurology, Elsevier, Edinburgh
3. Birklein F, Sittl R, Spitzer A, et al (1997) Sudomotor function in sympathetic reflex dystrophy. Pain 69:49–54
4. Birklein F, Riedl B, Neundörfer B, et al (1998) Sympathetic vasoconstrictor reflex pattern in patients with complex regional pain syndrome. Pain 75:93–100
5. Birklein F, Walther D, Bigalke H, et al (2002) Sudomotor testing predicts the presence of neutralizing botulinum A toxin antibodies. Ann Neurol 52:68–73
6. Bligh J (1979) The central neurology of mammalian thermoregulation. Neuroscience 4:1213–1216
7. Bolton B, Carmichael EA, Stürup G (1936) Vasoconstriction following deep inspiration. J Physiol 86:83–94
8. Boric MP, Figueroa XF, Donoso MV, et al (1999) Rise in endothelium-derived NO after stimulation of rat perivascular sympathetic mesenteric nerves. Am J Physiol 277: H1027–H1035
9. Calder JS, Holten I, McAllister RM (1998) Evidence for immune system involvement in reflex sympathetic dystrophy. J Hand Surg Br 23:147–150
10. Carvajal JA, Germain AM, Huidobro-Toro JP, et al (2000) Molecular mechanism of cGMP-mediated smooth muscle relaxation. J Cell Physiol 184:409–420
11. Clough GF (1999) Role of nitric oxide in the regulation of microvascular perfusion in human skin in vivo. J Physiol 516(Pt 2):549–557
12. Clough GF, Church MK (2002) Vascular responses in the skin: an accessible model of inflammation. News Physiol Sci 17:170–174
13. Di Angelantonio S, Giniatullin R, Costa V, et al (2003) Modulation of neuronal nicotinic receptor function by the

neuropeptides CGRP and substance P on autonomic nerve cells. Br J Pharmacol 139:1061–1073
14. Duran WN, Seyama A, Yoshimura K, et al (2000) Stimulation of NO production and of eNOS phosphorylation in the microcirculation in vivo. Microvasc Res 60:104–111
15. Fagius J, Karhuvaara S, Sundlöf G (1989) The cold pressor test: effects on sympathetic nerve activity in human muscle and skin nerve fascicles. Acta Physiol Scand 137:325–334
16. Figueroa XF, Poblete MI, Boric MP, et al (2001) Clonidine-induced nitric oxide-dependent vasorelaxation mediated by endothelial alpha(2)-adrenoceptor activation. Br J Pharmacol 134:957–968
17. Hecht MJ, Birklein F, Winterholler M (2004) Successful treatment of axillary hyperhidrosis with very low doses of botulinum toxin B: a pilot study. Arch Dermatol Res 295:318–319
18. Heijnen CJ, Rouppe vd, V, Wulffraat N, et al (1996) Functional alpha 1-adrenergic receptors on leukocytes of patients with polyarticular juvenile rheumatoid arthritis. J Neuroimmunol 71:223–226
19. Henriksen O (1991) Sympathetic reflex control of blood flow in human peripheral tissues. Acta Physiol Scand Suppl 603:33–39
20. Hermann D, Schlereth T, Vogt T, et al (2005) Clonidine induces nitric oxide- and prostaglandin-mediated vasodilation in healthy human skin. J Appl Physiol 99:2266–2270
21. Howarth D, Burstal R, Hayes C, et al (1999) Autonomic regulation of lymphatic flow in the lower extremity demonstrated on lymphoscintigraphy in patients with reflex sympathetic dystrophy. Clin Nucl Med 24:383–387
22. Inoue S, Murakami N (1976) Unit responses in the medulla oblongata of rabbit to changes in local and cutaneous temperature. J Physiol 259:339–356
23. Jänig W (1990) Functions of the sympathetic innervation of the skin. In: Loewy AD, Spyer KM (eds) Central Regulation of Autonomic Functions, Oxford University Press, New York
24. Jarajapu YP, Johnston F, Berry C, et al (2001) Functional characterization of alpha1-adrenoceptor subtypes in human subcutaneous resistance arteries. J Pharmacol Exp Ther 299:729–734
25. Kennedy WR, Wendelschafer-Crabb G, Brelje TC (1994) Innervation and vasculature of human sweat glands: an immunohistochemistry-laser scanning confocal fluorescence microscopy study. J Neurosci 14:6825–6833
26. Lewis T (1937) The nocifensor system of nerves and its reactions. Br Med J 1:431–437
27. Low PA, Kennedy WR (1997) Cutaneous effectors as indicators of abnormal sympathetic function. In: Morris JL, Gibbins IJ (eds) Autonomic innervation of the skin, Harwood academic publishers, Amsterdam
28. Low PA, Caskey PE, Tuck RR, et al (1983) Quantitative sudomotor axon reflex test in normal and neuropathic subjects. Ann Neurol 14:573–580
29. Low PA, Kihara M, Cardone C (1993) Pharmacology and morphometry of the eccrine sweat gland in vivo. In: Low PA (ed) Clinical Autonomic Disorders, Little, Brown and Company, Boston
30. Maestroni GJ (2004) Modulation of skin norepinephrine turnover by allergen sensitization: impact on contact hypersensitivity and T helper priming. J Invest Dermatol 122:119–124
31. Maestroni GJ (2006) Sympathetic nervous system influence on the innate immune response. Ann N Y Acad Sci 1069:195–207
32. Marriott I, Marshall JM, Johns EJ (1990) Cutaneous vascular responses evoked in the hand by the cold pressor test and by mental arithmetic. Clin Sci Colch 79:43–50
33. McQuay HJ, Tramer M, Nye BA, et al (1996) A systematic review of antidepressants in neuropathic pain. Pain 68:217–227
34. Minor V (1927) Ein Neues Verfahren zu der klinischen Untersuchung der Schweissabsonderung. Z Neurologie 101:302–308
35. Moncada S, Vane JR (1979) The role of prostacyclin in vascular tissue. Fed Proc 38:66–71
36. Nadel ER, Bullard RW, Stolwijk JAE (1971) Importance of skin temperature in the regulation of sweating. J Appl Physiol 31:80–87
37. Nadel ER, Mitchell JW, Saltin B, et al (1971) Peripheral modifications to the central drive for sweating. J Appl Physiol 31:828–833
38. Namer B, Bickel A, Kramer H, et al (2004) Chemically and electrically induced sweating and flare reaction. Auton Neurosci 114:72–82
39. Nolano M, Crisci C, Santoro L, et al (2000) Absent innervation of skin and sweat glands in congenital insensitivity to pain with anhidrosis. Clin Neurophysiol 111:1596–1601
40. Oberle J, Elam M, Karlsson T, et al (1988) Temperature-dependent interaction between vasoconstrictor and vasodilatator mechanisms in human skin. Acta Physiol Scand 132:459–469
41. Ogawa T, Low PA (1993) Autonomic regulation of temperature and sweating. In: Low PA (ed) Clinical Autonomic Disorders, Little, Brown and Company, Boston
42. Ogawa T, Asayama M, Ito M, et al (1979) Significance of skin pressure in body heat balance. Jpn J Physiol 29:805–816
43. Riedl B, Nischik M, Birklein F, et al (1998) Spatial extension of sudomotor axon reflex sweating in human skin. J Aut Nerv Syst 69:83–88
44. Riedl B, Beckmann T, Neundorfer B, et al (2001) Autonomic failure after stroke – is it indicative for pathophysiology of complex regional pain syndrome? Acta Neurol Scand 103:27–34
45. Sato J, Perl ER (1991) Adrenergic excitation of cutaneous pain receptors induced by peripheral nerve injury. Science 251:1608–1610
46. Sato K, Ohtsuyama M, Sato F (1993) Normal and abnormal eccrine sweat gland function. In: Low PA (ed) Clinical Autonomic Disorders, Little, brown and Company, Boston
47. Schick CH, Fronek K, Held A, et al (2003) Differential effects of surgical sympathetic block on sudomotor and vasoconstrictor function. Neurology 60:1770–1776
48. Schlereth T, Brosda N, Birklein F (2005) Somatotopic arrangement of sudomotor axon reflex sweating in humans. Auton Neurosci 123:76–81
49. Schlereth T, Brosda N, Birklein F (2005) Spreading of sudomotor axon reflexes in human skin. Neurology 64:1417–1421
50. Schlereth T, Mouka I, Eisenbarth G, et al (2005) Botulinum toxin A (Botox(R)) and sweating-dose efficacy and comparison to other BoNT preparations. Auton Neurosci 117:120–126

51. Schlereth T, Dittmar JO, Seewald B, et al (2006) Peripheral amplification of sweating – a role for calcitonin gene-related peptide. J Physiol 576:823–832
52. Tarkowski E, Naver H, Wallin BG, et al (1995) Lateralization of T-lymphocyte responses in patients with stroke. Effect of sympathetic dysfunction? Stroke 26:57–62
53. Togel B, Greve B, Raulin C (2002) Current therapeutic strategies for hyperhidrosis: a review. Eur J Dermatol 12:219–223
54. Tooke JE, Ostergren J, Lins PE, et al (1987) Skin microvascular blood flow control in long duration diabetics with and without complications. Diabetes Res 5:189–192
55. Wasner G, Schattschneider J, Baron R (2002) Skin temperature side differences – a diagnostic tool for CRPS? Pain 98:19–26
56. White DG, Martin W (1989) Differential control and calcium-dependence of production of endothelium-derived relaxing factor and prostacyclin by pig aortic endothelial cells. Br J Pharmacol 97:683–690
57. Wise RA, Wigley FM, White B, et al (2004) Efficacy and tolerability of a selective alpha(2C)-adrenergic receptor blocker in recovery from cold-induced vasospasm in scleroderma patients: a single-center, double-blind, placebo-controlled, randomized crossover study. Arthritis Rheum 50:3994–4001
58. Yen LD, Bennett GJ, Ribeiro-Da-Silva A (2006) Sympathetic sprouting and changes in nociceptive sensory innervation in the glabrous skin of the rat hind paw following partial peripheral nerve injury. J Comp Neurol 495:679–690
59. Zahn S, Leis S, Schick C, et al (2004) No {alpha}-adrenoreceptor-induced C-fiber activation in healthy human skin. J Appl Physiol 96:1380–1384

Immune Circuits of the Skin

4

E. Weinstein and R.D. Granstein

Contents

Key Features

- Skin is the initial barrier to attack by exogenous elements.
- Cutaneous immunity, the active process by which immune homeostasis in the skin is maintained, is achieved through the interplay of innate and adaptive (acquired) arms.
- Innate immunity is a rapid, first-line defense that includes cellular (e.g., Langerhans cells, keratinocytes) and secreted elements (e.g., cytokines, antimicrobial peptides triggered by pattern-recognition receptors (e.g., toll-like receptors).
- The adaptive immune response is highly specific, robust, and capable of remembering encountered antigens; T lymphocytes are the principal effectors of adaptive immunity in the skin.
- Once thought to function independently, the innate and adaptive systems are now considered interdependent contributors to an incompletely understood immune circuit.

Synonyms Box: Integument, skin, including all derivatives of epidermis (e.g., hair, nails); T cell, T lymphocyte; B cell, B lymphocyte; cytokines, secreted hormone-like proteins involved in immune function and cellular communication; chemokines, cytokines involved in chemotaxis of leukocytes; interleukins, cytokines whose structure has been defined

4.1 Introduction to the Immune System and Its Components

In vertebrates, immune function is accomplished through two systems of host defenses. The innate component constitutes a rapid and stereotyped means of containing infection while the acquired, or adaptive, arm provides a more robust, targeted defense that develops more slowly after first exposure to pathogens [17]. Traditionally, these responses have been examined in isolation, a reflection of their complementary but apparently discrete processes. This paradigm of divided action has been challenged, however, as new research has elucidated important interplay between innate and adaptive systems. Nowhere else in the body is this interface more evident and relevant than in the skin, the principal barrier to exogenous attack.

Several models have been applied to describe the immunological environment of the integument, from *skin associated lymphoid tissue* (SALT) in 1978 to the

R.D. Granstein and T.A. Luger (eds.), *Neuroimmunology of the Skin*,

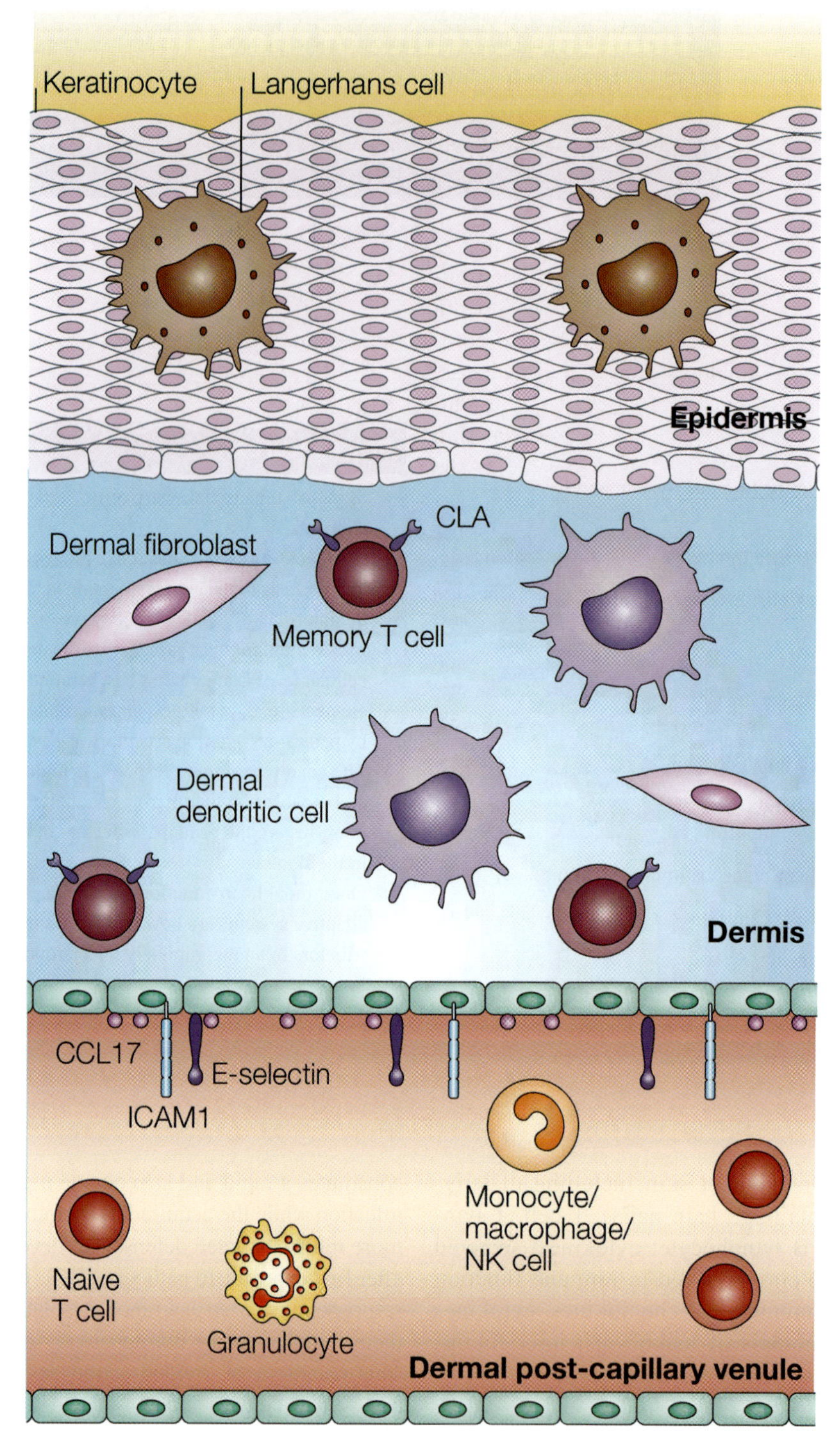

Fig. 4.1 Immune components of normal skin. Cells and cell-surface receptors of the epidermis, dermis, and cutaneous vasculature are shown. Reprinted by permission from Macmillan Publishers Ltd: Nature Reviews Immunology [41], copyright (2004)

more recent and inclusive *skin immune system* (SIS), among others [12,74]. This evolution reflects emerging insight into the immunological contributions made by an array of cellular and humoral elements. For innate immunity, surveillance is accomplished by a set of epidermal and dermal resident sentinels (Fig. 4.1). Under cover of the stratum corneum, these cells include Langerhans cells, dermal dendritic cells, keratinocytes, mast cells, and macrophages [41]. Recirculating neutrophils and natural killer cells are important, transient cellular effectors in this first line of defense. In addition, cytokines, antimicrobial peptides, and complement proteins are soluble activators and mediators that shape the innate response. Resident T cells represent the adaptive immune system in quiescent skin and are joined by recruited B and T lymphocytes in the setting of inflammation.

Regardless of the site of activity, the major distinctions between innate and acquired systems can be understood in terms of differences in the receptors they use and in their dissimilar capacities for immunological memory. While responses of the acquired system improve and mature with repeated exposure to the same antigen, those of innate immunity remain unchanged. Second, cells of the innate arm recognize antigens belonging to broad classes of pathogens. The adaptive system, on the other hand, somatically rearranges its receptors to bind highly specific targets [18,48]. Differences in their receptors also account, in part, for the general principle that innate effectors are triggered first, followed sometime later by the adaptive arm. Throughout the body, macrophages and dendritic cells are the prototypical early responders. In this group are the Langerhans cells, a subset of dendritic cells residing in the epidermis. Through the stimulation of receptors for pathogen-associated motifs, these cells initiate an inflammatory cascade that results in rapid containment of infection and in the activation of the adaptive system. It is increasingly evident, however, that the receptors of innate immunity are both more widely distributed than was previously thought and play a central role in controlling the adaptive response [48].

4.2 Receptors of Innate Immunity

The cells of the innate immune system rely on a series of germ line-encoded receptors to accomplish the tasks of rapid response to and eradication of invading pathogens. Far less numerous in variety than the organisms against which they protect, these receptors recognize highly conserved pathogen-associated molecular patterns common to many classes of microorganisms and are thus termed pattern-recognition receptors. The recognized antigens, such as lipopolysaccharide, peptidoglycan, and lipotechoic acid, are typically components essential to microbial function and are not natively produced by the host [48]. This ensures a broad but self-tolerant defense. The mechanism of this defense depends on the receptor subtype, the cell on which the receptor is expressed, and the receptor's response to the binding of its cognate antigen.

A useful functional classification distinguishes between secreted, endocytic, and intracellular signaling receptors [47,48]. Mannan-binding lectins, for example, are secreted serum proteins that opsonize pathogens and trigger the classical complement cascade. Macrophage mannose and scavenger receptors, by contrast, trigger endocytosis for pathogen clearance. Principal among the intracellular signaling group is the family of Toll-like receptors (TLRs) named for their structural homology to drosophila Toll transmembrane proteins (important in the fly's embryological development and immune function).

Since the identification of TLR-4 in 1997 [49], ten members of the TLR family have been reported. All share characteristic extra- and intracellular domains that function in the recognition of and response to a range of pathogenic components [76], affording the innate immune system a greater degree of antigen specificity than had been previously recognized. With the aid of adapter proteins, for example, TLR-4 has been shown to be critical in the recognition of Gram-negative lipopolysaccharide (LPS) [35]. Similarly, TLR-2-deficient mice have been shown to be hyporesponsive to components of Gram-positive organisms. The full repertoire of TLRs allows for the recognition of pathogenic ligands derived from bacteria, viruses, and fungi, as well as some nonpathogenic host proteins and synthetic compounds [47,76]. The import of these receptors lies in their ability to couple the pattern recognition of the innate immune system with the specificity and robustness of adaptive immunity. The activation of TLRs triggers a signaling cascade that culminates in increased expression of genes coding for key components of the inflammatory response.

This transduction pathway is closely related to those of drosophila Toll and mammalian IL-1 receptors, leading to the activation of NF-κB [49,53], a nuclear transcription factor responsible for the induction of chemokines, cytokines, costimulatory molecules, adhesion molecules, and major histocompatability complex (MHC) molecules [6,29,76]. Together, these inflammatory elements help to initiate and coordinate the adaptive immune response. The vital role of TLRs has been demonstrated in mutant mice lacking MyD88, an adapter protein in TLR signaling pathways. These mice were unable to mount effective responses to various TLR ligands (see [76] for summary) and were more tolerant of highly immunogenic skin allografts [81]. Though a MyD88-independent pathway for dendritic cell activation by LPS has been elucidated, this too depends on intact TLR signaling [84].

The expression of TLRs has now been demonstrated in several classes of resident skin cells. This distribution reflects both the diversity of functions mediated by these receptors, as well as the previously unrecognized immunological contributions made by a variety of cell types. In fact, the discovery of TLRs on Langerhans cells, dermal dendritic cells, keratinocytes, and on mast cells has helped to reshape the perspective on the interplay between innate and adaptive immunity, while the vital, active role of cutaneous cells has been illuminated.

4.3 TLR-Bearing Cells: Effectors of Innate Immunity

4.3.1 Dendritic Cells

Cutaneous dendritic cells are bone marrow-derived antigen presenting cells (APCs) located in the epidermis as Langerhans cells and in dermis as dermal dendritic cells. Though they account for only 2–4% of epidermal cells, Langerhans cells are widely regarded as the ambassadors of cutaneous immunity for their role in coordinating the immune response. Identified in 1868, these cells were first postulated to be neural agents and, almost a century later, exhausted melanocytes [61]. It was not until the late 1970s and early 1980s that molecules suggesting an immune function for Langerhans cells were detected on their surface membranes, including Fc receptors (that bind immunoglobulin constant regions), C3 complement receptors, and MHC class II molecules [38,62,73]. In the context of contact hypersensitivity, it had been further shown that Langerhans cells were phagocytic, capable of migration to skin-draining lymph nodes, and could be found closely apposed to lymphocytes [70,71]. By the mid-1980s, a class of splenic dendritic cells that functioned as potent stimulators of the mixed leukocyte reaction (MLR) had been described. Thus, when Schuler and Steinman demonstrated that cultured murine epidermal Langerhans cells could be coaxed to adopt a splenic dendritic cell phenotype, their position as cutaneous APCs was clarified [67].

Through a combination of TLR activation and antigen-independent responses to cytokines, Langerhans and dermal dendritic cells function to initiate and direct adaptive immunity. The coexpression of TLR types 1, 2, 4, 5, and 9 afford dendritic cells sensitivity to bacterial LPS, lipoproteins, glycolipids, flagellin, and CpG DNA [2,33,76,80]. Upon activation, dendritic cells are primed for antigen presentation while in transit to lymph nodes where naïve T lymphocytes are concentrated [41]. This system of antigen gathering and remote presentation solves the logistical problem of introducing pathogen to the small number of clonal T cells receptive to a given epitope. To achieve this, dendritic cells target endocytosed material for presentation on MHC molecules rather than for destruction (as in macrophages). Endowed with several classes of pattern-recognition and Fc receptors for antigen capture (in addition to constitutive macropinocytosis), these cells are alternately efficient foragers when immature and effective stimulators of T cells once activated [4]. Recent research has also revealed a modulatory role for neuropeptides and the sympathetic nervous system in the stimulation of lymphocytes by skin dendritic cells [69]. The details of the activation and presentation processes are discussed later.

Importantly, dendritic cells in the skin are phenotypically and functionally varied. For example, the distribution of TLRs has recently been shown to differ between dermal dendritic and Langerhans cells. Langerhans cells are relatively deficient in bacteria-binding TLR types 2, 4, and 5, suggesting a possible mechanism for tolerance to commensal organisms in the epidermis, and imiquimod-sensitive TLR-7 is notably absent from murine Langerhans cells [3,51,79]. In addition to its unique pattern of TLR expression, the Langerhans cell, on the one hand, is remarkable for the cytoplasmic inclusion of Birbeck granules, racket-shaped bodies associated with surface expression of C-type lectin Langerin/CD207 [78]. Dermal dendritic cells, on the other hand, are characterized by the presence of DC-SIGN/CD209, a lectin that distinguishes the dendritic cells from other dermal components [77]. Still, common to both classes of cells is their myeloid origin, demonstrated in vitro by the transformation of $CD14^{+}$ monocytes into Langerhans cells [30]. However, there exists a third subset called plasmacytoid dendritic cells that arise from lymphoid precursors and can be found in the dermal layer during active inflammation [77]. It remains unclear whether these plasmacytoid cells participate in antigen presentation.

The paradigm in which Langerhans cells serve as independent activators of cutaneous immunity has been challenged by the discovery of important phenotypic and functional differences between these cells and other dendritic-cell subsets. Specifically, Asahina and Tamaki demonstrated a subdued response by murine Langerhans cells to LPS and viral antigens relative to splenic dendritic cells and peritoneal macrophages [3]. Also, unlike their splenic counterparts, the Langerhans cells could not be stimulated to release interferon (IFN)-γ, important in recruiting type 1 helper (Th1) T cells, but were relatively prolific secretors of chemokines that promote the type 2 (Th2) response [3,26]. Thus, the authors propose that even in their earliest responses, Langerhans cells may be responding to the influence of their epidermal milieu (e.g., keratinocytes and cytokines) rather than initiating inflammation. That Langerhans cells may further play a modulatory role, downregulating the inflammatory response, has been suggested by the surprising finding that mice deficient in epidermal Langerhans cells were particularly susceptible to contact hypersensitivity by diphtheria toxin [36]. It is evident that the character of Langerhans cells has yet to be fully delineated.

4.3.2 Keratinocytes

Historically, keratinocytes were believed to make only a passive contribution to the skin's homeostatic function. Through a network of tight, intercellular adhesions and production of the stratum corneum, these cells form the initial barrier to exogenous injury and prevent dessication. This understanding has been expanded, however, as keratinocytes are now known to play an important, active role in cutaneous immunity. Their participation is triggered both mechanically (e.g., through trauma), photochemically (e.g., by ultraviolet radiation (UVR)), and chemically (via receptor-mediated response), serving to initiate inflammation and to couple the innate and adaptive arms of immunity.

In the absence of apparent provocation, keratinocytes are constitutive secretors of pro-inflammatory cytokines, including interleukin (IL)-1α and tumor necrosis factor (TNF)-α. In reaction to endotoxin and UVR, the release of TNF-α is enhanced [39]. The secretion of IL-1α [43] is similarly upregulated by physical injury, resulting in the nuclear factor (NF)-κB-mediated activation of more than 90 known gene targets in skin [52]. In autocrine/paracrine fashion, the IL-1R-bearing keratinocytes are responsive to the IL-1 surge and are further stimulated to release ganulocyte-macrophage colony stimulating factor (GM-CSF) and other mediators of inflammation [32]. Keratinocytes have been shown to secrete soluble factors IL-6, IL-7, IL-12, IL-15, and IL-18 [11,72], all of which can promote and shape the adaptive immune response.

The immunological role of keratinocytes is also suggested by the expression of IFN-γ-induced MHC class II peptides and intercellular adhesion molecule (ICAM)-1 on the keratinocyte cell surface [7,21]. The recent demonstration of TLR types 1–6 and 9 on keratinocytes [50] has revealed yet another mechanism by which the NF-κB-dependent inflammatory pathway may be invoked in these cells. However, the extent to which keratinocytes can effectively present antigen and independently stimulate lymphocyte activation remains unclear. In addition, these cells are thought to modulate the immune response through the production and secretion of downregulatory cytokines. IL-1 receptor antagonist (IL-1ra), for example, is known to competitively inhibit the binding of IL-1 to its receptor and can be found in high concentrations in keratinocytes. Mice deficient in IL-1ra have been shown to be more susceptible to lethal endotoxemia and this cytokine is thought to play a role in modifying the degree of inflammation [34]. Moreover, the anti-inflammatory cytokine IL-10 has also been shown to be inducible in keratinocytes exposed to UVB radiation, otherwise associated with a pro-inflammatory response [23]. The important immunological contribution of keratinocytes may thus represent a careful balance between activating and inhibiting factors.

4.3.3 Mast Cells

Mast cells are bone marrow-derived effectors of immunity. Well-recognized for their expression of high-affinity receptors for IgE (FcεR) [37], they comprise a diverse, migratory class of granule-bearing cells. Though widely dispersed throughout the body, mast cells are concentrated at sites of greatest potential exposure to pathogens and allergens, including the skin, respiratory tract, and intestine [28]. They are best understood for their role in triggering inflammation, degranulating in response to the cross-linking of FcεR receptors by IgE, and releasing preformed histamine and TNF-α, prostaglandins, and leukotrienes [17]. In this capacity, mast cells function both as defenders against parasites and as culprits in the development of pathological disorders such as anaphylaxis and asthma [28]. The proximity of these cells to blood vessels and vagal nerve terminals in the skin ensures that their secreted mediators of vascular permeability and immune-cell recruitment are likely to reach their targets [46,82].

In addition to granule-associated and arachadonic acid-derived products, mast cells prolifically secrete cytokines. The Th2-promoting interleukins, IL-4, IL-5, and IL-13, are characteristic among these. However, mast cells have also been shown to produce Th1-related factors, including IFN-γ, IL-12, and IL-18, as well as numerous chemokines and growth factors [46]. The precise mix of preformed factors, eicosanoids, cytokines, and chemokines released by mast-cell activation appears to depend on the quality of the stimulus and its receptor target. Recent studies on cultured human mast cells in an IFN-γ milieu have demonstrated an FcγRI-mediated response to IgG that includes secretion of all of these mediators [83]. IgE stimulation and Ig-binding superantigens give a similarly mixed response. In contrast, the chemotactic complement system may preferentially stimulate degranulation [28,54]. In addition to these indirect methods of activation, the presence of TLRs on the mast cell surface confers an ability to respond directly to pathogens. The currently recognized pattern of TLR expression in mast cells is broad, similar to that in keratinocytes, including most of the known subtypes (types 2 and 4 are the best studied to date) [75,76]. Their activation in cultured mast cells yields a burst of cytokines and chemokines, often without concomitant degranulation [16].

Whatever the stimulus, the activation of mast cells and the secretion of their chemical mediators may be most relevant to immunity for the resulting recruitment of its effectors. Mast cell-derived IL-1, TNF-α, and histamine have clear impact on the permeability of the vascular endothelium, as well as on its expression of adhesion molecules for immune cells (including ICAM-1, VCAM-1, and selectins). In addition, secreted chemokines are chemoattractive for these cells and, through various combinations, may be selective in their recruitment [46]. The extent to which mast cells can serve as APCs and help to coordinate the adaptive response in vivo remains uncertain [27,28]. However, their overall relevance to an effective immune response has been clearly demonstrated in animal models of mast cell-deficiency [45].

4.4 The Integrated Innate Response

Activated through receptor stimulation, direct injury to keratinocytes, or APC-mediated triggers, the innate immune system functions to contain infection. Inflammatory mediators originating from its effector cells diffuse into the cutaneous vasculature, upregulating the endothelial adhesion molecules required for diapedesis of circulating immune cells recruited by chemical signals [31]. This heightened expression is supplementary to the presence of ICAM-1, E-selectin, and CCL-17 on resting endothelial cells (shown in Fig. 4.1) [41]. Notably, intestinal fibroblasts have been shown to react in ways similar to vascular endothelial cells, responding to IL-1 and TNF-α by upregulating adhesion molecules and secreting additional cytokines [56]. It is likely that cutaneous fibroblasts have a comparable role. Further, vascular permeability is increased both by these inflammatory secretions and by neighboring nerve endings that release vasoactive neuropeptides such as substance P [20].

The recruited infiltrate is permitted passage across the dermal vasculature and includes neutrophils, natural killer cells, and non-resident T cells. The phagocytic neutrophils are reinforcing to the resident macrophages, engulfing antibody- and complement-opsonized pathogens and debris. Destruction of this material is accomplished through intracellular free radical species and catabolizing peptides [17]. The surrounding inflammatory environment, especially the presence of IL-1 and TNF-α, has been shown to be stimulatory to the respiratory burst necessary for the production of caustic radicals and augments the killing ability of these neutrophils [24]. Also protective are natural killer cells that act in the setting of viral infection to identify and destroy affected host cells. The putative mechanisms for their directly cytotoxic effect depend on the presence of both surface immunoglobulin receptors and molecules sensitive to the downregulation of MHC-I peptides on their host targets (a frequent sequela of viral activity) [8,42].

In addition to the cellular elements, the innate response is armed with a set of soluble assistants. Complement proteins are serum factors that bind invading microorganisms, either perforating their cellular membranes or coating them for subsequent phagocytosis. Complement fragments are also chemoattractive to neutrophils and stimulatory for degranulation of mast cells [17]. A second category of soluble factors has been the focus of recent research both for its dependence on TLR activation and for its import in initiating adaptive involvement. This class of antimicrobial peptides (AMPs) comprises a large family of genetically encoded molecules that contribute to the cutaneous "chemical barrier" [14,65,66] protecting the exposed skin surface. Chief among them are the human β-defensins (hBDs) and cathelicidins [5], two classes of AMPs constitutively expressed and rapidly induced in response to epithelial injury and inflammation [66]. Keratinocytes are not the sole producers of AMPs, however, as other immune cells of the skin such as mast cells have been implicated in producing them [19].

The spectra and mechanisms of antibacterial action have been shown to differ across and within subtypes of AMPs. For example, the antimicrobial and protease-inhibiting actions of the cathelicidins have been attributed to two proteolytically cleaved domains – a highly conserved cathelin-like domain and a variable LL-37 C-terminal peptide. Together, these confer broad-spectrum antibacterial activity to the cathelicidin molecule [85]. In contrast, the cationic hBDs demonstrate a range of killing activity, from preferentially Gram-negative (hBD-2) to broad-spectrum (hBD-3) [66]. Though named for their antibacterial properties, and previously regarded exclusively as components of innate immunity, the AMPs likely play additional roles of only recently elucidated complexity. As chemotactic factors, hBDs and cathelicidins have been shown to attract dendritic cells, T lymphocytes, neutrophils, monocytes/macrophages, and mast cells [65]. In addition, keratinocytes show a receptor-mediated response to both classes of AMPs, resulting in migration and proliferation (relevant in wound-healing), and the release of additional inflammatory cytokines [55].

4.5 The Interface between Innate and Adaptive Immunity

Concurrent with the protective actions of the nonspecific, early immune response is the engagement of the acquired immune system. To achieve this, the secreted pro-inflammatory mediators that recruit and activate the effectors of innate immunity simultaneously induce the activation and maturation of professional APCs. Dendritic cells activated by keratinocyte-derived TNF-α, IL-1α, and IL-18 are primed for effective antigen presentation to naïve T lymphocytes through a process that includes (1) the enhancement of phagocytic activity, (2) the upregulation of costimulatory molecules CD40, CD80, and CD86, (3) increased cell-surface expression of class II MHC molecules [58], (4) heightened expression of chemokine receptors responsible for localizing dendritic cells to T cell-rich areas, and (5) the production of cytokines that aid in directing the T cell response (e.g., IL-12, IL-18, and IL-10) [59]. As mentioned, this process may also be stimulated through the direct activation of TLRs on the APCs. Notably, both mechanisms rely on the NF-κB pathway. [9]

Unlike the dendritic cells to which they respond, T lymphocytes bear receptors that, when considered together, can bind almost any protein antigen. Their variety is achieved through a process of genetic recombination (analogous to that of immunoglobulin production by B cells) that occurs after the migration of the bone marrow-derived T cells to the thymus. The T cell receptors (TCRs) are membrane-bound heterodimers of either α/β or γ/δ subunits that recognize peptide fragments presented on MHC or MHC-like molecules [15]. Though the subpopulation of γ/δ-expressing cells is not MHC-restricted [10], the vast majority of T lymphocytes in skin (as in blood) express α/β TCRs and thus require contact with the MHC–antigen complex for activation [25]. It is not surprising, therefore, that cutaneous T cells can be found in close proximity to Langerhans cells in basal regions of the epidermis [22]. In normal skin, T lymphocytes have been shown to cluster near postcapillary venules high in the papillary dermis, just below the dermal–epidermal junction – only 2% were intraepidermal [13,22].

The pairing of TCR and MHC-antigen complexes is refereed by the expression of either CD4 or CD8 molecules on the surface of α/β T cells. The TCRs of $CD4^+$ cells interact exclusively with MHC class II molecules, and those of $CD8^+$ cells only with MHC-I. MHC class I molecules are expressed by all nucleated cells and are complexed with antigen in the setting of intracellular viral infection. This presentation of viral product targets the affected cells for destruction by cytotoxic $CD8^+$ T cells. The activation of $CD4^+$ T-helper cells, rather, is accomplished by MHC-II-bearing APCs [17]. In healthy epidermis, $CD8^+$ cells outnumber $CD4^+$ cells; in the dermis, these are more evenly distributed [13]. It is the $CD4^+$ cell, however, that enhances the phagocytic attack on microbes and guides the flexible adaptive immune response [1].

The activation of $CD4^+$ cells by Langerhans or dermal dendritic cells occurs in skin-draining lymph nodes or, less frequently, within the cutaneous layers. Kupper and Medzhitov describe a trilogy of *immune surveillance* measures by which the immune system introduces naïve T cells to their cognate antigens, directs the activated lymphocytes to the skin, and distributes clonally derived memory cells throughout the body [41]. In the first step, TLR- and cytokine-activated cutaneous dendritic cells downmodulate the expression of E-cadherin, a surface adhesion molecule with high avidity for keratinocytes [68]. Untethered, these cells undergo the maturation process described above while emigrating from the skin to lymph nodes via afferent lymphatics [40]. Here, the mature dendritic cells, now bearing the costimulatory molecules and MHC class II-antigen complexes necessary for T-cell activation, bind and prime naïve $CD4^+$ cells expressing appropriate TCRs.

4.6 Completing the Cutaneous Immune Circuit

In addition to their role in initiating the adaptive response, dendritic cells help to direct its course. Secreted IL-12 and IL-23 stimulate the differentiation of activated $CD4^+$ cells into Th1 cells. These, in turn, release factors IL-2, IFN-γ, and TNF-α that promote T cell proliferation, reinforce the antimicrobial capacity of phagocytes, and induce the differentiation of cytotoxic lymphocytes. Alternatively, a predominance of dendritic cell-derived IL-4 drives the conversion of activated T-helper cells into Th2 cells – these secrete IL-4, IL-5, IL-6, and IL-10, important in quelling inflammation, as well as in stimulating eosinophils and B cells (not present in healthy human skin) [4,15,18]. Th2 cells are the primary effectors in fighting helminthic parasites and the primary culprits in allergic reactions [1]. Not all the progeny of proliferating $CD4^+$ cells are destined to become effectors, however. Some develop into memory cells that survive long after their stimulating antigen is cleared and enable a rapid secondary immune response in the setting of reinfection [1]. Notably, the presence of memory T cells in the skin is not achieved at random and the mechanisms for this localization are now understood to be relevant to both adaptive immune function and dysfunction.

The trafficking of memory T cells to the skin layers has been associated with cutaneous lymphocyte-associated antigen (CLA), a carbohydrate cell–surface receptor found on roughly 90% of cutaneous T cells during inflammation and only sparsely expressed on non-skin-homing T lymphocytes [64]. Evidence for the importance of this molecule was suggested by the affinity of CLA for E-selectin, present in large numbers on activated endothelial cells [57]. More recently, studies have confirmed this function of CLA and demonstrated a role for a host of cytokines (e.g., CXCR2, CCR4, CCR6, CXCR3, and CCR10) in the cutaneous migration of memory T cells [64]. These findings relate directly to the second element in the model of *immune surveillance*. In addition, the actions of CLA have offered new insight into the mechanisms of T cell-mediated skin diseases such as psoriasis, allergic contact dermatitis, atopic dermatitis, cutaneous graft-vs.-host disease, and cutaneous T-cell lymphoma [60,64]. CLA-marked T cells are characteristically involved in each of these diatheses.

With the introduction of activated T cells into the skin (or their activation in situ), the immune circuit is complete. Inflammation and the immune response subside as pathogens are eliminated, pro-inflammatory cytokines are diluted, and suppressive factors such as IL-10 and TGF-β predominate. Memory T cells specific for the inciting antigen remain localized to the skin and are supplemented by a set of circulating central memory cells that confer similar immunity at other sites (the final component of *immune surveillance*) [41]. Even in immune quiescence, there is a role for Langerhans and dermal dendritic cells. Given the appropriate milieu, dendritic cells have been shown to support the differentiation of regulatory T-helper subtypes, including Treg and Tr1 cells. Unlike their Th1 and Th2 cousins, these regulatory cells inhibit inflammation and T cell proliferation through both direct contact and secretion of IL-10 and TGF-β. Recent research suggests that dendritic cells at various stages of maturation are competent to induce the development of these cells. Though the precise mechanisms by which this occurs remain uncertain, it is clear that the presence of IL-10 and cell-surface factors (such as MHC and costimulatory molecules) are involved [44,63]. Regardless, this interaction between dendritic and T cells establishes tolerance toward certain non-pathogenic and endogenous antigens, beyond that determined during immune-system development.

Whether activating or suppressive, the influence of innate immunity over the adaptive system is profound. By tracing the circuit of cutaneous immune activity, it becomes further evident that this influence is bi-directional. It is within this context of interdependence that the pathological mechanisms of cutaneous immune diseases can be understood and options for their treatment explored.

Summary for the Clinician

Skin is the initial barrier to attack by exogenous elements and is now understood to play an active role in initiating and supporting immune function. Immunity is achieved by the intricate interplay of innate and acquired (adaptive) arms. Innate immunity serves to contain infection by rapid, stereotyped methods targeted to broad classes of infectious agents. The adaptive response, rather, is slow on first exposure to a given antigen but is highly specific and increasingly robust with repeat antigenic encounters. Skin hosts the receptors (e.g., pattern-recognition receptors), cellular sentinels, and the molecular milieu involved in activating and coordinating these arms of immunity. Specifically, toll-like receptors (TLRs) comprise an important class of pattern-recognition receptors recently shown to function at the interface of innate and adaptive immunity. Upon activation, TLRs trigger inflammation and the upregulation of chemokines, cytokines, costimulatory molecules, and other molecular mediators of adaptive immunity. TLRs are distributed across numerous cell types in the skin, including antigen presenting cells (APCs), keratinocytes, and mast cells. Important among cutaneous APCs are Langerhans cells and dermal dendritic cells that, when stimulated, migrate to skin-draining lymph nodes and mature into potent activators of T lymphocytes (the cellular effectors of adaptive immunity). Recent research has demonstrated a complex immunomodulatory role for Langerhans cells, as well, supporting the notion that immune function reflects a careful balance of activating and inhibitory factors. Similarly, keratinocytes, once considered mere passive contributors to immune homeostasis are now thought to factor in shaping the nature and the extent of the immune response. However actuated, the cells and signals of innate immunity promote the inflammatory cascade through which phagocytic neutrophils are recruited and soluble assistants (e.g., antimicrobial peptides (AMPs)) are secreted at injured or infected sites. At the same time, the activation, maturation, and migration of APCs coupled with the secretion of chemotactic factors by innate cells support the mobilization of T cells, triggering the acquired immune system. It is by further influence of APCs that T lymphocytes differentiate into effector, memory, and regulatory subtypes. Cutaneous lymphocyte-associated antigen (CLA)-bearing T cells home to sites in the skin layers and reinforce the innate system in resolving the inciting injury. Through an increasingly sophisticated understanding of this immune circuit, both the pathophysiology of immune-mediated skin diseases and options for their treatment can be rationally explored.

Acknowledgment

This work is supported by NIH R01 AR42429.

References

1. Abbas AK, Lichtman AH (2003) Cellular and Molecular Immunology, 5th ed. Saunders, Philadelphia
2. Akira S, Takeda K, Kaisho T (2001) Toll-like receptors: critical proteins linking innate and acquired immunity. Nature Immunology, 2:675–680
3. Asahina A, Tamaki K (2006) Role of Langerhans cells in cutaneous protective immunity: is the reappraisal necessary? Journal of Dermatological Science, 44:1–9
4. Banchereau J, Steinman RM (1998) Dendritic cells and the control of immunity. Nature, 392:245–252
5. Bardan A, Nizet V, Gallo RL (2004) Antimicrobial peptides and the skin. Expert Opinion on Biological Therapy, 4:543–549
6. Barnes PJ (1997) Nuclear factor-kappa B. The International Journal of Biochemistry and Cell Biology, 29:867–870
7. Basham TY, Nickoloff BJ, Merigan TC, et al (1985) Recombinant gamma interferon differentially regulates class II antigen expression and biosynthesis on cultured normal human keratinocytes. Journal of Interferon Research, 5:23–32

8. Biron CA, Nguyen KB, Pien GC, et al (1999) Natural killer cells in antiviral defense: function and regulation by innate cytokines. Annual Review of Immunology, 17:189–220
9. Bonizzi G, Karin M (2004) The two NF-kappaB activation pathways and their role in innate and adaptive immunity. Trends in Immunology, 25:280–288
10. Born W, Cady C, Jones-Carson J, et al (1999) Immunoregulatory functions of gamma delta T cells. Advances in Immunology, 71:77–144
11. Bos JD (ed) (2005) Skin Immune System (SIS): Cutaneous Immunology and Clinical Immunodermatology, 3rd ed. CRC Press, Boca Raton
12. Bos JD, Kapsenberg ML (1986) The skin immune system (SIS): its cellular constituents and their interactions. Immunology Today, 7:235–240
13. Bos JD, Zonneveld I, Das PK, et al (1987) The skin immune system (SIS): distribution and immunophenotype of lymphocyte subpopulations in normal human skin. Journal of Investigative Dermatology, 88:569–573
14. Braff MH, Bardan A, Nizet V, et al (2005) Cutaneous defense mechanisms by antimicrobial peptides. Journal of Investigative Dermatology, 125:9–13
15. Clark R, Kupper T (2005) Old meets new: the interaction between innate and adaptive immunity. Journal of Investigative Dermatology, 125:629–637
16. Dawicki W, Marshall JS (2007) New and emerging roles for mast cells in host defence. Current Opinion in Immunology, 19:31–38
17. Delves PJ, Roitt IM (2000) The immune system. First of two parts. The New England Journal of Medicine, 343:37–49
18. Delves PJ, Roitt IM (2000) The immune system. Second of two parts. The New England Journal of Medicine, 343:108–117
19. Di Nardo A, Vitiello A, Gallo RL (2003) Cutting edge: mast cell antimicrobial activity is mediated by expression of cathelicidin antimicrobial peptide. The Journal of Immunology, 170:2274–2278
20. Dunnick CA, Gibran NS, Heimbach DM (1996) Substance P has a role in neurogenic mediation of human burn wound healing. The Journal of Burn Care and Rehabilitation, 17:390–396
21. Dustin ML, Singer KH, Tuck DT, et al (1988) Adhesion of T lymphoblasts to epidermal keratinocytes is regulated by interferon gamma and is mediated by intercellular adhesion molecule 1 (ICAM-1). The Journal of Experimental Medicine, 167:1323–1340
22. Elbe A, Foster CA, Stingl G (1996) T-cell receptor alpha beta and gamma delta T cells in rat and human skin – are they equivalent? Seminars in Immunology, 8:341–349
23. Enk CD, Sredni D, Blauvelt A, et al (1995) Induction of IL-10 gene expression in human keratinocytes by UVB exposure in vivo and in vitro. The Journal of Immunology, 154:4851–4856
24. Ferrante A, Nandoskar M, Walz A, et al (1988) Effects of tumour necrosis factor alpha and interleukin-1 alpha and beta on human neutrophil migration, respiratory burst and degranulation. International Archives of Allergy and Applied Immunology, 86:82–91
25. Foster CA, Yokozeki H, Rappersberger K, et al (1990) Human epidermal T cells predominantly belong to the lineage expressing alpha/beta T cell receptor. The Journal of Experimental Medicine, 171:997–1013
26. Fujita H, Asahina A, Sugaya M, et al (2005) Differential production of Th1- and Th2-type chemokines by mouse Langerhans cells and splenic dendritic cells. Journal of Investigative Dermatology, 124:343–350
27. Galli SJ, Maurer M, Lantz CS (1999) Mast cells as sentinels of innate immunity. Current Opinion in Immunology, 11:53–59
28. Galli SJ, Nakae S, Tsai M (2005) Mast cells in the development of adaptive immune responses. Nature Immunology, 6:135–142
29. Gay NJ, Keith FJ (1991) Drosophila Toll and IL-1 receptor. Nature, 351:355–356
30. Geissmann F, Prost C, Monnet JP, et al (1998) Transforming growth factor beta1, in the presence of granulocyte/macrophage colony-stimulating factor and interleukin 4, induces differentiation of human peripheral blood monocytes into dendritic Langerhans cells. The Journal of Experimental Medicine, 187:961–966
31. Groves RW, Allen MH, Ross EL, et al (1995) Tumour necrosis factor alpha is pro-inflammatory in normal human skin and modulates cutaneous adhesion molecule expression. The British Journal of Dermatology, 132:345–352
32. Groves RW, Rauschmayr T, Nakamura K, et al (1996) Inflammatory and hyperproliferative skin disease in mice that express elevated levels of the IL-1 receptor (type I) on epidermal keratinocytes. Evidence that IL-1-inducible secondary cytokines produced by keratinocytes in vivo can cause skin disease. The Journal of Clinical Investigation, 98:336–344
33. Hemmi H, Takeuchi O, Kawai T, et al (2000) A Toll-like receptor recognizes bacterial DNA. Nature, 408:740–745
34. Hirsch E, Irikura VM, Paul SM, et al (1996) Functions of interleukin 1 receptor antagonist in gene knockout and overproducing mice. Proceedings of the National Academy of Sciences of the United States of America, 93:11008–11013
35. Hoshino K, Takeuchi O, Kawai T, et al (1999) Cutting edge: toll-like receptor 4 (TLR4)-deficient mice are hyporesponsive to lipopolysaccharide: evidence for TLR4 as the Lps gene product. The Journal of Immunology, 162:3749–3752
36. Kaplan DH, Jenison MC, Saeland S, et al (2005) Epidermal langerhans cell-deficient mice develop enhanced contact hypersensitivity. Immunity, 23:611–620
37. Kinet JP (1999) The high-affinity IgE receptor (Fc epsilon RI): from physiology to pathology. Annual Review of Immunology, 17:931–972
38. Klareskog L, Tjernlund U, Forsum U, et al (1977) Epidermal Langerhans cells express Ia antigens. Nature, 268:248–250
39. Kock A, Schwarz T, Kirnbauer R, et al (1990) Human keratinocytes are a source for tumor necrosis factor alpha: evidence for synthesis and release upon stimulation with endotoxin or ultraviolet light. The Journal of Experimental Medicine, 172:1609–1614
40. Kripke ML, Munn CG, Jeevan A, et al (1990) Evidence that cutaneous antigen-presenting cells migrate to regional lymph nodes during contact sensitization. The Journal of Immunology, 145:2833–2838
41. Kupper TS, Fuhlbrigge RC (2004) Immune surveillance in the skin: mechanisms and clinical consequences. Nature Reviews, 4:211–222

42. Lanier LL (1998) NK cell receptors. Annual Review of Immunology, 16:359–393
43. Lee RT, Briggs WH, Cheng GC, et al (1997) Mechanical deformation promotes secretion of IL-1 alpha and IL-1 receptor antagonist. The Journal of Immunology, 159: 5084–5088
44. Levings MK, Gregori S, Tresoldi E, et al (2005) Differentiation of Tr1 cells by immature dendritic cells requires IL-10 but not CD25+CD4+ Tr cells. Blood, 105:1162–1169
45. Malaviya R, Abraham SN (2001) Mast cell modulation of immune responses to bacteria. Immunological Reviews, 179:16–24
46. Marshall JS (2004) Mast-cell responses to pathogens. Nature Reviews, 4:787–799
47. Medzhitov R (2001) Toll-like receptors and innate immunity. Nature Reviews, 1:135–145
48. Medzhitov R, Janeway C Jr. (2000) Innate immunity. The New England Journal of Medicine, 343:338–344
49. Medzhitov R, Preston-Hurlburt P, Janeway CA, Jr. (1997) A human homologue of the Drosophila Toll protein signals activation of adaptive immunity. Nature, 388:394–397
50. Miller LS, Modlin RL (2007) Human keratinocyte Toll-like receptors promote distinct immune responses. Journal of Investigative Dermatology, 127:262–263
51. Mitsui H, Watanabe T, Saeki H, et al (2004) Differential expression and function of Toll-like receptors in Langerhans cells: comparison with splenic dendritic cells. Journal of Investigative Dermatology, 122:95–102
52. Murphy JE, Robert C, Kupper TS (2000) Interleukin-1 and cutaneous inflammation: a crucial link between innate and acquired immunity. Journal of Investigative Dermatology, 114:602–608
53. Muzio M, Natoli G, Saccani S, et al (1998) The human toll signaling pathway: divergence of nuclear factor kappaB and JNK/SAPK activation upstream of tumor necrosis factor receptor-associated factor 6 (TRAF6). The Journal of Experimental Medicine, 187:2097–2101
54. Nilsson G, Johnell M, Hammer CH, et al (1996) C3a and C5a are chemotaxins for human mast cells and act through distinct receptors via a pertussis toxin-sensitive signal transduction pathway. The Journal of Immunology, 157:1693–1698
55. Niyonsaba F, Ushio H, Nakano N, et al (2007) Antimicrobial peptides human beta-defensins stimulate epidermal keratinocyte migration, proliferation and production of proinflammatory cytokines and chemokines. Journal of Investigative Dermatology, 127:594–604
56. Pang G, Couch L, Batey R, et al (1994) GM-CSF, IL-1 alpha, IL-1 beta, IL-6, IL-8, IL-10, ICAM-1 and VCAM-1 gene expression and cytokine production in human duodenal fibroblasts stimulated with lipopolysaccharide, IL-1 alpha and TNF-alpha. Clinical and Experimental Immunology, 96:437–443
57. Picker LJ, Kishimoto TK, Smith CW, et al (1991) ELAM-1 is an adhesion molecule for skin-homing T cells. Nature, 349:796–799
58. Pierre P, Turley SJ, Gatti E, et al (1997) Developmental regulation of MHC class II transport in mouse dendritic cells. Nature, 388:787–792
59. Reis e Sousa C (2001) Dendritic cells as sensors of infection. Immunity, 14:495–498
60. Robert C, Kupper TS (1999) Inflammatory skin diseases, T cells, and immune surveillance. The New England Journal of Medicine, 341:1817–1828
61. Romani N, Holzmann S, Tripp CH, et al (2003) Langerhans cells – dendritic cells of the epidermis. Apmis, 111:725–740
62. Rowden G, Lewis MG, Sullivan AK (1977) Ia antigen expression on human epidermal Langerhans cells. Nature, 268:247–248
63. Rutella S, Danese S, Leone G (2006) Tolerogenic dendritic cells: cytokine modulation comes of age. Blood, 108:1435–1440
64. Santamaria-Babi LF (2004) CLA(+) T cells in cutaneous diseases. European Journal of Dermatology, 14:13–18
65. Schauber J, Gallo RL (2007) Expanding the roles of antimicrobial peptides in skin: alarming and arming keratinocytes. Journal of Investigative Dermatology, 127:510–512
66. Schroder JM, Harder J (2006) Antimicrobial skin peptides and proteins. Cellular and Molecular Life Sciences, 63:469–486
67. Schuler G, Romani N, Steinman RM (1985) A comparison of murine epidermal Langerhans cells with spleen dendritic cells. Journal of Investigative Dermatology, 85:99s–106s
68. Schwarzenberger K, Udey MC (1996) Contact allergens and epidermal proinflammatory cytokines modulate Langerhans cell E-cadherin expression in situ. Journal of Investigative Dermatology, 106:553–558
69. Seiffert K, Granstein RD (2006) Neuroendocrine regulation of skin dendritic cells. Annals of the New York Academy of Sciences, 1088:195–206
70. Silberberg I (1973) Apposition of mononuclear cells to langerhans cells in contact allergic reactions. An ultrastructural study. Acta Dermato-Venereologica, 53:1–12
71. Silberberg-Sinakin I, Thorbecke GJ (1980) Contact hypersensitivity and Langerhans cells. Journal of Investigative Dermatology, 75:61–67
72. Steinhoff M, Brzoska T, Luger TA (2001) Keratinocytes in epidermal immune responses. Current Opinion in Allergy and Clinical Immunology, 1:469–476
73. Stingl G, Wolff-Schreiner EC, Pichler WJ, et al (1977) Epidermal Langerhans cells bear Fc and C3 receptors. Nature, 268:245–246
74. Streilein JW (1978) Lymphocyte traffic, T-cell malignancies and the skin. Journal of Investigative Dermatology, 71:167–171
75. Supajatura V, Ushio H, Nakao A, et al (2002) Differential responses of mast cell Toll-like receptors 2 and 4 in allergy and innate immunity. The Journal of Clinical Investigation, 109:1351–1359
76. Takeda K, Kaisho T, Akira S (2003) Toll-like receptors. Annual Review of Immunology, 21:335–376
77. Valladeau J, Saeland S (2005) Cutaneous dendritic cells. Seminars in Immunology, 17:273–283
78. Valladeau J, Ravel O, Dezutter-Dambuyant C, et al (2000) Langerin, a novel C-type lectin specific to Langerhans cells, is an endocytic receptor that induces the formation of Birbeck granules. Immunity, 12:71–81
79. van der Aar AM, Sylva-Steenland RM, Bos JD, et al (2007) Loss of TLR2, TLR4, and TLR5 on Langerhans cells abolishes bacterial recognition. The Journal of Immunology, 178:1986–1990
80. Visintin A, Mazzoni A, Spitzer JH, et al (2001) Regulation of Toll-like receptors in human monocytes and dendritic cells. The Journal of Immunology, 166:249–255

81. Walker WE, Nasr IW, Camirand G, et al (2006) Absence of innate MyD88 signaling promotes inducible allograft acceptance. The Journal of Immunology, 177: 5307–5316
82. Williams RM, Berthoud HR, Stead RH (1997) Vagal afferent nerve fibres contact mast cells in rat small intestinal mucosa. Neuroimmunomodulation, 4:266–270
83. Woolhiser MR, Brockow K, Metcalfe DD (2004) Activation of human mast cells by aggregated IgG through FcgammaRI: additive effects of C3a. Clinical Immunology, 110:172–180
84. Yamamoto M, Sato S, Hemmi H, et al (2003) Role of adaptor TRIF in the MyD88-independent toll-like receptor signaling pathway. Science, 301:640–643
85. Zaiou M, Nizet V, Gallo RL (2003) Antimicrobial and protease inhibitory functions of the human cathelicidin (hCAP18/LL-37) prosequence. Journal of Investigative Dermatology, 120:810–816

Modulation of Immune Cells by Products of Nerves

5

A.M. Bender and R.D. Granstein

Contents

Synonyms Box: CD–80, B7–1; CD–86, B7–2

Abbreviations: *APC* Antigen presenting cell, *BDNF* Brain-derived neurotrophic factor, *cAMP* Cyclic adenosine monophosphate, *CGRP* Calcitonin gene related peptide, *CHS* Contact hypersensitivity, *DC* Dendritic cell, *DTH* Delayed type hypersensitivity, *Ig* Immunoglobulin, *IL* Interleukin, *LC* Langerhans cell, *MHC* Major histocompatibility complex, *NGF* Nerve growth factor, *NK-1* Neurokinin-1 receptor, *NK-2* Neurokinin-2 receptor, *NT-3/4/5* Neurotrophin-3/4/5, *PACAP* Pituitary adenylate cyclase activating peptide, *SP* Substance P, *TNF* Tumor necrosis factor, *UVR* Ultraviolet radiation, *VIP* Vasoactive intestinal peptide

Key Features

- Neuropeptides released by sensory nerves alter cutaneous immune cell functions.
- Contact hypersensitivity (CHS) and delayed type hypersensitivity (DTH) are T-cell mediated immune reactions modulated by neuropeptides such as substance P, CGRP, VIP, and PACAP.
- Antigen presentation is modulated by local neuropeptides and is an important step in initiating a cellular immune response. The specific role of Langerhans cells in CHS, however, is controversial.
- Neuropeptides participate in ultraviolet radiation-induced immunosuppression.
- Neuropeptides exert their actions through modulating immune cell surface molecule expression, cytokine secretion, and intracellular cascades.
- Mast cells play a role in bidirectional signaling between neurons and the immune system.
- T cell, B cell, and neutrophil activity are modulated by neuropeptides.
- Neurotrophins alter neuropeptide release and may play a role in inflammatory skin disease.

5.1 Introduction

The nervous and immune systems are intimately linked. These systems communicate via a common language of neuropeptides and cytokines to create a network that can be thought of as one “neuroimmunocutaneous system” [48]. The linkage between the nervous and immune systems is both physical and functional. Structurally, nerves innervate the epidermis, the outer-most layer of skin, and have been found in close relationship with Langerhans cells (LCs), Merkel cells, keratinocytes, and mast cells [13,29]. Similar to a synapse, nerves may be able to release preformed neurosecretory vesicles into the vicinity of cutaneous and immune cells residing in

R.D. Granstein and T.A. Luger (eds.), *Neuroimmunology of the Skin*,

the skin to alter their functions. Inflammatory diseases such as atopic dermatitis and psoriasis are associated with altered neuropeptide expression [1,17], further supporting the link between inflammation and products of nerves.

This chapter will focus on the immunomodulating effect of products released from primary afferent sensory neurons. Discussion of autonomic fibers and their products, specifically catecholamines, will be covered in Chap. 8. Neuropeptides associated with sensory nerves include substance P (SP), calcitonin gene-related peptide (CGRP), vasoactive intestinal peptide (VIP), and pituitary adenylate cyclase-activating polypeptide (PACAP) among others [72]. Neuropeptides affect the function of immune cells in multiple manners [24,72]. The immune functions discussed in this chapter will include contact hypersensitivity (CHS) and delayed type hypersensitivity (DTH) reactions, antigen presentation function, and ultraviolet radiation (UVR) induced immunosuppression. The detailed effects of neuropeptides on modulating cell surface markers, cytokine expression, and intracellular signaling cascades will also be examined. In addition, a summary of neuropeptide actions on T and B lymphocytes, mast cells, and other immune cells will be briefly explored (Table 5.1).

5.2 Contact Hypersensitivity (CHS) and Delayed-Type Hypersensitivity (DTH)

Contact hypersensitivity (CHS) and delayed-type hypersensitivity (DTH) are T cell mediated immune reactions. After sensitization with hapten, LCs (and/or dermal dendritic cells) migrate to regional lymph nodes where they present antigen to T cells. Later, elicitation mobilizes memory T cells for a specific immune response on subsequent challenge with the antigen [57].

The CHS response to a contact sensitizer can be abolished by destroying the nerve fibers in the draining lymph nodes after contact sensitization, supporting the critical role of nerves in the CHS response. CHS can then be restored in denervated lymph nodes by directly supplying the LN with injected substance P [59]. These findings suggest that the CHS response is controlled by products of local nerves, either through intact nerve fiber signaling or through direct application of SP to the lymph node.

SP augments CHS responses while CGRP inhibits CHS responses in most assays. When SP is injected intradermally, CHS is enhanced [51]. This effect appears to be specific as administration of a specific SP antagonist blocks the ability of SP to enhance CHS [51]. Mice deficient in neurokinin-1 (NK-1) receptors, the primary receptor for SP, show diminished allergic contact dermatitis, suggesting that SP is needed for a robust inflammatory response. In contrast to SP, pre-treatment with CGRP inhibits CHS. Intradermal injection of CGRP prior to sensitization with an allergic contact sensitizer inhibits the induction phase of CHS. This effect was found to be a local phenomenon as hapten applied at a distant site from CGRP administration shows no inhibition of CHS [4]. Similar to pre-treatment with CGRP, if mice are injected intradermally with PACAP and sensitized at the injected site, induction of CHS is inhibited [12,35,36].

DTH responses are also inhibited by CGRP. If murine LCs are pre-treated with CGRP, their ability to elicit a DTH response to antigen is inhibited [65]. Interestingly, when cells are cultured in the presence of anti-IL-10 along with CGRP, they are able to elicit a full DTH response, suggesting a key role for IL-10 in mediating the functional effects of CGRP [65]. In a similar manner, PACAP and VIP inhibit the ability of murine epidermal cells enriched for LCs to elicit DTH in prior immunized mice [35,36].

5.3 In Vitro Antigen Presenting Function

CGRP inhibits the antigen-presenting function of Langerhans cells and macrophages and this may partially explain its ability to inhibit CHS and DTH. Using a mixed epidermal cell/lymphocyte reaction, CGRP inhibits alloantigen presentation by LC-enriched cultures of murine epidermal cells [4,29]. In a similar manner, CGRP, PACAP, and VIP inhibit presentation of antigen to an antigen-specific Th1 clone in enriched murine LC cultures [5,35,36,66]. CGRP also inhibits antigen presentation by human peripheral blood mononuclear cells [21].

5.4 Langerhans Cells and CHS

Langerhans cells (LCs) are antigen presenting cells residing in the epidermis and have been thought to be key in CHS responses. Both CHS reactions require effective presentation of hapten to T cells in order for sensitization and elicitation to occur. An unexpected finding concerning the role of LCs in the CHS response was recently observed in a study utilizing transgenic mice that constitutively lack epidermal LCs. In these transgenic mice, the regulatory elements for human Langerin were used to drive expression of diphtheria toxin, resulting in mice that lacked epidermal LCs. In these mice, CHS was found to be augmented instead of abrogated. In this manner, it was hypothesized that epidermal LCs may be dispensable for CHS and that LCs may actually play a suppressive role [32]. Further studies using a variety of model systems are needed in order to support these results and further delineate the role of LCs in hypersensitivity reactions. If LCs do indeed play a suppressive role in CHS, further understanding the role of neuropeptides in this process may be beneficial in developing novel therapeutic targets for inflammatory diseases.

5.5 Ultraviolet Radiation-Induced Immunosuppression

Modulation of the neuroimmune system by UVR will be further discussed in Chap. 14. Briefly, it should be noted that UVR suppresses induction of CHS and that CGRP has been implicated in the mechanism of UVR-induced immunosuppression [23]. It is thought that CGRP is locally released after cutaneous UVR and may decrease the number of LCs in the skin after UVR exposure [23,38]. In addition, CGRP antagonists restore CHS induction after UVR, suggesting that CGRP participates in the process by which UVR impairs CHS induction [50]. Release of TNF-alpha from mast cells, induced by CGRP, may be involved here [50]. Intradermal injection of a SP agonist, which acts on the NK-1 receptor, reverses UVR-dependent tolerance and augments CHS. In this way, SP may act as an adjuvant [51].

5.6 Surface Molecule Expression

It is thought that CGRP acts to inhibit antigen presentation through inhibition of CD86 surface molecule expression. CD86 (also known as B7-2) is an important co-stimulatory molecule necessary for antigen presentation by major histocompatibility complex (MHC) class II molecules. CGRP inhibits the induction of B7-2 expression on murine LCs and macrophages [5,10,21,29]. If IL-10 neutralizing antibodies are present during the period of exposure to CGRP, the inhibition of the induction of B7-2 expression by CGRP is abrogated [21,65]. This suggests that IL-10 may be involved in the mechanism by which B7-2 expression is inhibited after CGRP exposure. PACAP also downregulates CD86 expression in LPS/GM-CSF-stimulated XS106 cells, a LC-like cell line [35]. In addition, PACAP downregulates B7.1 and B7.2 expression in LPS/IFN-γ-activated macrophages [14].

SP may exert its proinflammatory actions by altering surface molecule expression on dermal microvascular endothelial cells. The ability of SP to augment expression of intracellular adhesion molecule 1 (ICAM-1), vascular cell adhesion molecule 1 (VCAM-1), and endothelial-leukocyte adhesion molecule-1 (ELAM-1) may be important in promoting leukocyte migration to sites of inflammation [25,41,44].

Chemokine receptor expression is also affected by neuropeptides. VIP downregulates CCR7 expression in LPS-stimulated mature dendritic cells [73]. In this manner, inflammatory immune responses may be prevented by hindering CCR7 expression, which is required for dendritic cell migration to lymph nodes. Further elucidation of the relationship between neuropeptides and chemokine surface receptor expression is needed to better understand the effects of neuropeptides on immune cell trafficking.

5.7 Cytokine Secretion

The ability of CGRP to suppress cell mediated immune responses is thought to be mediated in part by favoring cytokine expression that leads to a reduced T_H1-type immune response. CGRP induces upregulation of IL-10, an anti-inflammatory cytokine, and downregulation of IL-1β, a proinflammatory cytokine, in the LC-like line XS52 [65]. CGRP also augments IL-10 production in human monocyte cultures [21]. In contrast, SP is thought to augment antigen presentation and T cell responses through boosting inflammatory cytokine release and downregulating anti-inflammatory cytokine release to favor immune activation [26,27,33,42].

IL-10 suppresses T_H1 cell-mediated immunity, in part, through inhibition of antigen presentation to T_H1 cells [18]. Injection of IL-10 into the skin inhibits induction of contact hypersensitivity at that site [18]. The immunosuppressive effects of PACAP and VIP may be partly due to augmented IL-10 secretion in LCs, as it was found that IL-10 secretion was augmented after PACAP and VIP application in the XS106 Langerhans cell-like line [35,36].

IL-1β is a cytokine that is selectively expressed in the epidermis by LCs and its mRNA increases after application of contact sensitizers [18,28,45]. Expression of IL-1β is critical for initiation of primary immune responses in the skin and is found to be inhibited by CGRP [65]. In addition, PACAP and VIP inhibit LPS/GM-CSF-induced stimulation of IL-1β secretion in the Langerhans cell-like line XS106, which may partially explain the suppressive actions of these neuropeptides on CHS [35,36]. SP, on the other hand, can directly bind macrophages and augment production of IL-1 along with other pro-inflammatory molecules [26,27,33,42].

IL-12 and IL-23 are cytokines that strongly favor Th1-type cell-mediated immunity. CGRP has been found to inhibit expression and release of the p40 subunit of IL-12/23 from LPS-stimulated Langerhans cell-like line XS52, murine macrophages, and human PBMC [21,64,65]. These effects are abrogated in the presence of a CGRP antagonist. In addition, VIP is able to inhibit IL-12/23 p40 release from LPS-stimulated XS106 cells [36].

TNFα may also participate in the process of CHS induction impairment. When neutralizing antibodies to TNFα are injected prior to intradermal administration of CGRP, CHS induction is restored and the reduction in DC density is abrogated [50]. Mast cells induce release of TNFα upon CGRP application, suggesting a link between these two mediators. Interestingly, CGRP fails to impair CHS induction in mast cell-deficient mice, suggesting that mast cells are a necessary player in the process of CGRP-induced CHS suppression. The authors of this report speculate that release of TNFα in response to CGRP may interfere with antigen-presenting function in the skin, resulting in a suppressed CHS response [50].

5.8 Intracellular Mechanisms

Many neuropeptides exert their influence by acting on specific G protein-coupled receptors, which activate cAMP-protein kinase A or the phosphatidylinositol-protein kinase C pathways. Experimental evidence has shown that CGRP, VIP, and PACAP act through increasing intracellular cAMP levels [45,50,55,66]. The effect of CGRP on cAMP is thought to be specific as it can be inhibited by coculture with the truncated form of CGRP [CGRP-(8–37)], which acts as a specific competitive inhibitor [5]. The adenylate cyclase-cAMP-protein kinase A pathway is generally thought of as a pathway that downregulates or inhibits the amplification and effector phases of immune responses in T cells, B cells, and macrophages [31]. In addition to its cAMP-dependent pathway, VIP and PACAP can act via a cAMP-independent pathway, which inhibits NF-κB activation in some cell types [14,37].

SP is thought to primarily act through activation of NF-κB [41,43,57]. In addition to NF-κB, other transcriptional gene regulators modulated by SP include nuclear factor of activated T-cells (NFAT), cAMP responsive element (CRE), and activator protein-1 (AP-1) [3,11,39]. As the complexity of intracellular signaling mechanisms becomes apparent, it should be noted that a single neuropeptide may act through multiple transcription factor pathways depending on the specificity of the particular receptors, cellular subtypes, tissue, or species investigated.

5.9 Mast Cells

Neuropeptides can also modulate mast cell functions [6,26]. Mast cells are located in close proximity to SP-releasing C-fibers and endothelial cells [61]. Substance P released from neurons and keratinocytes can participate in mast-cell-neurite communication, supporting the notion of bidirectional signaling between neurons and immune cells [62]. Indeed, an increased number of dermal contacts between nerves and mast cells have been observed in lesional and nonlesional skin of patients with atopic dermatitis and nummular eczema. In addition, SP and CGRP fibers were found to be considerably increased in such lesional samples [30,49]. These clinical findings support the notion of the role of mast cells in promoting neurogenic inflammation.

SP may augment inflammatory responses in the skin by inducing mast cell TNFα mRNA expression and TNFα secretion [2]. SP also increases mast cell degranulation and release of histamine [16,19,55]. It is thought that tryptase from degranulated mast cells may cleave proteinase-activated receptor 2 (PAR2) at the plasma membrane of peripheral neurons and stimulate the release of local SP and CGRP, perpetuating the cycle of neurogenic inflammation [60].

5.10 T Lymphocyte Function

Supporting its immunosuppressive role, CGRP can inhibit T cell functions by suppressing T cell proliferation and IL-2 production [8,69,71]. In addition, CGRP inhibits proliferation of human PBMC [21]. In contrast to CGRP, SP plays an immunostimulatory role on

T cells. SP augments proliferation of T lymphocytes and enhances IL-2 expression in activated human T cells [9,54].

In line with its anti-inflammatory functions, VIP can inhibit T cell IL-2 secretion and induce differentiation and survival of T_H2 cells [14]. PACAP also promotes a shift from T_H1 towards T_H2 differentiation in vivo and in vitro [14]. Recently, it has been shown that VIP may augment populations of T regulatory cells in vivo [14]. Injection of VIP into T-cell receptor transgenic mice results in an expansion of T regulatory cells expressing Foxp3 and the CD4+CD25+ phenotype. These regulatory T cells are able to inhibit responder T cell proliferation, impair DTH responses, and prevent graft-vs.-host disease in transgenic mice [14]. Further understanding of the ability of VIP to augment the T regulatory cell population may be important in developing novel treatments for transplant medicine in addition to autoimmune and inflammatory disease processes.

5.11 B Lymphocyte Function

In addition to altering T cell functions, neuropeptides modulate B cell differentiation and immunoglobulin (Ig) expression. CGRP inhibits surface Ig expression in a LPS-stimulated murine pre-B cell line [46,47]. SP, on the other hand, augments B cell Ig secretion in stimulated murine leukocyte cultures and in a murine B cell line [53,61].

5.12 Other Immune Cell Functions

CGRP also plays an inhibitory role on neutrophils and other immune cell types. CGRP inhibits natural killer cell activity [68] and inhibits neutrophil accumulation [22]. In contrast, SP plays a stimulatory role on neutrophils. SP augments neutrophil chemotaxis, induces neutrophil lysosomal enzyme release, and augments neutrophil and macrophage phagocytosis [14,58]. High dose SP also induces eosinophil activation, adherence, and migration [20].

5.13 Neurotrophins

In addition to the neuropeptides mentioned previously, it should be noted that immune cells also produce and respond to neurotrophins such as nerve growth factor (NGF), brain-derived neurotrophic factor (BDNF), neurotrophin-3 (NT-3), and neurotrophin-4/5 (NT-4/5) [52]. Neurotrophins were originally discovered as growth factors supporting neuron maintenance and survival. More recently, it has been discovered that the target cells of neurotrophins are not limited to neurons. Immune cells express neurotrophins and their receptors [7]. Serum NGF levels are augmented in patients with atopic dermatitis and are positively correlated with inflammatory disease activity [67]. It is thought that augmented levels of neurotrophins in the circulation may be from increased local production within inflamed areas of skin. NGF can stimulate substance P expression in nociceptive sensory neurons [40] and can regulate mRNA expression of SP and CGRP [15,56]. In a cyclical manner, local products of nerves such as SP and neurokinin A can regulate NGF synthesis [34] and NGF in turn can regulate the synthesis of other products of nerves, namely SP and CGRP [32,40]. It is thought that enhanced NGF synthesis by keratinocytes may elicit abnormal sensory innervation and further the vicious cycle of allergic inflammation. Indeed, inflamed skin has shown increased numbers of sensory nerves and enhanced neuropeptide synthesis [15,49,63,70]. In this way, neurotrophins add another layer of complexity to the neuroimmunocutaneous system. Further research in the area of neurotrophins promises to enlighten understanding concerning inflammatory skin disease and the mechanisms by which the nervous system affects the immune system.

5.14 Conclusion

In conclusion, these data support that SP generally augments cellular immunity. The pro-inflammatory effects of SP lead to increased T cell proliferation, pro-inflammatory cytokine release, and leukocyte migration. CGRP, VIP, and PACAP, on the other hand, appear to exert mainly an inhibitory role on CHS and DTH. The known effect of some neuropeptides on immune cell function is summarized in Table 5.1. Additional studies involving the mechanisms of the influences of the neuroimmune axis on the skin would be of great interest. A further understanding of these pathways may have the potential to lead to new therapeutic approaches based on influencing neuroimmune pathways and activities.

Table 5.1 Modulation of Immune Cell by Products of Nerves

	CHS and DTH	APC function	Surface molecule expression	Cytokine secretion	Intracellular mechanisms	T and B cell function	Other immune cell functions
SP	*Augments CHS [51] *Reverses UVR-dependent tolerance [51]	—	*Augments endothelial cell adhesion molecule expression [41, 44]	*Augments IL-1 and TNFα [26,27,33,42]	*Augments NF-κB activation [41,43,57] *Also regulates NFAT, CRE, and AP-1 [3,11,39]	*Augments T cell proliferation and IL-2 production [9,54] *Augments B cell Ig secretion [53,61]	*Augments neutrophil chemotaxis and lysosomal enzyme release *Augments macrophage phagocytosis [14,58]
CGRP	*Inhibits CHS [4] and DTH [65] *Involved in UVR-dependent tolerance [23,38,50]	*Inhibits APC function [4,29]	*Inhibits CD86 expression [5,10,21,29]	*Inhibits IL-1 and IL-12 p40 [21,64,65] *Augments IL-10 [15, 18, 36]	*Augments intracellular cAMP [5]	*Inhibits T cell proliferation and IL-2 production [8,69] *Inhibits B cell Ig secretion [46]	*Inhibits NK cell activity [68] *Inhibits neutrophil accumulation [22]
VIP	*Inhibits DTH [36]	*Inhibits APC function [36]	*Inhibits CCR7 expression in mature DCs [73]	*Inhibits IL-12 [36] *Augments IL-10 [36]	*Augments intracellular cAMP [4,36] *Also inhibits NF-κB activation [14,37]	*Inhibits T cell proliferation [8] *Promotes T_H2 differentiation [14] *Augments T regulatory cells [14]	—
PACAP	*Inhibits CHS and DTH [35]	*Inhibits APC function [35]	*Inhibits CD86 expression [35]	*Augments IL-10 [35]	*Augments intracellular cAMP [14] *Also inhibits NF-κB activation [14,37]	*Promotes T_H2 differentiation [14]	—

Summary for the Clinician

- The nervous system and the immune systems communicate via a common language of neuropeptides.
- Substance P augments contact hypersensitivity (CHS), while calcitonin gene related peptide (CGRP) and pituitary adenylate cyclase activating peptide (PACAP) inhibit CHS in most assays. CGRP, vasointestinal peptide (VIP), and PACAP also inhibit delayed type hypersensitivity (DTH).
- CGRP, VIP, and PACAP mediate their anti-inflammatory effects by inhibiting Langerhans cell antigen presentation, suppressing B7 surface expression, upregulating IL-10 expression, and downregulating IL-1β and IL-12 p40 expression.
- Ultraviolet radiation-induced immunosuppression is thought to involve CGRP release from local nerves to impair CHS induction.
- Substance P mediates its pro-inflammatory effects by augmenting IL-1 and TNFα production in macrophages. Substance P also is able to upregulate leukocyte adhesion molecule expression on endothelial cells.
- Neuropeptides act through specific G protein-coupled receptors. CGRP primarily augments cAMP expression and adenylate cyclase activation. Substance P activates NF-κB.
- Mast cells are located in close proximity to neuropeptide-releasing nerves and may participate in mast-cell-neurite communication.
- T lymphocyte proliferation and B lymphocyte immunoglobulin secretion are inhibited by CGRP and are augmented by substance P.
- Immune cells can produce and respond to neurotrophins which may enhance neurogenic inflammation.

Acknowledgment

This work is supported by NIH R01 AR42429.

References

1. Anand P, et al (1991) Neuropeptides in skin disease: increased VIP in eczema and psoriasis but not axillary hyperhidrosis. Br J Dermatol 124(6):547–549
2. Ansel JC, et al (1993) Substance P selectively activates TNF-alpha gene expression in murine mast cells. J Immunol 150(10):4478–4485
3. Ansel JC, et al, (1997) Interactions of the skin and nervous system. J Investig Dermatol Symp Proc 2(1):23–26
4. Asahina A, et al (1995) Inhibition of the induction of delayed-type and contact hypersensitivity by calcitonin gene-related peptide. J Immunol 154(7):3056–3061
5. Asahina A, et al (1995) Specific induction of cAMP in Langerhans cells by calcitonin gene-related peptide: relevance to functional effects. Proc Natl Acad Sci U S A 92(18):8323–8327
6. Baxter JH, Adamik R (1978) Differences in requirements and actions of various histamine-releasing agents. Biochem Pharmacol 27(4):497–503
7. Bonini S, et al (2003) Nerve growth factor: neurotrophin or cytokine? Int Arch Allergy Immunol 131(2):80–84
8. Boudard F, Bastide M (1991) Inhibition of mouse T-cell proliferation by CGRP and VIP: effects of these neuropeptides on IL-2 production and cAMP synthesis. J Neurosci Res 29(1):29–41
9. Calvo CF, Chavanel G, Senik A (1992) Substance P enhances IL-2 expression in activated human T cells. J Immunol 148(11):3498–3504
10. Carucci JA, et al (2000) Calcitonin gene-related peptide decreases expression of HLA-DR and CD86 by human dendritic cells and dampens dendritic cell-driven T cell-proliferative responses via the type I calcitonin gene-related peptide receptor. J Immunol 164(7):3494–3499
11. Christian C, Gilbert M, Payan DG (1994) Stimulation of transcriptional regulatory activity by substance P. Neuroimmunomodulation 1(3):159–164
12. Dai R, Streilein JW (1997) Ultraviolet B-exposed and soluble factor-pre-incubated epidermal Langerhans cells fail to induce contact hypersensitivity and promote DNP-specific tolerance. J Invest Dermatol 108(5):721–726
13. Dalsgaard CJ, Rydh M, Haegerstrand A (1989) Cutaneous innervation in man visualized with protein gene product 9.5 (PGP 9.5) antibodies. Histochemistry 92(5):385–390
14. Delgado AV, McManus AT, Chambers JP (2005) Exogenous administration of Substance P enhances wound healing in a novel skin-injury model. Exp Biol Med (Maywood) 230(4):271–280
15. Donnerer J, Schuligoi R, Stein C, Amann R (1993) Upregulation, release, and axonal transport of substance P and calcitonin gene-related peptide in adjuvant inflammation and regulatory function of nerve growth factor. Regulatory Peptides 46:150–154
16. Ebertz JM, et al (1987) Substance P-induced histamine release in human cutaneous mast cells. J Invest Dermatol 88(6):682–685
17. Eedy DJ, et al (1991) Neuropeptides in psoriasis: an immunocytochemical and radioimmunoassay study. J Invest Dermatol 96(4):434–438
18. Enk AH, et al (1994) Induction of hapten-specific tolerance by interleukin 10 in vivo. J Exp Med 179(4):1397–1402
19. Fewtrell CM, et al (1982) The effects of substance P on histamine and 5-hydroxytryptamine release in the rat. J Physiol 330:393–411
20. Foster AP, Cunningham FM (2003) Substance P induces activation, adherence and migration of equine eosinophils. J Vet Pharmacol Ther 26(2):131–138
21. Fox FE, et al (1997) Calcitonin gene-related peptide inhibits proliferation and antigen presentation by human peripheral blood mononuclear cells: effects on B7, interleukin 10, and interleukin 12. J Invest Dermatol 108(1):43–48
22. Gherardini G, et al (1998) Calcitonin gene-related peptide improves skin flap survival and tissue inflammation. Neuropeptides 32(3):269–273

23. Gillardon F, et al (1995) Calcitonin gene-related peptide and nitric oxide are involved in ultraviolet radiation-induced immunosuppression. Eur J Pharmacol 293(4):395–400
24. Girolomoni G, Tigelaar RE (1990) Capsaicin-sensitive primary sensory neurons are potent modulators of murine delayed-type hypersensitivity reactions. J Immunol 145(4): 1105–1112
25. Goeddel DV, et al (1986) Tumor necrosis factors: gene structure and biological activities. Cold Spring Harbor Symp Quantitat Biol 51(Pt 1):597–609
26. Hartung HP, Toyka KV (1983) Activation of macrophages by substance P: induction of oxidative burst and thromboxane release. Eur J Pharmacol 89(3–4):301–305
27. Hartung HP, Wolters K, Toyka KV (1986) Substance P: binding properties and studies on cellular responses in guinea pig macrophages. J Immunol 136(10):3856–3863
28. Heufler C, et al (1992) Cytokine gene expression in murine epidermal cell suspensions: interleukin 1 beta and macrophage inflammatory protein 1 alpha are selectively expressed in Langerhans cells but are differentially regulated in culture. J Exp Med 176(4):1221–1226
29. Hosoi J, et al (1993) Regulation of Langerhans cell function by nerves containing calcitonin gene-related peptide. Nature 363(6425):159–163
30. Jarvikallio A, Harvima IT, Naukkarinen A (2003) Mast cells, nerves and neuropeptides in atopic dermatitis and nummular eczema. Arch Dermatol Res 295(1):2–7
31. Kammer GM (1988) The adenylate cyclase-cAMP-protein kinase A pathway and regulation of the immune response. Immunol Today 9(7–8):222–229
32. Kaplan DH, et al (2005) Epidermal langerhans cell-deficient mice develop enhanced contact hypersensitivity. Immunity 23(6):611–620
33. Kimball ES, Persico FJ, Vaught JL (1988) Neurokinin-induced generation of interleukin-1 in a macrophage cell line. Ann N Y Acad Sci 540:688–690
34. Kinkelin I, et al (2000) Increase in NGF content and nerve fiber sprouting in human allergic contact eczema. Cell Tissue Res 302(1):31–37
35. Kodali S, et al (2003) Pituitary adenylate cyclase-activating polypeptide inhibits cutaneous immune function. Eur J Immunol 33(11):3070–3079
36. Kodali S, et al (2004) Vasoactive intestinal peptide modulates Langerhans cell immune function. J Immunol 173(10):6082–6088
37. Leceta J, et al (2000) Receptors and transcriptional factors involved in the anti-inflammatory activity of VIP and PACAP. Ann N Y Acad Sci 921:92–102
38. Legat FJ, et al (2004) The role of calcitonin gene-related peptide in cutaneous immunosuppression induced by repeated subinflammatory ultraviolet irradiation exposure. Exp Dermatol 13(4):242–250
39. Lieb K, et al (1997) The neuropeptide substance P activates transcription factor NF-kappa B and kappa B-dependent gene expression in human astrocytoma cells. J Immunol 159(10):4952–4958
40. Lindsay RM, Harmar AJ (1989) Nerve growth factor regulates expression of neuropeptide genes in adult sensory neurons. Nature 337(6205):362–364
41. Lindsey KQ, et al (2000) Neural regulation of endothelial cell-mediated inflammation. J Investig Dermatol Symp Proc 5(1):74–78
42. Lotz M, Vaughan JH, Carson DA (1988) Effect of neuropeptides on production of inflammatory cytokines by human monocytes. Science 241(4870):1218–1221
43. Marriott I, et al (2000) Substance P activates NF-kappaB independent of elevations in intracellular calcium in murine macrophages and dendritic cells. J Neuroimmunol 102(2):163–171
44. Matis WL, Lavker RM, Murphy GF (1990) Substance P induces the expression of an endothelial-leukocyte adhesion molecule by microvascular endothelium. J Invest Dermatol 94(4):492–495
45. Matsue H, et al (1992) Langerhans cells are the major source of mRNA for IL-1 beta and MIP-1 alpha among unstimulated mouse epidermal cells. J Invest Dermatol 99(5):537–541
46. McGillis JP, et al (1993) Modulation of B lymphocyte differentiation by calcitonin gene-related peptide (CGRP). II. Inhibition of LPS-induced kappa light chain expression by CGRP. Cell Immunol 150(2):405–416
47. McGillis JP, Rangnekar V, Ciallella JR (1995) A role for calcitonin gene related peptide (CGRP) in the regulation of early B lymphocyte differentiation. Can J Physiol Pharmacol 73(7):1057–1064
48. Misery L (1997) Skin, immunity and the nervous system. Br J Dermatol 137(6):843–850
49. Naukkarinen A, Nickoloff BJ, Farber EM (1989) Quantification of cutaneous sensory nerves and their substance P content in psoriasis. J Invest Dermatol 92(1):126–129
50. Niizeki H, Alard P, Streilein JW (1997) Calcitonin gene-related peptide is necessary for ultraviolet B-impaired induction of contact hypersensitivity. J Immunol 159(11):5183–5186
51. Niizeki H, Kurimoto I, Streilein JW (1999) A substance p agonist acts as an adjuvant to promote hapten-specific skin immunity. J Invest Dermatol 112(4):437–442
52. Nockher WA, Renz H (2006) Neurotrophins in allergic diseases: from neuronal growth factors to intercellular signaling molecules. J Allergy Clin Immunol 117(3):583–589
53. Pascual DW, et al (1991) Substance P acts directly upon cloned B lymphoma cells to enhance IgA and IgM production. J Immunol 146(7):2130–2136
54. Payan DG, Brewster DR, Goetzl EJ (1983) Specific stimulation of human T lymphocytes by substance P. J Immunol 131(4):1613–1615
55. Payan DG, Levine JD, Goetzl EJ (1984) Modulation of immunity and hypersensitivity by sensory neuropeptides. J Immunol 132(4):1601–1604
56. Pincelli C, Fantini F, Giannetti A (1993) Neuropeptides and skin inflammation. Dermatology 187(3):153–158
57. Scholzen T, et al (1998) Neuropeptides in the skin: interactions between the neuroendocrine and the skin immune systems. Exp Dermatol 7(2–3):81–96
58. Scholzen TE, et al (2004) Cutaneous allergic contact dermatitis responses are diminished in mice deficient in neurokinin 1 receptors and augmented by neurokinin 2 receptor blockage. FASEB J 18(9):1007–1009
59. Shepherd AJ, et al (2005) Mobilisation of specific T cells from lymph nodes in contact sensitivity requires substance P. J Neuroimmunol 164(1–2):115–123
60. Song IS, et al (2000) Substance P induction of murine keratinocyte PAM 212 interleukin 1 production is mediated by the neurokinin 2 receptor (NK-2R). Exp Dermatol 9(1):42–52

61. Stead RH, Bienenstock J, Stanisz AM (1987) Neuropeptide regulation of mucosal immunity. Immunol Rev 100: 333–359
62. Suzuki R, et al (1999) Direct neurite-mast cell communication in vitro occurs via the neuropeptide substance P. J Immunol 163(5):2410–2415
63. Tobin D, Nabarro G, Baart de la Faille H, van Vloten WA, van der Putte SC, Schu HJ (1992) Increased number of immunoreactive nerve fibers in atopic dermatitis. J Allergy Clin Immunol 90:613–614
64. Torii H, et al (1997) Expression of neurotrophic factors and neuropeptide receptors by Langerhans cells and the Langerhans cell-like cell line XS52: further support for a functional relationship between Langerhans cells and epidermal nerves. J Invest Dermatol 109(4):586–591
65. Torii H, et al (1997) Regulation of cytokine expression in macrophages and the Langerhans cell--like line XS52 by calcitonin gene--related peptide. J Leukoc Biol 61(2):216–223
66. Torii H, Tamaki K, Granstein RD, The effect of neuropeptides/hormones on Langerhans cells. J Dermatol Sci 20(1):21–28
67. Toyoda M, et al (2002) Nerve growth factor and substance P are useful plasma markers of disease activity in atopic dermatitis. Br J Dermatol 147(1):71–79
68. Umeda Y, Arisawa M, Inhibition of natural killer activity by calcitonin gene--related peptide. Immunopharmacol Immunotoxicol 11(2–3):309–320
69. Umeda Y, et al (1988) Inhibition of mitogen-stimulated T lymphocyte proliferation by calcitonin gene-related peptide. Biochem Biophys Res Commun 154(1):227–235
70. Urashima R, Mihara M (1998) Cutaneous nerves in atopic dermatitis: a histological, immunohistochemical, and electron microscopic study. Virchows Arch 432:363–370
71. Wang F, et al (1992) Calcitonin gene-related peptide inhibits interleukin 2 production by murine T lymphocytes. J Biol Chem 267(29):21052–21057
72. Weihe E, Hartschuh W (1988) Multiple peptides in cutaneous nerves: regulators under physiological conditions and a pathogenetic role in skin disease? Semin Dermatol 7(4):284–300
73. Weng Y, et al (2007) Regulatory effects of vasoactive intestinal peptide on the migration of mature dendritic cells. J Neuroimmunol 182(1–2):48–54

Regulation of Immune Cells by POMC Peptides

6

T.A. Luger, T. Brzoska, K. Loser, and M. Böhm

Contents

Synonyms Box: Adrenocorticotropin (ACTH); Agouti signalling protein (ASIP); Clusters of Differentiation (CD); Corticotropin-Releasing Hormone (CRH); Dendritic cells (DC); Dextran sodium sulphate (DSS); Experimental autoimmune enecephalitis (EAE); Intercellular adhesion molecule-1 (ICAM); Interleukin-1 (IL); Interleukin-1 receptor (IL-IR); Interleukin-1 receptor-associated kinase 1 (IRAKI); Lipopolysaccharide (LPS); Major histocompatibility (MHC); Mitogen-activated protein kinase (MAPK); Melanocortin receptor (MC-R); melanocyte-stimulting hormone (MSH); Nuclear factor-kB (NF-kB); Peripheral blood mononuclear cells (PBMC); Prohormone convertase (PC); Proopiomelanocortin (POMC); Protein kinase A (PKA); Toll-like receptor (TLR); tumor necrosis factor (TNF); Vascular cell adhesion molecule-1 (VCAM-1).

Key Features

- Cells of the immune system express receptors for melanocortins, especially MC-IR.
- POMC-derived peptides such as α-MSH or ACTH have potent immunomodulatory effects on cells of the immune system including downregulation of proinflammatory cytokines and adhesion molecules, induction of IL-10 and generation of regulatory T cells.
- The immuno-modulatory in vitro effects are recapitulated in a diversity of animal models in which α-MSH has anti-inflammatory effects.

6.1 Introduction

Alpha-melanocyte stimulating hormone (α-MSH) is a tridecapeptide derived from proopiomelanocortin (POMC) by posttranslational processing [12]. Several studies have provided ample evidence that α-MSH in vitro as well as in vivo has potent anti-inflammatory and immuno-regulatory effects [16,32]. Most of these effects are mediated via melanocortin receptors expressed on cells of the immune system as well as on resident non-immune cell types of peripheral tissues [16]. Moreover, the C-terminal tripeptide of α-MSH, KPV, and a related tripeptide K(D)PT both exhibit anti-inflammatory properties as seen for α-MSH [16]. Therefore, the emphasis of this brief review is on the spectrum of immuno-modulatory and anti-inflammatory activities of α-MSH and related peptides as well as their therapeutic potential for the treatment of immune-mediated inflammatory diseases.

6.2 α-MSH and Melanocortin Receptors

α-MSH is generated from a precursor hormone called proopiomelanocortin (POMC), which functions as the source for several peptide hormones such as the

R.D. Granstein and T.A. Luger (eds.), *Neuroimmunology of the Skin*,

melanocortins, adrenocorticotropin (ACTH), α-, β-, and γ-MSH, as well as the endogenous opioid β-endorphin [26]. POMC is proteolytically cleaved by prohormone convertases (PCs), which belong to the family of serine proteases of the subtilisin/kexin type [67]. Although POMC peptides were originally considered as neurohormones, it is now well established that POMC expression and processing may occur in many peripheral tissues. In addition to their originally described pigmentation inducing capacity, melanocortins are now well known to exert a variety of other biological activities, including the regulation of immunity and inflammation [12,32]. The generation of melanocortins is controlled by endogenous mediators such as corticotropin releasing hormone (CRH), pro-inflammatory cytokines such as interleukin-1 (IL-1) and tumor necrosis factor-α (TNF-α) as well as exogenous noxious stimuli such as ultraviolet irradiation [12].

Upon investigation of the anti-inflammatory capacities of α-MSH, it turned out that these appear to be restricted to smaller fragments such as the C-terminal peptide fragment of α-MSH (KPV). Other small molecular weight peptides include the N-acetylated and C-amidated tripeptide KPV as well as several stereoisomers [38]. A structurally related derivate is K(D)PT in which the hydrophobic amino acid valine of KPV is substituted by the more polar amino acid threonine. The all L-form of K(D)PT has first been described as a part of IL-1β and seems to be capable of interacting with the IL-1 receptor type I (IL-1RI) [16]. Whether KPV and K(D)PT bind to MC-Rs and utilize the same signaling pathways as the natural ligands is not yet clear [16].

Melanocortins elicit their biological effects via binding to specific surface receptors – melanocortin receptors (MC-Rs) – expressed on target cells. These receptors belong to the superfamily of G-protein coupled receptors with seven transmembrane domains and the different MC-Rs bind melanocortins with differential affinity [17]. Five MC-R subtypes, MC-1R to MC-5R, have been identified and cloned. Human MC-1R and MC-4R discriminate poorly between ACTH and α-MSH, while MC-2R is selective for ACTH. α-MSH is the preferred, though not exclusive, MC-5R ligand, and MC-3R is the least selective receptor of the family [65]. MC-Rs are more widely expressed throughout the body. In particular, MC-1R has been detected not only on melanocytes but also on many other cell types, including inflammatory and immunocompetent cells [12]. Recent evidence exists for non-melanocortin signaling through MC-1R. Accordingly, components of the innate immunity such as β-defensins have been identified as additional ligands for MC-1R on melanocytes and thereby to participate in melanogenesis [18]. Moreover, there is strong support for MC-3R being expressed on macrophages and thereby contributing to the immunomodulatory potential of melanocortins [27].

6.3 Anti-Inflammatory and Immuno-Modulatory Capacities of α-MSH

Several investigations have provided ample evidence for a broad spectrum of anti-inflammatory and immuno-modulatory activities of α-MSH, mainly due to its capacity to regulate the production of cytokines and chemokines as well as the expression of surface molecules required for adhesion and co-stimulation of immunocytes (Table 6.1) [16]. The anti-inflammatory effects of α-MSH have been observed for extremely low, for example, subpicomolar, concentrations, where based upon the ligand affinity of MC-1R only few receptors would be occupied [12,43]. It is therefore possible that the anti-inflammatory effects of α-MSH are not only mediated by MC-Rs but also by additional pathways. Accordingly, previous studies could indeed demonstrate that α-MSH potently and selectively reduces surface binding of radiolabeled IL-1β to the T-cells [53].

Many studies have addressed the question whether α-MSH affects the function of dendritic cells (DC), monocytes, and macrophages. Accordingly, human as well as murine peripheral blood-derived monocytes and macrophages have been shown to express the MC-1R. Usually MC-1R expression on monocytic cells is low but significantly enhanced upon encountering inflammatory stimuli such as bacterial endotoxins and mitogens [6,8,60,64,69]. There is also recent evidence for the expression of MC-3R on murine and human macrophages and its involvement in mediating the anti-inflammatory effects of α-MSH [29,28]. Moreover, human peripheral blood DCs when cultured in the presence of GM-CSF, IL-4, and monocyte conditioned medium were found to express MC-1R. In comparison to immature DCs after 5 days of culture matured DC at day 8 expressed the highest levels of MC-1R [6]. In addition, MC-1R also was detected on other cells being involved in the regulation of immune and inflammatory responses such as neutrophils, basophils, eosinophils, mast cells, fibroblasts, keratinocytes, and endothelial cells [5,9,10,13,14,20,37,59]. The finding of immunocompetent cells being able to express MC-1R indicates that α-MSH may affect their

Table 6.1 Anti-inflammatory effects of α-MSH in vitro

	Effect	Reference
Cytokines		
– TNFα	Inhibition	[71,76]
– IL-1β	Inhibition	[31]
– IL-6	Inhibition	[41]
– IL-8	Inhibition	[10,15]
– Groα	Inhibition	[15]
– IFNγ	Inhibition	[47,74]
– IL-4	Inhibition	[13]
– IL-12 (p70)	Inhibition	[72]
– IL-10	Induction	[7]
Other Mediators		
– iNOS	Inhibition	[35]
– NO_2^-	Inhibition	[69]
– Neopterin	Inhibition	[60]
– Histamine	Inhibition (mouse)	[1]
	Induction (human)	[34]
– PGE	Inhibition	[19]
– PGF_2	Inhibition	[55]
Adhesion Molecules		
– ICAM 1	Inhibition	[11,37]
– VCAM	Inhibition	[37]
– E-selectin	Inhibition	[37]
Costimulatory Molecules		
– CD11b	Inhibition	[59]
– CD86	Inhibition	[8]
– CD80	Inhibition	[6]
– CD40	Inhibition	[6]
– CD14	Inhibition	[64]
Toll like Receptor: TLR4	Inhibition	[72]
Neutrophil Migration	Inhibition (CXCR)	[20,51]
Antigen Presentation	Inhibition	[6,8]
Suppressor T-cells	Induction	[45]
Regulatory T-cells	Induction	[46,54]
IgE production	Inhibition	[2]

functions and thereby play a role in the regulation of immuneresponses. In contrast, melanocortins such as β-MSH and γ-MSH, which require the expression of other MC-Rs, exert a less pronounced immunomodulating activity. The binding of melanocortins to MC-1R is known to be blocked by the agouti signaling protein (ASIP), which is encoded by the agouti locus. Accordingly, ASIP inhibits the effects of α-MSH on melanocytes, but it is not yet fully elucidated whether it also neutralizes α-MSH-mediated effects on the function of immunocompetent cells [70].

There is evidence from several studies that α-MSH is capable to modulate the function of MC-1R expressing monocytes, macrophages, and DCs. In macrophages, α-MSH downregulates the synthesis and release of pro-inflammatory cytokines such as IL-1, IL-6, and TNF-α, the production of pro-inflammatory nitric oxide and neopterin, as well as the expression of the endotoxin receptor CD14 [44,60,64]. In macrophages, α-MSH also downregulates the production of IL-12, which plays a crucial role in the development of naïve T-cells into Th_1-cells [72]. α-MSH also turned out to be a potent inhibitor of IL-1-mediated effects such as thymocyte proliferation and fever induction [41]. Via modulating the secretion of chemokines such as IL-8 and Gro-α, α-MSH is able to regulate the migration of monocytes and neutrophils and to inhibit IL-8-mediated chemotaxis via down-regulation of the IL-8 receptor CXCR1/2 on neutrophils [15,16,51]. Moreover, exposure of peripheral blood mononuclear cells or human keratinocytes to α-MSH increased both IL-10 mRNA and protein [7,62]. Induction of IL-10 is currently believed to be a key event responsible for immuno-modulatory and anti-inflammatory activities of α-MSH.

There is evidence that α-MSH affects the expression of MHC antigens and costimulatory molecules, which are required to mount an effective immuneresponse. Accordingly, α-MSH in a dose-dependent manner downregulates the expression of major histocompatibility (MHC) class I molecules on monocytes and significantly suppresses the expression of CD86 and CD40 on monocytes and DCs, whereas CD80 expression was not substantially altered [6,8]. Thus α-MSH via blocking costimulatory signals and inducing IL-10 appears to prevent DC maturation and may serve as one of the signals required for immunosuppression and possibly tolerance induction (Fig. 6.1). This is further supported by investigations showing that a subpopulation of regulatory T-lymphocytes is generated upon treatment of DC with α-MSH. They are characterized by the expression of $CD25^+$, $CD4^+$, $CTLA4^+$, Foxp3 and the production of increased levels of transforming growth factor (TGF-β) [46,54].

The overall expression of MC-Rs in several lymphocyte subsets currently is believed to be low or undetectable [4]. However, there is recent evidence for MC-1R being expressed on T-cells and a direct effect of α-MSH on lymphocyte functions. Accordingly, $CD8^+$ but not $CD4^+$ T cells upon stimulation were found to express MC-1R but none of the other MC-Rs. Moreover, α-MSH induced a suppressor phenotype associated with increased cytotoxic activity in $CD8^+$ T cells, and α-MSH-treated $CD8^+$ T cells were able to suppress contact allergy in vivo, indicating a suppressive activity of α-MSH-stimulated $CD8^+$ T-cells [45]. There is also evidence for MC-1R and MC-3R

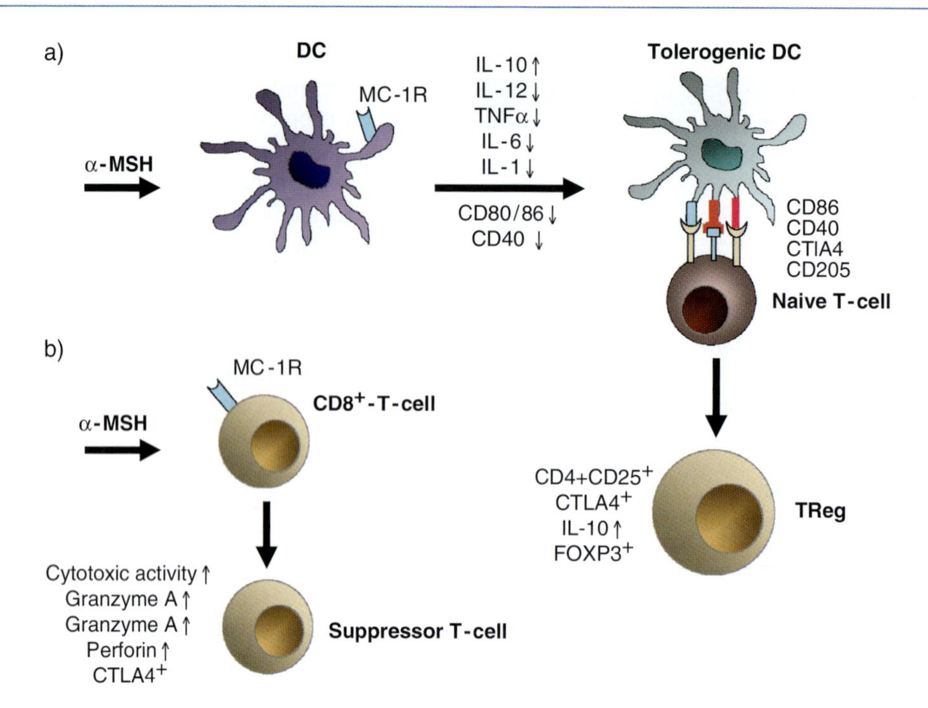

Fig. 6.1 α-MSH generates suppressor T-cells and regulatory T-cells. **a** α-MSH via downregulating costimulatory molecule expression (CD80/86, CD40, CD205) as well as cytokines (IL-1β, IL-6, TNF-α, IL-12) and upregulating IL-10 and CD205 (molecules that have previously been associated with the DC-mediated generation/expansion of regulatory T cells) generates tolerogenic DC, which have the capacity to induce $CD4^+CD25^+$, $CTLA4^+$, $IL\text{-}10^+$, $FOXP3^+$ Regulatory T-cells from naïve T-cells. **b** α-MSH upregulates granzyme A, granzyme B, perforin expression, and cytotoxic activity in MC-1R expressing $CD8^+$, $CTLA4^+$ T-cells and thereby favors the generation of suppressor T-cells

expression by B-lymphocytes and a role of α-MSH in mediating B-cell functions [23]. Accordingly, α-MSH and ACTH were found to modulate the IL-4 and anti CD40 mediated IgE release by human peripheral blood mononuclear cells (PBMC). At low physiological concentrations, α-MSH increases IgE synthesis, whereas higher concentrations significantly inhibit IgE production by B-cells [2,3].

In addition to the effects on DCs, T-, and B-cells, α-MSH also has been demonstrated to mediate the functions of inflammatory cells such as mast-cells, basophils, and eosinophils. Accordingly, mast cells express MC-R1 as well as MC-R5, and histamine release is regulated in human mast cells by ACTH and α-MSH [5,34]. Human basophils also express MC-1R and α-MSH suppresses basophil functions as well as IL-4 production [13]. Similarly, human eosinophils express MC-1R and upon exposure to α-MSH the IL-3 mediated expression of CD69 and CD11b as well as the release of superoxide anion was inhibited [59]. Furthermore, α-MSH is a potent inhibitor of the IL-1, TNF-α, or LPS mediated expression of adhesion molecules such as intercellular adhesion molecule-1 (ICAM-1) and P-selectin on dermal vascular endothelial cells. Ultimately, the inhibition of inflammatory cell adhesion and transmigration may contribute to the anti-inflammatory potential of α-MSH [37].

An important component of the intracellular signaling cascade underlying the anti-inflammatory effects of α-MSH is the cAMP-PKA pathway [44]. By increasing intracellular cAMP, α-MSH was found to inhibit cytokine (IL-1β, IL-6, TNFα, and IL-8) or LPS mediated nuclear factor-κB (NF-κB) nuclear translocation and DNA binding via inhibiting phosphorylation and subsequent degradation of IκB, the inhibitor of NF-κB [15,49,50,52]. Moreover, α-MSH inhibits the p38 mitogen activated protein kinase (p38MAPK) pathway, resulting in a diminished binding of the TATA-box binding protein to both DNA and NF-κB. This leads to an ineffective generation of RNA polymerase II, which is required for an efficient transcription [77]. Recently it has been reported that α-MSH also regulates p38MAPK as well as NF-κB activation via inducing the binding of IL-1R-associated kinase 1 (IRAK1) – the proximal intracellular signal molecule of endotoxin-bound TLR4 – to its inhibitor IRAK-M in activated macrophages [72]. In addition to the inhibition of the NF-κB signaling pathway, α-MSH also has been shown to alter signaling via the AP1 and the interferon-regulatory factor 1 pathway [10,44,72].

6.4 Therapeutic Potential of α-MSH and Related Peptides

Several animal models of inflammatory immune mediated diseases have been used to confirm the in vitro anti-inflammatory effects of α-MSH (Table 6.2). Using mouse models of contact hypersensitivity (CHS) intravenous as well as epicutaneous application of α-MSH significantly suppressed both the sensitization and elicitation phase of the cutaneous immune response and induced hapten-specific tolerance [63]. Induction of immune tolerance by α-MSH could be abrogated by application of an antibody against the IL-10 strongly, suggesting that this cytokine plays a key role in mediating the molecular anti-inflammatory mechanisms of α-MSH [33]. Moreover, the α-MSH related peptides KPV as well as K(D)PT like the parent molecule α-MSH were able to exert a similar anti-inflammatory activity and to induce hapten specific tolerance [16]. The relevance of the anti-inflammatory potential of α-MSH in murine contact dermatitis was further supported by preliminary findings, demonstrating that α-MSH topically applied in a cream formulation reduced nickel-induced contact eczema in man [68].

Using a murine model, α-MSH applied intraperitoneally was able to inhibit allergic airway inflammation induced by aerosol sensitization and subsequent challenges with ovalbumin. In addition, levels of both IL-4 and IL-13, two important pro-allergic cytokines, were found to be decreased in the bronchoalveolar lavage fluid of allergic mice treated with α-MSH. In accordance with the role of IL-10 in α-MSH-mediated suppression of CHS, the anti-inflammatory action of the peptide in allergic airway inflammation was dependent on the presence of IL-10 [58].

The in vitro effects of α-MSH on dermal microvascular endothelial cells encouraged the investigation of its effects in vivo using a model of LPS-induced cutaneous vasculitis (local Shwartzman reaction). Accordingly, a single intraperitoneal injection of α-MSH was able to diminish the vascular damage and hemorrhage via downregulating the sustained expression of vascular E-selectin and vascular cell adhesion molecule-1 (VCAM-1), two adhesion molecules crucially required for the diapedesis and activation of leukocytes, ultimately resulting in extravasation, inflammation, and hemorrhagic vascular damage [66].

α-MSH has also been tested for its anti-inflammatory potential using animal models of autoimmune uveitis as well as corneal injury. In a murine model of experimental autoimmune uveitis, α-MSH given intravenously significantly suppressed endotoxin-induced uveitis [56,75]. The anti-inflammatory mechanism of α-MSH in these models appears to be linked to the induction of regulatory T cells, since adoptively transferred T cells generated by α-MSH and TGF-β_2 in vitro also were found to suppress experimental autoimmune uveoretinitis [54].

Table 6.2 Therapeutic potential of αMSH and fragments for animal models of immune mediated inflammatory diseases

Model		Application of α-MSH	Species	Reference
CHS	Urate crystals	s.c.	Rat	[25]
	DNFB, Oxazolone	Epicutaneous	Mouse	[63]
	DNFB	i.p.	Mouse	[39]
	TNCB	i.v.	Mouse	[33]
	λ-carrageenan	i.p.	Mouse	[39]
	IL-1, IL-6, TNFα	i.p.	Mouse	[40]
	IL-8, LTB4, PAF	i.c.v.	Mouse	[21]
Vasculitis	LPS	i.p.	Mouse	[66]
Allergic airway inflammation	OVA	i.v.	Mouse	[58]
Gastrointestinal inflammation	DSS	i.p.	Mouse	[48]
	$CD45RB^{high}$-T-cells	i.p.	Mouse	[42]
Autoimmune uveitis	M.Tb. cell wall	i.v.	Mouse	[56,75]
Arthritis	M.Tb. cell wall	i.p.	Rat	[22]
EAE	Ig-proteolipid protein	i.v.	Mouse	[36,73]

In animal models of arthritis, repeated intraperitoneal administration of α-MSH significantly attenuated the clinical and histological signs of adjuvant-induced experimental arthritis as compared to control animals. α-MSH was similarly effective as prednisolone but did not cause significant weight loss [22]. Furthermore, in a rat model of gouty arthritis elicited by intraarticular injection of monosodium urate monohydrate crystals, the MC-3R antagonist SHU9119 blocked the anti-inflammatory action of the α-MSH precursor and structurally related peptide ACTH [30]. These data indicate that MC-3R may be a relevant target for the treatment of arthritis.

The effect of α-MSH in experimentally induced colitis was investigated using mouse models of dextran sodium sulphate (DSS)-induced and $CD45RB^{high}$ transfer colitis. α-MSH profoundly inhibited weight loss, and prevented disintegration of the general condition of the animals [42,61]. In a rat model of DSS-induced colitis, intraperitoneal injection of α-MSH likewise reduced the colonic macroscopic lesions compared to untreated ones in both acute and chronic colitis groups [57]. Similar effects have been observed when the mice were treated with the tripeptides KPV or K(D)PT [42,48]. There is evidence for an important role of the MC-1R in the regulation of inflammatory responses of the gut, since in mice with a frame shift mutation in the MC-1R gene (MC-1Re/e) [48] DSS-induced colitis mice was significantly aggravated in comparison to the wild type (C57BL/6wt).

To investigate the possibility whether α-MSH may suppress autoimmune diseases, a murine model of experimental autoimmune encephalomyelitis (EAE) was investigated. In comparison to untreated mice injection of α-MSH at the onset of paralysis resulted in a significantly reduced severity and tempo of EAE. This effect seems to be due to the induction of TGF-β producing regulatory T-cells and the reduction of IFNγ producing Th_1-cells [73]. In a recent study, it was demonstrated that α-MSH transduced PLP specific T-cells had a preventive and therapeutic effect on active relapse-remitting EAE in an antigen-inducible manner [36]. Therefore, α-MSH as well as its fragments may serve as promising new therapeutic tools to treat autoimmune diseases of the central nervous system.

6.5 Conclusions

α-MSH and related peptides in vitro exert several immuno-regulatory and anti-inflammatory activities. In particular, via affecting the function of DC and T-cells they apparently participate in the generation of regulatory T-cells and the induction of immune tolerance. The broad spectrum of anti-inflammatory effects is further mediated by the capacity of α-MSH and its fragments to regulate the function of almost any cell participating in inflammation such as mast cells, basophils, eosinophils, and endothelial cells. The molecular mechanism being responsible for the anti-inflammatory effects of α-MSH is the downregulation of transcription factor activation seen for NFκB. The effects of α-MSH on inflammatory and immunocompetent cells largely seem to depend on the expression of MC-1R and to some degree also on MC-3R. Furthermore, there is recent evidence that α-MSH-derived tripeptides may function via a di/tripeptide transporter (PepT1) being expressed in immune and intestinal epithelial cells [24]. Therefore, α-MSH-related tripeptides such as KPV and K(D)PT appear to be well suited for being developed as novel compounds for the treatment of immune-mediated inflammatory diseases.

Summary for the Clinician

- α-MSH can be regarded as a lead substance for testing the anti-inflammatory action of melanocortin peptides in various models of inflammation.
- Truncated tripeptides and derivatives corresponding to the C-terminal domain of α-MSH have preserved anti-inflammatory effects.
- Due to their physiochemical nature such tripeptides may become novel disease-modifying agents in the future treatment of immune-mediated inflammatory diseases.

References

1. Adachi S, Nakano T, Vliagoftis H, Metcalfe DD (1999) Receptor-mediated modulation of murine mast cell function by alpha-melanocyte stimulating hormone. J Immunol 163:3363–3368
2. Aebischer I, Stampfli MR, Zurcher A, et al (1994) Neuropeptides are potent modulators of human in vitro immunoglobulin Esynthesis. Eur J Immunol 24:1908–1913
3. Aebischer I, Stämpfli M, Miescher S, Horn M, Zürcher AW, Stadler BM (1995) Neuropeptides accentuate interleukin-4 induced human immunoglobulin E synthesis in vitro. Exp Dermatol 4:418–423
4. Andersen GN, Hagglund M, Nagaeva O, et al (2005) Quantitative measurement of the levels of melanocortin

receptor subtype 1, 2, 3 and 5 and pro-opio-melanocortin peptide gene expression in subsets of human peripheral blood leucocytes. Scand J Immunol 61:279–284
5. Artuc M, Grutzkau A, Luger T, Henz BM (1999) Expression of MC1- and MC5-receptors on the human mast cell line HMC-1. Ann N Y Acad Sci 885:364–367
6. Becher E, Mahnke K, Brzoska T, Kalden DH, Grabbe S, Luger TA (1999) Human peripheral blood-derived dendritic cells express functional melanocortin receptor MC-1R. Ann N Y Acad Sci 885:188–195
7. Bhardwaj RS, Schwarz A, Becher E, et al (1996) Pro-opiomelanocortin-derived peptides induce IL-10 production in human monocytes. J Immunol 156:2517–2521
8. Bhardwaj RS, Becher E, Mahnke K, et al (1997) Evidence for the differential expression of the functional alpha melanocyte stimulating hormone receptor MC-1 on human monocytes. J Immunol 158:3378–3384
9. Böhm M, Raghunath M, Sunderkötter C et al (2004) Collagen metabolism is novel target of the neuropeptide α-melanocyte stimulating hormone. J Biol Chem 279:6959–6966
10. Böhm M, Schulte U, Kalden DH, Luger TA (1999) Alpha-melanocyte-stimulating hormone modulates activation of NF-κB and AP-1 and secretion of IL-8 in human dermal fibroblasts. Ann N Y Acad Sci 885:277–286
11. Bohm M, Eickelmann M, Li Z, et al (2005) Detection of functionally active melanocortin receptors and evidence for an immunoregulatory activity of alpha-melanocyte-stimulating hormone in human dermal papilla cells. Endocrinology 146:4635–4646
12. Böhm M, Luger TA, Tobin DJ, Garcia-Borron JC (2006) Melanocortin receptor ligands: new horizons for skin biology and clinical dermatology. J Invest Dermatol 126:1966–1975
13. Böhm M, Raap U, Schiller M, Straub RH, Haas H, Slominski A, Kapp A, Luger TA (2007) Expression and functional significance of HPA axis components in human basophils. Exp Dermatol 16:371–372 (abstract)
14. Brzoska T, Scholzen T, Becher E, Luger TA (1997) Effect of UV light on the production of proopiomelanocortin-derived peptides and melanocortin receptors in the skin. In: Altmeyer P, Hoffmann K, Stücker M (eds) Skin Cancer and UV-Radiation. Springer-Verlag, Berlin, pp 227–237
15. Brzoska T, Kalden DH, Scholzen T, Luger TA (1999) Molecular basis of α-MSH/IL-1 antagonism. Ann N Y Acad Sci 885:230–238
16. Brzoska T, Luger TA, Maaser C, Abels C, Böhm M (2008) α-Melanocyte-stimulating hormone and related tripeptides. Biochemistry, anti-inflammatory effects in vitro and in vivo and future perspectives for the treatment of immune-mediated inflammatory diseases. Endocr Rev 29:581–602
17. Buzzetti R, McLoughlin L, Lavender PM, Clark AJ, Rees LH (1989) Expression of pro-opiomelanocortin gene and quantification of adrenocorticotropic hormone-like immunoreactivity in human normal peripheral mononuclear cells and lymphoid and myeloid malignancies. J Clin Invest 83:733–737
18. Candille SI, Kaelin CB, Cattanach BM, et al (2007) A defensin mutation causes black coat color in domestic dogs. Science 318:1418–1423
19. Cannon JG, Tatro JB, Reichlin S, Dinarello CA (1986) Alpha melanocyte stimulating hormone inhibits immunostimulatory and inflammatory actions of interleukin 1. J Immunol 137:2232–2236
20. Catania A, Rajora N, Capsoni F, Minonzio F, Star RA, Lipton JM (1996) The neuropeptide alpha-MSH has specific receptors on neutrophils and reduces chemotaxis in vitro. Peptides 17:675–679
21. Ceriani G, Macaluso A, Catania A, Lipton JM (1994) Central neurogenic antiinflammatory action of alpha-MSH: modulation of peripheral inflammation induced by cytokines and other mediators of inflammation. Neuroendocrinology 59:138–143
22. Ceriani G, Diaz J, Murphree S, Catania A, Lipton JM (1994) The neuropeptide alpha-melanocyte-stimulating hormone inhibits experimental arthritis in rats. Neuroimmunomodulation 1:28–32
23. Cooper A, Robinson SJ, Pickard C, Jackson CL, Friedmann PS, Healy E (2005) Alpha-melanocyte-stimulating hormone suppresses antigen-induced lymphocyte proliferation in humans independently of melanocortin 1 receptor gene status. J Immunol 175:4806–4813
24. Dalmasso G, Charrier-Hisamuddin L, Thu Nguyen HT, Yan Y, Sitaraman S, Merlin D (2008) PepT1-mediated tripeptide KPV uptake reduces intestinal inflammation. Gastroenterology 134:166–178
25. Denko CW, Gabriel P (1985) Effects of peptide hormones in urate crystal inflammation. J Rheumatol 12:971–975
26. Eipper BA, Mains RE (1980) Structure and biosynthesis of pro-adrenocorticotropin/endorphin and related peptides. Endocr Rev 1:1–27
27. Getting SJ, Flower RJ, Perretti M (1999) Agonism at melanocortin receptor type 3 on macrophages inhibits neutrophil influx. Inflamm Res 48(Suppl 2):S140–S141
28. Getting SJ, Gibbs L, Clark AJ, Flower RJ, Perretti M (1999) POMC gene-derived peptides activate melanocortin type 3 receptor on murine macrophages, suppress cytokine release, and inhibit neutrophil migration in acute experimental inflammation. J Immunol 162:7446–7453
29. Getting SJ, Allcock GH, Flower R, Perretti M (2001) Natural and synthetic agonists of the melanocortin receptor type 3 possess anti-inflammatory properties. J Leukoc Biol 69:98–104
30. Getting SJ, Christian HC, Flower RJ, Perretti M (2002) Activation of melanocortin type 3 receptor as a molecular mechanism for adrenocorticotropic hormone efficacy in gouty arthritis. Arthritis Rheum 46:2765–2775
31. Getting SJ, Christian HC, Lam CW, et al (2003) Redundancy of a functional melanocortin 1 receptor in the anti-inflammatory actions of melanocortin peptides: studies in the recessive yellow (e/e) mouse suggest an important role for melanocortin 3 receptor. J Immunol 170:3323–3330
32. Gonzalez-Rey E, Chorny A, Delgado M (2007) Regulation of immune tolerance by anti-inflammatory neuropeptides. Nat Rev Immunol 7:52–63
33. Grabbe S, Bhardwaj RS, Steinert M, et al (1996) Alpha-melanocyte stimulating hormone induces hapten-specific tolerance in mice. J Immunol 156:473–478
34. Grutzkau A, Henz BM, Kirchhof L, Luger T, Artuc M (2000) Alpha-melanocyte stimulating hormone acts as

a selective inducer of secretory functions in human mast cells. Biochem Biophys Res Commun 278:14–19
35. Gupta AK, Diaz RA, Higham S, Kone BC (2000) Alpha-MSH inhibits induction of C/EBPbeta-DNA binding activity and NOS2 gene transcription in macrophages. Kidney Int 57:2239–2248
36. Han D, Tian Y, Zhang M, Zhou Z, Lu J (2007) Prevention and treatment of experimental autoimmune encephalomyelitis with recombinant adeno-associated virus-mediated alpha-melanocyte-stimulating hormone-transduced PLP139–151-specific T cells. Gene Ther 14:383–395
37. Hartmeyer M, Scholzen T, Becher E, et al (1997) Human microvascular endothelial cells (HMEC-1) express the melanocortin receptor type 1 and produce increased levels of IL-8 upon stimulation with αMSH. J Immunol 159:1930–1937
38. Haskell-Luevano C, Sawyer TK, Hendrata S, et al (1996) Truncation studies of alpha-melanotropin peptides identify tripeptide analogues exhibiting prolonged agonist bioactivity. Peptides 17:995–1002
39. Hiltz ME, Lipton JM (1990) Alpha-MSH peptides inhibit acute inflammation and contact sensitivity. Peptides 11:979–982
40. Hiltz ME, Catania A, Lipton JM (1992) Alpha-MSH peptides inhibit acute inflammation induced in mice by rIL-1 beta, rIL-6, rTNF-alpha and endogenous pyrogen but not that caused by LTB4, PAF and rIL-8. Cytokine 4:320–328
41. Huang QH, Hruby VJ, Tatro JB (1998) Systemic alpha-MSH suppresses LPS fever via central melanocortin receptors independently of its suppression of corticosterone and IL-6 release. Am J Physiol 275:R524–R530
42. Kannengiesser K, Maaser C, Heidemann J, et al (2008) Melanocortin-derived tripeptide KPV has anti-inflammatory potential in murine models of inflammatory bowel disease. Inflamm Bowel Dis 14:324–331
43. Lam CW, Getting SJ (2004) Melanocortin receptor type 3 as a potential target for anti-inflammatory therapy. Curr Drug Targets Inflamm Allergy 3:311–315
44. Lipton JM, Catania A (1997) Antiinflammatory actions of the neuroimmunomodulator α-MSH. Immunol Today 18:140–145
45. Loser K, Brzoska T, Beissert S, Luger TA (2006) α-Melanocyte stimulating hormone treated CD8+ T cells control contact allergy and anti-tumoral immunity. J Invest Dermatol 126:109 (Abstract)
46. Loser K, Brzoska T, Kupas V, Höcker N, Voskort M, Beissert S, Abels Ch, Luger TA (2008) α-Melanocyte stimulating hormone induces tolerogenic dendritic cells that generate CD4+ regulatory T cells. J Invest Dermatol 128:iii (Abstract)
47. Luger TA, Schauer E, Trautinger F, et al (1993) Production of immunosuppressing melanotropins by keratinocytes. Ann N Y Acad Sci 680:567–570
48. Maaser C, Kannengiesser K, Specht C, et al (2006) Crucial role of the melanocortin receptor MC1R in experimental colitis. Gut 55:1415–1422
49. Manna SK, Aggarwal BB (1998) α-melanocyte-stimulating hormone inhibits the nuclear transcription factor NF-kB activation induced by various inflammatory agents. J Immunol 161:2873–2880
50. Manna SK, Ramesh GT (2005) Interleukin-8 induces nuclear transcription factor-kappaB through a TRAF6-dependent pathway. J Biol Chem 280:7010–7021
51. Manna SK, Sarkar A, Sreenivasan Y (2006) Alpha-melanocyte-stimulating hormone down-regulates CXC receptors through activation of neutrophil elastase. Eur J Immunol 36:754–769
52. Moustafa M, Szabo M, Ghanem GE, et al (2002) Inhibition of tumor necrosis factor-alpha stimulated NFkappaB/p65 in human keratinocytes by alpha-melanocyte stimulating hormone and adrenocorticotropic hormone peptides. J Invest Dermatol 119:1244–1253
53. Mugridge KG, Perretti M, Ghiara P, Parente L (1991) Alpha-melanocyte-stimulating hormone reduces interleukin-1 beta effects on rat stomach preparations possibly through interference with a type I receptor. Eur J Pharmacol 197:151–155
54. Namba K, Kitaichi N, Nishida T, Taylor AW (2002) Induction of regulatory T cells by the immunomodulating cytokines alpha-melanocyte-stimulating hormone and transforming growth factor-beta2. J Leukoc Biol 72:946–952
55. Nicolaou A, Estdale SE, Tsatmali M, Herrero DP, Thody AJ (2004) Prostaglandin production by melanocytic cells and the effect of alpha-melanocyte stimulating hormone. FEBS Lett 570:223–226
56. Nishida T, Miyata S, Itoh Y, et al (2004) Anti-inflammatory effects of alpha-melanocyte-stimulating hormone against rat endotoxin-induced uveitis and the time course of inflammatory agents in aqueous humor. Int Immunopharmacol 4:1059–1066
57. Oktar BK, Ercan F, Yegen BC, Alican I (2000) The effect of alpha-melanocyte stimulating hormone on colonic inflammation in the rat. Peptides 21:1271–1277
58. Raap U, Brzoska T, Sohl S, et al (2003) Alpha-melanocyte-stimulating hormone inhibits allergic airway inflammation. J Immunol 171:353–359
59. Raap U, Schefzyk M, Bruder M, Schiller M, Luger TA, Kapp A, Böhm M (2007) The eosinophil – a novel target for the neuropeptide alpha-melanocyte-stimulating hormone. Exp Dermatol 16:377–378 (Abstract)
60. Rajora N, Ceriani G, Catania A, Star RA, Murphy MT, Lipton JM (1996) Alpha-MSH production, receptors, and influence on neopterin in a human monocyte/macrophage cell line. J Leukoc Biol 59:248–253
61. Rajora N, Boccoli G, Catania A, Lipton JM (1997) alpha-MSH modulates experimental inflammatory bowel disease. Peptides 18:381–385
62. Redondo P, Garcia-Foncillas J, Okroujnov I, Bandres E (1998) α-MSH regulates interleukin-10 expression by human keratinocytes. Arch Dermatol Res 290:425–428
63. Rheins LA, Cotleur AL, Kleier RS, Hoppenjans WB, Sauder DN, Nordlund JJ (1989) Alpha-melanocyte stimulating hormone modulates contact hypersensitivity responsiveness in C57/BL6 mice. J Invest Dermatol 93:511–517
64. Sarkar A, Sreenivasan Y, Manna SK (2003) Alpha-melanocyte-stimulating hormone inhibits lipopolysaccharide-induced biological responses by downregulating CD14 from macrophages. FEBS Lett 553:286–294
65. Schioth HB, Muceniece R, Larsson M, et al (1997) Binding of cyclic and linear MSH core peptides to the melanocortin receptor subtypes. Eur J Pharmacol 319:369–373
66. Scholzen TE, Sunderkotter C, Kalden DH, et al (2003) Alpha-melanocyte stimulating hormone prevents lipopolysaccharide-induced vasculitis by down-regulating endothelial cell adhesion molecule expression. Endocrinology 144:360–370

67. Seidah NG, Benjannet S, Hamelin J, et al (1999) The subtilisin/kexin family of precursor convertases. Emphasis on PC1, PC2/7B2, POMC and the novel enzyme SKI-1. Ann N Y Acad Sci 885:57–74
68. Slominski A, Wortsman J, Luger T, Paus R, Solomon S (2000) Corticotropin releasing hormone and proopiomelanocortin involvement in the cutaneous response to stress. Physiol Rev 80:979–1020
69. Star RA, Rajora N, Huang J, Chavez R, Catania A, Lipton JM (1995) Evidence of autocrine modulation of macrophage nitric oxide synthase by alpha-MSH. Proc Natl Acad Sci U S A 92:8016–8020
70. Suzuki I, Tada A, Ollmann MM, et al (1997) Agouti signaling protein inhibits melanogenesis and the response of human melanocytes to alpha-melanotropin. J Invest Dermatol 108:838–842
71. Taherzadeh S, Sharma S, Chhajlani V, et al (1999) Alpha-MSH and its receptors in regulation of tumor necrosis factor-alpha production by human monocyte/macrophages. Am J Physiol 276:R1289–R1294
72. Taylor AW (2005) The immunomodulating neuropeptide alpha-melanocyte-stimulating hormone (alpha-MSH) suppresses LPS-stimulated TLR4 with IRAK-M in macrophages. J Neuroimmunol 162:43–50
73. Taylor AW, Kitaichi N (2008) The diminishment of experimental autoimmune encephalomyelitis (EAE) by neuropeptide alpha-melanocyte stimulating hormone (alpha-MSH) therapy. Brain Behav Immun 22:639–646
74. Taylor AW, Streilein JW, Cousins SW (1994) Alpha-melanocyte-stimulating hormone suppresses antigen-stimulated T cell production of gamma-interferon. Neuroimmunomodulation 1:188–194
75. Taylor AW, Yee DG, Nishida T, Namba K (2000) Neuropeptide regulation of immunity. The immunosuppressive activity of alpha-melanocyte-stimulating hormone (alpha-MSH). Ann N Y Acad Sci 917:239–247
76. Wong KY, Rajora N, Boccoli G, Catania A, Lipton JM (1997) A potential mechanism of local anti-inflammatory action of alpha-melanocyte-stimulating hormone within the brain: modulation of tumor necrosis factor-alpha production by human astrocytic cells. Neuroimmunomodulation 4:37–41
77. Yoon SW, Goh SH, Chun JS, et al (2003) Alpha-melanocyte-stimulating hormone inhibits lipopolysaccharide-induced tumor necrosis factor-alpha production in leukocytes by modulating protein kinase A, p38 kinase, and nuclear factor kappa B signaling pathways. J Biol Chem 278:32914–32920

Regulation of Cutaneous Immunity by Catecholamines

7

K. Seiffert

Contents

Synonyms Box: adrenergic receptor = adrenoceptor = catecholamine receptor; epinephrine = adrenaline; norepinephrine = noradrenaline

Abbreviations: *AD* Atopic dermatitis, *AR* Adrenergic receptor, *BMDDC* Bone marrow-derived dendritic cells, *CHS* Contact hypersensitivity, *DNFB* Dinitroflourobenzene, *DTH* Delayed type hypersensitivity, *FITC* Flourescein isothiocyanate, *IFN* Interferon, *IL* Interleukin, *LC* Langerhans cells, *LPS* Lipopolysaccharide, *NE* Norepinephrine, *SNS* Sympathetic nervous system

Key Features

- *The Sympathetic Nervous System and Stress Responses:* An activation of the sympathetic nervous system (SNS), one of the main pathways involved in the stress response, produces an immediate, widespread "fight or flight" response. Transmitters of the SNS triggered during stress also shape the general and the cutaneous immune response.
- *Catecholamines:* The catecholamines epinephrine and norepinephrine act as hormones in the blood circulation and as neurotransmitters in the central and peripheral nervous system. Catecholamine (or adrenergic) receptors are abundantly expressed on resident and infiltrating skin cells.
- *Cutaneous Immunity:* Cutaneous immune function is regulated by multiple neuroendocrine pathways, among them are the hypothalamic pituitary axis, glucocorticoids, neuropeptides, endogenous opioids, and, prominently, the SNS. Langerhans cells in the skin have been shown to respond to adrenergic signals by altering their ability to present antigens, produce cytokines, express co-stimulatory molecules and migrate to draining lymph nodes.
- *Atopic Dermatitis:* Flares of atopic dermatitis are often reported to be triggered by stressful life events, and a dysregulation of the adrenergic response has been linked to immunological abnormalities. The relevance of psychosocial stress as a trigger factor is emphasized by the therapeutic effect of psychosocial interventions.
- *Psoriasis:* β-Adrenergic receptor blocking drugs can induce the first manifestation or exacerbation of psoriasis. Catecholamines suppress Th1 responses and skew the immune response towards Th2 through β-adrenergic receptors. The pharmacologic blockade of β-receptors may, thus, lead to a predominance of Th1 pathways and initiation or exacerbation of disease in genetically predisposed individuals.

R.D. Granstein and T.A. Luger (eds.), *Neuroimmunology of the Skin*,

7.1 Introduction

Although it is deeply ingrained in the public perception that stress will affect one's emotional and physical well being, the scientific community has been slow to accept this concept. Despite some early studies showing that academic stress or chronic overexertion and fatigue will affect disease susceptibility [27], only recently has it been accepted that an organism perceiving a "stressor" will create a behavioral response that includes transmitting hormonal or neural messages to the periphery to affect the immune system. Accumulating knowledge over the past decades indicates that the nervous system may communicate with the immune system to affect the health status of an individual in general [7] and the cutaneous immune response in particular [49]. Not only has it been reported that inflammatory skin conditions such as atopic dermatitis, psoriasis, and acne worsen with stress, but stress hormones, such as catecholamines and glucocorticoids, as well as neuropeptides have been extensively studied in regards to their immunomodulating function in the skin. This review will focus on the role of mediators of the sympathetic nervous system (SNS), one of the main pathways activated in the stress response, in cutaneous immune function.

7.2 The Sympathetic Nervous System and the Stress Response

Under normal, resting conditions, the autonomic nervous system regulates functions such as heart rate, vascular tone, gastrointestinal motility, and respiratory rate through the interaction of its three arms, the sympathetic, the parasympathetic, and the enteric nervous system. It functions to constantly adjust the organism to external and internal influences. Any stressor or threat to the stability or homeostasis of the internal milieu is counteracted by adaptive forces of the organism, and leads to the activation of the SNS. The activation of the SNS produces an immediate, widespread response that has been called the "fight or flight" response. Signals from the central nervous system trigger the release of the catecholamines epinephrine and, to a lesser extent, norepinephrine (NE) from the adrenal medulla and also of NE from sympathetic neurons. This activation is intended to prepare the individual for imminent danger by increasing heart rate and cardiac output, skeletal muscle vasodilation, cutaneous and gastrointestinal vasoconstriction, pupillary dilation, bronchial dilation, and piloerection. But neuroendocrine hormones triggered during stress also shape the immune response: while acute stress, as an evolutionary adaptive psychophysiological survival mechanism, may temporarily enhance general immune functions [10], chronic stress or depression leads to decreased host defenses, decreased response to vaccines, and viral susceptibility, possibly resulting in atopic syndromes, autoimmune diseases, or malignancy [8,26]. Acute and chronic stress can influence *cutaneous* immune responses in a similar manner, as outlined in the following sections.

7.3 Sympathetic Innervation and Catecholamine Production in the Skin

The sympathetic nervous system originates from the brainstem, from where preganglionic fibers travel down the spinal column and terminate in pre-and paravertebral ganglia [8]. Postganglionic fibers lead to multiple innervated tissues, such as smooth muscles of the vasculature, heart, skeletal muscles, kidney, gut, fat, and also primary and secondary lymphoid tissues, and the skin, where they release NE as their primary neurotransmitter. The body's main sources of epinephrine are the chromaffine cells of the adrenal medulla, which is directly innervated by preganglionic neurons. These cells are embryologically and anatomically homologous to cells of sympathetic ganglia, and release epinephrine and NE at a ratio of 4:1 [7].

There are several ways in which catecholamines can reach the skin. Most prominent is the release of NE from sympathetic nerve fibers that travel together with sensory nerves to innervate blood vessels, sweat glands, and hair follicles and finally appear as single nerve fibers in dermis and epidermis [4,58]. It is well established that NE serves as the main neurotransmitter in the periphery, but circulating epinephrine as well as some NE can reach target organs humoraly after being produced in the adrenal medulla. Apart from the classical ways by which catecholamines can reach the skin, as a neurotransmitter or circulating hormone, it has become apparent that skin cells themselves hold the full capacity for catecholamine synthesis [19]. Keratinocytes express key enzymes of catecholamine biosynthesis [44], and upregulated de novo epinephrine synthesis as well as higher levels of epinephrine have been found in undifferentiated epidermal keratinocytes when compared to differentiated ones [43]. Interestingly, keratinocytes also contain catecholamine-degrading enzymes and, thus, seem to be not only a source, but also a regulator of this important signaling system. It has been speculated that this system significantly contributes in situ to the epidermal immune response. Further research is needed to examine under which circumstances epidermal keratinocytes release epinephrine, and if this local machinery exists mainly for homeostatic purposes or if it is part of the "stress system" as well.

7.4 Adrenergic Receptor Expression in the Skin

A number of infectious and non-infectious stimuli can lead to release of NE from sympathetic nerve terminals [28] that lie in close proximity to cells within most organ systems [27]. Most cells in the body express receptors for catecholamines [27] and, given the rich innervation of the skin by sympathetic nerves, it is no surprise that several cell types within the skin express adrenergic receptors (AR) as well. AR are of the α or β type, which are subdivided into nine subtypes termed α_{1A}, α_{1B}, α_{1D}, α_{2A}, α_{2B}, α_{2C}, and β_1, β_2, and β_3 [22]. In human and murine skin, β_2-AR seems to be the most abundant adrenoceptor [54,56], although α-ARs are present as well [13]. Among the AR-expressing cell populations within the skin are keratinocytes, melanocytes, fibroblasts, and mast cells [35,43,56]. Specifically, freshly isolated human keratinocytes as well as the keratinocyte cell line HaCaT have been shown to express β_2-AR and practically no β_1-AR by autoradiographic mapping and radioligand binding experiments [53]. In parallel with the decreasing ability of keratinocytes undergoing differentiation to produce catecholamines, the number of β_2-adrenoceptors per cell also decreases [45]. Human melanocytes have been shown to express α1- and β2-ARs as well, at least after stimulation with NE in the case of the α_1-receptor, and have been implied in melanogenesis [19]. Human skin fibroblasts express β_1, β_2, and β_3 ARs in increasing density [18]. There are no publications describing AR expression on sebocytes to date, but one study reports that α- and β-adrenergic catecholamines stimulate sebocyte growth [41]. Seiffert et al. [50] have shown that murine Langerhans cells, the potent antigen presenting cell population in the epidermis, and Langerhans cell-like cell lines express the adrenergic receptors β_2-AR and α_{1A}-AR, and that receptor engagement leads to a potent downregulation of antigen presentation and cutaneous immune function as discussed below.

Among the cells that may infiltrate the skin or at least circulate in the vicinity of the skin are T and B lymphocytes and dendritic cells other than Langerhans cells. Among those cells, Th1 and B lymphocytes almost exclusively express β_2-ARs [28]. α-AR expression seems to be limited to specific innate immune cell subsets and is primarily of the α_1 subtype [25]. Other types of dendritic cells, such as bone marrow-derived dendritic cells (BMDDC), have been shown to express receptors for β_1-, β_2-, α_{2A}-, and α_{2C} [34].

7.5 The Sympathetic Nervous System and Cellular and Humoral Immune Function

Catecholamines have been shown to affect numerous cell types involved in innate and adaptive immune responses [30]. Noradrenergic innervation is present early in development and the arrival of fibers generally precedes the development of the cellular component of the immune system, which suggests a role for products of the SNS in the development of this system [8]. Primary and secondary lymphoid organs are extensively innervated by noradrenergic sympathetic nerve fibers, and interestingly, noradrenergic innervation of spleen and lymph nodes is diminished progressively during aging, a time when cell-mediated immune function is also suppressed [30]. Sympathetic fibers are primarily found in zones rich in CD4+ T cells and macrophages [15,16], whereas nodular and follicular zones of developing or maturing B cells are poorly innervated [7]. Furthermore, T cells, macrophages, and mast cells are regularly seen in contact with the terminals of peripheral nerves from sympathetic and sensory ganglia [8], and adrenergic receptors are present on numerous inflammatory cells such as T-lymphocytes, mast cells, monocytes/macrophages, eosinophils, and neutrophils [24,30].

Given the close anatomical relationship of lymphocytes and sympathetic fibers, it is not surprising that catecholamines have been found to modulate immune cell function. Whether the direction of immune responses is geared towards enhancement or supression is still a hotly debated topic. Numerous studies have focused on the immunosuppressive effect on T lymphocytes [28]. But the effect of the SNS on the immune response has been revealed to be more complex, depending on the cell type involved, the state of cellular differentiation, the timing of the stressor, and the organ involved. Systemically, catecholamines seem to suppress Th1 responses and enhance Th2 responses. They drive a Th2 shift at the level of both APCs and Th1 cells by potently inhibiting the production of interleukin (IL)-12 and enhancing the production of IL-10 through stimulation of β-ARs, which are expressed on Th1 but not on Th2 cells [7,21,38]. The effect of catecholamines on T cells appears to be dependent on the state of cellular differentiation and the timing of AR stimulation in relation to the time of cell activation (reviewed in [27]). For example, when naïve T cells are stimulated to differentiate into Th1 cells in the presence of NE, these Th1 cells will produce more interferon (IFN)γ per cell than without NE, indicating that NE participates in the generation of optimal protective Th1-mediated immune responses. The effect of NE on Th1 effector cells is more complex; when added before stimulation, NE reduces the amount of IFNγ produced by Th1 cells. It has no effect when added at the time of T cell stimulation, but will increase Th1 IFNγ production when added after stimulation. Catecholamines have also been found to influence B cell, natural killer cell, neutrophil [2,47], and macrophage [8] distribution and function.

In addition to skewing the immune response towards Th2 profiles, it has been suggested that stress hormones lead to a redistribution of leukocytes from the blood to peripheral organs, including the skin, and that acute stress has an immune enhancing function, while chronic stress dampens the immune response in the long run (see Chap. 12 and reviews by Dhabhar [9,10]). Additionally, the direction of the ensuing immune response may be influenced by several factors such as the organ involved, the nature of the response, the presence or absence of antigen, and/or the presence and relative expression of particular receptor subtypes on the surface of immune cells [7].

7.6 Catecholamines in Cutaneous Immune Function: Basic Research Approaches

The effect of adrenergic stimulation on circulating lymphocytes is closely paralleled by its effect on resident immune cells in the skin. The skin is not only the largest human organ but also the primary immune defense barrier. Positioned at the interface between an organism's internal milieu and an external environment characterized by constant assault with potential microbial pathogens, the skin serves not only as a physical protective barrier but it is also able to defend the body by rapidly mounting an immune response to injury and microbial insult. When exploring the effects of catecholamines on cutaneous immune function, different logical approaches have been taken. While some groups have focused on the effect of isolated catecholamines on either cutaneous immune responses or specific cell types, others have looked at the consequences of stress in general on the whole organism or on distinct cell types/organs. While looking at stress responses in general will yield functionally relevant information on how the different stress response systems interact to change immune function generally and locally, it is often difficult to discern a direct effect of the SNS vs. multiple feedback and substitute mechanisms. An obvious problem of this approach is that all stressors lead to the activation of multiple intertwined stress response systems, such as the HPA axis and glucocorticoids, neuropeptides, endogenous opioids, and not just the SNS. Thus, another approach is to focus on the isolated effect of catecholamines on distinct cell populations.

7.6.1 Catecholamines Regulate Cutaneous Immune Cell Function

Langerhans cells (LC) are epidermal immune cells that will first encounter invading pathogens and/or sensitizing/irritating substances, and present their antigens to the effector cells of the immune system. Pretreatment of murine epidermal cell preparations as well as purified murine LCs with epinephrine or norepinephrine in vitro lead to a significant dose-dependent inhibition of their ability to present antigen to an antigen-specific Th1 clone, as measured by the reduced ability of the T cells to produce the pro-inflammatory cytokine IFNγ [50]. This reduced antigen-presenting capability could, in part, be due to a modulation of LC cytokine secretion by catecholamines, particularly cytokines that would normally promote a Th1 response, such as IL-12. As shown with the LC-like murine cell line XS106A, incubation with epinephrine prior to stimulation with LPS inhibited LPS-induced IL-12- and TNFα-production, and slightly upregulated IL-10 production [51]. These findings are supported by studies in BMDDC where NE inhibited lipopolysaccharide (LPS)-induced IL-12 production and promoted LPS-induced IL-10 production, an effect that was mediated by both β- and α_2-ARs [32].

These findings suggest that, in vitro, the direct effect of catecholamines on LC is to dampen their antigen presenting function. But do catecholamines influence LC function in vivo? To determine whether catecholamines affect antigen presentation in the *elicitation phase* of an immune response, a murine model of delayed type hypersensitivity (DTH), in which mice are immunized by subcutaneous injection of tumor fragments (the S1509 spindle cell tumor line) was used. Injection of antigen-pulsed dendritic cells into the hind footpad of previously immunized mice at a later time elicits a strong footpad swelling response in this model. If epidermal cells enriched for LC content were pretreated in vitro with epinephrine or NE prior to antigen pulsing, the ability of these cells to elicit DTH reactions in vivo was significantly suppressed. This effect could be blocked completely by the β_2-adrenergic receptor antagonist ICI 118,551, but not by the nonspecific α-adrenergic blocker phentolamine, indicating that the profound inhibitory effect of catecholamines on antigen presentation in the *elicitation phase* of an immune response is $\beta 2$-mediated. To test the ability of epinephrine to modify the *sensitization phase* of an immune response, a murine model of contact hypersensitivity (CHS) was used. In this prototype of a T cell mediated immune response, immunization is achieved by applying the potent contact sensitizer dinitroflourobenzene (DNFB) to the back skin of a mouse. When the same animal is challenged a week later on the ear, a strong immune response, characterized by inflammation, edema, and an increase in pinna thickness, ensues. Mice injected intradermally with epinephrine before application of DNFB at the injected site showed a markedly reduced ear swelling, and thus, reduced immune response upon DNFB challenge 7 days later. Interestingly, the immune inhibitory effect remained when epinephrine was injected at a site distant from the site of DNFB immunization [50]. The systemic effect of intradermally injected catecholamines may be due to rapid diffusion into surrounding tissues. Alternatively, catecholamines may be taken up systemically and influence immune cell trafficking and other related functions. In any case, it is evident that catecholamines suppress LC function when administered to naïve cells before antigen activation. The finding that the adrenergic system inhibits CHS sensitization via β-receptors is supported by findings of Maestroni [31], who showed that topical application of the adrenergic agonist prazosin at the time of sensitization with FITC leads to a decrease in ear swelling upon challenge 6 days later. Likewise, a transient inhibition of sympathetic activity during sensitization by injection of the ganglionic blocker pentolinium increases FITC-induced CHS reactions [32], indicating that *endogenous* catecholamines may also play a role in shaping the immune response.

Interestingly, ganglionic blockade during skin sensitization has also been shown to increase IFNγ production in draining lymph nodes [32], suggesting an increased Th1 response. This parallels findings by other groups that found Th2 polarization (decreased Th1 but increased Th2 cytokine profiles) after catecholamine treatment (see Sect. 7.5). Likewise, blocking β_2-receptors resulted in an increased production of IFNγ, and IL-2, but not IL-4 in draining lymph node cells [34]. This result indicates that adrenergic signaling in addition to antigen presentation may influence the extent of Th1 priming, but not necessarily cause a Th2 shift in the adaptive response.

Upon capture and processing of an antigen, LCs start a maturation process that includes upregulation of surface molecules, secretion of proinflammatory cytokines, and migration toward lymphoid organs, where they present antigen and interact with other lymphocytes [59]. Impaired migration has been suggested as one mechanism of catecholamine-induced inhibition of cutaneous immune function. Maestroni [31] showed that α-adrenergic blockers inhibited FITC-induced migration of LC (defined as being FITC+ CD86+) to draining lymph nodes 24 h after sensitization. While this finding may indicate that α-agonists would increase LC migration, it could also reflect a relative, compensatory, increase in β-adrenergic tone after α-blockade, and thus, β-induced inhibition of migration. Indeed, in vivo blockade of β_2-AR in murine skin during FITC-sensitization by topical application of the β_2-adrenergic antagonist ICI 118,551 enhanced LC-migration to draining lymph nodes (LC defined as FITC+ CD11c+ appearing in the draining lymph nodes after 24 h). IL-10 is a crucial cytokine in LC migration to lymph nodes, since IL-10 knockout mice show enhanced migration of LC [61]. The mechanism by which catecholamines may interfere with the LC migration process may involve enhancing IL-10 secretion, and thus, decreasing their chemotactic response to chemokines that are essential in the emigration pathway from the site of antigen deposition to regional lymph nodes, namely CCL-19 and CCL-21 [34]. The amount of NE present in the skin is determined through the balance between synthesis, release, and degradation or reuptake. Interestingly, recent evidence suggests that the type of antigen encountered by the skin may affect the amount of NE in the skin, and thus the direction of the ensuing immune response [33]. Contact sensitizers that induce a predominant

Th1 response, such as oxazolone, appear to do so by inhibiting local NE turnover, whereas contact sensitizers that induce a predominant Th2 response, such as FITC, do not affect NE turnover.

The consensus of the studies discussed above [31,32,34,50,51] indicates a downregulation of LC-related immune functions by adrenergic agonists. But it has become evident that the immune regulating effect of the adrenergic system on dendritic cell (and/or LC)-function crucially depends on the state of dendritic cell activation. NE decreases migration and CHS response when added to immature BMDDCs, but enhances migration and CHS when added to maturing BMDDCs 1 h after stimulation [34]. When given 24 h after stimulation, NE has no effect. Given the role β_2-AR plays in downregulating immature LC-immune function, it may be expected that this receptor is also instrumental in determining immune enhancement in maturing cells. Contrary to this assumption, the short acting β_2-agonists salbutamol decreases CHS when acting on immature cells, but has no effect after 1 and 24 h of maturation. This may suggest either a rapid downregulation of β_2-ARs, or it could indicate that the effect on maturing cells is mediated by ARs other than β_2. In summary, catecholamines seem to inhibit immature LC function by downregulating antigen presentation through restraining the production of inflammatory cytokines and migration to draining lymph nodes. But the effect of the SNS is crucially dependent on the state of dendritic cell maturation when encountering the stressor, with either no effect or even enhanced function when acting on maturing or mature cells, respectively.

7.6.2 Catecholamines Mediate the Cutaneous Stress Response

Although it is important to focus on the isolated effects of catecholamines on distinct cell populations, it is somewhat questionable how isolated effects will be integrated into the complete immune response to stressors that no doubt impact multiple intertwined stress response pathways. The general effect of stress on cutaneous immunity will be discussed in detail in Chap. 12 of this publication. This chapter will focus on the possible role the SNS is playing in cutaneous immune reactions to stress.

Two common stressors used in examining stress effects in the animal model are restraint stress or increased population densities. Hosoi et al. [23] showed that the elicitation of contact hypersensitivity was suppressed in mice that received either type of stressor. Epidermal sheets of stressed mice showed a decrease in I-A molecules (a member of class II major histocompatibility complex) as well as lower cell densities and numbers of dendrites, indicating diminished LC function. This effect was not observed in adrenalectomized mice, suggesting an inhibitory role of adrenergic hormones in regulating cutaneous immunity. But just as with the response of LCs to adrenergic signaling as discussed above, timing of a stressor is a crucial factor in the development of an enhanced or suppressed immune response. Dhabhar and colleagues showed that exposure of previously sensitized mice to an acute stressor immediately before re-exposure to antigen resulted in a significant, long-lasting increase of CHS and increased numbers of leucocytes at the site of antigen challenge [11]. Catecholamines as well as glucocorticoids are involved in this stress-induced enhancement of CHS, since adrenalectomy eliminated the stress-induced enhancement of DTH, while low-dose corticosterone and EPI administration restored DTH enhancement [12]. Flint et al. [17] convincingly demonstrated the importance of the timing of a stressor (whether it acts in a naïve or sensitized animal) in determining the direction of an immune response. When BALB/c mice were restrained before cutaneous sensitization with DNFB, chemically induced ear swelling and leukocyte infiltration were diminished upon challenge with antigen. In comparison, if animals were restrained before antigen challenge, the immune response was enhanced. This effect could be blocked in part by a glucocorticoid receptor antagonist, but it is very likely that other stress transmitters, such as neuropeptides and catecholamines, were involved as well. Nevertheless, some conflicting evidence whether stress (before sensitization) up- or downregulated the ensuing immune response remains in the literature. In recent experiments by Viswanathan et al. [60], restraint stress prior to DNFB immunization increased the magnitude of pinna swelling measured 6 and 24 h after DNFB administration, and enhanced CD11c$^+$ cell maturation and migration to regional lymph nodes. Of note, though, the authors investigated events at initiation phase of the immune response, but not at recall as done by Flint et al. [17].

It seems that not just the timing of the stressor, but also the amount and type of antigen encountered by the organism can influence the outcome of stress mediated immune responses. In experiments similar to Flint's by Saint-Mezard et al. [42], restraining mice before sensitization with DNFB led to enhanced DTH upon re-exposure to DNFB 5 days later. Since the two study designs were comparable (same animal model, same stressor, same contact sensitizer), but have opposing outcomes, it is tempting to speculate that the amount of contact sensitizer used (10 times higher in Saint-Mezard's experiments) may be the deciding factor. Future studies will be needed to unravel the exact mechanisms by which specific stressors mediate effects on cutaneous immunity and to delineate the variables that ultimately guide the directionality of this immune response.

7.7 Catecholamines in Cutaneous Immune Function: Clinical Relevance

As reviewed above, transmitters of the SNS have the ability to significantly influence cutaneous immune reactions *in vitro* and in *in vivo* animal models. But what is the evidence that the SNS also affects skin disorders seen by physicians on a daily basis? There are plenty of anecdotal reports linking stress to atopic dermatitis, psoriasis, and multiple other skin diseases, and conditions associated with increased sympathetic activity, like psychological pressure and anxiety, are often said to exacerbate these disorders. The following paragraphs will highlight the role of the SNS in two skin conditions, psoriasis and atopic dermatitis, that are exemplary for Th1 or Th2 dominant immune responses, respectively.

7.7.1 Catecholamines and Psoriasis

A well-known phenomenon in dermatologic practice is the induction or exacerbation of psoriasis by β-AR blocking drugs. Although the mechanisms involving both immunologic and non-immunologic factors have been examined in various studies, no consensus has been reached to date [36]. Psoriasis is generally seen as a Th1-driven disease. Given that catecholamines have been shown to suppress Th1 responses and skew the immune response towards Th2 through β-ARs, as discussed above, it is tempting to speculate that pharmacologic blockade of β-receptors may lead to a predominance of Th1 pathways and, thus, initiation or exacerbation of disease in genetically predisposed individuals.

Another possible theory is that there is a general dysregulation of the β-adrenergic reactivity. Increased constitutive blood levels of epinephrine and NE and increased urinary catecholamine excretion, as well as increased reactivity of the SNS to stress (as evidenced by higher blood pressure and pulse rate) has been reported in psoriasis [1,6]. Paradoxically, β_2-adrenergic receptor mRNAs in the involved epidermis are significantly decreased compared with uninvolved epidermis [55,57]. Recently, an association of the Arg16Gly polymorphism of the β_2-AR with psoriasis has been suggested [37], but it is not clear yet how these observations relate to pathomechanisms in psoriasis. Clearly, more work needs to be done to elucidate a hypothetic role of β-ARs in this condition.

7.7.2 Catecholamines and Atopic Dermatitis

Among skin diseases, the influence of stress has been most extensively explored in AD. This is in no small part due to frequent anecdotal patient reports of exacerbation after stressful life events. A growing number of reports also support the role of psychological stress [5]. Of note, in psychological tests, AD patients have highly increased levels of anxiety, a condition typically accompanied by persistent sympathetic overarousal [52]. Early observations linked the sympathetic nervous system to immunological abnormalities. A partial block in β-ARs in AD was postulated since leukocytes from AD patients show a decreased rise in intracellular levels of cAMP upon incubation with β- agonists due to increased phosphodiesterase levels [20,39]. This may trigger the release of mediators such as histamine, prostaglandin, and leukotrienes, in turn leading to itch and inflammation in atopic individuals [3]. Although not examined thoroughly, polymorphisms in the β_2-AR that could alter the structure and function have been suggested [40]. Interestingly, significantly higher NE levels in plasma have been reported in patients with AD than healthy controls. On the other hand, cell extracts from epidermal suction blister roofs revealed a threefold induction of the NE-degrading enzyme monoamine oxidase A [46], supporting earlier observations of defective catecholamine/adrenoceptor signaling.

The initiation of AD appears to be driven by an allergen-induced activation of Th2 cells [29], and, as discussed earlier, catecholamines can skew the immune response towards a Th2 phenotype. Additionally, an increase in $CD8^+$ T lymphocyte counts after mental stress in AD patients is accompanied by elevated NE plasma levels, further supporting the importance of neuroimmunological mechanisms in AD [48]. Although additional work is necessary to further clarify the role of the SNS in the initiation and exacerbation of disease, the relevance of psychosocial stress as a trigger factor is emphasized by the therapeutic effect of psychosocial interventions, including stress management or relaxation training, in AD [14].

Summary for the Clinician

It is widely recognized that stress affects immune responses in general and the skin in particular. A major pathway that regulates the stress response involves the sympathetic nervous system. The skin is extensively innervated by sympathetic nerve fibers, and almost all circulating and resident immune cells express adrenergic receptors. Although stress is generally perceived as immunosuppressive and thought to increase susceptibility to infection and cancer, it is paradoxically also thought to exacerbate certain inflammatory diseases. Recent work suggests that the directionality of the immune response is determined by multiple factors such as the duration, type and amount of the "stressor", as well as the type and state of maturation of the immune cell the adrenergic system is acting upon. In general, acute stress is found to be immuno-enhancing, whereas chronic stress is immunosuppressive. Systemically, catecholamines skew the immune response away from a Th1 towards a Th2 response by suppressing cytokines like IL-12 and TNFα, and enhancing IL-10 and IL-4. At the cellular level, catecholamines suppress antigen presentation, CHS, and DTH through β_2-adrenoceptor mediated pathways. The mechanisms of this inhibition may involve altered cytokine production and reduced migration of LCs to regional lymph nodes. In any case, the immune regulating effect of the adrenergic system crucially depends on the state of dendritic cell activation, as migration and CHS are enhanced when catecholamines are acting on maturing cells after antigen stimulation. Clearly, the neuroendocrine regulation of cutaneous immune function is a complex system involving not only the SNS, but also the hypothalamic–pituitary–adrenal axis, and numerous distinct neuropeptides (see Chaps. 6, 7, 11–13). Under certain circumstances this system can become unbalanced, and lead to the exacerbation of cutaneous disease, for example, atopic dermatitis and psoriasis. A dysregulation of the adrenergic response has been suggested in both conditions, and the clinician should take these mechanisms into consideration when treating the affected population.

References

1. Arnetz BB, Fjellner B, Eneroth P, et al (1985) Stress and psoriasis: psychoendocrine and metabolic reactions in psoriatic patients during standardized stressor exposure. Psychosom Med 47:528–541
2. Benschop RJ, Rodriguez-Feuerhahn M, Schedlowski M (1996) Catecholamine-induced leukocytosis: early observations, current research, and future directions. Brain Behav Immun 10:77–91
3. Bos JD, Kapsenberg ML, Smitt JH (1994) Pathogenesis of atopic eczema. Lancet 343:1338–1341
4. Botchkarev VA, Peters EM, Botchkareva NV, et al (1999) Hair cycle-dependent changes in adrenergic skin innervation, and hair growth modulation by adrenergic drugs. J Invest Dermatol 113:878–887
5. Buske-Kirschbaum A, Geiben A, Hellhammer D (2001) Psychobiological aspects of atopic dermatitis: an overview. Psychother Psychosom 70:6–16
6. Buske-Kirschbaum A, Ebrecht M, Kern S, et al (2006) Endocrine stress responses in TH1-mediated chronic inflammatory skin disease (psoriasis vulgaris) – do they parallel stress-induced endocrine changes in TH2-mediated inflammatory dermatoses (atopic dermatitis)? Psychoneuroendocrinology 31:439–446
7. Chrousos GP (2000) Stress, chronic inflammation, and emotional and physical well-being: concurrent effects and chronic sequelae. J Allergy Clin Immunol 106:S275–S291
8. Chrousos GP (2000) The stress response and immune function: clinical implications. The 1999 Novera H. Spector Lecture. Ann N Y Acad Sci 917:38–67
9. Dhabhar FS (2002) Stress-induced augmentation of immune function – the role of stress hormones, leukocyte trafficking, and cytokines. Brain Behav Immun 16:785–798
10. Dhabhar FS (2003) Stress, leukocyte trafficking, and the augmentation of skin immune function. Ann N Y Acad Sci 992:205–217
11. Dhabhar FS, McEwen BS (1996) Stress-induced enhancement of antigen-specific cell-mediated immunity. J Immunol 156:2608–2615
12. Dhabhar FS, McEwen BS (1999) Enhancing versus suppressive effects of stress hormones on skin immune function. Proc Natl Acad Sci U S A 96:1059–1064
13. Drummond PD, Skipworth S, Finch PM (1996) Alpha 1-adrenoceptors in normal and hyperalgesic human skin. Clin Sci (Lond) 91:73–77
14. Ehlers A, Stangier U, Gieler U (1995) Treatment of atopic dermatitis: a comparison of psychological and dermatological approaches to relapse prevention. J Consult Clin Psychol 63:624–635
15. Felten DL, Felten SY, Carlson SL, et al (1985) Noradrenergic and peptidergic innervation of lymphoid tissue. J Immunol 135:755s-765s
16. Felten SY, Madden KS, Bellinger DL, et al (1998) The role of the sympathetic nervous system in the modulation of immune responses. Adv Pharmacol 42:583–587
17. Flint MS, Valosen JM, Johnson EA, et al (2001) Restraint stress applied prior to chemical sensitization modulates the development of allergic contact dermatitis differently than restraint prior to challenge. J Neuroimmunol 113:72–80

18. Furlan C, Sterin-Borda L, Borda E (2005) Activation of beta3 adrenergic receptor decreases DNA synthesis in human skin fibroblasts via cyclic GMP/nitric oxide pathway. Cell Physiol Biochem 16:175–182
19. Grando SA, Pittelkow MR, Schallreuter KU (2006) Adrenergic and cholinergic control in the biology of epidermis: physiological and clinical significance. J Invest Dermatol 126:1948–1965
20. Grewe SR, Chan SC, Hanifin JM (1982) Elevated leukocyte cyclic AMP-phosphodiesterase in atopic disease: a possible mechanism for cyclic AMP-agonist hyporesponsiveness. J Allergy Clin Immunol 70:452–457
21. Hasko G, Szabo C, Nemeth ZH, et al (1998) Stimulation of beta-adrenoceptors inhibits endotoxin-induced IL-12 production in normal and IL-10 deficient mice. J Neuroimmunol 88:57–61
22. Hoffman BB (2001) Adrenoceptor-activating and other sympathomimetic drugs. In: Katzung BG (ed) Basic and Clinical Pharmacology, Lange Medical Books/McGraw-Hill, New York
23. Hosoi J, Tsuchiya T, Denda M, et al (1998) Modification of LC phenotype and suppression of contact hypersensitivity response by stress. J Cutan Med Surg 3:79–84
24. Johnson M (2002) Effects of beta2-agonists on resident and infiltrating inflammatory cells. J Allergy Clin Immunol 110:S282–S290
25. Kavelaars A (2002) Regulated expression of alpha-1 adrenergic receptors in the immune system. Brain Behav Immun 16:799–807
26. Kiecolt-Glaser JK, McGuire L, Robles TF, et al (2002) Psychoneuroimmunology: psychological influences on immune function and health. J Consult Clin Psychol 70:537–547
27. Kin NW, Sanders VM (2006) It takes nerve to tell T and B cells what to do. J Leukoc Biol 79:1093–1104
28. Kohm AP, Sanders VM (2001) Norepinephrine and beta 2-adrenergic receptor stimulation regulate $CD4^+$ T and B lymphocyte function in vitro and in vivo. Pharmacol Rev 53:487–525
29. Leung DY, Soter NA (2001) Cellular and immunologic mechanisms in atopic dermatitis. J Am Acad Dermatol 44: S1–S12
30. Madden KS, Sanders VM, Felten DL (1995) Catecholamine influences and sympathetic neural modulation of immune responsiveness. Annu Rev Pharmacol Toxicol 35:417–448
31. Maestroni GJ (2000) Dendritic cell migration controlled by alpha 1b-adrenergic receptors. J Immunol 165:6743–6747
32. Maestroni GJ (2002) Short exposure of maturing, bone marrow-derived dendritic cells to norepinephrine: impact on kinetics of cytokine production and Th development. J Neuroimmunol 129:106–114
33. Maestroni GJ (2004) Modulation of skin norepinephrine turnover by allergen sensitization: impact on contact hypersensitivity and T helper priming. J Invest Dermatol 122:119–124
34. Maestroni GJ, Mazzola P (2003) Langerhans cells beta 2-adrenoceptors: role in migration, cytokine production, Th priming and contact hypersensitivity. J Neuroimmunol 144:91–99
35. Moroni F, Fantozzi R, Masini E, et al (1977) The modulation of histamine release by alpha-adrenoceptors: evidences in murine neoplastic mast cells. Agents Actions 7:57–61
36. O'Brien M, Koo J (2006) The mechanism of lithium and beta-blocking agents in inducing and exacerbating psoriasis. J Drugs Dermatol 5:426–432
37. Ozkur M, Erbagci Z, Nacak M, et al (2004) Association of the Arg16Gly polymorphism of the beta-2-adrenergic receptor with psoriasis. J Dermatol Sci 35:162–164
38. Panina-Bordignon P, Mazzeo D, Lucia PD, et al (1997) Beta2-agonists prevent Th1 development by selective inhibition of interleukin 12. J Clin Invest 100:1513–1519
39. Reed CE, Busse WW, Lee TP (1976) Adrenergic mechanisms and the adenyl cyclase system in atopic dermatitis. J Invest Dermatol 67:333–338
40. Roguedas AM, Audrezet MP, Scotet V, et al (2006) Intrinsic atopic dermatitis is associated with a beta-2 adrenergic receptor polymorphism. Acta Derm Venereol 86:447–448
41. Rosenfield RL, Wu PP, Ciletti N (2002) Sebaceous epithelial cell differentiation requires cyclic adenosine monophosphate generation. In Vitro Cell Dev Biol Anim 38:54–57
42. Saint-Mezard P, Chavagnac C, Bosset S, et al (2003) Psychological stress exerts an adjuvant effect on skin dendritic cell functions in vivo. J Immunol 171:4073–4080
43. Schallreuter KU (1997) Epidermal adrenergic signal transduction as part of the neuronal network in the human epidermis. J Investig Dermatol Symp Proc 2:37–40
44. Schallreuter KU, Wood JM, Lemke R, et al (1992) Production of catecholamines in the human epidermis. Biochem Biophys Res Commun 189:72–78
45. Schallreuter KU, Wood JM, Pittelkow MR, et al (1993) Increased in vitro expression of beta 2-adrenoceptors in differentiating lesional keratinocytes of vitiligo patients. Arch Dermatol Res 285:216–220
46. Schallreuter KU, Pittelkow MR, Swanson NN, et al (1997) Altered catecholamine synthesis and degradation in the epidermis of patients with atopic eczema. Arch Dermatol Res 289:663–666
47. Schedlowski M, Falk A, Rohne A, et al (1993) Catecholamines induce alterations of distribution and activity of human natural killer (NK) cells. J Clin Immunol 13:344–351
48. Schmid-Ott G, Jaeger B, Adamek C, et al (2001) Levels of circulating CD8(+) T lymphocytes, natural killer cells, and eosinophils increase upon acute psychosocial stress in patients with atopic dermatitis. J Allergy Clin Immunol 107:171–177
49. Seiffert K, Granstein RD (2006) Neuroendocrine regulation of skin dendritic cells. Ann N Y Acad Sci 1088:195–206
50. Seiffert K, Hosoi J, Torii H, et al (2002) Catecholamines inhibit the antigen-presenting capability of epidermal Langerhans cells. J Immunol 168:6128–6135
51. Seiffert K, Hosoi J, Torii H, et al (2003) Epinephrine regulates TNF-α expression in the Langerhans cell-like cell line XS106A. J Invest Dermatol 121:abstract 896
52. Seiffert K, Hilbert E, Schaechinger H, et al (2005) Psychophysiological reactivity under mental stress in atopic dermatitis. Dermatology 210:286–293

53. Steinkraus V, Korner C, Steinfath M, et al (1991) High density of beta 2-adrenoceptors in a human keratinocyte cell line with complete epidermal differentiation capacity (HaCaT). Arch Dermatol Res 283:328–332
54. Steinkraus V, Steinfath M, Korner C, et al (1992) Binding of beta-adrenergic receptors in human skin. J Invest Dermatol 98:475–480
55. Steinkraus V, Steinfath M, Stove L, et al (1993) Beta-adrenergic receptors in psoriasis: evidence for down-regulation in lesional skin. Arch Dermatol Res 285: 300–304
56. Steinkraus V, Mak JC, Pichlmeier U, et al (1996) Autoradiographic mapping of beta-adrenoceptors in human skin. Arch Dermatol Res 288:549–553
57. Takahashi H, Kinouchi M, Tamura T, et al (1996) Decreased beta 2-adrenergic receptor-mRNA and loricrin-mRNA, and increased involucrin-mRNA transcripts in psoriatic epidermis: analysis by reverse transcription-polymerase chain reaction. Br J Dermatol 134:1065–1069
58. Tausk FA, Christian E, Johansson O, et al (1993) Neurobiology of the skin. In: Fitzpatrick TB, Eisen AZ, Wolff K, Freedberg IM, Austen FK (eds) Dermatology in General Medicine, McGraw-Hill, New York
59. Valladeau J, Saeland S (2005) Cutaneous dendritic cells. Semin Immunol 17:273–283
60. Viswanathan K, Daugherty C, Dhabhar FS (2005) Stress as an endogenous adjuvant: augmentation of the immunization phase of cell-mediated immunity. Int Immunol 17:1059–1069
61. Wang B, Zhuang L, Fujisawa H, et al (1999) Enhanced epidermal Langerhans cell migration in IL-10 knockout mice. J Immunol 162:277–283

The Role of Neuropeptide Endopeptidases in Cutaneous Immunity

8

T.E. Scholzen

Contents

Synonyms Box: Neprilysin, neutral endopeptidases, common acute lymphoblast leukemia antigen (CALLA) CD10, enkephalinase, CD10, NEP, EC 3.4.24.11; angiotensin-converting enzyme, dipeptidyl carboxypeptidase, kininase II, ACE, CD143, EC 3.4.15.1; Dipeptidyl peptidase IV, CD26, EC 3.4.14.5

Key Features

- Proteolytic processing and degradation plays an important role in modulating the generation and bioactivity of neuroendocrine peptide mediators, a class of key molecules in cutaneous biology.
- Accordingly, the cellular localization and expression, and the molecular biology and structural properties of selected intracellular prohormone convertases and ectopically expressed zinc-binding metalloendoproteases are discussed.
- A special reference will be made to the physiologic and pathophysiologic significance of these endopeptidases in cutaneous immunobiology.
- Because of the number of pathologically relevant changes in inflammation and tumor progression that can be directly attributed to neprilysin and angiotensin-converting enzyme, a particular focus will be on the role of these enzymes in modulating innate and adaptive immune responses in the skin.

Abbreviations: *ACE* Angiotensin-converting enzyme, *ACTH* Adrenocorticotropin, *Ag* Antigen, *Ang* Angiotensin, *BK* Bradykinin, *CGRP* Calcitonin gene-related peptide, *CTCL* Cutaneous T-cell lymphomas, *DC* Dendritic cell(s), *DPIV* Dipeptidyl peptidase IV, *EAE* Experimental autoimmune encephalomyelitis, *ECE* Endothelin-converting enzyme, *EC* Endothelial cells, *END* Endorphin, MC_x Melanocortin receptor, *MHC*

R.D. Granstein and T.A. Luger (eds.), *Neuroimmunology of the Skin*,

Major histocompatibility complex, *MSH* Melanocyte-stimulating hormone, *NEP* Neprilysin, *PACAP* Pituitary adenylate-cyclase-activating polypeptide, *PC* Prohormone convertase, *POMC* Proopiomelanocortin, *SP* Substance P, *TC* T-cell(s), T_H helper T-cells, T_{eff} Effector T-cells, *VIP* Vasoactive intestinal peptide

8.1 Introduction

Almost every aspect of cutaneous cellular and tissue function, including proliferation, differentiation, maturation, communication, antigen (Ag) presentation, and survival of cells as well as hair growths, eccrine gland function, wound healing, and tissue regeneration, is modulated by neuropeptides. It is thus quite comprehensible that a variety of mechanisms have evolved, which limit their temporal, spatial, and developmental bioactivity. These include temporally and spatially controlled mediator generation and release, the regulated expression of specific receptors on cellular targets, receptor desensitization and resensitization, and the clearance of excessive extracellular peptides. Proteases participate in several of the above mechanisms; thus taking a key regulatory role in cutaneous peptide mediator bioavailability. As such they serve to generate bioactive peptides from inactive prohormones in order to initialize inflammatory and trophic responses. Importantly, they also rapidly terminate the bioactivity of neuropeptides released from nerve terminals or endocrine cells and thus prevent the development of a neuropeptides-initialized or -augmented deleterious chronic inflammation. In addition, microbial invaders or parasites use peptidases as an evolutionarily successful strategy to manipulate the host immune defenses. Peptidases play an important role in cutaneous plasticity and wound healing by modulating trophic neuropeptide activities. Moreover, beyond a mere catalytic function, ectopeptidases trigger specific intracellular signal transduction, participate in cell–cell or cell–virus recognition, and mediate or modulate binding to extracellular matrix components. This chapter highlights some of the current knowledge on peptidase function in cutaneous immunity and outlines clinical and potential future research areas derived from key functions of these enzymes. In addition, as zinc metalloproteases are among the largest group of proteases relevant for the extracellular cleavage of neuroendocrine mediators, a special emphasis will be made on this class of proteases.

8.2 Intracellular Endoproteases Convert Inactive Prohormones to Bioactive Mediators

8.2.1 Cellular Localization and Expression

Despite the important role that intracellular endoproteolytic processing and activation of prohormones, particularly of proopiomelanocortin (POMC), by prohormone convertase (PC) plays for cutaneous physiologic and pathophysiologic responses, this chapter's focus centers on functions of extracellular proteases. Importantly, with respect to cutaneous melanocortin generation, some extracellular proteases may also be capable of processing larger precursors, resulting in bioactive melanocortin receptor-(MC-) activating POMC peptides. Neuroendocrine hormones such as adrenocorticotrophin (ACTH) or α-melanocyte-stimulating hormone (α-MSH) are released in the skin as part of an intrinsic cutaneous hypothalamus–pituitary–adrenal axis and mediate the cutaneous response to invasive and noninvasive exogenous stress via MC receptors [11,87]. PC are an evolutionary conserved class of secretory serine proteases of the subtilisin/kexin-type that comprise PC1/3, PC2, furin/PACE, PACE4, PC4, PC5/6, PC7, and SKI-1 [81]. POMC peptides and POMC processing enzymes including PC1, PC2, PACE4, or furin have been identified in a variety of skin cells, skin appendages, cutaneous carcinoma cells, and immune cells (reviewed in [10,42,81]).

8.2.2 Molecular Biology and Structural Properties

The conserved structure of the PC catalytic domain suggests that these proteases have evolved from a common ancestral precursor gene. Subtilisin/kex family PC are specialized for cleaving multiple hormones, growth factors, and receptor precursors by limited internal proteolysis at single or multiple basic recognition sites, within the general motive K/R-$(X)_n$-K/R↓ [81]. PC expression and activity is strictly regulated at the tissue, cell, or subcellular level.

Autocatalytic activation of PC zymogens is another means to control PC bioactivity within the secretory pathway. In some cases, that is, for PC2, cofactors such as the binding protein 7B2 are required for efficient zymogen activation of proPC and full functional activity [50,81].

8.2.3 Physiologic Significance

The primary function of cutaneous PC is the conversion of the POMC prohormone in various skin cells. The resulting generation of α-MSH, ACTH, or β-endorphin is highly relevant for skin immunity, stress response, and pigmentation (for more details see Chap. 6 and [42]). Studies of PC1/3- and PC2-deficient mice revealed that deletion of these enzymes impairs the processing of POMC and other prohormones, although some redundancies might exist [71]. The pathophysiologic consequences of a dysregulated PC expression for cutaneous immunity have not yet been fully explored. The simultaneous episodic expression of PC1, PC2, and POMC during the murine hair cycle suggests a regulatory function of PC for the pilosebaceous unit [49]. POMC, MC, and PC expression in some skin cells are synergistically regulated by UV light, melanocortins, and pro-inflammatory cytokines in vitro. These stimuli may simultaneously increase production and responsiveness of cutaneous cells to POMC peptides, although this has not been conclusively confirmed in vivo [73,76]. An upregulated PC1, PC2, and furin expression positively correlates with malignant neuroendocrine tumors and of several other cancers [39], suggesting that subtilisin/kex-like convertases may increase tumorigenesis and aggressiveness by augmenting processing and activation of mitogenic peptides. Thus, PC both serve as a prognostic marker for tumor progression and constitute an important pharmacologic target in cancer therapy, since PC inhibition drastically reduced the metastatic properties of certain tumor cells [39,71]. Interestingly, some bacterial toxin precursors (e.g., *Diphteria* toxin, *Botulinum* neurotoxin), as well as viral glycoproteins of HIV-1, Ebola, and others viruses need proteolytic PC activation for their toxic or infectious capacity and/or the cell–cell spreading. This demonstrates the high relevance of PC for the cutaneous response to infectious agents and, therefore, PC inhibition could be beneficial for abrogating microbial-induced cytopathicity (reviewed in [39]).

8.3 Dipeptidylpeptidase IV/CD26

8.3.1 Cellular Localization and Expression

Dipeptidyl peptidase (DP) IV (CD26, EC 3.4.14.5) is a multifunctional homodimeric glycoprotein with functional roles in hematology, endocrinology, immunology, endothelial cell (EC), and cancer biology and metabolism. DPIV is part of a six member gene family of enzymes that, in addition to DPIV, includes fibroblast activating protein (FAP), DP-like (DPL) 1, DPL2, DP8, DP9, and prolyloligopeptidase (POP) [13,27]. Human DPIV is ubiquitously expressed by capillary EC, activated lymphocytes, DC subpopulations, and on apical surfaces of epithelial cells [27]. In addition, soluble forms of the enzyme have been described. Cutaneous DPIV is expressed on keratinocytes [60], fibroblasts [59], melanocytes [55], the axon–Schwann cell interface [20], and TCs [41].

8.3.2 Molecular Biology and Structural Properties

The structural and biochemical properties of DPIV have been described in detail in [27]. The DPIV gene product is a 766 amino acid (AA) ectoprotease with an apparent monomeric molecular weight of about 110kDa. Characteristically, full functional DPIV peptidase activity requires homodimerization between one of two extracellular hydrolase domains. This results in a rather unique post-proline dipeptidyl aminopeptidase activity of DPIV by cutting off N-terminal X-P or X-A dipeptides from polypeptides. A variety of DPIV peptide substrates have functional relevance for skin (patho) physiology. These comprise at least 9 CCL and CXC chemokines (i.e., CCL5, RANTES, or CXCL10, IFNγ-induced protein), hormones (i.e., glucagon-like peptides (GLP), prolactin, leutinizing hormone α, chorionic gonadotropin α chain), enkephalins, and neuropeptides such as neuropeptide Y, pituitary-adenylate cyclase-activating polypeptide (PACAP) 38, vasoactive intestinal peptide (VIP), and SP.

8.3.3 Functional Roles of DPIV/CD26 in Immunity and Inflammation

There is compelling evidence that DPIV has a number of important physiologic functions in endocrinology and metabolism. For instance, DPIV degrades GLP and glucose-dependent insulinotropic peptide. The inhibition of DPIV results in accumulation of these peptides, which stimulates greater insulin production and is therefore beneficial for the treatment of insulin-independent *Diabetes mellitus* (see [27] for details). DPIV/CD26 has gained considerable interest as a T cell (TC) activation marker, and is also expressed by some dendritic cell (DC) subpopulations [24]. Accordingly, DPIV expression, together with other TC activation markers such as CD25, CD71 CD45RO, or CD29, increases significantly after antigenic and mitogenic stimulation, or treatment with the T helper (T_H)1 cytokine IL-12. Overexpression of human CD26 in TC of transgenic mice reduces thymus cellularity, impairs thymocyte proliferation, and increases the number of peripheral apoptotic $CD4^+$- or $CD8^+$-TC, indicating the importance of CD26 for peripheral T lymphocyte homeostasis [85]. Interestingly, DPIV upregulation increases degradation of VIP and PACAP, two neuropeptides known to trigger T_H2 immune responses via IL-4, IL-5, and IL-10 induction in $CD4^+$ TC [25,27]. Up- or downregulated CD26 expression may therefore shift the T_H1/T_H2 balance [72,82] with high relevance for psoriasis, atopic dermatitis [5], or rheumatoid arthritis (RA) [8]. However, studies of murine experimental autoimmune encephalomyelitis (EAE) and RA revealed surprising discrepancies between functional inhibition of DPIV by genetic knock-out and pharmacologic inhibitors. While DPIV-inhibiting drugs delayed the onset and severity of experimental EAE or RA, missing CD26 activity in knock-out mice or in human patients was inversely correlated to the severity of Ag-induced RA or EAE. Thus, DPIV inhibitors may have additional functional targets [8]. CD26 expressed in lipid rafts within the immunologic synapse also transduces intracellular signals that overlap with the TC receptor/CD3 signaling pathways. DPIV enzyme activity may therefore be dispensable for full immunologic activity of TC, since costimulatory activity of CD26 in vitro is retained in DPIV mutants that lack hydrolase activity [27].

8.3.4 DPIV/CD26 and the Development of Neoplasms

The skin is the host for a number of extranodal non-Hodgkin cutaneous TC lymphomas (CTCL) [41]. Strikingly, in the most relevant types of CTCL, mycosis fungoides (MF) and Sézary syndrome (SS), skin-homing malignant $CD3^+$ $CD4^+$ $CD7^{variable}CLA^+$ $CCR4^+$ TCs characteristically lack CD26 expression. DPIV/CD26 is an important diagnostic marker for SS or MF, and also highly relevant for the pathophysiology of CTCL [88]. Missing CD26 compromises the TC capability to degrade the constitutively expressed cutaneous stem cell factor 1, which may promote homing of malignant TC into the skin [58]. Infiltrating CTCL-$CD26^-$ TC then generate an immunosuppressive T_H2 environment with immature DC incapable of efficiently presenting phagocytozed material derived from apoptotic malignant TC. The resulting regulatory TC (T_{reg}) then contribute to immunosuppression and CTCL-TC tolerance [41]. Evidently, DPIV may also be relevant for growth, invasiveness, and metastasis of other tumors. For instance, DPIV expression is inversely correlated to progression of melanoma and even absent in metastatic melanoma [67,101]. Conversely, overexpression of DPIV in melanoma cells suppresses tumor progression in nude mice possibly independently of a DPIV enzymatic activity. Likewise, DPIV interacts with ECM components such as collagen, fibronectin, E-cadherin, or tissue inhibitors of matrix metalloproteases [103]. Consequently, a higher DPIV expression enhances adherence of tumor cells to the ECM, which may be anti-invasive by preventing detachment of tumor cells from the solid tumor. Alternatively, DPIV may also hamper basic fibroblast growth factor (bFGF) mitogenic signaling pathways [106]. In summary, DPIV/CD26 (patho-)physiologic effects in tumorigenesis possibly depend on a direct interaction with ECM components or an indirect degradation of important inflammatory mediators.

8.4 Ectopic Zinc Metalloendopeptidases: Neprilysin and Angiotensin-Converting Enzyme

8.4.1 Cellular Localization and Expression

Neprilysin (NEP, EC 3.4.24.11, CD10) and angiotensin-converting enzyme ACE (EC 3.4.15.1, CD143), two mechanistically related EC surface zinc metallopeptidases, are widely distributed in the body and highly expressed in the vascular endothelium, kidney epithelium, lung, or CNS [99,100]. NEP ("enkephalinase" or "common acute lymphoblastic leukemia antigen" (CALLA)) was initially isolated more than 30 years ago as an insulin B chain-degrading enzyme abundantly expressed in the renal brush border membrane [100]. The history of the ACE family has accomplished a journey from the original discovery and isolation of ACE 50 years ago as "hypertension-converting enzyme" until the recent identification of the related carboxypeptidase ACE2 as vasopeptidase and coreceptor for the severe acute respiratory syndrome corona virus (SARS-CoV) [30,31]. The somatic isoform of ACE and ACE2 are abundantly expressed in the vascular endothelium surface of the lung and in brush-border membranes of kidney, intestine, placenta, and the choroid plexus. A soluble isoform of somatic ACE is present in the plasma, and a smaller isoenzyme essential for male fertility is expressed in the testis [30,99]. Cutaneous NEP as well as components of the renin-angiotensin system (RAS) including ACE and ACE2 are expressed in basal keratinocytes, hair follicles, eccrine and sebaceous glands, the microvascular endothelium, and large nerves or nerve-ensheathing Schwann cells (reviewed in [31,75,89]). NEP and ACE, but not ACE2, are expressed in DC, macrophages, or TC and in bone marrow stromal cells, suggesting functional roles in hematopoiesis and immunity [15,31,38,45,47]. A recently identified secreted homologue NEP2 related to *D. melanogaster* NEP is expressed in the renal tube and in testis, but its relevance for skin immunity still has to be determined [96].

8.4.2 Molecular Biology and Structural Properties

Despite structural similarities between ACE and NEP, NEP is evolutionarily closer related to the bacterial Zn protease thermolysin. NEP homologues were identified in all organisms from simple prokaryotes to higher vertebrates, including men [100]. The mammalian NEP family now comprises at least seven members, with NEP and endothelin-converting enzyme (ECE, EC 3.4.24.71) as best characterized. The molecular biology of NEP is described in [69]. As a type II integral membrane protein ectopeptidase, NEP (90–110 kDa) consists of a short N-terminal intracellular domain, a transmembrane anchor, and a large C-terminal extracellular domain that contains a Zn-coordinating active site constituted by a HExxH and an ExxA/GD sequence [100]. Two closely related ACE isoenzymes (somatic and germinal) have been identified in mammalian cells. Somatic ACE (150–180 kDa) is a type I C-terminally membrane-anchored glycoprotein that contains two highly homologous extracellular domains (N-domain, C-domain), each bearing a zinc-coordinating catalytic site. By contrast, the smaller testicular ACE involved in male fertility contains a single catalytic site identical to the C-terminal domain of somatic ACE [44,99].

8.4.3 Physiologic Roles of NEP and ACE

Much help in understanding NEP and ACE function is derived from using mercaptoalkanoyl inhibitors such as captopril [99], or selective NEP inhibitors such as thiorphan and phosphoramidon [69]. The latter is a *streptomyces tanashiensis* product suggesting an evolutionary old relationship between zinc metalloproteases-expressing prokaryotes and eukaryotes [100]. In addition to its widespread role in turning off neuronal signals transmitted via SP or enkephalins, the panel of today's known NEP substrates includes, but is not limited to, vasoactive peptides such as bradykinin (BK) or angiotensin (Ang) I, atrial natriuretic peptide (ANP), growth factors such as bombesin, chemotactic peptides such as fMLP, and most recently β-amyloid (Aβ) peptide, the key initiator of Alzheimer's disease (AD) [35,86,100]. The structure of the mammalian NEP extracellular domain limits accessibility of the catalytic site for substrates. Therefore, and in contrast to the protease activity of thermolysin, NEP is an oligopeptidase cleaving peptides predominantly non-terminal before hydrophobic AA residues [69]. ACE's active sites display endopeptidase and dipeptidyl carboxypeptidase activity, which differ in their pH/chloride dependency and substrate specificity [36,105]. A prominent physiological feature of NEP and ACE is their overlap in competitively cleaved substrates, resulting in opposing roles in renal and cardiovascular regulation (reviewed in [9,86]). The successful 30 year use of ACE inhibitors to therapeutically intervene with hypertension and cardiovascular dysfunction has encouraged attempts to additionally inhibit related enzymes such as NEP and recently ECE with a single drug. Advantage of such dual ACE/NEP- or triple ACE/NEP/ECE-specific "vasopeptidase inhibitors" is a limited generation of proinflammatory, blood pressure-rising vasoconstrictors (Ang II, ET-1), the accumulation of vasodilators (BK, Ang_{1-7}), and additionally of the diuretic ANP. Consequently, vasopeptidase inhibition lowers blood pressure, diminishes cardiac hypertrophy and fibrosis, promotes renal natriuresis and diuresis, but also bears the risk of serious adverse effects [3,9,91].

8.4.4 Zinc Metalloproteases Terminate "Danger Signals"

NEP and ACE are ancient components of the innate and adaptive immune response of higher vertebrates. First, they participate in the immediate host defense – unfortunately with initial advances for a microbial intruder. Bacterial thermolysin-like peptidases facilitate entry into the host by degrading peptides with antimicrobial properties, for instance nerve-derived host SP, but also α-MSH or adrenomedullin [93] [Fig. 8.1(**1**)]. In parallel, peptidases enhance the bacterial invasiveness by degrading ECM (collagen IV) and facilitate the entry into the host's circulation [Fig. 8.1(**2**,**3**)] [51,52]. The complex peptidase–substrate interplay is reflected by the fact that – from the defendant's point of view – limited degradation of vasoactive and neuropeptides may be advantageous in early inflammation. Neuronal and cellular-derived SP, CGRP, or BK induce a vicious cycle of releasing IL-1 or TNFα from cutaneous cells that conversely promote the release and axonal transport of SP and CGRP in sensory neurons ([65,68,74,75], and included references.). Indeed, a downregulated NEP expression by irritants or endotoxins as observed in respiratory tract or intestinal inflammation suggests that prolonged activity of SP and BK may be advantageous for kick-off and progression of (neurogenic) inflammation and subsequent events, such as a full Ag-specific inflammatory response at the site of Ag challenge.

Numerous studies using vasopeptidase inhibitors or NEP-/ACE-deficient mice confirm the significance of NEP and ACE for cutaneous innate immunity [65,68,74,75]. The lack of NEP increased inflammatory responses and lethality in various murine inflammatory models (reviewed in [75]), and functional deletion of NEP and/or ACE markedly exacerbated murine allergic contact dermatitis (ACD) [23,77,78]. Missing NEP or ACE particularly on, or in the vicinity of, hematopoietic and immune cells modulates hematopoiesis [38], profoundly disturbs the local immune cell-activating cyto- and chemokine microenvironment, and imbalances pro- and anti-inflammatory neuropeptides. The latter fine-tune the DC:TC interface and T_H differentiation in adaptive immunity. Initialization of cutaneous delayed-type hypersensitivity requires activation and migration

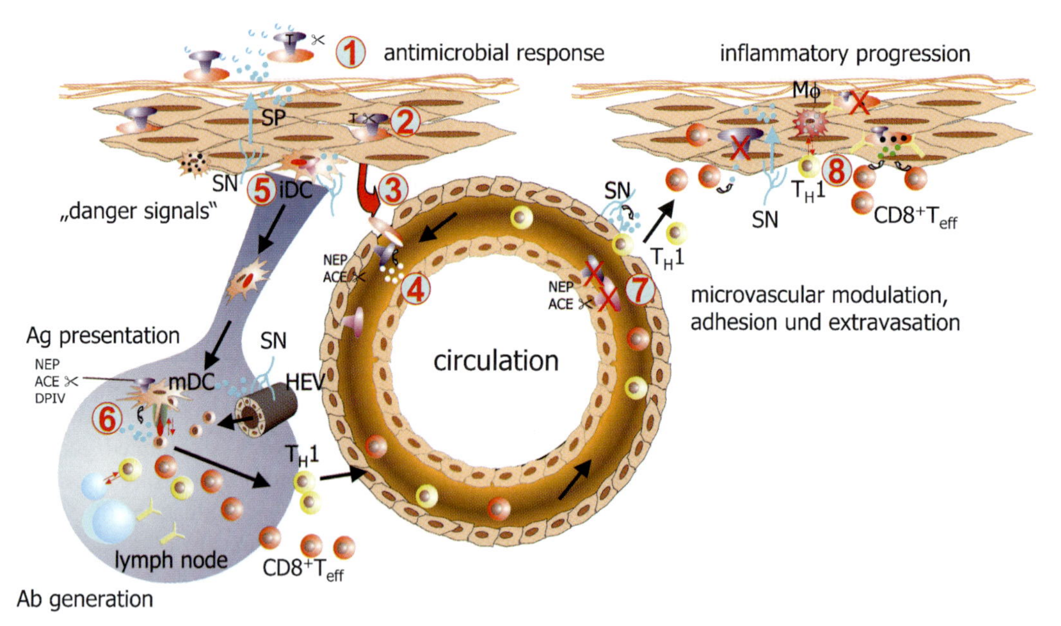

Fig. 8.1 Endopeptidases control multiple steps in inflammation and initialization of adaptive immunity. Intruding bacterial pathogens utilize thermolysin-like peptidases (T) to partly degrade antimicrobial neuropeptides, for example, substance P (SP) derived from sensory neurons (SN) or keratinocytes (**1**). By degrading extracellular matrix, peptidases facilitate microbial entry into the host (**2**) and accession to the vascular lumen (**3**). In some cases, host peptidases (NEP) may liberate anti-inflammatory peptides (α-MSH) from pathogen precursor molecules to compromise host's immunity (**4**). Released SP and other pro-inflammatory peptides serve as "danger signals" to activate residing immature dendritic cells (iDC) and induce DC maturation and migration into draining lymph node (LN) (**5**). In the LN, neuropeptides (SP) released from sensory nerves contacting high endothelial venules (HEV) or from matured DC (mDC) interacting with naïve T-cells (TC) promote T-helper (T_H) 1 polarization and clonal expansion of T_H1 and CD8+ effector TC ($CD8^+ T_{eff}$), particularly in the absence of NEP or ACE (**6**). Inflammation-downregulated peptidases at the site of ongoing inflammation promote vascular responses to released neuropeptides, as well as the recruitment and extravasation of inflammatory cells (**7**). In environment lacking functional NEP, prolonged activity of pro-inflammatory neuropeptides released from SN, macrophages (Mϕ), or TC promote multiple leukocyte effector functions, and in parallel, the temporarily increased availability of growth factors facilitates recovery of the damaged tissue (**8**)

[Fig. 8.1(**5**)] of Ag-laden dermal DC or Langerhans cells and contact to naïve $CD4^+$ TC in skin-draining lymph nodes [Fig. 8.1(**6**)]. Vasoactive peptides such as SP, BK, and Ang II via NK_1, B_2, and AT1/AT2 receptors, respectively, profoundly modulate myeloid DC precursor differentiation [38] and boost important DC and TC functions (Fig. 8.2) [45]. APC (DC or macrophages) and TC express NK_1 and B_2 as well as RAS components (ACE, AT1/AT2 receptors) and constitute extra-neuronal sources of SP [2,17,45,46,56]. Importantly, SP and BK activate DC NF-κB, a central transcription factor involved in DC maturation marker (CD11c, MHC) and costimulatory molecule (CD40, B7 molecules) expression, and thus NEP/ACE-deficiency may promote functional maturation of $CD11c^+$ DC [2,45]. Vice versa, antagonist blocking of B_2 or NK_1 receptors impairs Ag sensitization in mice lacking functional ACE or NEP, which could be mimicked by adoptive transfer of NK_1 receptor antagonist-treated, Ag-pulsed wild type bone-marrow DC generated in vitro [23,77–79]. Exogenous or endogenous SP contributes to mitogen- or Ag-induced TC proliferation and promotes a T_H1 response characterized by reduced IL-4/IL-5 and increased IL-2/IFNγ expression [45,74,75,79]. Hence, prolonged bioavailability

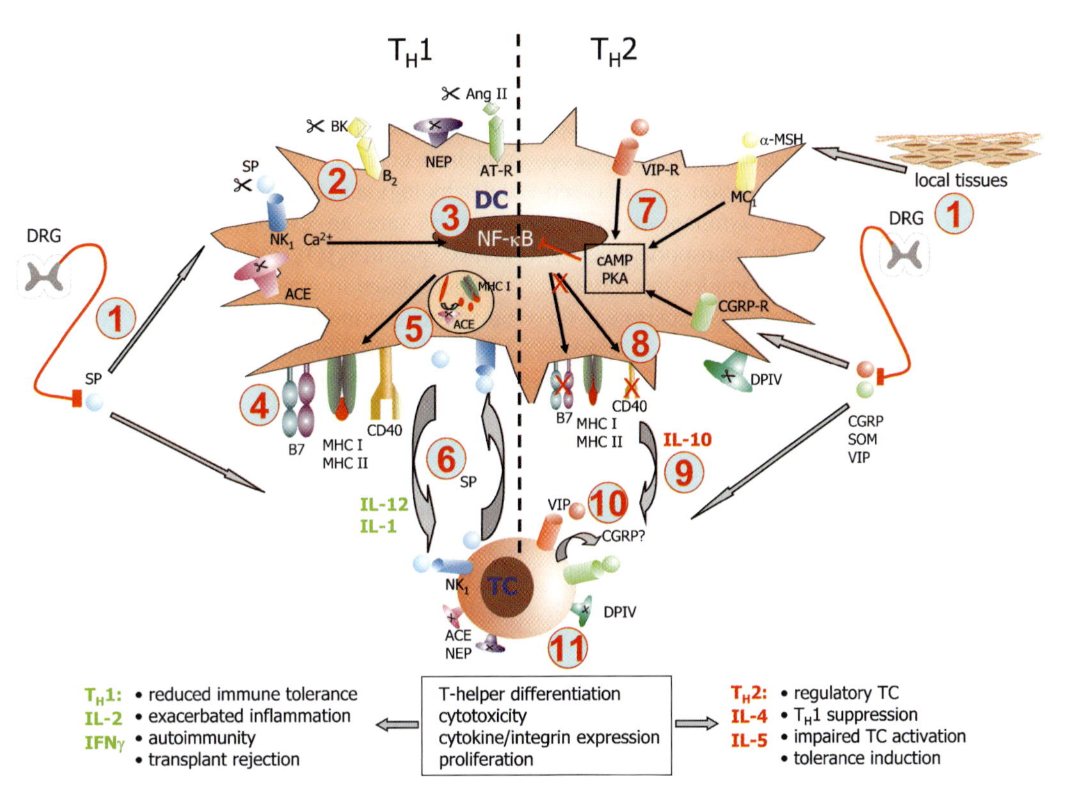

Fig. 8.2 Endopeptidases control the neuroendocrine hormone-balanced immune response transmitted by the immunologic synapse. Neuroendocrine and vasoactive peptides are released from cutaneous sensory neurons or cutaneous or lymphoid tissues (**1**). Ectopic (endo-) peptidase (NEP, ACE) mediator degradation controls the cellular accessibility of substance P (SP), bradykinin (BK), and angiotensin (Ang) II and activation of specific neurokinin (NK_1), bradykinin (B_2), and Ang (AT) receptors expressed by dendritic cells (DC, (**2**)). SP, BK, and Ang II constitute endogenous danger signals that stimulate DC pro-inflammatory signal transduction pathways such as the release of Ca^{2+} and subsequent activation of NF-κB (**3**). NF-κB plays a central role in the induction of DC costimulatory B7 molecules, CD40 or major histocompatibility complex (MHC) expression, and pro-inflammatory T-helper 1 (T_H1) cytokine release (**4**). Endopeptidases (e.g., ACE) improve the efficacy of antigen presentation to T-cells (TC) by intra- or extracellular trimming of peptides for MHC class I or II presentation (**5**). Endogenous SP from DC and TC may in an auto/paracrine manner drive T_H1 polarization under control of DC- or TC-expressed NEP and ACE (**6**). Via specific receptors expressed by DC and TC, α-MSH, calcitonin-gene related peptide (CGRP), somatostatin (SOM), and VIP trigger intracellular cAMP/protein kinase A (PKA) signal transduction (**7**). This inhibits endotoxin, cytokine-, or SP/BK-induced NF-κB activation and may promote a tolerogenic DC phenotype characterized by reduced TC costimulation (**8**), and release of anti-inflammatory cytokines (IL-10) (**9**). As demonstrated for VIP, such DC drive T_H2 polarization, regulatory TC (T_{reg}) development, and suppression of T_H1 immune responses. T_H2-inducing neuropeptides such as VIP (**10**) under the control of TC-expressed dipeptidyl peptidase IV/CD26 (DPIV, (**11**)) may also directly trigger T_{reg} development from naïve $CD4^+$ $CD25^+$ TC, which dampens T_H1 and $CD8^+$ T_{eff} responses

of "proinflammatory" tachykinins and kinins constitutes an endogenous "danger signal" per se that generate fully matured MHC class II/I expressing DC, which drive a T_H1 polarization and efficiently prime T_H and effector cell (T_{eff}) responses (Fig. 8.2). This may be beneficial for fighting intracellular pathogens or malignant cells, but could also result in uncontrolled inflammation or even organ-specific autoimmunity. In contrast, "T_H2" neuropeptides such as VIP/PACAP, CGRP, or α-MSH oppose the above by impairing Ag-presentation, and generating tolerogenic DC that attenuate TC activation. In addition, this may cause a T_H2 differentiation and give rise to regulatory CD4 and IL-10-producing $CD8^+$ $CD28^-CTLA4^+$ TC that subsequently suppress Ag-specific T_H1 mediated immune responses [25]. However, nature has predetermined additional roles for ACE and related peptidases for APC function in adaptive immunity. DC ACE is an important chaperone in the processing and MHC I-restricted presentation of viral and other peptides to $CD8^+$ cytotoxic T lymphocytes (CTL) [83,108]. ACE interacts with the transporter associated with Ag presentation (TAP) complex that shuttles cytosolic peptides into the exocytic compartment for association with nascent MHC I molecules. Carboxy-terminal ACE trimming of certain peptides improves their intracellular transport, and thus rescue their otherwise inefficient presentation to CTL [108]. Cathepsins, or a recently described asparaginyl endopeptidase, play a similar role in the endocytotic lysosomal pathway of DC and B-cells, with tremendous impact on MHC II-restricted Ag presentation [104]. Importantly, glycosylation or deamidation of certain self-proteins prevents cleavage by these proteases. Thus, a loss of these initially present post-translational modifications over time may suddenly render a self-protein susceptible for APC processing and recognition by TC, with fatal consequences for the maintenance of tolerance against self-antigens [19,104]. Thus, expression and specificity of endopeptidases in the MHC I or II compartment is essential for the control of peptide antigenicity.

8.4.5 Regulated NEP and ACE: A Protective Role Against Ultraviolet Irradiation?

Recent observations nurtured the hypothesis that NEP indirectly may have UV-protecting properties. UVB irradiation has been demonstrated to downregulate the NEP activity in human melanocytes. Pharmacologic NEP inhibition increased the efficacy of α-MSH or ACTH to induce tyrosinase activity and microphtalia expression, suggesting that NEP plays a role in melanogenesis [1]. However, the role of UV in regulating NEP and ACE expression may depend on the cellular and functional context. UV light or cytokine-exposed microvascular EC displayed a time-dependent loss of cell surface ACE in vitro, whereas the initially low NEP expression increased [75]. This complex reciprocal regulatory system of vasopeptidases, which has also been observed in hypertensive rats [37], may be of particular significance for EC function. Recent bioanalytical studies demonstrated that endothelial NEP and ACE are highly relevant for processing stress (ACTH) and anti-inflammatory hormone (α-MSH) [42]. However, instead of a mere removal of extracellular POMC peptides, NEP and ACE peptidolysis generated bioactive novel MC_x agonists or antagonists distinct from the parental peptide. This phenomenon may have currently undefined functional roles for vascular biology and cutaneous inflammatory responses, since EC are an established source and target of POMC peptides. Of note, invertebrate parasites use vasopeptidases to generate anti-inflammatory immunosuppressive melanocortins from POMC precursors, suggesting an evolutionary conserved mechanism to compromise a host's immune system [42].

8.4.6 Role of NEP and ACE in Cutaneous Wound Healing and Plasticity

It is now widely appreciated that NEP expression down-regulated by irritants, pro-inflammatory cytokines, endotoxins, or phorbol ester, respectively, increases the cellular accessibility of mediators and may promote inflammation and support cell growths as well as the development of neoplasms [34,95]. For instance, in nonhealing skin of *diabetes mellitus* patients or of mutant diabetic *db/db* mice, NEP expression and activity are increased ([75] and included references.). SP and CGRP may contribute to wound healing, as they promote keratinocyte, endothelial cell, and fibroblast proliferation and migration in vitro. Like bFGF, they also enhance angiogenesis and neovascularization in vivo (reviewed in [6,74]). Thus, increased NEP compromises re-epithelialization and wound healing, potentially by degrading growth factors such as bFGF-2 [26] and SP, which may facilitate development of diabetic ulcers [62]. Despite an improvement of extra-cutaneous wound healing after concomitantly applied SP and NEP inhibitors [7], these agents may exacerbate neuroinflammation and contribute to hypertrophic scar formation [80]. However, in pathologic cutaneous scar tissue, an increased expression of ACE in comparison to normal or wounded skin, particularly by myofibroblasts, was detected. Potentially, by generating abundant Ang II, ACE may have profibrotic and, thus, adverse effects on cutaneous wound healing similar to remodeling in the heart after myocardial infarct [54,92].

8.4.7 Development of Neoplasms: Shutting Off Growth-Promoting Signals is the Key

Allergens or irritants such as cigarette smoke impair NEP expression and activity in the lung, which may increase respiratory distress and neurogenic inflammation. Moreover, reduced NEP activity in certain forms of prostate or lung carcinoma prevents hydrolysis of mitogens such as bombesins, ET-1, or SP, which may promote peptide-driven tumor growths [84,95]. Likewise, reduced NEP activity hampers generation of the growth inhibitory fragment SP_{1-7} from NEP cleavage of SP [22]. One of the discussed mechanisms is an indirect phosphorylation and activation of the insulin growth factor 1 receptor (IGF-1R) that, further downstream, activates protein kinase B (Akt) anti-apoptotic cell survival pathways [95]. In line with this hypothesis, T and B cells become CD10-positive when undergoing apoptosis [16,53]. Hence, NEP expression represents a safety device that protects cells from mediators released from apoptotic cells and assures apoptotic cell-death by preventing access of cell survival-assuring mitogens. Interestingly, an apparent positive correlation of higher NEP levels with the malignancy of melanoma have been observed [4,102]. This increased NEP expression was accompanied by downregulated anti-apoptotic and upregulated pro-apoptotic proteins (Bcl-2, and Bax, respectively) [4]. Thus, NEP serves as a marker for apoptosis and progression of melanoma. However, the divergent expression pattern of NEP in metastatic melanoma, in contrast to other cancers, suggests that NEP may have diverse and tumor-specific modulating properties that require detailed future analysis. Noteworthy in this context, not all effects of vasopeptidases can be attributed to changes in extracellular mediator levels. Very recently, an ACE inhibitor-induced dimerization, phosphorylation, and signaling of ACE has been demonstrated in EC that enhanced endothelial ACE expression [43]. Likewise, as demonstrated for ACE, NEP, and tachykinin or BK receptors, the cellular co-expression [61] or a sterically close receptor-–peptidase association is important for receptor function and resensitization [18,21,48]. Therefore, direct vasopeptidase outside-in signaling or modulation of G protein-coupled receptor (GPCR) signal transduction independent of ligand cleavage may account for some effects of NEP or ACE inhibitors that require future analysis.

8.4.8 NEP and ACE: Are There Functional Roles in Psoriasis, Alopecia Areata, and Acne?

Innervation of the pilosebaceous unit plays a pivotal role for sebocyte function and hair growth. Stress-induced SP promotes both proliferation and differentiation of sebaceous glands, resulting in an increased size and lipogenesis. Surprisingly, in skin of acne patients in contrast to normal skin, the subcellular expression of SP-induced NEP is increased in germinal sebocytes [97]. This may argue against a stress-induced inflammation, although the exact role of increased NEP awaits future investigation. Likewise, in acute and late stage chronic alopecia areata, NEP is strongly expressed in follicular structures of the affected area, whereas the perivascular expression of ACE in alopecia areata lesions is diminished [98]. This may prevent the local generation of keratinocyte/fibroblast proliferation-inducing Ang II [90]. As SP is capable of inducing significant anagen hair growth, limited bioavailability of SP, Ang II, and growth factors in alopecia areata may attenuate hair growth and increase hair follicle regression and apoptosis [64,65]. Uncontrolled cellular access to sensory neuropeptides and growth factors may also contribute to the pathogenesis of psoriasis. A reduced NEP expression in acute psoriatic lesions, but not in healthy skin, may locally increase SP levels that contribute to keratinocyte hyperproliferation, inflammation, and psoriatic pruritus [12,57,75]. In conjunction with a reportedly enhanced neuronal density and NK_1 expression, this may explain, in part, the susceptibility of psoriasis to exogenous stress and, thus, NK_1 antagonists might be beneficial in psoriasis therapy [40]. Likewise, ACE has been associated with this disease [63], although the exact role of the enzyme is somewhat obscure. Some studies proposed elevated serum ACE levels in psoriasis as a diagnostic marker and suggested that increased amounts of the pro-inflammatory Ang II may contribute to the pathogenesis of this disease [32,70]. Several other reports, however, have associated pharmacologic ACE inhibition with the induction or exacerbation of psoriasis [14,33,107]. Thus, one adverse side-effect of vasopeptidase inhibitors may be the worsening of psoriasis due to prolonged availability of SP or BK.

Summary for the Clinician

As outlined in this chapter, endopeptidases have multiple substrates and different modes of action in the skin and elsewhere. Particularly, NEP is an excellent example that endopeptidases in addition to their metabolizing/degrading function mediate biotransformation of inert precursor peptides to active mediators, and act as convertases by converting one bioactive receptor ligand into a second peptide that activates another receptor. Finally, competitive cleavage of a given substrate generates a peptide with opposing physiological effects to that produced by a closely related enzyme (i.e., NEP: Ang I $\rightarrow$ Ang_{1-7}, hypotensive vs. ACE: Ang I $\rightarrow$ Ang II = Ang_{1-8}, hypertensive). Evidently, a multiplicity of functions arises from these activities, and thus endopeptidases constitute important milestones in cardiovascular regulation, in neurogenic inflammation and immune recognition, or in tumorigenesis. Although it is tempting to drug-target neprilysin and others to achieve expressional or enzymatic regulation, one may likely replace one evil with another. This dilemma is best exemplified by attempts to simultaneously inhibit multiple vasopeptidases (ACE, NEP, and ECE) with one inhibitor. Although this is a promising strategy for the treatment of cardiovascular disorders, recent large-scale phase IV clinical studies have questioned this benefit [9]. Evidently, the ACE/NEP inhibitor-dependently increased kinin, SP, and endothelin levels elevated the risk of potentially life-threatening angioedema, with an even higher incidence particularly in African-Americans [66] Vasopeptidase inhibition may also worsen cutaneous disorders with a neurogenic component [91], promote Ag sensitization, or trigger elicitation of allergic inflammation in already sensitized individuals. As explained above, NEP may also have (neuro-) protective roles in cancer and Alzheimer's disease [35]. Thus, future attempts of inhibiting peptidases with Aβ-catabolizing properties essentially require the design of drugs incapable of crossing the blood–brain-barrier, and critical validation of efficacy and safety of this promising therapeutic principle is mandatory. Vice versa, since downregulated NEP expression or activity frequently accompanies acute or chronic inflammation and the malignancy of tumors, an artificial substitution seems to be desirable. Thus, a tissue-specific use of recombinant enzymes [94] or a local enhancement of NEP expression and activity may provide a novel mechanism-based therapeutic approach towards management of (neurogenic) inflammatory skin disorders. Unfortunately, despite a revived interest in NEP regulation due to the impact of NEP on the CNS Aβ level, only few data are available demonstrating an upregulation of NEP, that is, by glucocorticoids, thrombin, or cAMP [28,29,35]. Obviously, the future challenge remains to develop tools to control peptidase expression and activity in a highly selective cell and tissue-specific manner.

References

1. Aberdam E, Auberger P, Ortonne JP, Ballotti R (2000) Neprilysin, a novel target for ultraviolet B regulation of melanogenesis via melanocortins. J Invest Dermatol 115:381–387
2. Aliberti J, Viola JP, Vieira-de-Abreu A, Bozza PT, Sher A, Scharfstein J (2003) Cutting edge: bradykinin induces IL-12 production by dendritic cells: a danger signal that drives Th1 polarization. J Immunol 170:5349–5353
3. Battistini B, Daull P, Jeng AY (2005) CGS 35601, a triple inhibitor of angiotensin converting enzyme, neutral endopeptidase and endothelin converting enzyme. Cardiovasc Drug Rev 23:317–330
4. Bilalovic N, Sandstad B, Golouh R, Nesland JM, Selak I, Torlakovic EE (2004) CD10 protein expression in tumor and stromal cells of malignant melanoma is associated with tumor progression. Mod Pathol 17:1251–1258
5. Bock O, Kreiselmeyer I, Mrowietz U (2001) Expression of dipeptidyl-peptidase IV (CD26) on CD8 + T cells is significantly decreased in patients with psoriasis vulgaris and atopic dermatitis. Exp Dermatol 10:414–419
6. Brain SD (1997) Sensory neuropeptides: their role in inflammation and wound healing. Immunopharmacology 37:133–152
7. Burssens P, Steyaert A, Forsyth R, van Ovost EJ, De PY, Verdonk R (2005) Exogenously administered substance P and neutral endopeptidase inhibitors stimulate fibroblast proliferation, angiogenesis and collagen organization during Achilles tendon healing. Foot Ankle Int 26:832–839
8. Busso N, Wagtmann N, Herling C, Chobaz-Peclat V, Bischof-Delaloye A, So A, et al. (2005) Circulating CD26 is negatively associated with inflammation in human and experimental arthritis. Am J Pathol 166:433–442
9. Campbell DJ (2003) Vasopeptidase inhibition: a double-edged sword? Hypertension 41:383–389
10. Castro MG, Morrison E (1997) Post-translational processing of proopiomelanocortin in the pituitary and in the brain. Crit Rev Neurobiol 11:35–57
11. Catania A, Gatti S, Colombo G, Lipton JM (2004) Targeting melanocortin receptors as a novel strategy to control inflammation. Pharmacol Rev 56:1–29
12. Chan J, Smoller BR, Raychauduri SP, Jiang WY, Farber EM (1997) Intraepidermal nerve fiber expression of calcitonin gene-related peptide, vasoactive intestinal peptide and substance P in psoriasis. Arch Dermatol Res 289:611–616
13. Chen T, Ajami K, McCaughan GW, Gorrell MD, Abbott CA (2003) Dipeptidyl peptidase IV gene family. The DPIV family. Adv Exp Med Biol 524:79–86
14. Cohen AD, Bonneh DY, Reuveni H, Vardy DA, Naggan L, Halevy S (2005) Drug exposure and psoriasis vulgaris: case–control and case–crossover studies. Acta Derm Venereol 85:299–303
15. Costerousse O, Allegrini J, Lopez M, Alhenc-Gelas F (1993) Angiotensin I-converting enzyme in human circulating mononuclear cells: genetic polymorphism of expression in T-lymphocytes. Biochem J 290(Pt 1):33–40
16. Cutrona G, Ferrarini M (2001) Expression of CD10 by human T cells that undergo apoptosis both in vitro and in vivo. Blood 97:2528
17. Danilov SM, Sadovnikova E, Scharenborg N, Balyasnikova IV, Svinareva DA, Semikina EL, et al. (2003) Angiotensin-converting enzyme (CD143) is abundantly expressed by

dendritic cells and discriminates human monocyte-derived dendritic cells from acute myeloid leukemia-derived dendritic cells. Exp Hematol 31:1301–1309
18. Deddish PA, Marcic B, Tan F, Jackman HL, Chen ZZ, Erdos EG (2002) Neprilysin inhibitors potentiate effects of bradykinin on B2 receptor. Hypertension 39:619–623
19. Doyle HA, Mamula MJ (2001) Post-translational protein modifications in antigen recognition and autoimmunity. Trends Immunol 22:443–449
20. Dubovy P (1987) Histochemical evidence for the presence of dipeptidylpeptidase IV in the Schwann cells of skin unmyelinated axons. Experientia 43:883–884
21. Erdos EG, Marcic BM (2001) Kinins, receptors, kininases and inhibitors – where did they lead us? Biol Chem 382:43–47
22. Erin N, Zhao W, Bylander J, Chase G, Clawson G (2006) Capsaicin-induced inactivation of sensory neurons promotes a more aggressive gene expression phenotype in breast cancer cells. Breast Cancer Res Treat 99:351–364
23. Fastrich M, Fabritz L, Luger TA, Scholzen TE (2006) Neprilysin/angiotensin-converting enzyme double-deficient mice: a mouse model to study inflammatory skin disease. J Invest Dermatol 126:1675
24. Gliddon DR, Howard CJ (2002) CD26 is expressed on a restricted subpopulation of dendritic cells in vivo. Eur J Immunol 32:1472–1481
25. Gonzalez-Rey E, Chorny A, Delgado M (2007) Regulation of immune tolerance by anti-inflammatory neuropeptides. Nat Rev Immunol 7:52–63
26. Goodman OB, Jr., Febbraio M, Simantov R, Zheng R, Shen R, Silverstein RL, et al. (2006) Neprilysin inhibits angiogenesis via proteolysis of fibroblast growth factor-2. J Biol Chem 281:33597–33605
27. Gorrell MD (2005) Dipeptidyl peptidase IV and related enzymes in cell biology and liver disorders. Clin Sci (Lond) 108:277–292
28. Graf K, Kunkel K, Zhang M, Grafe M, Schultz K, Schudt C, et al. (1995) Activation of adenylate cyclase and phosphodiesterase inhibition enhance neutral endopeptidase activity in human endothelial cells. Peptides 16: 1273–1278
29. Graf K, Koehne P, Grafe M, Zhang M, Auch-Schwelk W, Fleck E (1995) Regulation and differential expression of neutral endopeptidase 24.11 in human endothelial cells. Hypertension 26:230–235
30. Guy JL, Lambert DW, Warner FJ, Hooper NM, Turner AJ (2005) Membrane-associated zinc peptidase families: comparing ACE and ACE2. Biochim Biophys Acta 1751:2–8
31. Hamming I, Timens W, Bulthuis ML, Lely AT, Navis GJ, van GH (2004) Tissue distribution of ACE2 protein, the functional receptor for SARS coronavirus. A first step in understanding SARS pathogenesis. J Pathol 203:631–637
32. Huskic J, Alendar F, Matavulj A, Ostoic L (2004) Serum angiotensin converting enzyme in patients with psoriasis. Med Arh 58:202–205
33. Ikai K (1995) Exacerbation and induction of psoriasis by angiotensin-converting enzyme inhibitors. J Am Acad Dermatol 32:819
34. Ishimaru F, Potter NS, Shipp MA (1996) Phorbol ester-mediated regulation of CD10/neutral endopeptidase transcripts in acute lymphoblastic leukemias. Exp Hematol 24:43–48
35. Iwata N, Higuchi M, Saido TC (2005) Metabolism of amyloid-beta peptide and Alzheimer's disease. Pharmacol Ther 108:129–148
36. Jaspard E, Wei L, Alhenc-Gelas F (1993) Differences in the properties and enzymatic specificities of the two active sites of angiotensin I-converting enzyme (kininase II). Studies with bradykinin and other natural peptides. J Biol Chem 268:9496–9503
37. Jongun L (2004) Reciprocal regulation of angiotensin converting enzyme and neutral endopeptidase in rats with experimental hypertension. Physiol Res 53:365–368
38. Joshi DD, Dang A, Yadav P, Qian J, Bandari PS, Chen K, et al. (2001) Negative feedback on the effects of stem cell factor on hematopoiesis is partly mediated through neutral endopeptidase activity on substance P: a combined functional and proteomic study. Blood 98:2697–2706
39. Khatib AM, Siegfried G, Chretien M, Metrakos P, Seidah NG (2002) Proprotein convertases in tumor progression and malignancy: novel targets in cancer therapy. Am J Pathol 160:1921–1935
40. Kikwai L, Babu RJ, Kanikkannan N, Singh M (2004) Preformulation stability of Spantide II, a promising topical anti-inflammatory agent for the treatment of psoriasis and contact dermatitis. J Pharm Pharmacol 56:19–25
41. Kim EJ, Hess S, Richardson SK, Newton S, Showe LC, Benoit BM, et al. (2005) Immunopathogenesis and therapy of cutaneous T cell lymphoma. J Clin Invest 115:798–812
42. Koenig S, Luger TA, Scholzen TE (2006) Monitoring neuropeptide-specific proteases: processing of the proopiomelanocortin peptides adrenocorticotropin and α-melanocyte-stimulating hormone in the skin. Exp Dermatol 15:751–761
43. Kohlstedt K, Brandes RP, Muller-Esterl W, Busse R, Fleming I (2004) Angiotensin-converting enzyme is involved in outside-in signaling in endothelial cells. Circ Res 94:60–67
44. Krege JH, John SW, Langenbach LL, Hodgin JB, Hagaman JR, Bachman ES, et al. (1995) Male-female differences in fertility and blood pressure in ACE-deficient mice. Nature 375:146–148
45. Lambrecht BN (2001) Immunologists getting nervous: neuropeptides, dendritic cells and T cell activation. Respir Res 2:133–138
46. Lapteva N, Nieda M, Ando Y, Ide K, Hatta-Ohashi Y, Dymshits G, et al. (2001) Expression of renin-angiotensin system genes in immature and mature dendritic cells identified using human cDNA microarray. Biochem Biophys Res Commun 285:1059–1065
47. Lapteva N, Ide K, Nieda M, Ando Y, Hatta-Ohashi Y, Minami M, et al. (2002) Activation and suppression of renin-angiotensin system in human dendritic cells. Biochem Biophys Res Commun 296:194–200
48. Marcic B, Deddish PA, Skidgel RA, Erdos EG, Minshall R, Tan F (2000) Replacement of the transmembrane anchor in angiotensin I-converting enzyme (ACE) with a glycosylphosphatidylinositol tail affects activation of the B2 bradykinin receptor by ACE inhibitors. J Biol Chem 275:16110–16118
49. Mazurkiewicz JE, Corliss D, Slominski A (1999) Differential temporal and spatial expression of POMC mRNA and of the production of POMC peptides during the murine hair cycle. Ann N Y Acad Sci 885:427–429

50. Mbikay M, Seidah NG, Chretien M (2001) Neuroendocrine secretory protein 7B2: structure, expression and functions. Biochem J 357:329–342
51. Miyoshi S, Nakazawa H, Kawata K, Tomochika K, Tobe K, Shinoda S (1998) Characterization of the hemorrhagic reaction caused by Vibrio vulnificus metalloprotease, a member of the thermolysin family. Infect Immun 66:4851–4855
52. Miyoshi S, Watanabe H, Kawase T, Yamada H, Shinoda S (2004) Generation of active fragments from human zymogens in the bradykinin-generating cascade by extracellular proteases from *Vibrio vulnificus* and *V. parahaemolyticus*. Toxicon 44:887–893
53. Morabito F, Mangiola M, Rapezzi D, Zupo S, Oliva BM, Ferraris AM, et al. (2003) Expression of CD10 by B-chronic lymphocytic leukemia cells undergoing apoptosis in vivo and in vitro. Haematologica 88:864–873
54. Morihara K, Takai S, Takenaka H, Sakaguchi M, Okamoto Y, Morihara T, et al. (2006) Cutaneous tissue angiotensin-converting enzyme may participate in pathologic scar formation in human skin. J Am Acad Dermatol 54:251–257
55. Morrison ME, Vijayasaradhi S, Engelstein D, Albino AP, Houghton AN (1993) A marker for neoplastic progression of human melanocytes is a cell surface ectopeptidase. J Exp Med 177:1135–1143
56. Nahmod KA, Vermeulen ME, Raiden S, Salamone G, Gamberale R, Fernandez-Calotti P, et al (2003) Control of dendritic cell differentiation by angiotensin II. FASEB J 17:491–493
57. Nakamura M, Toyoda M, Morohashi M (2003) Pruritogenic mediators in psoriasis vulgaris: comparative evaluation of itch-associated cutaneous factors. Br J Dermatol 149:718–730
58. Narducci MG, Scala E, Bresin A, Caprini E, Picchio MC, Remotti D, et al. (2006) Skin homing of Sezary cells involves SDF-1-CXCR4 signaling and down-regulation of CD26/dipeptidylpeptidase IV. Blood 107:1108–1115
59. Nemoto E, Sugawara S, Takada H, Shoji S, Horiuch H (1999) Increase of CD26/dipeptidyl peptidase IV expression on human gingival fibroblasts upon stimulation with cytokines and bacterial components. Infect Immun 67:6225–6233
60. Novelli M, Savoia P, Fierro MT, Verrone A, Quaglino P, Bernengo MG (1996) Keratinocytes express dipeptidyl-peptidase IV (CD26) in benign and malignant skin diseases. Br J Dermatol 134:1052–1056
61. Okamoto A, Lovett M, Payan DG, Bunnett NW (1994) Interactions between neutral endopeptidase (EC 3.4.24.11) and the substance P (NK1) receptor expressed in mammalian cells. Biochem J 299:683–693
62. Olerud JE, Usui ML, Seckin D, Chiu DS, Haycox CL, Song IS, et al. (1999) Neutral endopeptidase expression and distribution in human skin and wounds. J Invest Dermatol 112:873–881
63. Ozkur M, Erbagci Z, Nacak M, Tuncel AA, Alasehirli B, Aynacioglu AS (2004) Association of insertion/deletion polymorphism of the angiotensin-converting enzyme gene with psoriasis. Br J Dermatol 151:792–795
64. Paus R, Heinzelmann T, Schultz KD, Furkert J, Fechner K, Czarnetzki BM (1994) Hair growth induction by substance P. Lab Invest 71:134–140
65. Peters EMJ, Ericson ME, Hosoi J, Seiffert K, Hordinsky MK, Ansel JC, et al. (2006) Neuropeptide control mechanisms in cutaneous biology: physiological and clinical significance. J Invest Dermatol 126:1937–1947
66. Pickering TG (2002) The rise and fall of omapatrilat. J Clin Hypertens 4:371–373
67. Roesch A, Wittschier S, Becker B, Landthaler M, Vogt T (2006) Loss of dipeptidyl peptidase IV immunostaining discriminates malignant melanomas from deep penetrating nevi. Mod Pathol 19:1378–1385
68. Roosterman D, Goerge T, Schneider SW, Bunnett NW, Steinhoff M (2006) Neuronal control of skin function: the skin as a neuroimmunoendocrine organ. Physiol Rev 86:1309–1379
69. Roques BP, Noble F, Dauge V, Fournie-Zaluski MC, Beaumont A (1993) Neutral endopeptidase 24.11: structure, inhibition, and experimental and clinical pharmacology. Pharmacol Rev 45:87–146
70. Ryder KW, Epinette WW, Jay SJ, Ransburg RC, Glick MR (1985) Serum angiotensin converting enzyme activity in patients with psoriasis. Clin Chim Acta 153:143–146
71. Scamuffa N, Calvo F, Chretien M, Seidah NG, Khatib AM (2006) Proprotein convertases: lessons from knockouts. FASEB J 20:1954–1963
72. Scheel-Toellner D, Richter E, Toellner KM, Reiling N, Wacker HH, Flad HD, et al. (1995) CD26 expression in leprosy and other granulomatous diseases correlates with the production of interferon-gamma. Lab Invest 73:685–690
73. Schiller M, Brzoska T, Bohm M, Metze D, Scholzen TE, Rougier A, et al. (2004) Solar-simulated ultraviolet radiation-induced upregulation of the melanocortin-1 receptor, proopiomelanocortin, and {alpha}-melanocyte-stimulating hormone in human epidermis in vivo. J Invest Dermatol 122:468–476
74. Scholzen T, Armstrong CA, Bunnett NW, Luger TA, Olerud JE, Ansel JC (1998) Neuropeptides in the skin: interactions between the neuroendocrine and the skin immune systems. Exp Dermatol 7:81–96
75. Scholzen TE, Luger TA (2004) Neutral endopeptidase and angiotensin-converting enzyme – key enzymes terminating the action of neuroendocrine mediators. Exp Dermatol 13:22–26
76. Scholzen TE, Kalden D-H, Brzoska T, Fastrich M, Schwarz T, Schiller M, et al. (2000) Expression of proopiomelanocortin peptides in human dermal microvascular endothelial cells: evidence for a regulation by ultraviolet light and interleukin-1. J Invest Dermatol 115:1021–1028
77. Scholzen TE, Steinhoff M, Bonaccorsi P, Klein R, Amadesi S, Geppetti P, et al. (2001) Neutral endopeptidase terminates substance P-induced inflammation in allergic contact dermatitis. J Immunol 166:1285–1291
78. Scholzen TE, Stander S, Riemann H, Brzoska T, Luger TA (2003) Modulation of cutaneous inflammation by angiotensin-converting enzyme. J Immunol 170:3866–3873
79. Scholzen TE, Steinhoff M, Sindrilaru A, Schwarz A, Bunnett NW, Luger TA, et al. (2004) Cutaneous allergic contact dermatitis responses are diminished in mice deficient in neurokinin 1 receptors and augmented by neurokinin 2 receptor blockage. FASEB J 18:1007–1009
80. Scott JR, Muangman PR, Tamura RN, Zhu KQ, Liang Z, Anthony J, et al. (2005) Substance P levels and neutral endopeptidase activity in acute burn wounds and hypertrophic scar. Plast Reconstr Surg 115:1095–1102
81. Seidah NG, Benjannet S, Hamelin J, Marmabachi AM, Basak A, Marcinkiewicz J, et al. (1999) The subtilisin/kexin family of precursor convertases: emphasis on PC1,

PC2/7B2, POMC and the novel enzyme SKI-1. Ann N Y Acad Sci 885:57–74
82. Seitzer U, Scheel-Toellner D, Hahn M, Heinemann G, Mattern T, Flad HD, et al. (1997) Comparative study of CD26 as a Th1-like and CD30 as a potential Th2-like operational marker in leprosy. Adv Exp Med Biol 421:217–221
83. Sherman LA, Burke TA, Biggs JA (1992) Extracellular processing of peptide antigens that bind class I major histocompatibility molecules. J Exp Med 175:1221–1226
84. Shipp MA, Look AT (1993) Hematopoietic differentiation antigens that are membrane-associated enzymes: cutting is the key! Blood 82:1052–1070
85. Simeoni L, Rufini A, Moretti T, Forte P, Aiuti A, Fantoni A (2002) Human CD26 expression in transgenic mice affects murine T-cell populations and modifies their subset distribution. Hum Immunol 63:719–730
86. Skidgel RA, Erdos EG (2004) Angiotensin converting enzyme (ACE) and neprilysin hydrolyze neuropeptides: a brief history, the beginning and follow-ups to early studies. Peptides 25:521–525
87. Slominski A, Wortsman J, Luger TA, Paus R, Solomon SG (2000) Corticotropin releasing hormone and proopiomelanocortin involvement in the cutaneous response to stress. Physiol Rev 80:979–1020
88. Sokolowska-Wojdylo M, Wenzel J, Gaffal E, Steitz J, Roszkiewicz J, Bieber T, et al. (2005) Absence of CD26 expression on skin-homing CLA+ CD4+ T lymphocytes in peripheral blood is a highly sensitive marker for early diagnosis and therapeutic monitoring of patients with Sezary syndrome. Clin Exp Dermatol 30:702–706
89. Steckelings UM, Czarnetzki BM (1995) The renin-angiotensin-system in the skin. Evidence for its presence and possible functional implications. Exp Dermatol 4:329–334
90. Steckelings UM, Artuc M, Paul M, Stoll M, Henz BM (1996) Angiotensin II stimulates proliferation of primary human keratinocytes via a non-AT1, non-AT2 angiotensin receptor. Biochem Biophys Res Commun 229:329–333
91. Steckelings UM, Artuc M, Wollschlager T, Wiehstutz S, Henz BM (2001) Angiotensin-converting enzyme inhibitors as inducers of adverse cutaneous reactions. Acta Derm Venereol 81:321–325
92. Steckelings UM, Wollschlager T, Peters J, Henz BM, Hermes B, Artuc M (2004) Human skin: source of and target organ for angiotensin II. Exp Dermatol 13:148–154
93. Steinman L (2004) Elaborate interactions between the immune and nervous systems. Nat Immunol 5:575–581
94. Sturiale S, Barbara G, Qui B, Figini M, Geppetti P, Gerard N, et al. (1999) Neutral endopeptidase (EC 3.4.24.11) terminates colitis by degrading SP. Proc Natl Acad Sci U S A 96:11653–11658
95. Sumitomo M, Shen R, Nanus DM (2005) Involvement of neutral endopeptidase in neoplastic progression. Biochim Biophys Acta 1751:52–59
96. Thomas JE, Rylett CM, Carhan A, Bland ND, Bingham RJ, Shirras AD, et al. (2005) Drosophila melanogaster NEP2 is a new soluble member of the neprilysin family of endopeptidases with implications for reproduction and renal function. Biochem J 386:357–366
97. Toyoda M, Morohashi M (2003) New aspects in acne inflammation. Dermatology 206:17–23
98. Toyoda M, Makino T, Kagoura M, Morohashi M (2001) Expression of neuropeptide-degrading enzymes in alopecia areata: an immunohistochemical study. Br J Dermatol 144:46–54
99. Turner AJ, Hooper NM (2002) The angiotensin-converting enzyme gene family: genomics and pharmacology. Trends Pharmacol Sci 23:177–183
100. Turner AJ, Isaac RE, Coates D (2001) The neprilysin (NEP) family of zinc metalloendopeptidases: genomics and function. Bioessays 23:261–269
101. Van den Oord JJ (1998) Expression of CD26/dipeptidyl-peptidase IV in benign and malignant pigment-cell lesions of the skin. Br J Dermatol 138:615–621
102. Velazquez EF, Yancovitz M, Pavlick A, Berman R, Shapiro R, Bogunovic D, et al. (2007) Clinical relevance of neutral endopeptidase (NEP/CD10) in melanoma. J Transl Med 5:2
103. Wang XM, Yu DM, McCaughan GW, Gorrell MD (2006) Extra-enzymatic roles of DPIV and FAP in cell adhesion and migration on collagen and fibronectin. Adv Exp Med Biol 575:213–222
104. Watts C, Matthews SP, Mazzeo D, Manoury B, Moss CX (2005) Asparaginyl endopeptidase: case history of a class II MHC compartment protease. Immunol Rev 207:218–228
105. Wei L, Clauser E, henc-Gelas F, Corvol P (1992) The two homologous domains of human angiotensin I-converting enzyme interact differently with competitive inhibitors. J Biol Chem 267:13398–13405
106. Wesley UV, McGroarty M, Homoyouni A (2005) Dipeptidyl peptidase inhibits malignant phenotype of prostate cancer cells by blocking basic fibroblast growth factor signaling pathway. Cancer Res 65:1325–1334
107. Wolf R, Tamir A, Brenner S (1990) Psoriasis related to angiotensin-converting enzyme inhibitors. Dermatologica 181:51–53
108. Yellen-Shaw AJ, Laughlin CE, Metrione RM, Eisenlohr LC (1997) Murine transporter associated with antigen presentation (TAP) preferences influence class I-restricted T cell responses. J Exp Med 186:1655–1662

Neuroinflammation and Toll-Like Receptors in the Skin

9

B. Rothschild, Y. Lu, H. Chen, P.I. Song, C.A. Armstrong, and J.C. Ansel

Contents

Key Features

- Toll-like receptors are key facilitators of innate immunity within the skin.
- Different TLRs are activated by various exogenous and endogenous ligands.
- TLR signaling provides a mechanism of activating rapid and directed immune responses in defense of the host.
- Evidence for TLR expression and function has recently been discovered in skin and neuronal tissues.
- The nervous system and specific neuromediators play a role in activation and subsequent response of the TLRs.
- Directed therapy that modifies TLR signaling may provide new avenues of treating auto-immune disease, skin cancer, and conditions of inappropriate inflammation.

Synonyms Box: TLR expression in the skin, TLR expression in the nervous system, neural regulation of TLRs, pattern-recognition receptors, innate immunity, cutaneous neuromediators of TLR

9.1 Host Innate and Adaptive Immunity

The mammalian immune system detects and eliminates pathogenic microorganisms by the activation of a highly organized and complex humoral and cellular system that discriminates between self and non-self. Immune responses have been traditionally divided into two overlapping and complimentary components: the innate immune system and the adaptive immune system [50]. The adaptive immune system evolved with the advent of jawed vertebrates and is characterized by the ability to detect and respond to a wide range of antigens by activating specific receptors on the cell surfaces of T and B cells. T and B cells are capable of expressing more than 10^{18} different antigen receptor specificities that allows these cells to respond to a vast number of foreign and sometimes self antigens. Activation of the antigen-specific receptors triggers selective host immunologic responses that include the clonal expansion of T cells, B cells, cytokine production, and immunoglobulin secretion.

Unlike the adaptive immune system, the innate immune system is expressed in organisms ranging from primitive eukaryotes to humans [50]. Although

R.D. Granstein and T.A. Luger (eds.), *Neuroimmunology of the Skin*,

the existence of the innate immune system has been appreciated for many years, its role and importance in the mammalian immune system has often been overshadowed by rapid advances in our understanding of the adaptive immune system in the past 30 years. The innate immune system is composed of both soluble and cellular components. Recently, in one of the major advances in the field of immunology, a number of cell-associated receptors termed pattern-recognition receptors (PRRs) were identified that recognize common molecular structures found in different types of microbial pathogens. This observation led to a dramatic change in our view of innate immunity and resulted in an explosion of publications examining the great variety and the surprising specificity of innate immune responses [58]. PRRs are capable of recognizing different microbe structures termed pathogen-associated molecular patterns (PAMPs); thus allowing the rapid identification and response to a wide variety of environmental pathogens. Many PRRs are located on the outer cell surface of leukocytes where interaction with different pathogens results in a series of intracellular signaling events that initiate directed host effector proinflammatory responses. Some PRRs can also be secreted or localized within the cell where they respond to either blood-borne or intracellular pathogens, respectively [54]. PRR mediated innate immune responses can include the production of acute phase proteins, complement, cytokines, chemokines, neuropeptides, antimicrobial peptides, and cell membrane costimulatory molecules. Cellular components of the innate immune system include NK cells, neutrophils, eosinophils, γδT cells, macrophages, and dendritic cells [58]. The rapid innate immune response to the foreign antigens or pathogens also triggers the activation of the slower antigen-specific adaptive T cell and B cell mediated immune response. In most cases, the initial innate immune response neutralizes and eliminates potential pathogens before the development of progressive infectious process. B and T cell adaptive immune responses both reinforce and amplify these innate immune responses [5].

9.2 Neuronal Regulation of Cutaneous Immunity

There is significant experimental evidence that the peripheral and central nervous systems participate in cutaneous innate and adaptive immune responses by the local and systemic release of neuropeptides and neurohormones that can act directly on a variety of epidermal and dermal cells, which express specific receptors for these neuromediators [100]. The neurologic system is capable of modulating both proinflammatory as well as immunosuppressive cutaneous immune responses depending on the biological properties of the released neuropeptides and the specific neuroreceptors expressed on the different target cells in the skin. It can be argued that the cutaneous neurological system is the first component of the innate immune system to be activated in the skin after exposure to a variety of environmental insults. Cutaneous injury not only triggers the release of multiple neuropeptides but also a variety of other neural activating agents such as cytokines, proteases, histamine, leukotrienes, nitric oxide, and transient receptor potential ion channel (TRP) mediators [105].

The sensory nerve fibers in the skin are composed of fine unmyelinated C fibers and myelinated A delta fibers, which are capable of synthesizing neuropeptides such as substance P (SP) and calcitonin gene-related peptide (CGRP) [94,107]. Cutaneous C-fiber neurons release these neuropeptides in response to a wide variety of noxious stimuli and activate not only regional sensory nerves but also epidermal and dermal cells expressing SP and CGRP receptors such as keratinocytes, Langerhans cells, melanocytes, dermal microvascular endothelial cells (DMEC), fibroblasts, resident leukocytes, and hair follicle cells [7,73,108]. SP and CGRP can modulate a variety of innate immune responses in the skin, including the expression of cytokines, chemokines, nitric oxide, and other soluble proinflammatory factors as well as cell associated immune proteins such as cellular adhesion molecules (CAMs) and class II MHC molecules [11,82]. These neuropeptide induced responses are believed to contribute to the cutaneous changes observed in many inflammatory diseases of the skin. In contrast to SP, CGRP demonstrates immunosuppressive activities such as inhibition of proinflammatory cytokine production and Langerhans cell responses [53].

Like cutaneous sensory nerves, the autonomic nervous system (sympathetic and parasympathetic nerves) can also regulate a number of biological processes in the skin [105]. Cutaneous autonomic nerve fibers are localized in the dermis where they innervate blood vessels, lymphatic vessels, hair follicles, eccrine glands, and apocrine glands [20,132]. The primary function of autonomic nerves in the skin is to regulate cutaneous blood and lymphatic vessel function as well as modulation of sweat gland, apocrine gland, and hair follicle activities. These autonomic neuronal effects are mediated by release of acetylcholine with activation of muscarinic and nicotinergic receptors on these cutaneous target tissues [35,106]. Acetylcholine has been reported to regulate keratinocyte proliferation, adhesion, migration, and differentiation as well as modulating the production of cellular cytokine and chemokine production [35,36,125]. In addition to acetylcholine, cutaneous autonomic nerves have been reported to release CGRP, VIP, and galanin, which may have immunosuppressive activities in the skin [53,67].

There is increasing evidence that the skin is both a target and source of neuroendocrine mediators. Different types of physical and psychological stress can activate the hypothalamus/pituitary system to release neurohormones such as CRH, ACTH, αMSH, and other specialized bioactive peptides that can act in an endocrine fashion on distal tissues such as the skin to modulate a variety of biological responses, including inflammation [105]. Adrenal glands can likewise produce catecholamines and cortisol in response to stressful stimuli, which are capable of modulating inflammatory responses. In recent years, it has become apparent that neurohormones such as CRH and αMSH in addition to opioids, endocannabinoids, endothelin, and TRP activating agents can be produced by various cutaneous cells [100,109]. An increasing number of studies indicate that this diverse group of mediators can act in an autocrine, paracrine, or endocrine manner to modulate inflammatory and neuronal responses in the skin.

The cutaneous neurological system is believed to play an important role in the pathogenesis of a variety of skin disorders such as urticaria, psoriasis, atopic dermatitis, contact dermatitis, prurigo, and wound healing. This concept is supported by experimental evidence that demonstrated altered expression of neuropeptides, neuropeptide receptors, and nerves, as well as symptoms such as itching, burning, and pain associated with these cutaneous disorders [12,122,123].

Although this discussion has focused primarily on the role of the neurological system on cutaneous innate immune responses, there is much evidence that many of these same neuropeptides and neurohormones can directly modulate T and B cell adaptive immune system responses [73]. In addition, neuromodulation of cutaneous innate immune responses can also initiate the activation of the antigen-specific adaptive immune system [124].

9.3 Mammalian TLRs: Structure, Expression, Localization, Ligands, Regulation, and Function

Our current understanding of the importance of PRRs in the innate immune system was greatly accelerated by the identification and characterization of a cell surface molecule in *Drosophila* called Toll. It was demonstrated that Toll has an essential role in the establishment of dorso-ventral polarity in the development of *Drosophila* as well as in the initiation of protective responses against pathogenic fungal infections [43]. Subsequently, a mammalian homologue of *Drosophila* Toll was identified [87] and designated Toll-like receptor (TLR). It was soon appreciated that TLR family members may have a critical role in mammalian immune responses to a wide range of infectious agents. TLR biology has become one of the most prominent areas of immunology research.

9.3.1 TLR Structure

To date, researchers have identified at least 11 members of the TLR family in humans and mice that appear to be derived from a common ancestral gene and share characteristic extracellular and cytoplasmic domains [58,140] (Fig. 9.1). The identified TLR proteins have been designated TLR1 through TLR11 and have been localized to specific chromosomes in man and mice [128]. TLRs belong to a family of type I transmembrane receptors [14] that can be divided into five subfamilies based on similarities in genomic structure and amino acid sequence [128]. Members of the TLR family share certain structural features such as several amino-terminal leucine rich repeats (LRRs) that appear to be important for both ligand binding and TLR dimerization after ligand binding [14], whereas the carboxy-terminal domains of TLRs contain a highly conserved region that is similar to the cytoplasmic portion of the interleukin 1 receptor (IL-1R) and is referred to as the Toll/IL-1R homologous region (TIR region) [86]. The TIR domain mediates protein–protein interactions with specific intracellular signal transduction proteins, which initiates the transcription of multiple cellular proinflammatory genes [14].

9.3.2 TLR Expression and Localization

TLRs are expressed by a variety of mammalian leukocytes and non-leukocytes. TLR expressing leukocytes include cells that often are the first line of defense

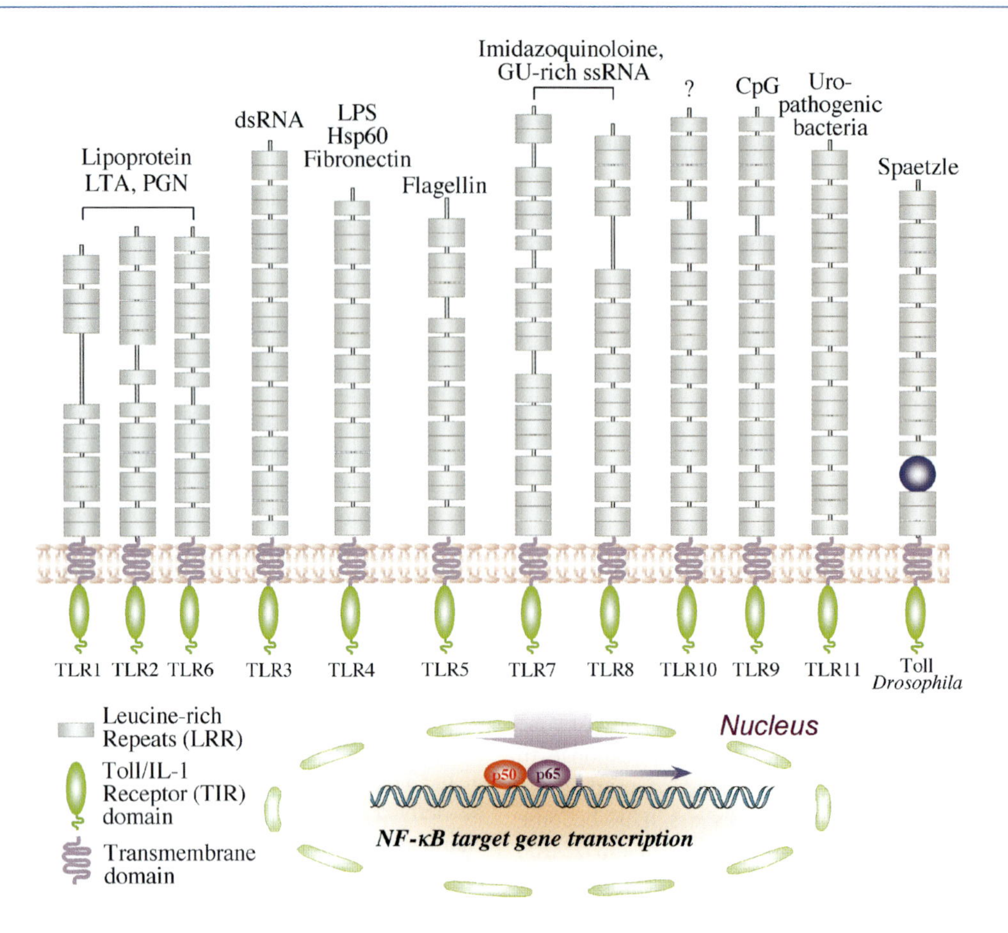

Fig. 9.1 Mammalian TLR structure and representative ligands. Toll receptors are characterized by both an extracellular amino-terminal leucine-rich repeat (LRR) domain that participates in ligand binding, and an intracellular carboxy-terminal domain termed Toll/interleukin-1 receptor (TIR) that initiates intracellular signal transduction. Representative ligands of different TLRs are indicated

against pathogens such as macrophages, dendritic cells, neutrophils, and mast cells [128]. Monocytes and macrophages have been shown to express most types of TLRs with the exception of TLR3 [95]. Differential expression of TLRs in distinct populations of dendritic cells has been observed [60]. For example, myeloid dendritic cells have been reported to express TLR1, TLR2, TLR4, TLR5, and TLR8, while plasmacytoid-type dendritic cells express predominantly TLR7 and TLR9 [55,59,60,71]. Recent studies indicate that the differential expression of TLRs by these dendritic cell subsets may significantly influence and direct the type of host immunological response to different foreign antigens or pathogens [48,88,92,130]. Mast cells have been reported to express TLR2, TLR4, TLR6, and TLR8, but not TLR5 [84,126]. Although the role of mast cell TLR in host immunity remains to be determined, it was observed that TLR4 mutated mice have defective mast cell proinflammatory responses to gram (−) bacteria derived LPS [84]. Additionally, it was found that mast cell accumulation in the skin in response to gram (+) derived PGN is TLR2 dependent [126]. It has been proposed that mast cell TLRs may have an important role in mediating inflammatory responses observed in atopic disease [126].

Although the majority of previous studies have focused on the biology of leukocyte TLRs, there is much evidence demonstrating the importance of TLRs in mediating local innate immune responses in non-leukocytes. The expression and function of TLRs in non-leukocytes may differ from TLR responses in leukocytes [30,117,118]. Recent work by our group and other laboratories have demonstrated that functional TLRs are present on corneal epithelial cells as well as human and murine keratinocytes [15,101,114–120]. In other studies, the basolateral portion of intestinal epithelial cells was found to express TLR5 that responds to bacterial flagellin [32], whereas intracellular intestinal epithelial TLR4 is activated by LPS. These TLRs are normally sequestered from the gut luminal surface to avoid activation by intestinal bacterial flora and are activated only if intestinal bacteria invade the outer gut luminal epithelial barrier [1,51,96]. Interestingly, recent studies also indicate that some interactions between gut epithelial TLRs and normal commensal bacteria may actually have some beneficial effects in the maintenance of intestinal epithelial homeostasis [104]. TLRs are also expressed on respiratory tract epithelium where they can respond to a variety of inhaled pathogens to initiate both protective and pathological immune responses [16]. Similar to gut epithelium, we observed that TLRs are not expressed on the outer surface of the cornea epithelium and the epidermis, but are expressed on epithelial cells located in the mid or basal portion of these tissues [117,118]. As a result, pathogen engagement and activation of cornea epithelial cells and keratinocyte TLR would occur only after epithelial compromise or injury.

9.3.3 TLR Ligands

The cellular recognition of a vast number of pathogens by a limited number of TLRs is possible because these microbes share similar essential structural components or combinations of components that are not expressed by mammalian cells. A number of TLR ligands have been identified by studies using specific microbial components as well as TLR knockout mice [61]. While most PRR ligands are derived from different microbes, there is also evidence that certain endogenous cellular proteins such as heat shock protein 70 (HSP70) and fibrinogen may also activate TLRs [112,131]. These endogenous cellular TLR ligands are released after cell injury and may act as a local "danger signal" that triggers a rapid regional host immune response to counter the agent causing the injury or infection. The identification and characterization of these endogenous cellular TLR ligands may lead to important new insights in our understanding of certain chronic inflammatory states, autoimmune diseases, and the development of novel vaccines. Recent studies have demonstrated the complexity of TLR biology. TLR family members are capable of initiating both distinct and overlapping cellular biological responses that are dependent on a number of factors, including the specific TLR ligand, TLR family member, intracellular signaling response pathway, and other locally released inflammatory mediators.

TLRs recognize unique PAMPs derived from a variety of pathogens. For example, TLR3 recognizes double-stranded RNA derived from intracellular viruses [8], whereas TLR5 recognizes flagellin, a 55 kDa monomer obtained from the flagella on the outer membrane of a number of gram negative bacteria [32,44]. In contrast to other TLRs, the first TLR7 ligands identified were not microbial-derived agents but the pharmaceutical compounds imiquimod and resiquimod [33,47]. These low molecular weight compounds are believed to mediate their therapeutic effects by inducing the production of cellular IL-12 and interferon gamma. Currently imiquimod is used topically on the skin to treat an expanding list of cutaneous diseases, including warts, dysplastic skin lesions, and even early skin cancers [81]. More recently, it was reported that TLR7 and TLR8 also respond to single stranded viral RNA [27,46,80]. Other studies found that TLR9 specifically responds to unmethylated CpG DNA motifs that are found in bacterial and viral genomes but not in vertebrate genetic material [17]. For a number of years, unmethylated CpG dinucleotides have been utilized as adjuvants for vaccines due to potent immunostimulatory activities of these compounds [65,70,139]. To date, TLR10 ligands have not been characterized although TLR10 is known to be a functional receptor that can mediate TLR signaling responses [42]. TLR11 was identified on murine uroepithelium, where it may function to trigger host defenses against urinary pathogens [140]. Interestingly, in contrast to rodents, humans have been found to express a nonfunctional form of uroepithelium TLR11, which may account for increased uropathogen susceptibility of humans compared to rodents [140].

9.3.4 TLR Signaling

Recent studies have elucidated many of the details of the TLR signaling pathway [4,6,69,97] (Fig. 9.2, Table 9.1). TLRs can form homodimers or heterodimers with other TLRs, which then bind by the cytoplasmic TIR region to corresponding homodimers of MyD88. This interaction initiates the recruitment of IRAK4 and IRAK1 to the TLR complex. IRAK4 subsequently phosphorylates IRAK1, which results in the recruitment of TRAF6. The IRAK1-IRAK4-TRAF6 complex then dissociates and acts upon another membrane protein complex consisting of TAK1, TAB1, and TAB2. This results in the formation of TAK1-TAB1-TAB2-TRAF6, which subsequently activates IKK. IKK then phosphorylates IkB, which is then degraded, allowing NF-κB translocation to the nucleus [3].

An alternative TLR3 and TLR4 signaling pathway independent of MyD88 has also recently been described [52]. Two additional TLR adaptors called TIR-adaptor protein (TIRAP or MAL) and TIR-domain

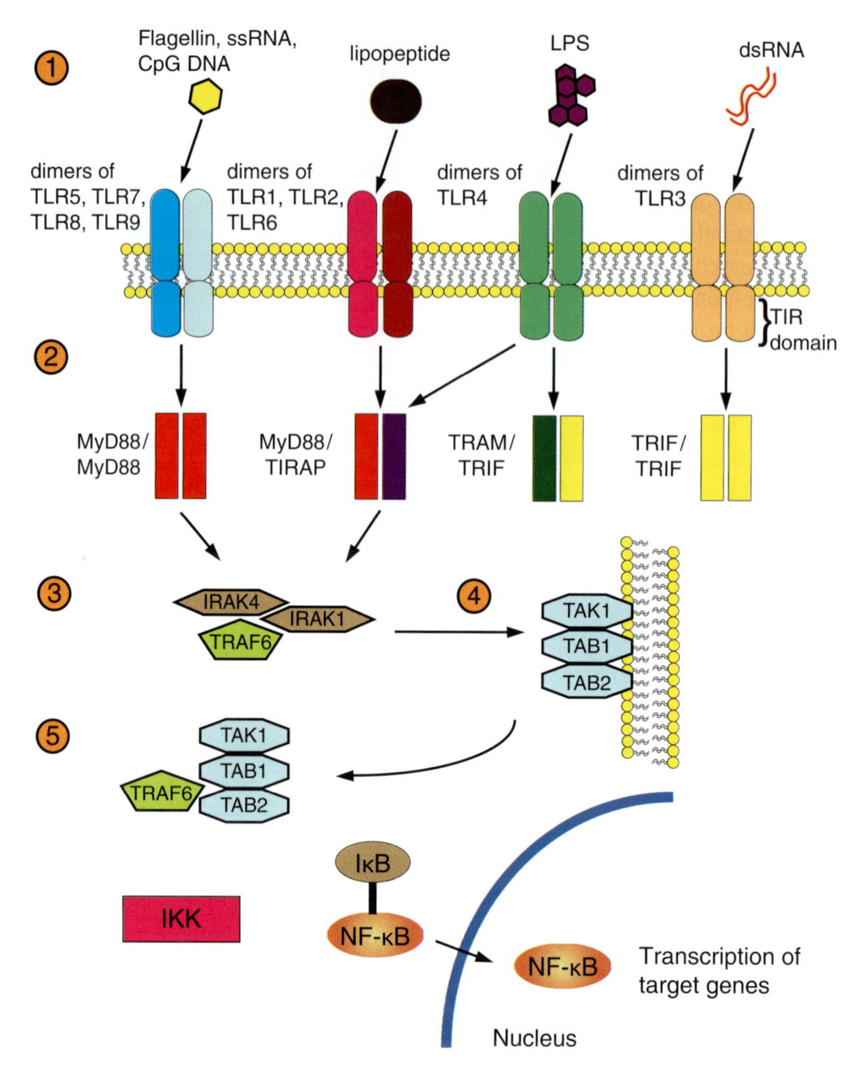

Fig. 9.2 TLR signaling response overview. (**1**) Ligands bind and activate TLRs, which form homodimers or heterodimers with other TLRs. (**2**) TIR adaptor proteins such as MyD88, TIRAP, TRAM, and TRIF form complexes with activated TLRs to initiate cell signaling. (**3**) IRAK4 recruits and phosphorylates IRAK1 and associates with TRAF6. (**4**) IRAK1-IRAK4-TRAF6 dissociates and interacts with TAK1-TAB1-TAB2. (**5**) A TAK1-TAB1-TAB2-TRAF6 complex phosphorylates IKK, which allows cytoplasmic NF-κB translocation to the nucleus and induction of transcription of multiple proinflammatory genes

adaptor inducing IFN-beta (TRIF or TICAM-1) have been identified and linked to the initial intracellular signal responses of these particular Toll-like receptors [52,137]. It appears that TLR3 signaling responses require TRIF, while TLR4 mediated responses involve both TRIF and a fourth adaptor molecule called TRIF-related adapter molecule (TRAM or TICAM-2) [138]. Alternatively, TLR4 is also able to signal through an interaction of MyD88 with the TIRAP adaptor protein [2,136].

Another TLR adaptor molecule has recently been identified and is termed SARM (sterile alpha and armadillo motif). SARM functions as a negative regulator of TRIF and therefore suppresses the TLR3 and TLR4 signaling pathways [24].

In addition to SARM, several other negative intracellular TLR regulators have been characterized [77] (Fig. 9.3). One such protein termed MyD88s, which is a shorter form of MyD88, forms inactive heterodimers with MyD88, which prevents MyD88 homodimerization and therefore blocks TLR downstream signaling [23]. Another cytosolic protein, IRAK-M, also serves as a TLR signaling inhibitor by binding to the MyD88 complex and preventing subsequent downstream events.

Table 9.1 Definitions of proteins involved in TLR Signaling

Name	Full Name	Function
FADD	FAS-associated via death domain	Activates apoptosis
IKK	IkappaB kinase	Phosphorylates IkappaB
IRAK	Interleukin-1-receptor-associated-kinase	Signaling downstream of adaptors
IRAK-M	Interleukin-1-receptor-associated-kinase-M	Negative intracellular regulator
IκB	IkappaB	Prevent NF-κB translocation
MyD88	Myeloid differentiation primary-response protein 88	Adaptor Protein, TLR1, 2, 4–10
MyD88s	Short form of MyD88	Negative intracellular regulator
NF-κB	Nuclear factor KappaB	Proinflammatory transcription
SARM	Sterile alpha and armadillo motif	Adaptor protein, suppress TRIF
Smad6	Sma- and Mad-related protein 6	Negative intracellular regulator
TAB	TAK1 binding protein	Signaling downstream of adaptors
TAK	Transforming growth factor β activated kinase	Signaling downstream of adaptors
TGF-β1	Transforming growth factor-beta 1	Anti-inflammatory cascade
TIR	Toll/IL-1R homologous region	Binds TLRs and adaptor proteins
TIRAP	TIR-adaptor protein	Adaptor protein, TLR 1, 2, 4, 6
TOLLIP	Toll-interacting protein	Negative intracellular regulator
TRAF	TNF receptor-associated factor	Signaling downstream of adaptors
TRAM	Toll-receptor-associated molecule	Adaptor protein, TLR4
TRIAD3A	[TLR binding RING finger protein – no full name]	Negative intracellular regulator
TRIF	TIR domain-containing adaptor induction IFN- β	Adaptor protein, TLR 3, 4

Mice deficient in IRAK-M demonstrate increased inflammation in response to LPS compared to wild-type animals [66]. Another protein termed TOLLIP (Toll-interacting protein) has been reported to decrease both TLR2 and TLR4-induced NF-κB activation, possibly by inhibition of IRAK1 function [141]. Likewise Smad6, a protein in the TGF-β1-BMP anti-inflammatory pathway, has recently been demonstrated to also abrogate IRAK1 signaling and therefore inhibit TLR responses [26]. Additional mechanisms exist to modulate TLR responses. For example, TRIAD3A is a protein that binds to the cytoplasmic domain of TLR9 and TLR4 and promotes their ubiquitylation and degradation. TLR-induced apoptosis can likewise inhibit TLR mediated responses. This occurs by the interaction of MyD88 with FADD (FAS associated death domain), which initiates activation of caspase-8 induced apoptosis [9,10]. It is unclear why some TLR ligands trigger apoptosis, whereas others upregulate TLR expression. Soluble TLR-like proteins have been identified in the serum of mice and humans. These soluble Toll-like receptor proteins can bind and sequester TLR ligands extracellularly and therefore prevent their interaction with cellular TLR [56,74]. Soluble isoforms of TLR2 and TLR4 are currently under investigation as anti-inflammtory agents. Therefore, as in other inflammatory pathways, TLR-induced responses are tightly regulated to prevent unwanted inflammation by prolonged activation, which could threaten the health of the host organism.

9.4 Microbial Modulation of TLR Expression and Function

Commensal organisms may have a beneficial role in host epithelial immune responses through modulation of cellular PRR function [63,129]. For example, normal bacterial flora activates PRRs to produce low levels of antimicrobial peptides and other proinflammatory proteins that are believed to inhibit colonization or infection by pathogenic bacteria. Recently it has been appreciated that certain pathogens are capable of producing factors that dampen cellular TLR inflammatory responses and therefore bypass normal host immune recognition and activation [38]. For example, vaccinia viral derived proteins such as A46R have been identified, which inhibit TLR signaling proteins MyD88 and TRIF, thus suppressing TLR and IL-1R induced NF-kB activation [19,121]. Another vaccinia protein called N1L antagonizes IKK activation and therefore blocks NF-kB mediated responses [28]. Likewise, the hepatitis C virus (HCV) expresses a protease that cleaves the intracellular TRIF protein, which, in turn, inhibits I: C-mediated TLR3-induced INF responses [76]. A number of RNA virus-derived proteins have also been reported to inhibit the RIG RNA helicase PRR system [21,91]. DNA viruses such as HSV have evolved mechanisms to block TLR mediated IRF and type I interferon production [89]. Bacteria and fungi have also developed ways to inhibit cellular TLR immune responses. For

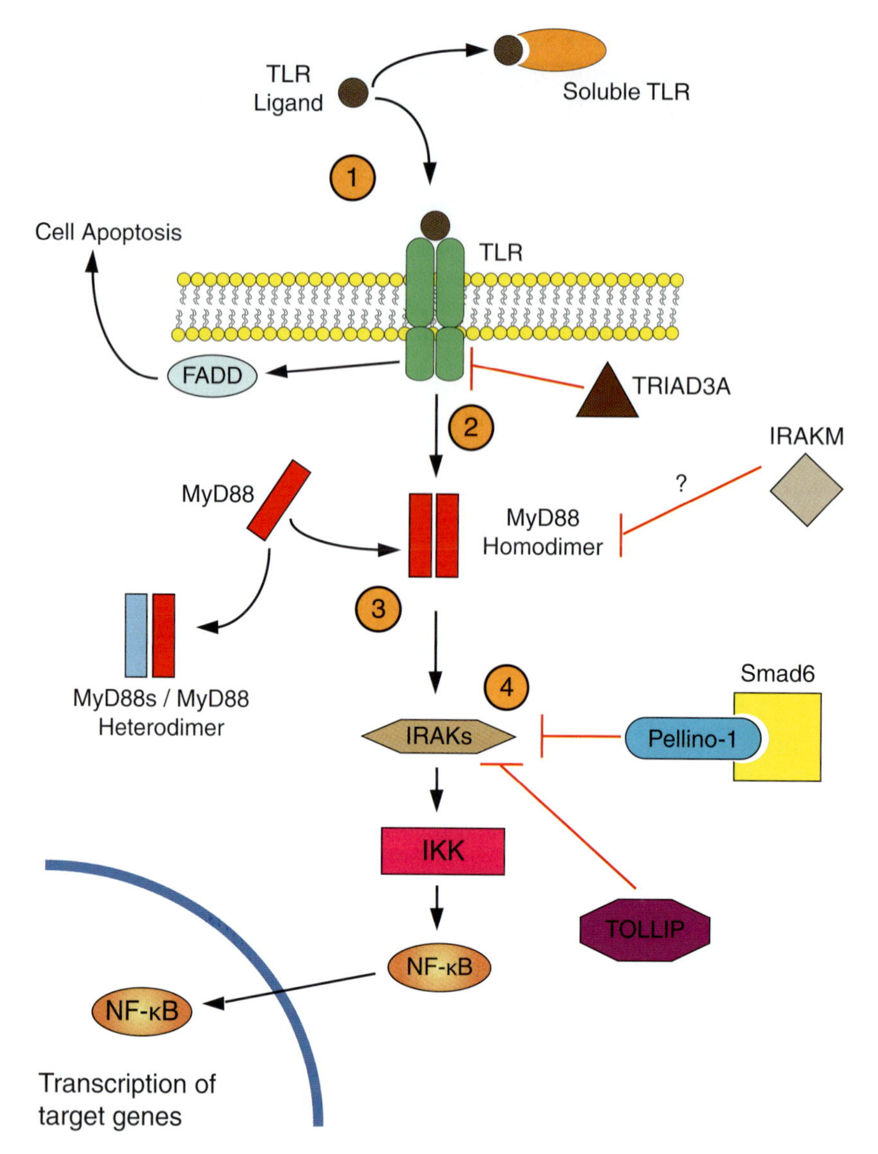

Fig. 9.3 Negative regulators of TLR signaling. There are multiple TLR signaling response inhibitors. (**1**) Soluble TLRs act as decoys to divert TLR ligands from cellular TLRs. (**2**) TRIAD3A binds to the cytoplasmic domain of TLR9 and TLR4 to inhibit signaling, whereas FADD can bind the death domain of TLR to initiate cellular apoptosis. (**3**) MyD88s and possibly IRAK-M bind to MyD88 and inhibit TLR signaling responses. (**4**) Smad6 and TOLLIP act as TLR signaling inhibitors by blocking IRAK kinase function

example, bacteria such as *H. pylori*, *P. gingivalis*, and *L. pneumonia* express modified LPS molecules that are less reactive to TLR compared with LPS molecules expressed by other gram (−) bacteria [49,113]. In addition, *Mycobacterium tuberculosis* has been found to suppress leukocyte INF-γ response by producing a TLR2-suppressing lipoprotein that allows persistent infection with this pathogen [99], whereas Yersinia bacteria has been reported to induce the production of immunosuppressive IL-10 by TLR2 activation [45].

Cellular TLR expression can also be upregulated in response to various microbial derived antigens. Exposure of macrophages to LPS initially leads to increased cell surface TLR2 expression, whereas repeated LPS exposure can result in desensitization of cells to the effects of LPS [78,83]. Likewise, we have recently observed that short exposures to LPS induces increased TLR4 expression in both cultured human keratinocytes as well as corneal epithelial cells [117,118]. *Mycobacterium avium* [134] and *Hemophilus influenza* have also been reported to increase TLR2 mRNA expression in leukocytes [110,111]. Like bacteria, certain viruses such as influenza A and Sendai virus were reported to up-regulate TLR1, TLR2, TLR3, and TLR7 expression in macrophages [80,93]. Finally, proinflammatory cytokines such as INF-γ, IL-2, IL-15, and TNF-α have been demonstrated to increase both monocyte and macrophage TLR expression [18,83].

9.5 Cutaneous TLRs

TLRs have been proposed to play an important role in the pathogenesis of a number of cutaneous diseases [64,85]. The epidermis is one of the primary sites of pathogen–host interaction. Keratinocytes are the predominant cell type in the epidermis and act not only as an effective physical barrier to harmful environmental agents, but when activated can initiate a surprising range of innate immune responses. In response to foreign antigens or pathogens, keratinocytes are capable of producing nitric oxide, defensins, chemokines, and cytokines [13,31,39,40,135]. Activated keratinocytes can also express cellular immunoreactive proteins such as ICAM-1 and class II MHC molecules. Until recently little was known about how keratinocytes recognize and respond to various pathogens. This concept changed dramatically after the identification of specific keratinocyte PRRs such as TLR. We have reported that human corneal epithelial cells, human keratinocytes, and more recently, murine keratinocytes expressed functional TLRs and TLR-associated molecules [114–118,120].

Other investigators have also detected functional keratinocyte TLRs. Kawai et al. found that cultured normal human keratinocytes (NHKC) constitutively express functional TLR2 and can be induced to express TLR4 [62]. Both TLR2 and TLR4 were detected in the epidermis of freshly isolated human skin samples. Activation of human keratinocyte TLRs was reported to result in the production of beta defensins and IL-8. Pretreatment of human keratinocytes with TLR2 ligands was found to induce hyporesponsiveness to TLR2 and TLR4 ligands. Pivarcsi et al. also reported that cultured human keratinocytes express TLR2, TLR4, and MyD88 [101]. They found that human keratinocyte TLR2 and TLR4 expression was augmented by LPS and interferon-gamma. In another study, Baker et al reported that human keratinocytes constitutively expressed mRNA for TLR1, TLR2, and TLR5, whereas TLR3 and TLR4 were barely detectable [15]. Mempel et a. also recently reported that *Staphylococcus aureus* was able to active human keratinocytes in a TLR2-specific fashion [90]. Exposure to *Candida albicans* or heat-killed *Mycobacterium tuberculosis* was found to activate human keratinocyte NF-kB [101]. In another recent report, Lebre et al. found that supernatants from TLR stimulated human keratinocytes induced the maturation of human dendritic cells, which promoted Th1 responses, indicating that keratinocytes may play an important role in directing T cell responses in the skin [75]. PRRs are believed to have a significant role in the host response and pathogenesis of cutaneous disorders such as acne, atopic dermatitis, candidiasis, psoriasis, syphilis, and leprosy.

9.6 Neuroregulation of Toll-Like Receptors

9.6.1 Neuronal TLR Expression and Function

TLRs have been identified in different types of neuronal cells including human glial cells, which have an important supportive role for the function of neurons in both the PNS and CNS [22]. Astrocytes and microglia in the CNS and Schwann cells in the PNS are considered subtypes of glial cells. TLRs and downstream adaptor proteins such as MyD88, TIRAP, and TRIF have been recently identified in human astrocytes [29].

Several reports indicate that primary CNS neurons also express TLR. For example, TLR3 expression was identified in a human neuron cell line (NT2-N) when infected with the herpes simplex virus [102]. Furthermore, Perkinje cells were found to express TLR3 in stroke and Alzheimer's disease patients [57]. In another study, human neurons were reported to express TLR3 and release inflammatory cytokines, including TNFα and IL-6 in response to TLR3 activation by double-stranded RNA [72]. Finally, a single recent report indicated that TLR4 could be detected in oral trigeminal neurons [133]. Therefore, these initial studies support the concept that both CNS and PNS may express TLRs.

TLR may also play an important role in some inflammatory and pathological disorders of the CNS. Mice deficient in the MyD88 have an impaired CNS inflammatory response to infection with *Streptococcus pneumoniae* [68]. Another study indicated that *Mycobacterium leprae* activation of TLR3 induced apoptosis of Schwann cells [98]. In contrast, MyD88 and TLR9 deficient mice were reported to be resistant to experimentally induced autoimmune encephalitis

(EAE) [103]. In another study, TLR4 deficient mice had decreased brain clearance of a protein implicated in the pathogenesis of Alzheimer's disease [127]. Although suggestive, these few animal studies demonstrate the current lack of information regarding the biological role of TLR in the CNS and PNS.

9.6.2 Neuropeptide Control of TLR Expression and Function

To date only a few studies have examined neuroregulation of TLR expression and function. One study reported that VIP released from intestinal neurons down-regulated TLR2 and TLR4 expression in intestinal epithelial cells, which was proposed to be secondary to VIP inhibition of IL1-β or IFN-γ production [34]. Another study indicated that VIP reduced human rheumatoid fibroblast TLR4 expression, although the mechanism of this effect was not known [37]. A recent paper likewise reported inhibition of TLR function in murine dendritic cells by CGRP, which was believed to be mediated by a protein called inducible cAMP early repressor (ICER), which participates in the cAMP/protein kinase A cell signaling pathway [41].

Our laboratory has recently initiated studies to determine if cutaneous neuropeptides can directly modulate the expression and function of TLR in keratinocytes. We observed that α-MSH was capable of inhibiting the expression and function of TLR2 in human keratinocytes [79]. In this study, we found that *Staphoylococcus aureus*-derived lipoteichoic acid (LTA) augmented both TLR2 and IL-8 expression in human keratinocytes. Pretreatment of these cells with α-MSH effectively inhibited this response to LTA, which was reversed by pretreatment of cells with the Agouti protein MC-1 receptor antagonist prior to the addition of α-MSH. In another study, we demonstrated that substance P can upregulate TLR expression and function in both murine and human keratinocytes [25]. This effect was inhibited by pretreatment of cells with neurokinin-1 (NK-1) or neurokinin-2 (NK-2) SP receptor antagonists. These initial studies therefore demonstrate another potential mechanism by which the neurological system may modulate immune responses in the skin (Fig. 9.4).

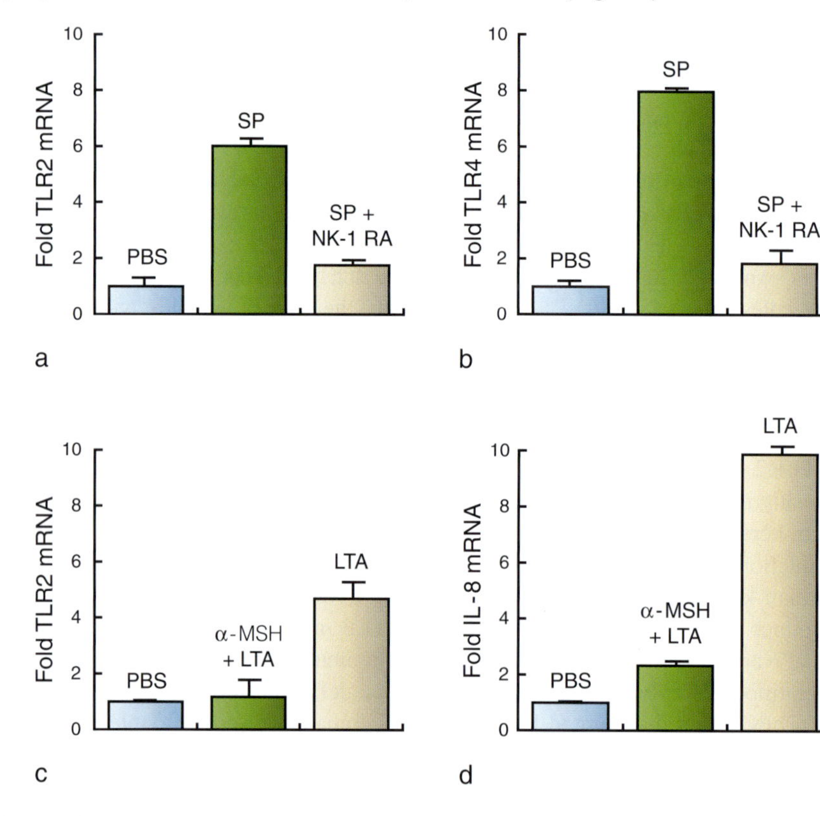

Fig. 9.4 TLR2 (a) and TLR4 (b) mRNA expression was measured in keratinocytes by quantitative RT-PCR 3 hours after the addition of 50 nM SP with or without pre-treating cells with a SP receptor antagonist (NK-1 RA) 2 hours prior to SP. SP induced an increase in the TLR mRNA expression in NHK. This response could be inhibited by pre-treating cells with 1 μM NK-1 RA. Similarly, TLR2 (c) and IL8 (d) mRNA expression was measured in keratinocytes 3 hours after the addition of 10 mg/ml LTA with or without pre-treating cells with α-MSH 2 hours prior to LTA. *Staphylococcus aureus* derived LTA induced an increase in the TLR2 or IL8 expression in NHK. This response could be inhibited by pre-treating cells with 10 nM α-MSH

Summary for the Clinician

In spite of the great excitement in recent years about the potential of new TLR-based therapies, we are only beginning to understand their mechanism of action and biological activities, including a range of beneficial and potentially harmful effects. In some situations, TLR agonists may be helpful as immune stimulants for the treatment of certain neoplastic, infectious, or immunodeficient conditions, whereas in other situations, TLR antagonists may prove to be useful as anti-inflammatory agents for inhibiting the detrimental effects of systemic infections. Since TLR-mediated innate responses are triggered by essential pathogen structural components not found in eukaryotes, pathogens rarely develop selective resistance to TLR-initiated immune activities. Therefore, PRRs such as TLR may be exciting new targets for the development of novel therapies to treat a wide range of human diseases.

In this chapter we highlighted some important interactions between the nervous system and cutaneous immunity. The skin – in conjunction with the cutaneous nervous system – serves a unique role in responding to most environmental noxious agents before they cause significant injury. Our recent studies also indicate that the neurological system can uniquely modulate cutaneous immune responses by modulating keratinocyte TLR expression and function.

References

1. Abreu MT, et al (2001) Decreased expression of Toll-like receptor-4 and MD-2 correlates with intestinal epithelial cell protection against dysregulated proinflammatory gene expression in response to bacterial lipopolysaccharide. J Immunol 167(3):1609–1616
2. Adachi O, et al (1998) Targeted disruption of the MyD88 gene results in loss of IL-1- and IL-18-mediated function. Immunity 9(1):143–150
3. Akira S (2003) Toll-like receptor signaling. J Biol Chem 278(40):38105–38108
4. Akira S, Takeda K (2004) Toll-like receptor signalling. Nat Rev Immunol 4(7):499–511
5. Akira S, Takeda K, Kaisho T (2001) Toll-like receptors: critical proteins linking innate and acquired immunity. Nat Immunol 2(8):675–680
6. Akira S, Yamamoto M, Takeda K (2003) Role of adapters in Toll-like receptor signalling. Biochem Soc Trans 31(Pt 3) : 637–642
7. Albertin G, et al (2003) Human skin keratinocytes and fibroblasts express adrenomedullin and its receptors, and adrenomedullin enhances their growth in vitro by stimulating proliferation and inhibiting apoptosis. Int J Mol Med 11(5):635–639
8. Alexopoulou L, et al (2001) Recognition of double-stranded RNA and activation of NF-kappaB by Toll-like receptor 3. Nature 413(6857):732–738
9. Aliprantis AO, et al (1999) Cell activation and apoptosis by bacterial lipoproteins through toll-like receptor-2. Science 285(5428):736–739
10. Aliprantis AO, et al (2000) The apoptotic signaling pathway activated by Toll-like receptor-2. EMBO J 19(13): 3325–3336
11. Ansel JC, et al (1993) Substance P selectively activates TNF-alpha gene expression in murine mast cells. J Immunol 150(10):4478–4485
12. Ansel JC, et al (1997) Interactions of the skin and nervous system. J Investig Dermatol Symp Proc 2(1):23–26
13. Arany I, et al (1996) Regulation of inducible nitric oxide synthase mRNA levels by differentiation and cytokines in human keratinocytes. Biochem Biophys Res Commun 220(3):618–622
14. Armant MA, Fenton MJ (2002) Toll-like receptors: a family of pattern-recognition receptors in mammals. Genome Biol 3(8):REVIEWS3011
15. Baker BS, et al (2003) Normal keratinocytes express Toll-like receptors (TLRs) 1, 2 and 5: modulation of TLR expression in chronic plaque psoriasis. Br J Dermatol 148(4):670–679
16. Basu S, Fenton MJ (2004) Toll-like receptors: function and roles in lung disease. Am J Physiol Lung Cell Mol Physiol 286(5):L887–L892
17. Bauer S, et al (2001) Human TLR9 confers responsiveness to bacterial DNA via species-specific CpG motif recognition. Proc Natl Acad Sci U S A 98(16):9237–9242
18. Bosisio D, et al (2002) Stimulation of toll-like receptor 4 expression in human mononuclear phagocytes by interferon-gamma: a molecular basis for priming and synergism with bacterial lipopolysaccharide. Blood 99(9):3427–3431
19. Bowie A, et al (2000) A46R and A52R from vaccinia virus are antagonists of host IL-1 and toll-like receptor signaling. Proc Natl Acad Sci U S A 97(18):10162–10167
20. Brain SD, Moore PK (1999) Pain and neurogenic inflammation. In: Parnham MJ (ed) Progress in Inflammation Research, Birkhauser, Basel
21. Breiman A, et al (2005) Inhibition of RIG-I-dependent signaling to the interferon pathway during hepatitis C virus expression and restoration of signaling by IKKepsilon. J Virol 79(7):3969–3978
22. Bsibsi M, et al (2002) Broad expression of Toll-like receptors in the human central nervous system. J Neuropathol Exp Neurol 61(11):1013–1021
23. Burns K, et al (2003) Inhibition of interleukin 1 receptor/Toll-like receptor signaling through the alternatively spliced, short form of MyD88 is due to its failure to recruit IRAK-4. J Exp Med 197(2):263–268
24. Carty M, et al (2006) The human adaptor SARM negatively regulates adaptor protein TRIF-dependent Toll-like receptor signaling. Nat Immunol 7(10):1074–1081
25. Chen H, et al (2006) Modulation of Toll-like receptor expression in murine and human keratinocytes by substance P: another link between the seurosensory and innate immune systems in the skin. J Invest Dermatol 127:124
26. Choi KC, et al (2006) Smad6 negatively regulates interleukin 1-receptor-Toll-like receptor signaling through direct interaction with the adaptor Pellino-1. Nat Immunol 7(10):1057–1065

27. Diebold SS, et al (2004) Innate antiviral responses by means of TLR7-mediated recognition of single-stranded RNA. Science 303(5663):1529–1531
28. DiPerna G, et al (2004) Poxvirus protein N1L targets the I-kappaB kinase complex, inhibits signaling to NF-kappaB by the tumor necrosis factor superfamily of receptors, and inhibits NF-kappaB and IRF3 signaling by toll-like receptors. J Biol Chem 279(35):36570–36578
29. Farina C, et al (2005) Preferential expression and function of Toll-like receptor 3 in human astrocytes. J Neuroimmunol 159(1–2):12–19
30. Faure E, et al (2000) Bacterial lipopolysaccharide activates NF-kappaB through toll-like receptor 4 (TLR-4) in cultured human dermal endothelial cells. Differential expression of TLR-4 and TLR-2 in endothelial cells. J Biol Chem 275(15):11058–11063
31. Frohm M, et al (1997) The expression of the gene coding for the antibacterial peptide LL-37 is induced in human keratinocytes during inflammatory disorders. J Biol Chem 272(24):15258–15263
32. Gewirtz AT, et al (2001) Cutting edge: bacterial flagellin activates basolaterally expressed TLR5 to induce epithelial proinflammatory gene expression. J Immunol 167(4):1882–1885
33. Gibson SJ, et al (2002) Plasmacytoid dendritic cells produce cytokines and mature in response to the TLR7 agonists, imiquimod and resiquimod. Cell Immunol 218(1–2):74–86
34. Gomariz RP, et al (2005) Time-course expression of Toll-like receptors 2 and 4 in inflammatory bowel disease and homeostatic effect of VIP. J Leukoc Biol 78(2):491–502
35. Grando SA (1997) Biological functions of keratinocyte cholinergic receptors. J Investig Dermatol Symp Proc 2(1):41–48
36. Grando SA, et al (1993) Human keratinocytes synthesize, secrete, and degrade acetylcholine. J Invest Dermatol 101(1):32–36
37. Gutierrez-Canas I, et al (2006) VIP down-regulates TLR4 expression and TLR4-mediated chemokine production in human rheumatoid synovial fibroblasts. Rheumatology (Oxford) 45(5):527–532
38. Haga IR, Bowie AG (2005) Evasion of innate immunity by vaccinia virus. Parasitology 130(Suppl):S11–S25
39. Harder J, et al (1997) A peptide antibiotic from human skin. Nature 387(6636):861
40. Harder J, et al (2001) Isolation and characterization of human beta -defensin-3, a novel human inducible peptide antibiotic. J Biol Chem 276(8):5707–5713
41. Harzenetter MD, et al (2007) Negative regulation of TLR responses by the neuropeptide CGRP is mediated by the transcriptional repressor ICER. J Immunol 179(1):607–615
42. Hasan U, et al (2005) Human TLR10 is a functional receptor, expressed by B cells and plasmacytoid dendritic cells, which activates gene transcription through MyD88. J Immunol 174(5):2942–2950
43. Hashimoto C, Hudson KL, Anderson KV (1988) The Toll gene of Drosophila, required for dorsal-ventral embryonic polarity, appears to encode a transmembrane protein. Cell 52(2):269–279
44. Hayashi F, et al (2001) The innate immune response to bacterial flagellin is mediated by Toll-like receptor 5. Nature 410(6832):1099–1103
45. Heesemann J, Sing A, Trulzsch K (2006) Yersinia's stratagem: targeting innate and adaptive immune defense. Curr Opin Microbiol 9(1):55–61
46. Heil F, et al (2004) Species-specific recognition of single-stranded RNA via toll-like receptor 7 and 8. Science 303(5663):1526–1529
47. Hemmi H, et al (2002) Small anti-viral compounds activate immune cells via the TLR7 MyD88-dependent signaling pathway. Nat Immunol 3(2):196–200
48. Hertz CJ, et al (2001) Microbial lipopeptides stimulate dendritic cell maturation via Toll-like receptor 2. J Immunol 166(4):2444–2450
49. Hirschfeld M, et al (2001) Signaling by toll-like receptor 2 and 4 agonists results in differential gene expression in murine macrophages. Infect Immun 69(3):1477–1482
50. Hoffmann JA, et al (1999) Phylogenetic perspectives in innate immunity. Science 284(5418):1313–1318
51. Hornef MW, et al (2002) Toll-like receptor 4 resides in the Golgi apparatus and colocalizes with internalized lipopolysaccharide in intestinal epithelial cells. J Exp Med 195(5):559–570
52. Horng T, Barton GM, Medzhitov R (2001) TIRAP: an adapter molecule in the Toll signaling pathway. Nat Immunol 2(9):835–841
53. Hosoi J, et al (1993) Regulation of Langerhans cell function by nerves containing calcitonin gene-related peptide. Nature 363(6425):159–163
54. Inohara N, Nunez G (2001) The NOD: a signaling module that regulates apoptosis and host defense against pathogens. Oncogene 20(44):6473–6481
55. Ito T, et al (2002) Interferon-alpha and interleukin-12 are induced differentially by Toll-like receptor 7 ligands in human blood dendritic cell subsets. J Exp Med 195(11):1507–1512
56. Iwami KI, et al (2000) Cutting edge: naturally occurring soluble form of mouse Toll-like receptor 4 inhibits lipopolysaccharide signaling. J Immunol 165(12):6682–6686
57. Jackson AC, Rossiter JP, Lafon M (2006) Expression of Toll-like receptor 3 in the human cerebellar cortex in rabies, herpes simplex encephalitis, and other neurological diseases. J Neurovirol 12(3):229–234
58. Janeway CA, Jr., Medzhitov R (2002) Innate immune recognition. Annu Rev Immunol 20:197–216
59. Jarrossay D, et al (2001) Specialization and complementarity in microbial molecule recognition by human myeloid and plasmacytoid dendritic cells. Eur J Immunol 31(11):3388–3393
60. Kadowaki N, et al (2001) Subsets of human dendritic cell precursors express different toll-like receptors and respond to different microbial antigens. J Exp Med 194(6):863–869
61. Kaisho T, Akira S (2002) Toll-like receptors as adjuvant receptors. Biochim Biophys Acta 1589(1):1–13
62. Kawai K, et al (2002) Expression of functional Toll-like receptor 2 on human epidermal keratinocytes. J Dermatol Sci 30(3):185–194
63. Kelly D, Conway S, Bacterial modulation of mucosal innate immunity. Mol Immunol 42(8):895–901
64. Kim J, et al (2002) Activation of toll-like receptor 2 in acne triggers inflammatory cytokine responses. J Immunol 169(3):1535–1541

65. Klinman DM, et al (2004) Use of CpG oligodeoxynucleotides as immune adjuvants. Immunol Rev 199:201–216
66. Kobayashi K, et al (2002) IRAK-M is a negative regulator of Toll-like receptor signaling. Cell 110(2):191–202
67. Kodali S, et al (2004) Vasoactive intestinal peptide modulates Langerhans cell immune function. J Immunol 173(10):6082–6088
68. Koedel U, et al (2004) MyD88 is required for mounting a robust host immune response to Streptococcus pneumoniae in the CNS. Brain 127(Pt 6):1437–1445
69. Kopp E, Medzhitov R (2003) Recognition of microbial infection by Toll-like receptors. Curr Opin Immunol 15(4):396–401
70. Krieg AM (2000) The role of CpG motifs in innate immunity. Curr Opin Immunol 12(1):35–43
71. Krug A, et al (2001) Toll-like receptor expression reveals CpG DNA as a unique microbial stimulus for plasmacytoid dendritic cells which synergizes with CD40 ligand to induce high amounts of IL-12. Eur J Immunol 31(10):3026–3037
72. Lafon M, et al (2006) The innate immune facet of brain: human neurons express TLR-3 and sense viral dsRNA. J Mol Neurosci 29(3):185–194
73. Lambrecht BN (2001) Immunologists getting nervous: neuropeptides, dendritic cells and T cell activation. Respir Res 2(3):133–138
74. LeBouder E, et al (2003) Soluble forms of Toll-like receptor (TLR)2 capable of modulating TLR2 signaling are present in human plasma and breast milk. J Immunol 171(12):6680–6689
75. Lebre MC, et al (2003) Double-stranded RNA-exposed human keratinocytes promote Th1 responses by inducing a Type-1 polarized phenotype in dendritic cells: role of keratinocyte-derived tumor necrosis factor alpha, type I interferons, and interleukin-18. J Invest Dermatol 120(6):990–997
76. Li K, et al (2005) Immune evasion by hepatitis C virus NS3/4A protease-mediated cleavage of the Toll-like receptor 3 adaptor protein TRIF. Proc Natl Acad Sci U S A 102(8):2992–2997
77. Liew FY, et al (2005) Negative regulation of toll-like receptor-mediated immune responses. Nat Rev Immunol 5(6):446–458
78. Lin Y, et al (2000) The lipopolysaccharide-activated toll-like receptor (TLR)-4 induces synthesis of the closely related receptor TLR-2 in adipocytes. J Biol Chem 275(32):24255–24263
79. Lu Y, et al (2007) Alpha-melanocyte-stimulating hormone inhibits the expression and function of keratinocyte TLR-2. J Invest Dermatol 127:S133
80. Lund JM, et al (2004) Recognition of single-stranded RNA viruses by Toll-like receptor 7. Proc Natl Acad Sci U S A 101(15):5598–5603
81. Mackenzie-Wood A, et al (2001) Imiquimod 5% cream in the treatment of Bowen's disease. J Am Acad Dermatol 44(3):462–470
82. Marriott I, Bost KL (2001) Expression of authentic substance P receptors in murine and human dendritic cells. J Neuroimmunol 114(1–2):131–141
83. Matsuguchi T, et al (2000) Gene expressions of Toll-like receptor 2, but not Toll-like receptor 4, is induced by LPS and inflammatory cytokines in mouse macrophages. J Immunol 165(10):5767–5772
84. McCurdy JD, Lin TJ, Marshall JS (2001) Toll-like receptor 4-mediated activation of murine mast cells. J Leukoc Biol 70(6):977–984
85. McInturff JE, Modlin RL, Kim J (2005) The role of toll-like receptors in the pathogenesis and treatment of dermatological disease. J Invest Dermatol 125(1):1–8
86. Medzhitov R (2001) Toll-like receptors and innate immunity. Nat Rev Immunol 1(2):135–145
87. Medzhitov R, Preston-Hurlburt P, Janeway CA, Jr. (1997) A human homologue of the Drosophila Toll protein signals activation of adaptive immunity. Nature 388(6640):394–397
88. Mellman I, Steinman RM (2001) Dendritic cells: specialized and regulated antigen processing machines. Cell 106(3):255–258
89. Melroe GT, DeLuca NA, Knipe DM (2004) Herpes simplex virus 1 has multiple mechanisms for blocking virus-induced interferon production. J Virol 78(16):8411–8420
90. Mempel M, et al (2003) Toll-like receptor expression in human keratinocytes: nuclear factor kappaB controlled gene activation by *Staphylococcus aureus* is toll-like receptor 2 but not toll-like receptor 4 or platelet activating factor receptor dependent. J Invest Dermatol 121(6):1389–1396
91. Mibayashi M, et al (2007) Inhibition of retinoic acid-inducible gene I-mediated induction of beta interferon by the NS1 protein of influenza A virus. J Virol 81(2):514–524
92. Michelsen KS, et al (2001) The role of toll-like receptors (TLRs) in bacteria-induced maturation of murine dendritic cells (DCS). Peptidoglycan and lipoteichoic acid are inducers of DC maturation and require TLR2. J Biol Chem 276(28):25680–15686
93. Miettinen M, et al (2001) IFNs activate toll-like receptor gene expression in viral infections. Genes Immun 2(6):349–355
94. Milner P, et al (2004) Regulation of substance P mRNA expression in human dermal microvascular endothelial cells. Clin Exp Rheumatol 22(3 Suppl 33):S24–S27
95. Muzio M, et al (2000) Differential expression and regulation of toll-like receptors (TLR) in human leukocytes: selective expression of TLR3 in dendritic cells. J Immunol 164(11):5998–6004
96. Naik S, et al (2001) Absence of Toll-like receptor 4 explains endotoxin hyporesponsiveness in human intestinal epithelium. J Pediatr Gastroenterol Nutr 32(4):449–453
97. O'Neill LA (2003) The role of MyD88-like adapters in Toll-like receptor signal transduction. Biochem Soc Trans 31(Pt 3):643–647
98. Oliveira RB, et al (2003) Expression of Toll-like receptor 2 on human Schwann cells: a mechanism of nerve damage in leprosy. Infect Immun 71(3):1427–1433
99. Pai RK, et al (2003) Inhibition of IFN-gamma-induced class II transactivator expression by a 19-kDa lipoprotein from Mycobacterium tuberculosis: a potential mechanism for immune evasion. J Immunol 171(1):175–184
100. Peters EM, et al (2006) Neuropeptide control mechanisms in cutaneous biology: physiological and clinical significance. J Invest Dermatol 126(9):1937–1947
101. Pivarcsi A, et al (2003) Expression and function of Toll-like receptors 2 and 4 in human keratinocytes. Int Immunol 15(6):721–730
102. Prehaud C, et al (2005) Virus infection switches TLR-3-positive human neurons to become strong producers of beta interferon. J Virol 79(20):12893–12904

103. Prinz M, et al (2006) Innate immunity mediated by TLR9 modulates pathogenicity in an animal model of multiple sclerosis. J Clin Invest 116(2):456–464
104. Rakoff-Nahoum S, et al (2004) Recognition of commensal microflora by toll-like receptors is required for intestinal homeostasis. Cell 118(2):229–241
105. Roosterman D, et al (2006) Neuronal control of skin function: the skin as a neuroimmunoendocrine organ. Physiol Rev 86(4):1309–1379
106. Schallreuter KU (1997) Epidermal adrenergic signal transduction as part of the neuronal network in the human epidermis. J Investig Dermatol Symp Proc 2(1):37–40
107. Schmidt R, et al (1995) Novel classes of responsive and unresponsive C nociceptors in human skin. J Neurosci 15(1 Pt 1):333–341
108. Scholzen T, et al (1998) Neuropeptides in the skin: interactions between the neuroendocrine and the skin immune systems. Exp Dermatol 7(2–3):81–96
109. Scholzen TE, et al (1999) Expression of proopiomelanocortin peptides and prohormone convertases by human dermal microvascular endothelial cells. Ann N Y Acad Sci 885:444–447
110. Shuto T, et al (2001) Activation of NF-kappa B by nontypeable Hemophilus influenzae is mediated by toll-like receptor 2-TAK1-dependent NIK-IKK alpha /beta-I kappa B alpha and MKK3/6-p38 MAP kinase signaling pathways in epithelial cells. Proc Natl Acad Sci U S A 98(15):8774–8779
111. Shuto T, et al (2002) Glucocorticoids synergistically enhance nontypeable Haemophilus influenzae-induced Toll-like receptor 2 expression via a negative cross-talk with p38 MAP kinase. J Biol Chem 277(19):17263–17270
112. Smiley ST, King JA, Hancock WW (2001) Fibrinogen stimulates macrophage chemokine secretion through toll-like receptor 4. J Immunol 167(5):2887–2894
113. Smith MF, Jr., et al (2003) Toll-like receptor (TLR) 2 and TLR5, but not TLR4, are required for Helicobacter pylori-induced NF-kappa B activation and chemokine expression by epithelial cells. J Biol Chem 278(35):32552–32560
114. Song IS, et al (1998) CD14 expression in rabbit and human corneas. Invest Ophthalmol Vis Sci 39:S773
115. Song IS, et al (1999) The expression and function of the LPS, CD14/Toll-like receptor 2 (TLR2) complex in human cornea. Invest Ophthalmol Vis Sci 40:S794
116. Song IS, et al (1999) The identity and function of CD14 LPS receptor and toll-like receptor 2 in a human epithelial cell line. J Invest Dermatol 112:546
117. Song PI, et al (2001) The expression of functional LPS receptor proteins CD14 and toll-like receptor 4 in human corneal cells. Invest Ophthalmol Vis Sci 42(12):2867–2877
118. Song PI, et al (2002) Human keratinocytes express functional CD14 and toll-like receptor 4. J Invest Dermatol 119(2):424–432
119. Song PI, et al (2002) Lipoteichoic acid-induced keratinocyte activation is mediated by Toll-like receptor 4. J Invest Dermatol 119:300
120. Song PI, et al (2003) The expression of Toll-like receptors, MD-2 and CD14 in mouse keratinocytes. J Invest Dermatol 121:Abstract 0945
121. Stack J, et al (2005) Vaccinia virus protein A46R targets multiple Toll-like-interleukin-1 receptor adaptors and contributes to virulence. J Exp Med 201(6):1007–1018
122. Steinhoff M, et al (2003) Modern aspects of cutaneous neurogenic inflammation. Arch Dermatol 139(11):1479–1488
123. Steinhoff M, et al (2006) Neurophysiological, neuroimmunological, and neuroendocrine basis of pruritus. J Invest Dermatol 126(8):1705–1718
124. Steinman L (2004) Elaborate interactions between the immune and nervous systems. Nat Immunol 5(6):575–581
125. Summers AE, Whelan CJ, Parsons ME (2003) Nicotinic acetylcholine receptor subunits and receptor activity in the epithelial cell line HT29. Life Sci 72(18–19):2091–2094
126. Supajatura V, et al (2002) Differential responses of mast cell Toll-like receptors 2 and 4 in allergy and innate immunity. J Clin Invest 109(10):1351–1359
127. Tahara K, et al (2006) Role of toll-like receptor signalling in Abeta uptake and clearance. Brain 129(Pt 11):3006–3019
128. Takeda K, Kaisho T, Akira S (2003) Toll-like receptors. Annu Rev Immunol 21:335–376
129. Tlaskalova-Hogenova, H, et al (2004) Commensal bacteria (normal microflora), mucosal immunity and chronic inflammatory and autoimmune diseases. Immunol Lett 93(2–3):97–108
130. Tsuji S, et al (2000) Maturation of human dendritic cells by cell wall skeleton of Mycobacterium bovis bacillus Calmette-Guerin: involvement of toll-like receptors. Infect Immun 68(12):6883–6890
131. Vabulas RM, et al (2002) HSP70 as endogenous stimulus of the Toll/interleukin-1 receptor signal pathway. J Biol Chem 277(17):15107–15112
132. Vetrugno R, et al (2003) Sympathetic skin response: basic mechanisms and clinical applications. Clin Auton Res 13(4):256–270
133. Wadachi R, Hargreaves KM (2006) Trigeminal nociceptors express TLR-4 and CD14: a mechanism for pain due to infection. J Dent Res 85(1):49–53
134. Wang T, Lafuse WP, Zwilling BS (2000) Regulation of toll-like receptor 2 expression by macrophages following Mycobacterium avium infection. J Immunol 165(11):6308–6313
135. Weller R (2003) Nitric oxide: a key mediator in cutaneous physiology. Clin Exp Dermatol 28(5):511–514
136. Yamamoto M, et al (2002) Essential role for TIRAP in activation of the signalling cascade shared by TLR2 and TLR4. Nature 420(6913):324–329
137. Yamamoto M, et al (2003) Role of adaptor TRIF in the MyD88-independent toll-like receptor signaling pathway. Science 301(5633):640–643
138. Yamamoto M, et al (2003) TRAM is specifically involved in the Toll-like receptor 4-mediated MyD88-independent signaling pathway. Nat Immunol 4(11):1144–1150
139. Yamamoto S, et al (1992) Unique palindromic sequences in synthetic oligonucleotides are required to induce IFN [correction of INF] and augment IFN-mediated [correction of INF] natural killer activity. J Immunol 148(12):4072–4076
140. Zhang D, et al (2004) A toll-like receptor that prevents infection by uropathogenic bacteria. Science 303(5663):1522–1526
141. Zhang G, Ghosh S (2002) Negative regulation of toll-like receptor-mediated signaling by Tollip. J Biol Chem 277(9):7059–7065

Section II

Stress and Cutaneous Immunity

Neuroendocrine Regulation of Skin Immune Response

10

G. Maestroni

Contents

Synonyms Box: Dendritic cell, antigen presenting cell; Th priming, antigen specific activation of T helper cells; adaptive immune response, acquired immune response; adrenergic, sympathetic; contact hypersensitivity, contact allergy; atopic dermatitis, neurodermitis

Key Features

- Disorders of skin immune activity are implicated in the pathogenesis of cutaneous infections, skin malignancies, and acquired inflammatory skin disorders.
- Immune cells of the skin include epidermal keratinocytes, which function both as a physical barrier and early warning system, Langerhans cells (LCs), intraepithelial lymphocytes, dermal dendritic cells (DCs), mast cells, and cutaneous memory T cells.
- The sympathetic nervous system (SNS) affects skin DC function and modulates their migration and Th1 priming ability. The DC receptors involved are the α1b- and β2-adrenergic receptors (ARs). These exert opposing functions with α1b-ARs stimulating and β2-ARs inhibiting DC (LC) migration. In particular, the effect of β2-ARs seems predominant and is mainly exerted by enhancing interleukin (IL)-10 production in TLR-activated DCs.
- DCs express the cannabinoid receptor CB2 and that its physiological ligand, that is, the endogenous endocannabinoid 2-arachydonoylglycerol (2-AG), may act as a chemoattractant for DCs. In vivo, 2-AG may recruit CD8+ DCs to the skin and act as an adjuvant for a Th1 adaptive response to a soluble protein. These findings may improve our understanding of the pathogenesis of skin diseases. In fact, alterations of skin ARs expression or function have been associated with skin inflammatory and autoimmune disorders. The involvement of the endogenous cannabinoid system in skin diseases is also possible but current evidence calls for further studies.

10.1 Introduction

The skin is the largest organ of the body and plays a central role in host defense. The epidermis is composed of keratinocytes that function both as physical barrier and early warning system. Immune cells of the epidermis include Langerhans cells (LCs) and intraepithelial lymphocytes. The dermis is composed of connective tissue produced by fibroblasts. Immune cells resident in the dermis include dermal dendritic cells (DCs), mast cells, and cutaneous lymphocyte antigen-positive memory T cells. Situated at the frontier with the external environment, the skin is well equipped for the innate immune response, which mediates nonspecific host defense. Recent studies have shown that the innate immune system is endowed with a highly sophisticated ability to discriminate between self and foreign pathogens. This

R.D. Granstein and T.A. Luger (eds.), *Neuroimmunology of the Skin*,

discrimination relies on a family of receptors, known as Toll-like receptors (TLRs), which play a crucial role in early host defense mechanisms. TLRs recognize pathogen-associated molecular patterns (PAMPs) and the consequent activation of innate immunity is necessary for the induction of acquired immunity, in particular, for Th1 priming. TLRs differ from each other in ligand specificities, expression pattern, and presumably in the target genes they can affect.

An ideal partner with the innate immune system of the skin is the nervous and neuroendocrine system, which may react rapidly (from milliseconds to minutes) to many types of nonspecific environmental stimuli. Neurotransmitters and neuropeptides bind to G-protein coupled receptors that activate the same signaling pathways as those triggered by inflammatory immune mediators.

Here we focused our attention on the neural and neuroendocrine influence on a crucial player of the innate and adaptive immune response, that is, DCs. Interestingly, DCs are uniquely able to either induce immune responses or to maintain a state of self-tolerance. Recent evidence has shown that the ability of DCs to induce tolerance in the steady-state is critical to the prevention of an autoimmune response. Likewise, DCs have been shown to induce several types of regulatory T cells, depending on the maturation state of the DC and the local microenvironment.

On the other hand, TLRs and neural receptors such as those of the adrenergic system have been related to skin disorders but in a strictly separate approach [5,8,24,28].

Here we report the role of ARs in modulating DC function and the influence of the endogenous cannabinoid 2-arachidonoylglycerol (2-AG) and its specific CB2 receptors.

10.2 Adrenergic Regulation of Skin DCs Function

DCs are bone marrow-derived antigen-presenting cells distributed in both lymphoid and non-lymphoid tissues [22,27]. After antigen internalization and inflammation, DCs leave the tissues interfacing with the external environment and enter the lymphatic vessels to reach the lymphoid organs and undergo maturation [22,27,33]. While still immature, the primary function of the DC is to capture and process antigens, then to present the antigenic peptides, and activate specific T cells [22,27]. Activation of naive T-helper (Th) cells also results in their polarization toward the Th1 and/or Th2 type, which orchestrates the immune effector mechanism that is more appropriate for the invading pathogen. Th1 cells promote cellular immunity, protecting against intracellular infection and cancer but carry the risk of organ-specific autoimmunity. Th2 cells promote humoral immunity, are highly effective against extracellular pathogens, and are involved in tolerance mechanisms and allergic diseases. The initiation and type of adaptive immune responses is controlled by innate immune recognition, which is mediated by DCs that produce IL-12, a prerequisite for both the activation of innate immunity and the development of Th1 responses [3,25].

Recently, we and others have begun to investigate neuronal influences on DCs. LCs, which reside within the epidermis, often lie in apposition with epidermal nerves, including both sensory and sympathetic fibers [4,30]. Chemokines expressed by endothelial cells in lymphatics and lymph node venules and chemokine receptors in LCs seem to contribute to LCs migration as chemoattractants and by triggering integrin-dependent adhesive interactions [3,9,32]. However, the mechanisms that drive LCs migration are not completely understood.

We investigated whether the sympathetic neurotransmitter norepinephrine (NE) could affect LCs migration and/or exert a chemotactic/chemokinetic effect on bone marrow-derived DCs in vitro. Fluorescein isothiocyanate (FITC) was used as an antigen to induce migration of LCs to regional lymph nodes. The results obtained showed that local application of the α1-AR antagonist, prazosin, inhibited migration of LCs to the draining lymph nodes. Furthermore, enhancement of DCs emigration from dorsal halves of ear skin was noted when NE was added in organ cultures, and prazosin inhibited the NE effect [12]. These results confirmed that NE can mobilize skin DCs via α1-ARs. To investigate whether the adrenergic influence on skin LCs migration in vivo resulted in an altered development of LC-dependent immune response, we measured the contact hypersensitivity (CHS) response to FITC after sensitization in presence of prazosin. Prazosin treatment during sensitization inhibited the CHS response expressed as net ear swelling after FITC challenge 6 days later [12]. This indicated that the prazosin-induced inhibition of DC migration resulted in a reduced sensitization to FITC. Other experiments performed in vitro revealed that NE is indeed a chemotactic and chemokinetic factor in bone marrow-derived DCs via α1-AR [12].

Other authors have confirmed the expression of mRNA coding for various ARs in LC; purified LC as well as LC-like cell lines expressed mRNA for β1, β2, and $\alpha1_A$-ARs [26]. Furthermore, pretreatment of LCs with NE or epinephrine suppressed their antigen-presenting ability and inhibited their capacity to elicit a CHS response. This effect was neutralized by the β2-AR blocker, ICI 118,551, suggesting the involvement of β2-ARs [26]. More recently, we also showed that bone-marrow-derived DC express β1-, β2-, $\alpha2_A$-, and $\alpha2_C$-ARs genes. Both β2- and $\alpha2_A$-ARs inhibited IL-12, while IL-10 was stimulated by β2-ARs only. In addition, LC migration, the CHS response, and production of IFN-γ and IL-2 in draining lymph node cells are increased in mice treated topically with the β2-AR antagonist, ICI 118,55, during FITC sensitization. Activation of β2-ARs in DCs before adoptive transfer could reduce both migration and CHS response to FITC. Finally, preincubation of DCs with LPS in the presence of the specific β2-AR agonist, salbutamol, impaired their chemotactic response to CCL19 and CCL21, and this effect was neutralized by anti-IL-10 mAb. Thus it appears that the physiological activation of β2-ARs in LCs results in stimulation of IL-10, which in turn restrains LCs migration, influencing antigen presentation and the consequent CHS response [15]. The reported adrenergic stimulation of LC migration via α1-ARs is seemingly in contrast with the present finding. A reasonable explanation may be that, physiologically, the final NE effect on LC migration results from two opposing effects: chemotaxis/chemokinesis mediated by α1-ARs and inhibition mediated by β2-ARs. The selective blockade of these two ARs results, in fact, in divergent effects on both LCs migration and CHS response [12,13,15]. The binding affinity of NE for α1-ARs is considerably higher than that for β2-ARs; this means that concentrations of NE that may activate α1-ARs do not activate β2-ARs and this might account for the contrasting results. In any case, our results suggested that the sympathetic nervous system is involved in regulating LCs migration and the consequent induction of the CHS response to a chemical allergen through modulation of IL-12 and IL-10 production. The sympathetic nervous system might thus play a modulating role during the innate immune response and this would contribute in determining the type and intensity of the adaptive response.

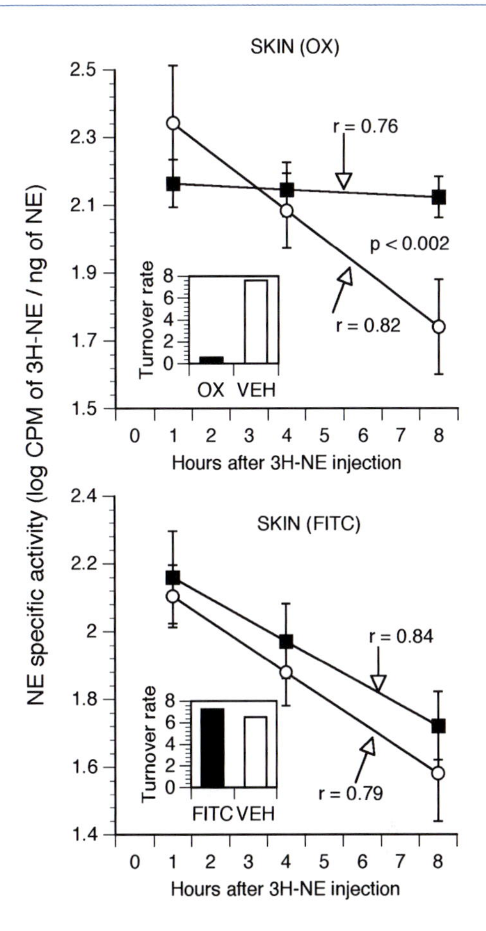

Fig. 10.1 Effect of allergen sensitization on the rate of NE release in the skin. Either 1, 4, and 8 h (**a**) or 18, 21, and 25 h (**b**) after application of oxazolone (OX), FITC, or vehicle (VEH), skin samples from 3H-NE-treated mice were collected for determination of the rate of NE release as NE turnover rate (nanograms per gram per hour; inset). Each point represents the mean ± standard deviation for specific activity from 21 mice per experiment; *r* values are the least square regression coefficients using raw data points for each line. The *p* values are indicated when significant. The specific activity of the 3H-NE (cpm per ng of NE) was calculated as the quotient of 3H-NE present in the tissue and the total tissue NE content. To determine NE turnover rates (rate constant × tissue NE) the specific activity of tissue NE was plotted as a function of time. The decay of specific activity is a first-order function, a straight line with a negative slope, and decay lines were calculated by the method of least squares. The rate constant represents the fraction of the NE pool replaced per unit time (h^{-1}). Rate constants (*k*) were calculated from the slope of logarithm of the specific activity vs. time relationship (0.434 [k] = slope). Differences between the slopes of the regression lines were tested by student's *t* test, using the pooled standard errors of sample regression. From [14]

We also reported that oxazolone, which induces a predominant Th1-type CHS response, but not FITC, which induces a prevailing Th2-type response, inhibits the local NE turnover in the skin of mice during the first 8 h of sensitization (Fig. 10.1). Oxazolone also induced higher expression of the inflammatory cytokines IL-1 and IL-6 mRNA in the skin [14]. Thus, the extent of Th1 priming in the adaptive response to a sensitizing agent seems to depend also on its ability to modulate the local sympathetic nervous activity during the innate immune response. This suggests the existence of a reciprocal cross-talk between the skin immune system and the SNS aimed at shaping the most appropriate immune responses.

10.3 The Endogenous Cannabinoid 2-AG as In Vivo Chemoattractant for DCs

Endogenous cannabinoids such as anandamide and 2-AG may be produced in most tissues upon a variety of stimuli, including exposure to agonists of TLRs and traumatic injury [18,29]. These substances act as physiological ligands for the G protein-coupled cannabinoid receptors CB1 and CB2, with CB2 receptors being expressed mostly in immunocompetent cells [29]. Interestingly, human myeloid DCs express both the cannabinoid CB1 and CB2 receptors and produce 2-AG [16].

Relevant to our previous work, peripheral sympathetic nervous system activity may be inhibited by activation of presynaptic cannabinoid CB1 receptors [6,17]. Therefore, we asked whether the presence of 2-AG during the innate immune response to a soluble protein could influence the ensuing adaptive response.

We studied the influence of 2-AG in a model in which a soluble foreign protein was injected in combination with peptidoglycan (PGN) from *S. aureus* to provide a danger signal to DCs. TLR2 recognizes the highest number of microbial products and PGN is a TLR2 agonist that has been reported to instruct myeloid DCs to induce a Th2 response [1,19]. Mice were injected with KLH and PGN in presence or absence of 2-AG (50 μM). The concentration of 2-AG used was similar to those induced in tissues by inflammatory stimuli or traumatic injury [18,29]. Ten days later the mice were boosted subcutaneously with KLH (50 μg) in complete Freund's adjuvant (CFA). After 7 days the mice were challenged with KLH (50 μg in 20 μl PBS) in the right hind footpad to evaluate the DTH response and cytokine production in draining popliteal lymph nodes. Addition of PGN to KLH during primary immunization resulted in a mixed Th1/Th2 response as assessed by the DTH response, IFN-γ, and IL-4 production in draining lymph node cells. The presence of 2-AG resulted in a significantly higher DTH response upon the recall memory response and in a strong inhibition of IL-4 production. These effects were neutralized when the specific CB2 antagonist SR 144528 was injected together with 2-AG in the primary immunization. Thus, 2-AG exerted a Th1 skewing effect during primary immunization via CB2 receptors.

We used real time RT-PCR to investigate CB2 receptor mRNA expression in lymphoid CD8+, CD11c+ and CD8−, CD11c+ DCs populations that were purified from the spleen of C57BL/6 mice or obtained from bone marrow cell cultures, respectively. We performed a semi-quantitative analysis of the CB2 mRNA expression in immature and mature DCs subsets using brain RNA as reference. We found that CD8+ CD11c+ DCs showed the highest expression of the CB2 mRNA and that both subsets of immature DCs showed higher expression of the CB2 mRNA compared to the control brain tissue. The CB2 expression was, however, reduced to the brain level upon maturation by PGN.

2-AG has been reported to enhance the migration of cells expressing the CB2 receptor [7,10]. Therefore we thought that the effect of 2-AG could be related to an enhanced migration of DCs. To challenge this hypothesis we injected OVA-FITC (50 μg) as foreign protein ±2-AG and 24 h later we enumerated the CD11c+ FITC+ and the CD8+ FITC+ cells in the CD11c+-enriched cell population from the draining lymph nodes. 2-AG indeed increased the number of DCs loaded with OVA-FITC in the draining lymph nodes, with the CD8+ FITC+ subpopulation being more affected. When the CD11c+ FITC+ general population was considered, the percent increase in migration was 163% ($p < 0.03$) while for the CD8+ FITC+ subset only, the figure rose to 364% ($p < 0.01$). The presence of SR 144528 completely counteracted these effects, indicating that the increase in the antigen-loaded DCs was a CB2-mediated phenomenon. In addition, 2-AG could exert a powerful chemotactic activity on both immature and mature DCs in vitro. Checkerboard analysis showed that this effect was real chemotaxis, as 2-AG was ineffective in absence of a concentration gradient. Again, the CB2 antagonist SR 144528 inhibited completely the chemotactic effect of 2-AG, indicating that the effect observed was CB2-mediated.

The observation that 2-AG may especially increase the number of antigen-loaded CD8+ DCs in the draining lymph nodes suggests that its action is possibly exerted on circulating DC precursors. In fact, the skin does not contain CD8+ DCs and 2-AG was reported to act as chemoattractant for human monocytes and macrophages [10] and, in general, for hematopoietic cells expressing the CB2 receptor [7]. Once recruited, DCs engulf and process the foreign protein while maturing upon the TLR2 stimulation. Finally they migrate to the draining lymph nodes where they present the antigen to Th cells (Fig. 10.2). The quality and quantity and subset of the DCs may influence Th priming [2,23]. On the one hand, during in vivo T-cell responses, CD8− DCs mainly induce Th2 responses, whereas CD8+ DCs elicit Th1 responses due to their high capacity to produce IL-12 [2]. On the other hand, in vitro and in vivo studies indicate that the functional division between CD8+ and CD8− DCs can be altered by environmental factors that modulate the function of DCs, in particular, the ability to produce IL-12 [2,14,15]. Thus, both the increased migration and type of DC subset might explain the final effect on Th priming. Thus, we suggest that the endocannabinoid 2-AG may act as a chemotactic substance capable of recruiting DCs and/or their precursor cells during the innate immune response that, in presence of a TLR2 agonist, consequently instruct a Th1 adaptive (Fig. 10.2).

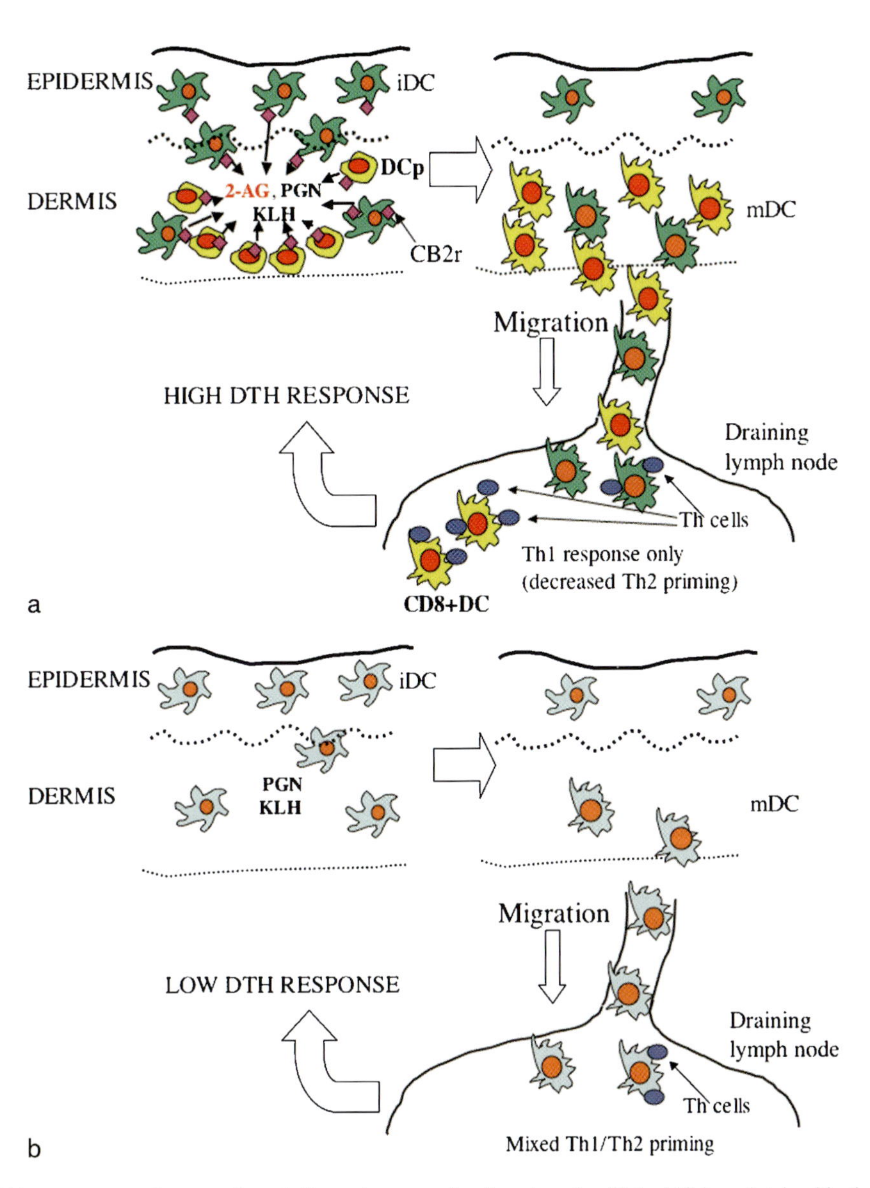

Fig. 10.2 AG (A) may exert a chemotactic activity on immature dendritic cells (iDC) and or on CD8+ dendritic cells precursors (DCp) via the cannabinoid CB2 receptor. In presence of the TLR2 agonist PGN and the foreign soluble protein KLH, the cells engulf the foreign protein and mature, ultimately leading to a Th1 shift of the adaptive response upon migration in the draining lymph nodes. This shift is maintained in the memory recall immune response that, in turn, results in a high delayed-type hypersensitivity (DTH) response. In the absence of 2-AG (B), no recruitment of dendritic cells or precursor cells takes place and the ensuing adaptive response is a mixed Th1/Th2 type that results in a low DTH response. From [11], with permission

10.4 Conclusion and Perspectives

Our studies on the SNS regulation of DC functions revealed that NE may shape the appropriate immune response to a foreign antigen by exerting a discrete influence on the cytokine production and migratory properties of the DC response to microbial products. NE acts mainly on β2-AR expressed in LCs (DCs) and results in the production of large amounts of IL-10 upon stimulation with allergens or TLR2 and TLR4, that is, the DC receptors for bacterial products. The presence of NE at the beginning of LC (DC) activation might serve to limit the inflammatory response and an excessive Th1 polarization, which may be potentially dangerous for the organism. By doing this, NE would also decrease the probability of autoimmune reactions. However, why the SNS should play such a role in the skin, that is, in an organ completely exposed to the external environment and its potential threats? This question seems important as all stimuli perceived as threats would activate the SNS in the so-called stress reaction. Other reports show that, indeed, stress may activate the skin immune responsiveness as a part of the homeostatic response to environmental challenges [31]. Other contributions in this book detail this topic. So we are in front of an apparent controversy. The existence of a controversy usually alludes to poor knowledge. The current dogma about the role of the innate immune response and DCs in the skin dictates that DCs must be alerted by danger signals (TLRs or other receptors) and undergo a complex cascade of events starting from antigen uptake, migration to the draining lymph nodes, and antigen presentation to naïve Th cells or cross presentation to CD8+ cells [21].

In the frame of this dogma we could consider that the discordant results might depend on the type of ARs that is activated. We might speculate that activation of α1-ARs by NE would result in increased migration and antigen presentation, while activation of β2-ARs by higher NE concentration or, better, by epinephrine would end in a down-regulation of DC migration and antigen presentation. However, are we sure that all immune reactions in the skin follow the above mentioned dogma? Some recent studies have started to question it. It has been reported that in certain cases migrating DCs do not present skin antigens in the draining lymph nodes but rather it is the lymph node resident DCs that do the job [20]. So the SNS influence that matters in term of type and strength of the adaptive immune response might depend, not only on events occurring locally in the skin at the infection site, but also in the draining lymph nodes, and this during the very early phase of the innate response.

This reasoning may apply also to endocannabinoids as well as for any other potential immunomodulating neuroendocrine factor. Cannabinoids have been, in fact, reported to exert both anti-inflammatory and pro-inflammatory effects, resulting in a complex and controversial picture. However, most in vivo studies are related to systemic and direct effects on T and B cell responses of pharmacological treatments, with the primary psychoactive component of marijuana, Δ^9-tetrahydrocannabinol, and related substances. In any case, we believe to have revealed a potentially important role of 2-AG, which might have several implications. For example, 2-AG is induced in the central nervous system by traumatic brain injury at tissue concentrations largely above that used in our study. Its pathophysiological relevance seems to be that of attenuating brain damage via a CB1-mediated mechanism in a model of closed head injury where no exogenous danger signals were involved. Perhaps, in presence of a danger signal (infection) or inflammation, the effect of 2-AG in the brain would promote neuroinflammatory diseases. As far as the skin is concerned, future studies should investigate the possible production of 2-AG in the skin and its physiopathological correlates.

Summary for the Clinician

The crucial role of the skin immune surveillance system is evident from the increased frequency of cutaneous infections and malignancies in patients with acquired immunodeficiency and in those receiving immunosuppressive drugs. Disorders of the skin immune activity are also implicated in the pathogenesis of acquired inflammatory skin disorders, including psoriasis, atopic dermatitis, contact hypersensitivity, lichen planus, alopecia areata, and vitiligo.

The present chapter describes studies showing that β-ARs in skin DCs play an important role in cutaneous innate and acquired immune responses. In particular, activation of these adrenoceptors inhibits DCs migration and Th1 priming. These findings should be considered together with the observation that most inflammatory skin disorders are associated with deranged local adrenergic mechanisms. In vitiligo, there is a dysregulation of catecholamine biosynthesis with increased plasma and epidermal norepinephrine levels associated with high numbers of β2-ARs in differentiating keratinocytes and with a defective calcium uptake in both keratinocytes and melanocytes. In atopic eczema, a point mutation in the β-AR gene could alter the structure and function of the receptor, thereby leading to a low density of receptors on both keratinocytes and peripheral blood lymphocytes. In psoriasis, β-ARs are downregulated and, interestingly, β-AR blockers may cause this inflammatory skin disease. [5,28] It should be noted that psoriasis is currently considered to be a Th1-mediated autoimmune disease.

Our finding that the endocannabinoid 2-AG may recruit DCs in the skin might have clinical relevance as inflammatory skin diseases are directly caused by massive infiltrate of immunocompetent cells. The involvement of the endocannabinoid system in skin diseases deserves to be investigated.

References

1. Agrawal S, et al. (2003) Cutting edge: different Toll-like receptor agonists instruct dendritic cells to induce distinct Th responses via differential modulation of extracellular signal-regulated kinase-mitogen-activated protein kinase and c-Fos. J Immunol 171:4984–4989
2. Ardavin C (2003) Origin, precursors and differentiation of mouse dendritic cells. Nat Rev Immunol 3:582–590
3. Bancherau J, Steinman RM (1998) Dendritic cells and the control of immunity. Nature 392:245–252
4. Botchkarev VA, Peters EM, Botchkareva NV, Maurer M, Paus R (1999) Hair cycle-dependent changes in adrenergic skin innervation, and hair growth modulation by adrenergic drugs. J Invest Dermatol 113:878–887
5. Halevy S, Livni E (1993) Beta-adrenergic blocking drugs and psoriasis: the role of an immunologic mechanism. J Am Acad Dermatol 29:504–505
6. Ishac EJ, Jiang L, Lake KD, Varga K, Abood ME, Kunos G (1996) Inhibition of exocytotic noradrenaline release by presynaptic cannabinoid CB1 receptors on peripheral sympathetic nerves. Br J Pharmacol 118:2023–2028
7. Jorda MA, et al. (2002) Hematopoietic cells expressing the peripheral cannabinoid receptor migrate in response to the endocannabinoid 2-arachidonoylglycerol. Blood 99:2786–2793
8. Kang SS, Kauls LS, Gaspari AA (2006) Toll-like receptors: applications to dermatologic disease. J Am Acad Dermatol 54:951–983; quiz 983–956
9. Kellermann S-A, Hudak S, Oldham ER, Liu Y-J, McEvoy LM (1999) The CC chemokine receptor-7 ligands 6Ckine and macrophage inflammatory protein-3b are potent chemoattractants for in vitro and in vivo-derived dendritic cells. J Immunol 162:3859–3864
10. Kishimoto S et al. (2003) 2-Arachidonoylglycerol induces the migration of HL-60 cells differentiated into macrophage-like cells and human peripheral blood monocytes through the cannabinoid CB2 receptor-dependent mechanism. J Biol Chem 278:24469–24475
11. Maestroni GJ (2004) The endogenous cannabinoid 2-arachidonoyl glycerol as in vivo chemoattractant for dendritic cells and adjuvant for Th1 response to a soluble protein. FASEB J 18:1914–1916
12. Maestroni GJM (2000) Dendritic cells migration controlled by a1b-adrenergic receptors. J Immunol 165:6743–6747
13. Maestroni GJM (2002) Short exposure of antigen-stimulated dendritic cells to norepinephrine: impact on kinetic of cytokine production and Th polarization. J Neuroimmunol 129:106–114
14. Maestroni GJM (2004) Modulation of skin norepinephrine turnover by allergen sensitization: impact on contact hypersensitivity and Th priming. J Invest Dermatol 122:119–124
15. Maestroni GJM, Mazzola P (2003) Langerhans cells b2-adrenoceptors: role in migration, cytokine production, Th priming and contact hypersensitivity. J Neuroimmunol 144:91–99
16. Matias I, Pochard P, Orlando P, Salzet M, Pestel J, Di Marzo V (2002) Presence and regulation of the endocannabinoid system in human dendritic cells. Eur J Biochem 269:3771–3778
17. Niederhoffer N, Schmid K, Szabo B (2003) The peripheral sympathetic nervous system is the major target of cannabinoids in eliciting cardiovascular depression. Naunyn Schmiedebergs Arch Pharmacol 367:434–443
18. Panikashvili D, et al. (2001) An endogenous cannabinoid (2-AG) is neuroprotective after brain injury. Nature 413:527–531
19. Qi H, Denning TL, Soong L (2003) Differential induction of interleukin-10 and interleukin-12 in dendritic cells by microbial toll-like receptor activators and skewing of T-cell cytokine profiles. Infect Immun 71:3337–3342
20. Randolph GJ (2006) Migratory dendritic cells: sometimes simply ferries? Immunity 25:15–18
21. Randolph GJ, Angeli V, Swartz MA (2005) Dendritic-cell trafficking to lymph nodes through lymphatic vessels. Nat Rev Immunol 5:617–628
22. Sallgaller ML, Lodge PA (1998) Use of cellular and cytokine adjuvants in the immunotherapy of cancer. J Surg Oncol 68:122–138
23. Sallusto F, Lanzavecchia A (1999) Mobilizing dendritic cells for tolerance, priming and chronic inflammation. J Exp Med 189:611–614
24. Schallreuter KU (1997) Epidermal adrenergic signal transduction as part of the neuronal network in the human epidermis. J Investig Dermatol Symp Proc 2:37–40
25. Schnare M, Barton GM, Holt AC, Takeda K, Akira S, Medzhitov M (2001) Toll-like receptors control activation of adaptive immune responses. Nat Immunol 2:947–950
26. Seiffert C, et al. (2002) Catecholamines inhibit the antigen-presenting capability of epidermal Langerhans cells. J Immunol 168:6128–6135
27. Shortman K, Caux C (1997) Dendritic cell development: multiple pathways to nature's adjuvants. Stem Cells 15:409–419
28. Steinkraus V, Steinfath M, Stove L, Korner C, Abeck D, Mensing H (1993) Beta-adrenergic receptors in psoriasis: evidence for down-regulation in lesional skin. Arch Dermatol Res 285:300–304
29. Sugiura T, Kobayashi Y, Oka S, Waku K (2002) Biosynthesis and degradation of anandamide and 2-arachidonoylglycerol

and their possible physiological significance. Prostaglandins Leukot Essent Fatty Acids 66:173–192
30. Torii H, Yan Z, Hosoi J, Granstein RD (1997) Expression of neurotrophic factors and neuropeptide receptors by Langerhans cells and the Langerhans cell-like cell line XS52: further support for a functional relationship between Langerhans cells and epidermal nerves. J Invest Dermatol 109:586–591
31. Viswanathan K, Daugherty C, Dhabhar FS (2005) Stress as an endogenous adjuvant: augmentation of the immunization phase of cell-mediated immunity. Int Immunol 17:1059–1069
32. Warnock RA, Campbell JJ, Deorf ME, Matsuzawa A, McEvoy LM, Butcher EC (2000) The role of chemokines in the microenvironmental control of T vrsus B cell arrest in Peyer's patch high endothelial venules. J Exp Med 191:77–88
33. Weinlich G, et al. (1998) Entry into lymphatics and maturation in situ of migrating murine cutaneous dendritic cells. J Invest Dermatol 110:441–448

Effects of Psychological Stress on Skin Immune Function: Implications for Immunoprotection Versus Immunopathology

11

F.S. Dhabhar

Contents

Key Features

- Psychological stress can have bidirectional effects on immune function, being immunosuppressive under certain conditions but immunoenhancing under others.
- Chronic stress (lasting for weeks, months, or years) is known to suppress or dysregulate immune function.
- Acute stress (lasting for minutes to hours) has been shown to enhance immune function, particularly in the skin.
- Acute stress experienced at the time of immune activation or antigen exposure significantly enhances primary/innate or secondary/adaptive immune responses.
- Acute stress induces a redistribution of leukocytes from the blood to compartments like the skin and sentinel lymph nodes, and this redistribution is thought to be one important mediator of the immunoenhancing effects of acute stress.
- The studies described here may provide substrates for interventions that could be designed to ameliorate stress-induced exacerbation of allergic, inflammatory, or autoimmune diseases.
- These studies also provide the framework for developing therapeutic interventions that may harness the body's endogenous adjuvant-like immunoenhancing mechanisms to enhance protective immunity during vaccination, infection, or wound healing.

R.D. Granstein and T.A. Luger (eds.), *Neuroimmunology of the Skin*,

Synonyms Box: CGRP, calcitonin gene related peptide; CHS, contact hypersensitivity; CMI, cell-mediated immunity; CRH, corticotropin-releasing hormone; CTACK, cutaneous T cell-attracting chemokine; CTL, cytolytic T cell; HPA, hypothalamic pituitary adrenal axis; IFN-gamma, interferon gamma; KLH, keyhole limpet hemocyanin; LTN, lymphotactin; MCP-1, monocyte chemotactic protein -1; MIP-1alpha, macrophage inflammatory protein 1 alpha; MSH, melanocyte-stimulating hormone; NK, natural killer cell; POMC, proopiomelanocortin; SCC, squamous cell carcinoma; SP, substance P; Th, helper T cell; TNF-alpha, tumor necrosis factor alpha; UV, ultraviolet; VIP, vasoactive intestinal peptide

11.1 Introduction

Psychological stress is known to suppress immune function and increase susceptibility to infections and cancer. Paradoxically, stress is also known to exacerbate autoimmune and inflammatory diseases, including psoriasis and numerous other inflammatory skin disorders. This suggests that stress may enhance immune function under certain conditions. It has recently been appreciated that while chronic stress suppresses or dysregulates immune function, acute stress often has immunoenhancing effects [28].

One of the most under-appreciated effects of stress on the immune system is its ability to induce significant changes in leukocyte distribution in the body [31]. Importantly, these changes in leukocyte distribution have significant effects on immune function in different body compartments, with the skin being a critical compartment in which the effects of different stressors are observed. Acute stress induces rapid, robust, and reversible changes in leukocyte distribution between the blood and other body compartments [31]. Acute stress can affect dendritic cell, neutrophil, macrophage, and lymphocyte trafficking, maturation, or function, in ways that can enhance innate and adaptive immunity [29,69,85,86]. Acute stress experienced prior to novel cutaneous antigen exposure increases memory T cell formation and results in a significant and long-lasting increase in immunity [29,69,86]. Similarly, acute stress experienced during antigen re-exposure enhances secondary immune responses [24]. This suggests that depending on the condition against which the immune response is directed, stress can enhance the acquisition as well as expression of immunoprotection and immunopathology.

In contrast to acute stress, chronic stress suppresses or dysregulates innate and adaptive immune responses through mechanisms that involve suppression of leukocyte numbers, trafficking, and function or changes in the Type 1-Type 2 cytokine balance [2,40]. Chronic stress has recently been shown to increase susceptibility to skin cancer by suppressing Type 1 cytokines and protective T cells while increasing suppressor T cell function [72].

We have suggested that the primary biological purpose of a physiologic stress response may be to promote survival, with stress hormones and neurotransmitters serving as beacons that prepare the immune system for potential challenges (e.g., wounding or infection) perceived by the brain (e.g., detection of an imminent attack) [28,31]. However, this same system may exacerbate immunopathology if the enhanced immune response is directed against innocuous or self-antigens, or if the system is over-activated as seen during chronic stress. In view of the ubiquitous nature of stress and its significant effects on immunoprotection and immunopathology, it is important to further elucidate the mechanisms mediating stress–immune interactions and to meaningfully translate findings from bench to bedside.

11.2 Stress

Although the word "stress" generally has negative connotations, stress is a familiar aspect of life, being a stimulant for some, but a burden for others. Numerous definitions have been proposed for the word stress. Each definition focuses on aspects of an internal or external challenge, disturbance, or stimulus; on perception of a stimulus by an organism; or on a physiological response of the organism to the stimulus [41,54,70]. Physical stressors have been defined as external challenges to homeostasis and psychological stressors as the "anticipation justified or not, that a challenge to homeostasis looms" [71]. An integrated definition states that stress is a constellation of events, consisting of a stimulus (stressor), that precipitates a reaction in the brain (stress perception), that activates physiologic fight or flight systems in the body (stress response) [25]. The physiologic stress response results in the release of neurotransmitters and hormones that serve as the brain's alarm signals to the body. It is often overlooked that a stress response has salubrious adaptive effects in the short run [28,29], although stress can be harmful when it is long-lasting [40,47,53,70].

An important distinguishing characteristic of stress is its duration and intensity. Thus, *acute stress* has been defined as stress that lasts for a period of minutes to hours, and *chronic stress* as stress that persists for several hours per day for weeks or months [25]. The intensity of stress may be gauged by the peak levels of stress hormones, neurotransmitters, and other physiological changes such as increases in heart rate and blood pressure, and by the amount of time for which these changes persist during stress and following the cessation of stress.

It is important to bear in mind that there exist significant individual differences in the manner and extent to which stress is perceived, processed, and coped with [28]. These differences become particularly relevant while studying human subjects because stress perception, processing, and coping mechanisms can have significant effects on the kinetics and peak levels of circulating stress hormones and on the duration for which these hormone levels are elevated. The magnitude and duration of catecholamine and glucocorticoid hormone exposure in turn can have significant effects on leukocyte distribution and function [27,64,74].

11.3 Stress-Induced Changes in Immune Cell Distribution

Effective immunoprotection requires rapid recruitment of leukocytes into sites of surgery, wounding, infection, or vaccination. Immune cells circulate continuously on surveillance pathways that take them from the blood, through various organs, and back into the blood. This circulation is essential for the maintenance of an effective immune defense network [80]. The numbers and proportions of leukocytes in the blood provide an important representation of the state of distribution of leukocytes in the body and of the state of activation of the immune system. The ability of acute stress to induce changes in leukocyte distribution within different body compartments is perhaps one of the most under-appreciated effects of stress and stress hormones on the immune system.

Numerous studies have shown that stress and stress hormones induce significant changes in absolute numbers and relative proportions of leukocytes in the blood. In fact, changes in blood leukocyte numbers were used as a measure of stress before methods were available to directly assay the hormone [44]. Studies have also shown that glucocorticoid [32,35,36] and catecholamine [7,8,17,57,58,65] hormones induce rapid and significant changes in leukocyte distribution and that these hormones are the major mediators of the effects of stress. Stress-induced changes in blood leukocyte numbers have been reported in fish [63], hamsters [9], mice [48], rats [30–32,67], rabbits [84], horses [79], non-human primates [60], and humans [14,43,57,66,73]. This suggests that the phenomenon of stress-induced leukocyte redistribution has a long evolutionary lineage, and that perhaps it has important functional significance.

Studies in rodents have shown that stress-induced changes in blood leukocyte numbers are characterized by a significant decrease in numbers and percentages of lymphocytes and monocytes, and by an increase in numbers and percentages of neutrophils [30,31]. Flow cytometric analyses revealed that absolute numbers of peripheral blood T cells, B cells, NK cells, and monocytes all show a rapid and significant decrease (40–70% lower than baseline) during stress [31]. Moreover, it has been shown that stress-induced changes in leukocyte numbers are rapidly reversed upon the cessation of stress [31]. In apparent contrast to animal studies, human studies have shown that stress can increase rather than decrease blood leukocyte numbers [14,15,56,61,73]. This apparent contradiction may be resolved by taking the following factors into consideration: First, stress-induced increases in blood leukocyte numbers are observed following stress conditions, which primarily result in the activation of the sympathetic nervous system. These stressors are often of a short duration (few minutes) or relatively mild (e.g., public speaking) [15,56,61,73]. Second, the increase in total leukocyte numbers may be accounted for by stress- or catecholamine-induced increases in granulocytes and NK cells [8,15,56,61,73]. Third, stress or pharmacologically induced increases in glucocorticoid hormones induce a significant decrease in blood lymphocyte and monocyte numbers [32,44,73,81]. Thus, stress conditions that result in a significant and sustained activation of the HPA axis result in a decrease in blood leukocyte numbers.

In view of the above discussion, it has been proposed that acute stress induces an initial increase followed by a decrease in blood leukocyte numbers. Stress conditions that result in activation of the sympathetic nervous system, especially conditions that induce high levels of norepinephrine, may induce an increase in circulating leukocyte numbers. These conditions may occur during the beginning of a stress response, very short duration stress (order of minutes), mild psychological stress, or during exercise. In contrast, stress conditions that result in the activation of the hypothalamic–pituitary–adrenal (HPA) axis induce a decrease in circulating leukocyte numbers. These conditions often occur during the later stages of a stress response, long duration acute stressors (order of hours), or during severe psychological, physical, or physiological stress. An elegant and interesting example in support of this hypothesis comes from Schedlowski et al. who measured changes in blood T cell and NK cell numbers as well as plasma catecholamine and cortisol levels in parachutists [73]. Measurements were made 2h before, immediately after, and 1h after the jump. Results showed a significant increase in T cell and NK cell numbers immediately (minutes) after the jump that was followed by a significant decrease 1 h after the jump. An early increase in plasma catecholamines preceded early increases in lymphocyte numbers, whereas the more delayed rise in plasma cortisol preceded the late decrease in lymphocyte numbers [73]. Importantly, changes in NK cell activity and antibody-dependent cell-mediated cytotoxicity closely paralleled changes in blood NK cell numbers; thus suggesting that changes in leukocyte numbers may be an important mediator of apparent changes in leukocyte "activity." Similarly, Rinner et al. have shown that a short stressor (1 min handling) induced an increase in mitogen-induced proliferation of T and B cells obtained from peripheral blood, while a longer stressor (2h immobilization) induced a decrease in the same proliferative responses [68]. In another example, Manuck et al. showed that acute psychological stress induced a significant increase in blood CTL numbers only in those subjects who showed hightened catecholamine and cardiovascular reactions to stress [52].

Thus, an acute stress response may induce biphasic changes in blood leukocyte numbers. Soon after the beginning of stress (order of minutes) or during mild acute stress, or exercise, catecholamine hormones and neurotransmitters induce the body's "soldiers" (leukocytes), to exit their "barracks" (spleen, lung, marginated pool, and other organs) and enter the "boulevards" (blood vessels and lymphatics). This results in an increase in blood leukocyte numbers, the effect being most prominent for NK cells and granulocytes. As the stress response continues, activation of the HPA axis results in the release of glucocorticoid hormones, which induce leukocytes to exit the blood and take position at potential "battle stations" (skin, mucosal lining of gastrointestinal and urinary-genital tracts, lung, liver, and lymph nodes) in preparation for immune challenges, which may be imposed by the actions of the stressor [24,27,31]. Such a redistribution of leukocytes results in a decrease in blood leukocyte numbers. Thus, acute stress may result in a redistribution of leukocytes from the barracks, through the boulevards, and to potential battle stations within the body.

Since the blood is the most accessible and commonly used compartment for human studies, it is important to carefully evaluate how changes in blood immune parameters might reflect in vivo immune function in the context of the specific experiments or study at hand. Moreover, since most blood collection procedures involve a certain amount of stress, since all patients or subjects will have experienced acute and chronic stress, and since many studies of psychophysiological effects on immune function focus on stress, the effects of stress on blood leukocyte distribution become a factor of considerable importance.

Dhabhar et al. were the first to propose that stress-induced changes in blood leukocyte distribution may represent an adaptive response [26,30]. They suggested that acute stress-induced changes in blood leukocyte numbers represent a redistribution of leukocytes from the blood to other organs such as the skin, draining sentinel lymph nodes, and other compartments [24,27]. They hypothesized that such a leukocyte redistribution may enhance immune function in compartments to which immune cells traffic during stress. In agreement with this hypothesis, it was demonstrated that a stress-induced redistribution of leukocytes from the blood to the skin is accompanied by a significant enhancement of skin immunity [24,26,33].

11.4 Functional Consequences of Stress-Induced Changes in Immune Cell Distribution

When interpreting data showing stress-induced changes in functional assays such as lymphocyte proliferation or NK activity, it may be important to bear in mind the effects of stress on the leukocyte composition of the compartment in which an immune parameter is being measured. For example, it has been shown that acute stress induces a redistribution of leukocytes from the blood to the skin and that this redistribution is accompanied by a significant enhancement of skin cell mediated immunity [24,29]. In what might at first glance appear to be contradicting results, acute stress has been shown to suppress splenic and peripheral blood responses to T cell mitogens [20] and splenic IgM production [89]. However, it is important to note that in contrast to the skin that is enriched in leukocytes during acute stress, peripheral blood and spleen are relatively depleted of leukocytes during acute stress [23]. This stress-induced decrease in blood and spleen leukocyte numbers may contribute to the acute stress-induced suppression of immune function in these compartments.

Moreover, in contrast to acute stress, chronic stress has been shown to suppress skin cell mediated immunity, and a chronic stress-induced suppression of blood leukocyte redistribution is thought to be one of the factors mediating the immunosuppressive effect of chronic stress [25]. Again, in what might appear to be contradicting results, chronic stress has been shown to enhance mitogen-induced proliferation of splenocytes [59] and splenic IgM production [89]. However, the spleen is relatively enriched in T cells during chronic glucocorticoid administration, suggesting that it may also be relatively enriched in T cells during chronic stress [55], and this increase in spleen leukocyte numbers may contribute to the chronic stress-induced enhancement of immune parameters measured in the spleen.

It is also important to bear in mind that the heterogeneity of the stress-induced changes in leukocyte distribution [31] suggests that using equal numbers of leukocytes in a functional assay may not account for stress-induced changes in relative percentages of different leukocyte subpopulations in the cell suspension being assayed. For example, samples that have been equalized for absolute numbers of total blood leukocytes from control vs. stressed animals may still contain different numbers of specific leukocyte subpopulations (e.g., T cells, B cells, or NK cells). Such changes in leukocyte composition may contribute to the effects of stress even in functional assays using equalized numbers of leukocytes from different treatment groups. This possibility needs to be taken into account before concluding that a given treatment changes an immune parameter on a "per cell" rather than a "per population" basis. Therefore, stress may affect immune function at a cellular level (e.g., phagocytosis, antigen presentation, killing, antibody production) and/or through leukocyte redistribution that could increase or decrease the number of cells with a specific functional capacity in the compartment being studied.

11.5 Effects of Acute Stress on Leukocyte Trafficking to a Site of Surgery or Immune Activation

Viswanathan and Dhabhar used a clinically relevant subcutaneously implanted surgical sponge model to elucidate the effects of stress on the kinetics, magnitude, subpopulation, and chemoattractant specificity of leukocyte trafficking to a site of immune activation or surgery [85]. Mice that were acutely stressed before subcutaneous implantation or the surgical sponge showed a two- to threefold higher neutrophil, macrophage, NK cell, and T cell infiltration than non-stressed animals. Leukocyte infiltration was evident as early as 6 h and peaked between 24 and 48 h. Importantly, sponges from non-stressed and acutely stressed mice had comparable and significantly lower leukocyte numbers at 72 h, indicating effective resolution of inflammation in both groups. These authors also examined the effects of stress on early (6 h) leukocyte infiltration in response to a predominantly pro-inflammatory cytokine, TNF-α, and lymphocyte-specific chemokine, lymphotactin (LTN). Acute stress significantly increased infiltration of macrophages, in response to saline, LTN, or TNF-α; neutrophils only in response to TNF-α; and NK and T cells only in response to LTN. These results showed that acute stress significantly enhances the kinetics and magnitude of leukocyte infiltration into a site of immune activation or surgery in a subpopulation and chemoattractant specific manner with tissue damage, antigen-, or pathogen-driven chemoattractants synergizing with acute stress to further determine the specific subpopulations that are recruited [85]. Thus, depending on the primary chemoattractants driving an immune response, acute stress may selectively mobilize specific leukocyte subpopulations into sites of surgery, wounding, or inflammation. Such a stress-induced increase in leukocyte trafficking may be an important mechanism by which acute stressors alter the course of different (innate vs. adaptive, early vs. late, acute vs. chronic) protective or pathological immune responses.

11.6 Acute Stress-Induced Enhancement of Innate/Primary Immune Responses in Skin

In view of the skin being one of the target organs to which leukocytes traffic during stress, studies were conducted to examine whether skin immunity is enhanced when immune activation/antigen exposure takes place following a stressful experience. Studies showed that acute stress experienced at the time of novel or primary antigen exposure results in a significant enhancement of the ensuing skin immune response [29]. Compared to controls, mice restrained for 2.5 h before primary immunization with keyhole limpet hemocyanin (KLH) showed a significantly enhanced immune response when re-exposed to KLH nine months later. This immunoenhancement was mediated by an increase in numbers of memory and effector helper T cells in sentinel lymph nodes at the time of primary immunization. Further analyses showed that the early stress-induced increase in T cell memory may have stimulated the robust increase in infiltrating lymphocyte and macrophage numbers observed months later at a novel site of antigen re-exposure. Enhanced leukocyte infiltration was driven by increased levels of the Type-1 cytokines, IL-2 and IFN-γ, and TNF-α, observed at the site of antigen re-exposure in animals that had been stressed at the time of primary immunization. Given the importance of inducing long-lasting increases in immunological memory during vaccination, we have suggested that the neuroendocrine stress response is nature's adjuvant that could be psychologically and/or pharmacologically manipulated to safely increase vaccine efficacy [28,29].

A similar enhancement of the sensitization/immunization/induction phase of cell-mediated immunity by different types of stressors administered at the time of antigen exposure has been observed in mice, rats, and non-human primates [12,18,88]. In a series of elegant experiments, Saint Mezard et al. showed that acute stress experienced at the time of sensitization resulted in a significant increase in the contact hypersensitivity (CHS) response [69]. These investigators showed that acute stress experienced during sensitization enhanced dendritic cell migration from skin to sentinel lymph nodes and also enhanced priming of lymph node CD8+ T cells. These CD8+ T cells responded in greater numbers at the site of antigen re-exposure during the recall phase of the CHS response. These studies also suggested that the effects of acute stress in this case were mediated primarily by norepinephrine [69].

Viswanathan et al. further elucidated the molecular and cellular mediators of the immunoenhancing effects of acute stress [86]. They showed that compared to nonstressed mice, acutely stressed animals showed significantly greater pinna swelling, leukocyte infiltration, and upregulated macrophage chemoattractant protein-1 (MCP-1), macrophage inflammatory protein-3α (MIP-3α), IL-1α, IL-1β, IL-6, TNF-α, and IFN-γ gene expression at the site of primary antigen exposure. Stressed animals also showed enhanced maturation and trafficking of dendritic cells from skin to lymph nodes, higher numbers of activated macrophages in skin and lymph nodes, increased T cell activation in lymph nodes, and enhanced recruitment of surveillance T cells to skin. These findings showed that important interactive components of innate (dendritic cells and macrophages) and adaptive (surveillance T cells) immunity are mediators of the stress-induced enhancement of a primary immune response. Such immunoenhancement during primary immunization may induce a long-term increase in immunologic memory, resulting in subsequent augmentation of the immune response during secondary antigen exposure.

In addition to elucidating mechanisms that could be targeted to reduce stress-induced exacerbation of proinflammatory reactions, the above-mentioned studies provide further support for the idea that a psychophysiological stress response is nature's fundamental survival mechanism that could be therapeutically harnessed to augment immune function during vaccination, wound healing, or infection.

11.7 Acute Stress-Induced Enhancement of Adaptive/Secondary Immune Responses in Skin

Studies have shown that in addition to enhancing primary cutaneous immune responses, acute stress experienced at the time of antigen re-exposure can also enhance secondary or recall responses in skin [24]. Compared to nonstressed controls, mice that were acutely stressed at the time of antigen re-exposure showed a significantly larger number of infiltrating leukocytes at the site of the immune reaction. These results demonstrated that a relatively mild behavioral manipulation can enhance an important class of immune responses that mediate harmful (allergic dermatitis) as well as beneficial (resistance to certain viruses, bacteria, and tumors) aspects of immune function.

Other groups have similarly shown enhancement of the elicitation/recall phase of cell-mediated immunity by different stressors administered at the time of antigen re-exposure, in mice, rats, hamsters, and non-human primates [9,12,18,88]. Flint et al. showed that acute stress enhanced CHS responses in both male and female mice [37]; however, they failed to observe the stress-induced enhancement of the sensitization phase of CHS [38] that has been reported by several independent groups as described above [10,12,18,29,69,86,88]. In contrast to the results of several studies (described above) from independent groups, Hosoi et al. have reported a suppression of skin CHS when stress is administered prior to elicitation [45].

While the mechanisms mediating a stress-induced enhancement of in vivo immune responses need to be further investigated, taken together, results from numerous studies show that acute stress can significantly enhance innate as well as adaptive immunity and can potentiate the immunization/sensitization/induction as well as the elicitation/recall phases of an immune response.

11.8 Hormone and Cytokine Mediators of Stress-Induced Enhancement of Immune Function

Although much work remains to be done to identify molecular, cellular, and physiological mechanisms mediating the adjuvant-like, immunoenhancing effects of acute stress, several studies have begun to identify endocrine and immune mediators of these effects. Studies have shown that corticosterone and epinephrine are important mediators of an acute stress-induced immunoenhancement [26]. Adrenalectomy, which eliminates the glucocorticoid and epinephrine stress response, eliminated the stress-induced enhancement of skin cell-mediated immunity (CMI). Low dose corticosterone or epinephrine administration significantly enhanced skin CMI [26]. In contrast, high dose corticosterone, chronic corticosterone, or low dose dexamethasone administration significantly suppressed skin CMI. These results suggested a novel role for adrenal stress hormones as endogenous immunoenhancing agents. They also showed that stress hormones released during a circumscribed or acute stress response may help prepare the immune system for potential challenges (e.g., wounding or infection) for which stress perception by the brain may serve as an early warning signal. Studies by Flint et al. have also suggested that corticosterone is a mediator of the stress-induced enhancement of skin CHS [37], while Saint-Mezard et al have suggested that the adjuvant-like effects of stress on dendritic cell and CD8+ T cell migration and function are mediated by norepinephrine [69].

Studies have also examined the immunological mediators of an acute stress-induced enhancement of skin immunity. Since gamma interferon (IFNγ) is a critical cytokine mediator of cell mediated immunity and delayed as well as contact hypersensitivity, studies were conducted to examine its role as a local mediator of the stress-induced enhancement of skin CMI [33]. The effect of acute stress on skin CMI was examined in wild-type and IFNγ receptor gene knockout mice (IFNγR–/–) that had been sensitized with 2,4-dinitro-1-fluorobenzene (DNFB). Acutely stressed wild-type mice showed a significantly larger CMI response than nonstressed mice. In contrast, IFNγR–/– mice failed to show a stress-induced enhancement of skin CMI. Immunoneutralization of IFNγ in wild-type mice significantly reduced the stress-induced enhancement of skin CMI. In addition, an inflammatory response to direct IFNγ-administration was significantly enhanced by acute stress. These results showed that IFNγ is an important local mediator of a stress-induced enhancement of skin CMI [33]. In addition to IFNγ, TNF-α, MCP-1, MIP-3α, IL-1, and IL-6 have also been associated with a stress-induced enhancement of the immunization/sensitization phase of skin cell-mediated immunity [29,86]. It is clear that further investigation is necessary in order to identify the most important molecular, cellular, and physiological mediators of a stress-induced enhancement of skin immunity.

11.9 Chronic Stress-Induced Suppression of Skin Immunity

In contrast to acute stressors, chronic stress has been shown to suppress or dysregulate immune function, including skin immunity (for review see [11,13,40,46, 49,50,76,90]). Dhabhar et al. conducted studies designed to examine the effects of increasing the intensity and duration of acute stress as well as the transition from acute to chronic stress on skin immune function [25]. These studies showed that acute stress administered for 2h prior to antigenic challenge significantly enhanced skin cell-mediated immunity [25]. Increasing the duration of stress from 2 to 5h produced the same magnitude immunoenhancement. Interestingly, increasing the intensity of acute stress produced a significantly larger enhancement of the CMI response, which was accompanied by increasing magnitudes of leukocyte redeployment. In contrast, these studies found suppression of the skin immune response when chronic stress exposure was begun 3 weeks before sensitization and either discontinued upon sensitization, or continued an additional week until challenge, or extended for 1 week after challenge [25]. Interestingly, acute stress-induced redistribution of peripheral blood lymphocytes was attenuated with increasing duration of stressor exposure and correlated with attenuated glucocorticoid responsivity. These results suggested that stress-induced alterations in lymphocyte redeployment may play an important role in mediating the bi-directional effects of stress on cutaneous cell-mediated immunity [25]. An association between chronic stress and reduced skin cell mediated immunity has also been reported in human subjects [78].

Given the importance of cutaneous cell-mediated immunity in elimination of immunoresponsive tumors like squamous cell carcinoma (SCC) [42,51], Saul et al. examined the effects of chronic stress on susceptibility to ultraviolet radiation (UV) induced SCC [72]. Mice were exposed to a minimal erythemal dose of UVB three times a week for 10 weeks. Half of the UVB-exposed mice were left nonstressed (i.e., they remained in their home cages) and the other half were chronically stressed (i.e., restrained during weeks 4–6). UV-induced tumors were measured weekly from week 11 through week 34, blood was collected at week 34, and tissues were collected at week 35. mRNA expression of IL-12p40, IFN-γ, IL-4, IL-10, CD3ε, and CCL27/CTACK, the skin T cell-homing chemokine, in dorsal skin was quantified using real-time polymerase chain reaction. CD4+, CD8+, and CD25+ leukocytes were counted using immunohistochemistry and flow cytometry.

Stressed mice had a shorter median time to first tumor (15 vs. 16.5 weeks) and reached 50% incidence earlier than controls (15 weeks vs. 21 weeks). Stressed mice also had lower IFN-γ, CCL27/CTACK, and CD3ε gene expression and lower CD4+ and CD8+ T cells infiltrating within and around tumors than nonstressed mice. In addition, stressed mice had higher numbers of tumor infiltrating and circulating CD4+ CD25+ suppressor cells than nonstressed mice. These studies showed that chronic stress increased susceptibility to UV-induced SCC by suppressing skin immunity, Type 1 cytokines, and protective T cells, and increasing active immunosuppression through regulatory/suppressor T cells [72].

11.10 Other Mediators and Effects of Stress on Skin Immune Function

While the above discussion has focused on the role of psychological stress and stress hormones in mediating changes in skin cell-mediated immunity, numerous other studies (many described in this volume) have examined other mediators of the effects of changes in central nervous system activity on skin immunity as well as cutaneous wound healing and changes in skin barrier function [5,19,22]. Although the principal stress hormones appear to play a major role, additional factors that could mediate the effects of stress and other emotional states on skin immunity include the actions of cutaneous nerves and the release of peptides like calcitonin gene related peptide (CGRP), substance P (SP), vasoactive intestinal peptide (VIP) [21,75], and proopiomelanocortin (POMC) peptides [62,77] like alpha-melanocyte stimulating hormone (alpha-MSH), local and systemic release of corticotropin releasing hormone (CRH) [3,34], mast cell factors [3,83], and mediators that also induce neurogenic inflammation [82]. Therefore, it is clear that much research remains to be done in order to identify the various psychophysiological factors that may mediate the effects of stress and other emotional states on skin immune reactivity.

11.11 Conclusion

An important function of physiological mediators released under conditions of acute psychological stress may be to ensure that appropriate leukocytes are present in the right place and at the right time to respond to an immune challenge which might be initiated by the stress-inducing agent (e.g., attack by a predator, invasion by a pathogen, etc.). The modulation of immune cell distribution by acute stress may be an adaptive response designed to enhance immune surveillance and increase the capacity of the immune system to respond to challenge in immune compartments (such as the skin), which serve as major defense barriers for the body. Thus, neurotransmitters and hormones released during stress may increase immune surveillance and help enhance immune preparedness for potential (or ongoing) immune challenge. Stress-induced immunoenhancement may increase immunoprotection during surgery, vaccination, or infection, but may also exacerbate immunopathology during inflammatory (dermatitis, cardiovascular disease, gingivitis) or autoimmune (psoriasis, arthritis, multiple sclerosis) diseases that are known to be negatively affected by stress [1,4,6,39].

Summary for the Clinician

The relationships between immune function and the physiological manifestations of stress are complex. Clinically, the important connection between stress and the exacerbation of skin disorders has been known for a long time [16,19,87]. The studies described here shed light on potential mechanisms that may mediate the bi-directional effects of stress on skin immune function, and provide substrates for clinical interventions that may be designed to dampen or eliminate stress-induced exacerbation of skin inflammation. While decades of research have examined the pathological effects of stress on immune function and on health, the study of salubrious or health-promoting effects of stress is relatively new [28,29]. Therefore, the studies presented here also provide a framework for developing therapeutic interventions that harness the mind and body's endogenous health-promoting mechanisms to enhance protective immunity during vaccination, infection, or wound healing. Much work remains to be done to elucidate the mechanisms mediating the salubrious vs. health-aversive effects of stress and to translate basic findings in the field from bench to bedside. However, this work is extremely important because stress is a ubiquitous aspect of life and is thought to play a role in the etiology of numerous diseases.

Acknowledgments

I thank current and previous members of my laboratory, particularly, Dr. Kavitha Viswanathan, Dr. Alison Saul, Kanika Ghai, Christine Daugherty, and Jean

Tillie, whose work and publications are among those discussed in this chapter. The work described here was supported by grants from the NIH (AI48995 and CA107498) and The Dana Foundation.

References

1. Ackerman KD, Heyman R, Rabin BS, Anderson BP, Houck PR, Frank E, Baum A (2002) Stressful life events precede exacerbations of multiple sclerosis. Psychosom Med 64(6):916–920
2. Ader R (2006) Psychoneuroimmunology IV. 4th ed, Academic Press, San Diego, p 1583
3. Akiyama H, Amano H, Bienenstock J (2005) Rat tracheal epithelial responses to water avoidance stress. J Allergy Clin Immunol 116(2):318–324
4. Al'Abadie MS, Kent GG, Gawkrodger DJ (1994) The relationship between stress and the onset and exacerbation of psoriasis and other skin conditions. Brit J Dermatol 130:199–203
5. Altemus M, Rao B, Dhabhar FS, Ding W, Granstein R (2001) Stress-induced changes in skin barrier function in healthy women. J Invest Dermatol 117:309–317
6. Amkraut AA, Solomon CF, Kraemer HC (1971) Stress, early experience and adjuvant-induced arthritis in the rat. Psychosom Med 33:203–214
7. Benschop RJ, Oostveen FG, Heijnen CJ, Ballieux RE (1993) Beta 2-adrenergic stimulation causes detachment of natural killer cells from cultured endothelium. Eur J Immunol 23(12):3242–3247
8. Benschop RJ, Rodriguez-Feuerhahn M, Schedlowski M (1996) Catecholamine-induced leukocytosis: early observations, current research, and future directions. Brain Behav Immun 10(2):77–91
9. Bilbo SD, Dhabhar FS, Viswanathan K, Saul A, Yellon SM, Nelson RJ (2002) Short day lengths augment stress-induced leukocyte trafficking and stress-induced enhancement of skin immune function. Proc Natl Acad Sci U S A 99(6):4067–4072
10. Bilbo SD, Hotchkiss AK, Chiavegatto S, Nelson RJ (2003) Blunted stress responses in delayed type hypersensitivity in mice lacking the neuronal isoform of nitric oxide synthase. J Neuroimmunol 140(1–2):41–48
11. Black PH (1994) Immune system-central nervous system interactions: effect and immunomodulatory consequences of immune system mediators on the brain. Antimicrobial Agents Chemother 38(1):7–12
12. Blecha F, Barry RA, Kelley KW (1982) Stress-induced alterations in delayed-type hypersensitivity to SRBC and contact sensitivity to DNFB in mice. Proc Soc Exp Biol Med 169:239–246
13. Borysenko M, Borysenko J (1982) Stress, behavior, and immunity: animal models and mediating mechanisms. General Hospital Psychiatr 4:59–67
14. Bosch JA, Berntson GG, Cacioppo JT, Dhabhar FS, Marucha PT (2003) Acute stress evokes selective mobilization of T cells that differ in chemokine receptor expression: a potential pathway linking immunologic reactivity to cardiovascular disease. Brain Behav Immun 17:251–259
15. Brosschot JF, Benschop RJ, Godaert GL, Olff M, De Smet M, Heijnen CJ, Ballieux RE (1994) Influence of life stress on immunological reactivity to mild psychological stress. Psychosom Med 56(3):216–224
16. Buske-Kirschbaum A, Geiben A, Hellhammer D (2001) Psychobiological aspects of atopic dermatitis: an overview. Psychother Psychosom 70(1):6–16
17. Carlson SL, Fox S, Abell KM (1997) Catecholamine modulation of lymphocyte homing to lymphoid tissues. Brain Behav Immun 11(4):307–320
18. Coe CL, Lubach G, Ershler WB (1989) Immunological consequences of maternal separation in infant primates, in Infant stress and coping, In: Lewis M, Worobey J (ed) Jossey-Bass Inc., New York, NY
19. Colavincenzo ML, Granstein RD (2006) Stress and the skin: a meeting report of the Weill Cornell symposium on the science of dermatology. J Invest Dermatol 126(12):2560–2561
20. Cunnick JE, Lysle DT, Kucinski BJ, Rabin BS (1990) Evidence that shock-induced immune suppression is mediated by adrenal hormones and peripheral beta-adrenergic receptors. Pharmacol Biochem Behav 36:645–651
21. Delgado M, Gonzalez-Rey E, Ganea D (2006) Vasoactive intestinal peptide: the dendritic cell → regulatory T cell axis. Ann N Y Acad Sci 1070:233–238
22. Denda M, Tsuchiya T, Hosoi J, Koyama J (1998) Immobilization-induced and crowded environment-induced stress delay barrier recovery in murine skin. Br J Dermatol 138(5):780–785
23. Dhabhar FS (1998) Stress-induced enhancement of cell-mediated immunity. Ann N Y Acad Sci 840:359–372
24. Dhabhar FS, McEwen BS (1996) Stress-induced enhancement of antigen-specific cell-mediated immunity. J Immunol 156:2608–2615
25. Dhabhar FS, McEwen BS (1997) Acute stress enhances while chronic stress suppresses immune function in vivo: a potential role for leukocyte trafficking. Brain Behav Immun 11(4):286–306
26. Dhabhar FS, McEwen BS (1999) Enhancing versus suppressive effects of stress hormones on skin immune function. Proc Natl Acad Sci U S A 96:1059–1064
27. Dhabhar FS, McEwen BS (2001) Bidirectional effects of stress and glucocorticoid hormones on immune function: possible explanations for paradoxical observations, In: Ader R, Felten DL, Cohen N (ed) Psychoneuroimmunology, 3rd ed, Academic Press, San Diego, pp 301–338
28. Dhabhar FS, McEwen BS (2006) Bidirectional effects of stress on immune function: possible explanations for salubrious as well as harmful effects, In: Ader R (ed) Psychoneuroimmunology IV, Elsevier, San Diego, pp 723–760
29. Dhabhar FS, Viswanathan K (2005) Short-term stress experienced at the time of immunization induces a long-lasting increase in immunological memory. Am J Physiol Regul Integr Comp Physiol 289(3):R738–744
30. Dhabhar FS, Miller AH, Stein M, McEwen BS, Spencer RL (1994) Diurnal and stress-induced changes in distribution of peripheral blood leukocyte subpopulations. Brain Behav Immun 8:66–79
31. Dhabhar FS, Miller AH, McEwen BS, Spencer RL (1995) Effects of stress on immune cell distribution – dynamics and hormonal mechanisms. J Immunol 154:5511–5527
32. Dhabhar FS, Miller AH, McEwen BS, Spencer RL (1996) Stress-induced changes in blood leukocyte distribution – role of adrenal steroid hormones. J Immunol 157:1638–1644

33. Dhabhar FS, Satoskar AR, Bluethmann H, David JR, McEwen BS (2000) Stress-Induced Enhancement of Skin Immune Function: A Role For IFNγ. Proc Natl Acad Sci U S A 97:2846–2851
34. Donelan J, Boucher W, Papadopoulou N, Lytinas M, Papaliodis D, Dobner P, Theoharides TC (2006) Corticotropin-releasing hormone induces skin vascular permeability through a neurotensin-dependent process. Proc Natl Acad Sci U S A 103(20):7759–7764
35. Fauci AS, Dale DC (1974) The effect of in vivo hydrocortisone on subpopulations of human lymphocytes. J Clin Invest 53:240–246
36. Fauci AS, Dale DC (1975) The effect of hydrocortisone on the kinetics of normal human lymphocytes. Blood 46:235–243
37. Flint MS, Miller DB, Tinkle SS (2000) Restraint-induced modulation of allergic and irritant contact dermatitis in male and female B6.129 mice. Brain Behav Immun 14(4):256–269
38. Flint MS, Valosen JM, Johnson EA, Miller DB, Tinkle SS (2001) Restraint stress applied prior to chemical sensitization modulates the development of allergic contact dermatitis differently than restraint prior to challenge. J Neuroimmunol 113(1):72–80
39. Garg A, Chren MM, Sands LP, Matsui MS, Marenus KD, Feingold KR, Elias PM (2001) Psychological stress perturbs epidermal permeability barrier homeostasis: implications for the pathogenesis of stress-associated skin disorders. Arch Dermatol 137(1):53–59
40. Glaser R, Kiecolt-Glaser JK (2005) Stress-induced immune dysfunction: implications for health. Nat Rev Immunol 5(3):243–251
41. Goldstein DS, McEwen B (2002) Allostasis, homeostats, and the nature of stress. Stress 5(1):55–58
42. Granstein RD, Matsui MS (2004) UV radiation-induced immunosuppression and skin cancer. Cutis 74(5 Suppl):4–9
43. Herbert TB, Cohen S (1993) Stress and immunity in humans: a meta-analytic review. Psychosom Med 55:364–379
44. Hoagland H, Elmadjian F, Pincus G (1946) Stressful psychomotor performance and adrenal cortical function as indicated by the lymphocyte reponse. J Clin Endocrinol 6:301–311
45. Hosoi J, Tsuchiya T, Denda M, Ashida Y, Takashima A, Granstein RD, Koyama J (1998) Modification of LC phenotype and suppression of contact hypersensitivity response by stress. J Cutan Med Surg 3(2):79–84
46. Irwin M (1994) Stress-induced immune suppression: role of brain corticotropin releasing hormone and autonomic nervous system mechanisms. Adv Neuroimmunol 4:29–47
47. Irwin M, Patterson T, Smith TL, Caldwell C, Brown SA, Gillin CJ, Grant I (1990) Reduction of immune function in life stress and depression. Biol Psychiatr. 27:22–30
48. Jensen MM (1969) Changes in leukocyte counts associated with various stressors. J Reticuloendothelial Soc 8:457–465
49. Khansari DN, Murgo AJ, Faith RE (1990) Effects of stress on the immune system. Immunol Today 11(5):170–175
50. Kort WJ (1994) The effect of chronic stress on the immune system. Adv Neuroimmunol 4:1–11
51. Kripke ML (1994) Ultraviolet radiation and immunology: something new under the sun – presidential address. Cancer Res 54(23):6102–6105
52. Manuck SB, Cohen S, Rabin BS, Muldoon MF, Bachen EA (1991) Individual differences in cellular immune response to stress. Psychol Sci 2(2):111–115
53. McEwen BS (1998) Protective and damaging effects of stress mediators: allostasis and allostatic load. NEJM 338(3): 171–179
54. McEwen BS (2002) The End of Stress As We Know It. Dana Press, Washington, DC, p 239
55. Miller AH, Spencer RL, Hasset J, Kim C, Rhee R, Cira D, Dhabhar FS, McEwen BS, Stein M (1994) Effects of selective Type I and Type II adrenal steroid receptor agonists on immune cell distribution. Endocrinology 135(5):1934–1944
56. Mills PJ, Berry CC, Dimsdale JE, Ziegler MG, Nelesen RA, Kennedy BP (1995) Lymphocyte subset redistribution in response to acute experimental stress: effects of gender, ethnicity, hypertension, and the sympathetic nervous system. Brain Behav Immun 9:61–69
57. Mills PJ, Ziegler MG, Rehman J, Maisel AS (1998) Catecholamines, catecholamine receptors, cell adhesion molecules, and acute stressor-related changes in cellular immunity. Adv Pharmacol 42:587–590
58. Mills PJ, Meck JV, Waters WW, D'Aunno D, Ziegler MG (2001) Peripheral leukocyte subpopulations and catecholamine levels in astronauts as a function of mission duration. Psychosom Med 63(6):886–890
59. Monjan AA, Collector MI (1977) Stress-induced modulation of the immune response. Science 196:307–308
60. Morrow-Tesch JL, McGlone JJ, Norman RL (1993) Consequences of restraint stress on natural killer cell activity, behavior, and hormone levels in Rhesus Macaques (*Macaca mulatta*). Psychoneuroendocrinol 18:383–395
61. Naliboff BD, Benton D, Solomon GF, Morley JE, Fahey JL, Bloom ET, Makinodan T, Gilmore SL (1991) Immunological changes in young and old adults during brief laboratory stress. Psychosom Med 53:121–132
62. Paus R, Botchkarev VA, Botchkareva NV, Mecklenburg L, Luger T, Slominski A (1999) The skin POMC system (SPS). Leads and lessons from the hair follicle. Ann N Y Acad Sci 885:350–363
63. Pickford GE, Srivastava AK, Slicher AM, Pang PKT (1971) The stress response in the abundance of circulating leukocytes in the Killifish, *Fundulus heteroclitus*. I. The cold-shock sequence and the effects of hypophysectomy. J Exp Zool 177:89–96
64. Pruett SB (2001) Quantitative aspects of stress-induced immunomodulation. Int Immunopharmacol 1(3):507–520
65. Redwine L, Snow S, Mills P, Irwin M (2003) Acute psychological stress: effects on chemotaxis and cellular adhesion molecule expression. Psychosom Med 65(4):598–603
66. Redwine L, Mills PJ, Sada M, Dimsdale J, Patterson T, Grant I (2004) Differential immune cell chemotaxis responses to acute psychological stress in Alzheimer caregivers compared to non-caregiver controls. Psychosom Med 66(5):770–775
67. Rinder CS, Mathew JP, Rinder HM, Tracey JB, Davis E, Smith BR (1997) Lymphocyte and monocyte subset changes during cardiopulmonary bypass: effects of aging and gender [see comments]. J Lab Clin Med 129(6):592–602
68. Rinner I, Schauenstein K, Mangge H, Porta S, Kvetnansky R (1992) Opposite effects of mild and severe stress on in vitro activation of rat peripheral blood lymphocytes. Brain Behav Immun 6:130–140
69. Saint-Mezard P, Chavagnac C, Bosset S, Ionescu M, Peyron E, Kaiserlian D, Nicolas JF, Berard F (2003) Psychological

stress exerts an adjuvant effect on skin dendritic cell functions in vivo. J Immunol 171(8):4073–4080
70. Sapolsky RM (2004) Why Zebras Don't Get Ulcers, 3rd ed. W.H. Freeman and Company, San Fransisco, CA, p 560
71. Sapolsky RM (2005) The influence of social hierarchy on primate health. Science 308:648–652
72. Saul AN, Oberyszyn TM, Daugherty C, Kusewitt D, Jones S, Jewell S, Malarkey WB, Lehman A, Lemeshow S, Dhabhar FS (2005) Chronic stress and susceptibility to skin cancer. J Nat Cancer Institute 97:1760–1767
73. Schedlowski M, Jacobs R, Stratman G, Richter S, Hädike A, Tewes U, Wagner TOF, Schmidt RE (1993) Changes of natural killer cells during acute psychological stress. J Clin Immunol 13(2):119–126
74. Schwab CL, Fan R, Zheng Q, Myers LP, Hebert P, Pruett SB (2005) Modeling and predicting stress-induced immunosuppression in mice using blood parameters. Toxicol Sci 83(1):101–113
75. Seiffert K, Granstein RD (2006) Neuroendocrine regulation of skin dendritic cells. Ann N Y Acad Sci 1088:195–206
76. Sklar LS, Anisman H (1981) Stress and cancer. Psychol Bull 89(3):369–406
77. Slominski A, Wortsman J, Luger T, Paus R, Solomon S (2000) Corticotropin releasing hormone and proopiomelanocortin involvement in the cutaneous response to stress. Physiol Rev 80(3):979–1020
78. Smith A, Vollmer-Conna U, Bennett B, Wakefield D, Hickie I, Lloyd A (2004) The relationship between distress and the development of a primary immune response to a novel antigen. Brain Behav Immun 18(1):65–75
79. Snow DH, Ricketts SW, Mason DK (1983) Hematological responses to racing and training exercise in Thoroughbred horses, with particular reference to the leukocyte response. Equine Vet J 15(2):149–154
80. Sprent J, Tough DF (1994) Lymphocyte life-span and memory. Science 265:1395–1400
81. Stein M, Ronzoni E, Gildea EF (1951) Physiological responses to heat stress and ACTH of normal and schizophrenic subjects. Am J Psychiatr 6:450–455
82. Steinhoff M, Stander S, Seeliger S, Ansel JC, Schmelz M, Luger T (2003) Modern aspects of cutaneous neurogenic inflammation. Arch Dermatol 139(11):1479–1488
83. Theoharides TC, Kalogeromitros D (2006) The critical role of mast cells in allergy and inflammation. Ann N Y Acad Sci 1088:78–99
84. Toft P, Svendsen P, Tonnesen E, Rasmussen JW, Christensen NJ (1993) Redistribution of lymphocytes after major surgical stress. Acta Anesthesiol Scand 37:245–249
85. Viswanathan K, Dhabhar FS (2005) Stress-induced enhancement of leukocyte trafficking into sites of surgery or immune activation. Proc Natl Acad Sci U S A 102(16): 5808–5813
86. Viswanathan K, Daugherty C, Dhabhar FS (2005) Stress as an endogenous adjuvant: augmentation of the immunization phase of cell-mediated immunity. Int Immunol 17(8):1059–1069
87. Weston WL, Huff JC (1976) Atopic dermatitis: etiology and pathogenesis. Pediatr Ann 5(12):759–762
88. Wood PG, Karol MH, Kusnecov AW, Rabin BS (1993) Enhancement of antigen-specific humoral and cell-mediated immunity by electric footshock stress in rats. Brain Behav Immun 7:121–134
89. Zalcman S, Anisman H (1993) Acute and chronic stressor effects on the antibody response to sheep red blood cells. Pharmacol Biochem Behav 46(2):445–452
90. Zwilling B (1992) Stress affects disease outcomes. Confronted with infectious disease agents, the nervous and immune systems interact in complex ways. ASM News 58(1):23–25

12 Photoneuroimmunology: Modulation of the Neuroimmune System by UV Radiation

P.H. Hart, J.J. Finlay-Jones, and S. Gorman

Contents

Synonyms Box: α-MSH, α-melanocyte stimulating hormone; CGRP, calcitonin gene-related protein; MC-1R, melanocortin-1 receptor; NGF, nerve growth factor; *cis*-UCA, *cis*-urocanic acid; *trans*-UCA, *trans*-urocanic acid; UVR, ultraviolet radiation

Key Features

- UVR component of sunlight is immunomodulatory with suppression of cellular immunity to antigens applied at both irradiated and non-irradiated sites.
- Via isomerization of *trans*-urocanic acid in the stratum corneum to its more soluble *cis* isomer, UVR activates peripheral sensory nerves for release of calcitonin gene-related protein and substance P.
- Dermal mast cells activated by neuropeptides contribute to the immunomodulatory properties of UV.
- UVR-induced keratinocyte nerve growth factor production augments activation of peripheral sensory nerves for neuropeptide release.
- UVR-induced keratinocyte α-melanocyte stimulating hormone is immunosuppressive by stimulating keratinocyte IL-10 production, as well as increasing melanogenesis and repair of UV-induced DNA damage in melanocytes and keratinocytes.
- There is debate upon whether UVA wavelengths (400–320 nm) have immunosuppressive properties similar to those of UVB (320–290 nm).

12.1 Introduction

Sunlight contains visible light (400–700 nm) as well as ultraviolet radiation (UVR); the latter is arguably one of the most important environmental insults and stressors to human skin. Environmental and dermatological photobiologists define UVA radiation as 400–320 nm, UVB radiation as 320–290 nm, and UVC radiation as 290–200 nm [15]. The UV component of terrestrial

R.D. Granstein and T.A. Luger (eds.), *Neuroimmunology of the Skin*,

sunlight is composed of approximately 5% UVB and 95% UVA radiation. UVC radiation is not present in sunlight reaching the earth's surface as it is effectively blocked by stratospheric ozone. UVR can be produced in the laboratory by use of optically filtered xenon arc lamps or more commonly by specialized fluorescent lamps. Most studies described in this chapter refer to the use of a UVR source enriched in UVB radiation (containing approximately 60% UVB radiation), although some studies have used solar-simulated light. In Sect. 12.7, investigations of the comparative immunoregulatory roles for UVA vs. UVB wavelengths are discussed.

As detailed many times in this book, strong evidence now exists that the immune and neuronal systems in the skin are a highly integrated and interactive unit with neuropeptides and neuroendocrine hormones of neuronal and non-neuronal origin, respectively, responsible for much of the cross-talk. In this chapter we present evidence that the neuroimmune system via neuropeptide–cell interactions plays an important role in the mechanisms by which UVR is immunomodulatory. Many laboratories [18], including our own [62], have shown that immune responses in neuropeptide-depleted mice (resulting from capsaicin treatment early in life) are not suppressed by UV irradiation. Further, capsaicin applied 48 and 24 h before exposure of human skin to erythemal UV reversed the ability of UVR to suppress Mantoux reactions in volunteers sensitive to tuberculin purified protein derivative [32]. Neuropeptides and neuroendocrine hormones both contribute to UV-induced immunomodulation, as well as to adaptive responses of skin to UV.

12.2 Photoneuroimmunology, Health, and Disease

The contribution of UV-induced immunomodulation to skin cancer development was first recognized two decades ago. Tumors develop in mice treated with a carcinogenic dose of UVR. If a UVR-induced tumor is transferred to a syngeneic mouse with a competent immune system, it is rejected [36]. However, if the recipient mouse is given a suberythemal dose of UVR before transfer, the tumor grows [36]. These, and other data, have demonstrated a dual role for UVR in skin cancer induction, where UVR induces DNA damage and also modulates the immune system. UVR-induced immunomodulation in humans for the progression of skin cancers has also been documented. Perhaps the strongest evidence relates to the increased skin cancer rate in transplant patients on immunosuppressive therapies [47]. Patients with non-melanoma skin cancers have a 30% increased risk of developing other cancers, which is consistent with the systemic nature of the immunomodulation induced by UVR [33]. It has been hypothesized that UVR-induced immunomodulation protects against over-zealous responses to nuclear antigens exposed by UVR-killed (sunburn) cells.

Further evidence of the immunomodulatory effects of UV radiation has been derived in the clinic, where UVR is used to treat inflammatory skin conditions such as psoriasis [40]. Suppression of the immune system by UVR may also affect responses to vaccines and to infectious agents [45]. In animal models, several experimental bacterial, viral, and protozoan infections have developed in UV-irradiated but not unirradiated hosts and suggests that individuals living at low latitudes may have an enhanced susceptibility to infectious diseases. Populations living at higher latitudes have an enhanced incidence of multiple sclerosis [65], and perhaps other autoimmune diseases, possibly due to a reduction in immunosuppressive UV radiation. Epidemiological studies are required to consolidate these proposals and to determine the contribution, if any, of UV-induced vitamin D status as a significant contributor to the disease patterns for populations dwelling at different latitudes. A potential interaction between the formation of 1,25 $(OH)_2$ vitamin D3 by UVR and photoneuroimmunology is currently unknown. At present, the contribution of vitamin D status to the immunomodulatory effects following both acute and chronic exposure to UV irradiation remains a subject of intense research.

Although studies with human skin are performed whenever possible, UVR-induced immunosuppression is most commonly studied in rodents by analysis of alterations to delayed or contact hypersensitivity responses to an experimental antigen or hapten. UVR modulates immune responses by multiple mechanisms. However, mechanisms clearly differ if the sensitizing antigen is applied to irradiated skin or to a non-irradiated site. "Local" UVR-induced immunosuppression is shown by application of the sensitizing antigen to the same site that has been irradiated and can be detected even with sub-erythemal amounts of UVR. The mechanisms of UVR-induced local immunosuppression involve modified antigen presentation by UVR-damaged Langerhans cells or immature dermal dendritic cells (for review, [7,64]).

The mechanisms behind "systemic" immunosuppression are less clear, particularly as the applied antigen will be handled by Langerhans cells or dendritic cells from unirradiated sites. For systemic immunosuppression, erythemal amounts of UVR are required, with UVR administered at least 3 days before antigen sensitization. The irradiation of skin may affect responses to antigens administered to distant skin, the airways, or the peritoneal cavity, to name but a few possible sites.

Dermal mast cells at the site of UV irradiation play an important role in the mechanism by which UV radiation is immunoregulatory to antigens applied to distant sites. From studies of UV-induced systemic suppression of contact hypersensitivity responses in mice, the following is known: (a) the prevalence of mast cells correlates directly with their susceptibility to UV-induced systemic immunosuppression, (b) BALB/c mice carrying Uvs1, a major locus for susceptibility to UV-induced immunosuppression, contain greater concentrations of dermal mast cells than BALB/c mice of the same parental origin, and (c) mast cell-depleted mice (Wf/Wf) are not susceptible to UV-induced systemic immunosuppression unless they were reconstituted on their dorsal skin with bone marrow-derived mast cells precursors from nonmutant mice, at least 6 weeks prior to UV irradiation at the same site [28]. Mast cells are also essential for UV-induced local immunosuppression [49] and suppression of delayed type hypersensitivity responses to allogeneic spleen cells [28].

Many products from mast cells may contribute to immunomodulatory events by production of TNFα and IL-10 [49]. Further, histamine working alone or together with other mast cell or keratinocyte products may initiate downstream immunomodulation. Histamine stimulates prostaglandin E_2 production by cultured keratinocytes, where prostaglandin E_2 may act locally on epidermal cells, in lymph nodes, or systemically [64]. Prostanoids are responsible for at least 50% of the systemic suppression of contact hypersensitivity responses caused by UV irradiation [27]. Histamine may modulate the maturation and activation of dendritic cells [44] that then contribute towards UVR-induced systemic immunomodulation. However, we have been unable to detect changes to CD11c-expressing antigen presenting cells in the lymph nodes of mice draining distant sites of antigen challenge [22]. As detailed below, neuroendocrine hormones from UVR-degranulating mast cells may also contribute to the adaptive response of skin to UVR, and the maintenance of epidermal health [3].

In both local and systemic immunosuppression caused by UVR, the neuroimmune system, including products of keratinocytes and nerves, has been implicated. Perhaps this is not surprising due to the important role of mast cells in UVR-immunoregulation, the close positioning of mast cells adjacent to nerves in skin [16] and the fact that neuropeptides are the major mechanism of IgE-independent mast cell activation.

12.3 Skin Photoreceptors for UVR. Links to the Neuroimmune System

Very little UVB radiation penetrates skin beneath the epidermis. Thus, photoreceptors in skin for UVR have been keenly investigated. UVR-initiated responses in the epidermis include cellular DNA damage, oxidation of membrane proteins, and alterations to intracellular signaling events, leading to deficient function or production of immunocompetent molecules. It has been hypothesized that UVR causes the oxidation of phosphatidylcholine in the membranes of irradiated skin cells [66]. In turn, oxidized phosphatidylcholine binds to, and activates, receptors for platelet activating factor on cells and stimulates a variety of downstream effects, including activation of MAP kinases, phospholipases, and gene transcription.

Several types of DNA damage may be induced by UVR, with cyclobutane pyrimidine dimers now recognized as the predominant DNA lesions in whole human skin exposed to UVA and UVB radiation [46]. Mechanisms of life and death and DNA repair are being investigated in UV-irradiated keratinocytes, melanocytes, and Langerhans cells and their role in UVR-immunoregulation studied. Langerhans cells and other dendritic cells with UVR-damaged DNA present antigen suboptimally in the skin-draining lymph nodes, which suppress immunity to the antigen, and induce regulatory cells and sometimes tolerance to the antigen [7]. As will be discussed later, UVR-induced neuroendocrine hormones may play an important role in the repair of UVR-induced pyrimidine dimers and the extent of UVR-induced immunomodulation, as well as the survival of precursor cells for keratinocyte and melanocyte tumors.

Another important process induced by UVR is the isomerization of *trans*-urocanic acid (*trans*-UCA) to its *cis* isomer [51]. *Trans*-UCA (deaminated histidine) is a molecular species located superficially in the stratum corneum of skin (comprising 0.5% of its dry weight). Upon irradiation *trans*-UCA is converted to its more soluble *cis* isomer. The action spectrum of UVR-induced suppression of contact hypersensitivity responses in mice closely follows the absorption spectrum of UCA [13]. Mice genetically deficient in histidase, and thus in skin UCA, are not susceptible to immunosuppression caused by UVR irradiation [51]. UCA isomers in human urine may be a biomarker of recent UVA/UVB radiation exposure [55]. A single exposure to approximately 70% of an MED of solar simulated UVR to 90% of skin area produced a five-fold increase in the ratio in urine of *cis*-UCA to *trans*-UCA relative to baseline [55]. The role of *cis*-UCA in UV-induced immunomodulation may vary with the response examined and the experimental model adopted but significant roles in delayed and contact hypersensitivity responses have been confirmed by neutralization studies with a *cis*-UCA monoclonal antibody (for review, [35]). The receptor and the target cells for the actions of *cis*-UCA remain a subject of investigation.

12.4 Stimulation by *cis*-Urocanic Acid of Neuropeptide Release from Peripheral Sensory Nerves

cis-UCA, but not *trans*-UCA, can activate sensory nerves in skin to release neuropeptides [35]. To investigate sensory nerve activation, a blister was induced on the hind footpad of rats, the surface epithelium removed, and a perspex chamber with inlet and outlet ports fixed over the blister base. A peristaltic pump controlled perfusion of UCA solutions, whilst the cutaneous blood flow was measured by an attached laser Doppler flowmeter probe [35]. *cis*-UCA at concentrations measured in UV-irradiated skin, but not *trans*-UCA, increased microvascular flow. Further, by use of receptor antagonists to substance P and CGRP, the action of *cis*-UCA largely depended on the combined activity of the neuropeptides, substance P, and calcitonin gene-related peptide (CGRP). Perfusion of *cis*-UCA over the base of blisters induced in rats depleted of sensory neuropeptides by subcutaneous injection of capsaicin when two days of age did not have any stimulatory effect above that seen with perfusion of *cis*-UCA together with neuropeptide receptor antagonists in control rats. Finally, the effect of *cis*-UCA on immunosuppression, not blood flow, was examined in sensory neuropeptide-depleted mice. Mice administered *cis*-UCA or UV to dorsal skin prior to hapten sensitization had a significantly reduced immune response to that hapten. However, if the mice were depleted of neuropeptides when 4 weeks of age, neither UV nor *cis*-UCA was immunoregulatory in adult mice [35]. Garssen and colleagues [18] first showed that UV-induced systemic immunosuppression was dependent on neuropeptides from sensory nerves, but this was the first study to determine a UV-induced mechanism for the stimulation of the peripheral terminals of primary afferent sensory nerves (Fig. 12.1).

The identity of the receptor for *cis*-UCA on the peripheral sensory nerves remains elusive. It had been reported that *cis*-UCA bound competitively to receptors for γ-amino-butyric acid (GABA) on rat cortex membranes [38]. However, GABA receptors are not generally found on peripheral nerves. Further, GABA at 10 mM did not alter microvascular flow in the rat blister model [35]. It was hypothesized that *cis*-UCA bound to a histamine receptor but quite distinct functions for *cis*-UCA and histamine have now been described on monocytes, keratinocytes, and fibroblasts (for review, [68]). Because of some structural similarities between *cis*-UCA and serotonin, there have also been suggestions that *cis*-UCA binds to the 5-hydroxytryptamine (5-HT)2A receptor. However, by studying the mechanism by which *cis*-UCA suppressed lipopolysaccharide-induced TNFα production by human monocytes, we have clearly shown that *cis*-UCA, serotonin, and a 5-HT2A receptor agonist use different membrane receptors [68].

12.5 Immunoregulatory Properties of UVR/ *cis*-Urocanic Acid-Induced Neuropeptides

12.5.1 Calcitonin Gene-Related Peptide

A link between UVR exposure and CGRP release from peripheral sensory nerves was first reported in rats in the early to mid 1990s, with CGRP implicated in UVR-induced vasodilation of skin [8,20]. An immune regulatory consequence of these CGRP changes was soon identified with numerous reports that UVR-induced CGRP was responsible for some of the suppressive properties of UVB on local contact hypersensitivity responses [4,21]. Topical application of CGRP had many of the same histological and functional consequences as UV, including a reduction of Ia-positive epidermal antigen presenting cells and suppressed contact hypersensitivity responses. Studies depending on the neutralizing properties of the CGRP receptor antagonist, $CGRP_{8-37}$, confirmed the functional involvement of CGRP in UVR-induced immunomodulation [18,57]. CGRP is

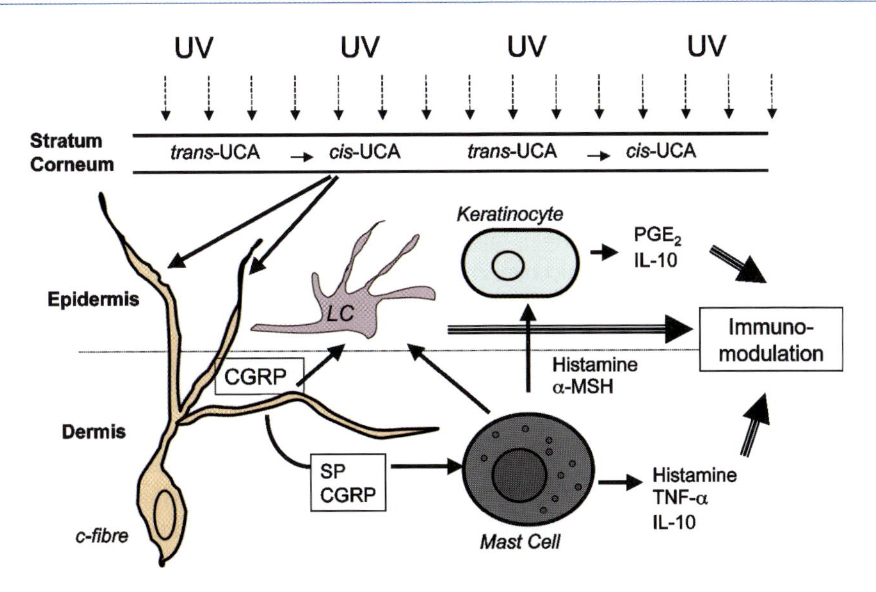

Fig. 12.1 Induction of immunomodulation by *cis*-urocanic acid (*cis*-UCA). *Trans*-urocanic acid in the stratum corneum of skin isomerizes to its *cis* form upon UV irradiation. *cis*-UCA stimulates peripheral sensory nerves by an as-yet-unidentified receptor for release of the neuropeptides CGRP and substance P (SP). Both CGRP and substance P degranulate dermal mast cells for release of the immunomodulatory mediators, histamine, α-MSH, IL-10, and TNF-α. In turn, these products of mast cells can contribute to the cross-talk between various skin cells, as well as more systemic target cells. CGRP as a product of activated sensory c-fibres can regulate the function of Langerhans cells (LC) migrating to draining lymph nodes

critical for not only local but also systemic suppression of contact hypersensitivity responses by UV radiation. Further, $CGRP_{8-37}$ reduced immunoregulation caused by a single erythemal dose of UV, and also to multiple suberythemal doses of UV administered daily for 4 days [18].

Most of the early studies (for review, [59]) hypothesized that CGRP was downregulating the activity of Langerhans cells, the major antigen presenting cells of the epidermis (Fig. 12.1). An intimate association of CGRP-containing epidermal nerves with Langerhans cells was shown in the early 1990s by confocal and transmission electron microscopy and further, CGRP was identified on the surface of some Langerhans cells [31]. Functionally, CGRP inhibited the ability of murine epidermal cells to stimulate allogeneic T cells and to present protein antigen to a T cell hybridoma [31]. Also, pre-exposure of Langerhans cells to CGRP before pulsing with tumor-associated antigens resulted in their reduced ability to elicit a delayed-type hypersensitivity response [31]. Activation of the CGRP receptor on Langerhans cells increases production of cyclic AMP by adenylate cyclase and augmentation of lipopolysaccharide- and granulocyte-macrophage colony stimulating factor-induced IL-10 production [5]. IL-10 downregulates CD86 and inhibits antigen presentation by Langerhans cells [39].

A further target cell for the functional activity of UV-induced CGRP is the dermal mast cell (Fig. 12.1). Neuropeptides are well characterized degranulating agents for rodent and human mast cells, with mast cell products such as histamine implicated in the vasodilatory effects of UV. The contribution of mast cells to UV-immunosuppression has been demonstrated by use of mast cell deficient mice and an inability of both UV and *cis*-UCA to be immunosuppressive in these mice [28,30,49]. In one murine study of hapten applied to the irradiated site, UVR-induced CGRP stimulated degranulation of mast cells causing the release of pre-formed TNFα which both reduced epidermal Ia density and suppressed a contact hypersensitivity response to the hapten. An anti-TNFα antibody administered intraperitoneally 6 h before the CGRP administration or UV irradiation of skin reversed CGRP- and UV-induced immunomodulation [49].

UVR-induced systemic immunosuppression is also reversed in mice given the CGRP antagonist, $CGRP_{8-37}$ (5×10^{-9} mol/mouse) when delivered intravenously 30 min before UV irradiation [49]. However, TNFα production by CGRP-activated mast cells is not required for UVB-induced systemic immunomodulation. We [29] and others [2], by use of anti-TNFα antibodies and experiments in TNFα−/− and TNF receptor−/− mice have clearly shown that TNF activity is not responsible for the systemic immunomodulatory effects of UVR. Our studies suggest that mast cell-derived histamine may contribute to the systemic effects of UVR [27,28]. The extent to which CGRP-activated mast cells are necessary for UVR-induced immunosuppression may vary for local and systemic models with a greater dependency in systemic immunosuppression [28]. Further, mast cells are immunomodulatory by different mechanisms depending on whether the experimental antigen is applied to the irradiated site or to a distant non-irradiated site.

The translation of these findings in rodents to UV-irradiated human skin has been addressed. Fifteen volunteers were exposed to UV for five consecutive days with one personally determined just perceptible minimal erythemal dose per day [60]. CGRP levels in human Finn chamber skin samples taken after UV exposure were significantly higher than those taken prior to UV irradiation. Further, CGRP levels of UVB-exposed volunteers positively correlated with the dose of UVR they received.

12.5.2 Substance P

Substance P immunoreactivity in rodent skin increases after UV irradiation [17]. Substance P is generally considered to have a proinflammatory role due to its involvement in the initial vascular inflammation that occurs in UV-irradiated skin, including vasodilation and plasma extravasation, leading to edema and erythema. In addition, substance P via the neurokinin-1 receptor can stimulate increased keratinocyte cytokine production of IL-1, IL-6, and TNFα to create an inflammatory epidermal environment [39,52,57]. In vitro studies have shown that substance P activates antigen presenting cells, leading to the initiation of antigen-specific responses and the production of Th1 cytokines [58]. Reports have differed with respect to the role of substance P in immune responses of experimental animals. A substance P agonist administered within 30 min of the sensitizing agent promoted hapten-specific skin immunity [50]. Further, substance P could reverse UVR-suppression of a local contact hypersensitivity response [61]. However, substance P has not as yet been implicated in UV-induced systemic suppression of contact hypersensitivity [18]. In human skin, topical capsaicin pretreatment had no effect on skin immunity as measured by a Mantoux response in tuberculin-sensitized volunteers [32].

Substance P and CGRP are colocalized in sensory nerves of the skin. From studies that used the rat blister model, both of these molecules were released upon exposure to *cis*-UCA [35]. The paradox of the differing roles for substance P and CGRP released from the same nerve terminals introduces the possibility that they act to reduce the immunoregulatory properties of the other. Further work is required to determine the extent of the interplay between sensory neuropeptides. One study has suggested that greater amounts of CGRP are produced by sensory nerves in chronically UV-irradiated skin [42]. Studies have implicated UV/*cis*-UCA-induced neuropeptides in mast cell degranulation, and release of suppressive mediators. Could substance P and CGRP differentially control mast cell activity? The response of mast cells enzymatically digested from human skin to substance P is mediated predominantly by neurokinin-1 receptors [26]. However, substance P can also act on mast cells by a unique pertussis-toxin-sensitive, cation-dependent, receptor-independent process [12]. Thus, the studies with a neurokinin-1 receptor antagonist may not have investigated all properties of substance P on mast cells. Substance P may have varied immunomodulatory properties on mast cells depending on the immune environment. Further, substance P triggered histamine release from purified human skin mast cells but did not stimulate the production of TNFα or IL-8, which was measured in response to FcεR1 activation. Substance P also suppressed IL-6 production by human skin mast cells [26]. These issues require further investigation particularly as substance P may cause immune enhancement or immune suppression depending on the other factors involved.

12.5.3 Bioavailability of UV/*cis*-Urocanic Acid-Induced Neuropeptides

UV irradiation of skin activates neuropeptide-degrading enzymes. Thus, the bioavailability of UVR-induced neuropeptides depends not only on their production but also on their breakdown by enzymes such as neutral endopeptidase, angiotensin-converting enzyme, and dipeptidyl peptidase IV (for review, [52]). CGRP is less sensitive than substance P to degradation by these zinc-metalloproteases [56]. Neutral endopeptidase and angiotensin-converting enzyme are regulated in an inversely correlated manner. For example, UV irradiation can dose-dependently reduce angiotensin-converting enzyme activity of epidermal cells after 24 and 48 h. Conversely, UV irradiation upregulates neutral endopeptidase expression and activity over time in cutaneous cells and reduces substance P activity (for review, [56]).

12.6 Immunoregulatory Properties of Neuroendocrine Hormones from UV-Irradiated Keratinocytes

Cross-talk between sensory nerves, keratinocytes, melanocytes, Langerhans cells, and other dendritic and lymphoid cells all contribute to the complex and integrated response of the epidermis to UV irradiation (Figs. 12.1 and 12.2). Two important contributors from UV-irradiated keratinocytes to the integrated neuroimmune response to UV are nerve growth factor (NGF) and α-melanocyte stimulating hormone (α-MSH).

12.6.1 Nerve Growth Factor

NGF belongs to the neurotrophin family of four structurally and functionally related proteins that not only control the development, maintenance, and apoptotic death of neurons but also regulate proliferation, differentiation, and apoptosis of non-neuronal cells (for review, [11]). In particular, neurotrophins contribute to the control of skin homeostasis and hair growth, with functions varying according to the ligand and the receptor and coreceptors expressed by the different cell types. Recent studies implicate UVR-induced keratinocyte NGF expression in UVR-induced systemic suppression of contact hypersensitivity responses in mice via effects on the neuroimmune system [62].

Increased expression of NGF was first shown immunohistochemically in the paws of UV-irradiated rats by Gillardon and colleagues [21], with peak expression after 12 h. In response to UV irradiation of dorsal skin of BALB/c mice, keratinocytes produce NGF with maximal expression detected immunohistochemically after 8 h [62]. Keratinocyte NGF expression may have several roles. NGF expression can protect keratinocytes and melanocytes from UVR-induced apoptosis via Bcl-2 expression [43]. NGF has also been reported to stimulate degranulation and cytokine release from skin mast cells [62]. However, we propose that the biologic action of UVR-induced, keratinocyte-derived NGF on dermal mast cells in vivo is indirect.

In our studies [62], UVR suppression of a contact hypersensitivity response was reversed by intraperitoneal administration of a polyclonal anti-NGF antibody to mice 3 h prior to UV irradiation. Further, NGF (20 μg/mouse) administered 5 days before hapten sensitization could suppress a contact hypersensitivity response but only in mice replete with mast cells and sensory nerves and not in mast cell-depleted mice

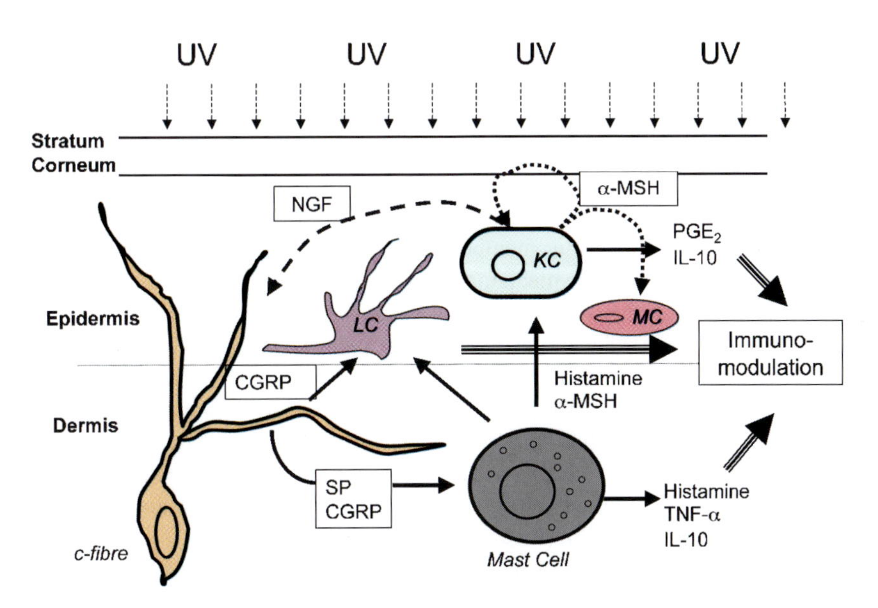

Fig. 12.2 Induction of immunomodulation by neuroimmune mediators from UV-irradiated keratinocytes (KC). UV-irradiated KC produce several neuroimmune mediators, including NGF and α-MSH. NGF stimulates release of the neuropeptides, CGRP, and substance P (SP), from peripheral sensory c-fibres, as well as providing growth and survival signals to nerves and mast cells. The immunomodulatory properties of CGRP and substance P, and the products from neuropeptide-activated mast cells, have been described in Fig. 12.1. α-MSH is anti-apoptotic for both KC and melanocytes (MC) and stimulates KC production of the regulatory cytokine IL-10

or mice treated with capsaicin at an early age. These data are consistent with UVR-induced NGF acting predominantly on the epidermal afferents of sensory c-fibers innervating the skin, with subsequent release of neuropeptides in the dermis and mast cell degranulation. This hypothesis is supported by the anatomical location of nerve terminals in the epidermis, with mast cells in murine dermis located remotely from the epidermis in the reticular layer.

The maximal expression of keratinocyte NGF 8–12 h after UV irradiation of skin provides a mechanism for sustained activation of sensory c-fibers for neuropeptide release. In support, increased CGRP immunoreactivity in c-fibers has been observed for up to 48 h after UV irradiation of rat skin [8]. This contrasts with a more immediate effect of *cis*-UCA on neuropeptide-induced immunoregulation that we detailed in Sect. 12.4, resulting in neuropeptide release within minutes of irradiation, and mast cell release of histamine and other immunomodulatory products. We attempted to determine a relationship between the biological actions of *cis*-UCA and keratinocyte NGF expression in UV-irradiated skin, but were unsuccessful due to the effect of the anti-NGF antibodies on mast cell homeostasis per se. Thus, it must be concluded that at separate times after UV-irradiation and in a potentially unrelated fashion, both *cis*-UCA and UVR-induced NGF independently activate sensory nerves for neuropeptide release, which in turn degranulates mast cells.

12.6.2 α-Melanocyte Stimulating Hormone

UV irradiation of human and murine skin stimulates increased expression of both α-MSH, a major member of the melanocortin family of neuroimmune mediators, and also the melanocortin-1 receptor (MC-1R) to which it binds (for review, [10]). Most skin cells are involved in the α-MSH-MC-1R network in that they can produce α-MSH and can respond to α-MSH. UVR-induced keratinocyte-derived α-MSH production has been considered part of an adaptive response to UVR exposure. α-MSH stimulates mammalian skin pigmentation and increased melanin synthesis by epidermal melanocytes. α–MSH can block UVR-induced apoptosis of normal human melanocytes and keratinocytes by stimulating nucleotide excision repair, thus reducing the amount of cyclobutane pyrimidine dimers induced by UVR and reducing the potential for melanocyte malignant transformation [9]. α-MSH can also reduce apoptosis by a mechanism that inactivates UVR-induced reactive oxygen species [37]. α-MSH stimulates keratinocyte proliferation and differentiation and thus may play a significant role in maintaining the functional integrity of the epidermis in response to UV radiation [9]. If shown to have similar effects on the repair of DNA lesions in UV-irradiated Langerhans cells and other antigen presenting cells in the epidermis, α-MSH may help to restore their optimal antigen presenting function, prevent the induction of regulatory T lymphocytes, and reverse consequent immunosuppression.

However, epicutaneously applied α-MSH can also inhibit both the induction and elicitation of a contact hypersensitivity response, as well as induce hapten-specific tolerance [23]. In these experiments, in vivo tolerance induction by α-MSH was abrogated by administration of an anti-IL-10 antibody at the site of sensitization [23] and suggested that some of the immunoregulatory functions of α-MSH were due to induction of keratinocyte IL-10 in concert with reduced production of inflammatory mediators by keratinocytes and other immune cells [53]. Degranulation of mast cells by α-MSH and release of histamine and TNFα may also contribute to UVR-induced immunomodulation [25], see Sect. 12.2. α-MSH-stimulated mast cells may also be a source of further regulatory α-MSH [3].

Further studies are required to better understand the immunoprotective and immunoregulatory properties of α-MSH and their contribution to photoneuroimmunology and to the development of both melanocytic and non-melanocytic skin cancers.

12.7 Role of Different UVR Wavelengths in Photoneuroimmunology

UVB radiation is generally more active than longer wavelength UVA radiation in photobiological processes. For example, neonatal UVB but not UVA radiation induces melanocytic neoplasms very similar to human cutaneous malignant melanomas in hepatocyte growth factor/scatter factor transgenic mice [14]. However, in the last decade studies from different laboratories have suggested that UVAII light (320–340 nm) may also have an immunomodulatory effect on immune responses.

In a murine study of delayed-type hypersensitivity responses, UVA (320–400 nm) radiation was as effective as solar-simulated light (UVA with UVB, 290–320 nm) in suppressing the elicitation of an established immune response [48]. Further, a sunscreen that filtered both UVA and UVB wavelengths, but not one filtering only UVB, prevented UVR-immune suppression [48]. In studies with 119 and 211 human volunteers, respectively [6,34], a sunscreen which was primarily a UVB radiation absorber protected against UVR-erythema but only partially against UVR-immunosuppression. This suggested that UVA radiation was immunosuppressive and that a broad-spectrum sunscreen with high UVA filtering capacity was optimal.

However, immunosuppression by UVA radiation remains a highly contested finding as others have reported a protective effect of UVA radiation against UVB-induced erythema/edema and systemic suppression of contact hypersensitivity [54]. Recently, the biological effects of varying ratios of UVA and UVB radiations were examined. With an increased UVA component, both the erythema/edema reaction and the degree of systemic immunosuppression were reduced. The photoprotection by UVA radiation has been largely attributed to the induction by UVA radiation of the heme-catabolizing enzyme, heme oxygenase-1, an enzyme that can neutralize some of the oxidative actions of UVA radiation [63]. The UVA component also prevented the UVB radiation-induced expression of IL-10 in irradiated skin [54].

Both UVA and UVB radiation can isomerize *trans*-UCA to its *cis* form [67]. In turn, *cis*-UCA stimulation of peripheral sensory nerves (see Sect. 12.4) may contribute to immunomodulation by both UVA and UVB radiation. However, there have been reports that UVA radiation can reverse the immunomodulatory effects of not only UVB radiation, but also *cis*-UCA [1,19]. Further study is urgently required to better understand the basis for these diverging reports of both an immunoprotective and an immunomodulatory effect of UVA radiation. This knowledge is critical for optimal sunscreen design.

12.8 Photoneuroimmunology in Chronically UV-Irradiated Skin

Most of this chapter has concentrated on the response of skin to a single experimental dose of UVR. After induction of various neuroimmunomodulators, their mechanisms of action have been studied by the measurement of immune responses and changes to these responses in the presence of antagonists. However, are these studies relevant to chronically UV-irradiated skin? Further, does the neuroimmune system adapt to chronic exposure to UVR?

In a murine study, neuropeptide levels, and in particular CGRP, increased with repeated UVR exposure [42]. Hairless mice (SKH-1) were exposed to subinflammatory UV three times per week for 4 weeks and then skin examined immunohistochemically 1, 3, and 7 days after the final UVR exposure. While the total number of epidermal nerve fibers did not significantly change, the number immunoreactive for CGRP significantly increased. Levels of CGRP in skin homogenates also increased [42]. In contrast, there was a relatively small increase in substance P immunoreactive epidermal nerve fibers, and in skin homogenates, there was no increase in immunoreactive substance P. The increased epidermal CGRP correlated with a reduced contact hypersensitivity response to DNFB applied to the chronically irradiated site, a suppression reversed by application of a commercial SPF20 sunscreen prior to each irradiation, or by intravenous $CGRP_{8-37}$ (200 nmol/kg) one day after the last of the chronic UV irradiations and 10 min before contact sensitization [42].

Human photoaged skin is also generally characterized by an increased number of epidermal and dermal nerve fibers. It has been suggested that UVR-stimulation of keratinocyte synthesis of NGF, a survival factor for developing sensory nerve fibers and a factor important in maintaining sensory nerve integrity, may be responsible. Chronically UV-irradiated skin also contains increased numbers of mast cells [24], which by crosstalk with cutaneous sensory nerve fibers are linked with the immunosuppressive properties of UVR and some of the tissue changes of photoaged skin [24]. UVR-induced NGF may also be linked with mast cell viability in skin [62]. These findings suggest photoaged skin may be highly sensitive to the immunomodulatory effects of UVR. However, this hypothesis requires further investigation.

The survival of skin cancer cells in photoaged skin may also be affected by an increased prevalence of neuropeptide and the major components of the neuroimmune system. Survival of skin cancer cells may be enhanced by the immunosuppressive and anti-apoptotic properties of the neuroimmune system (for review, [41]).

12.9 Conclusion

The epidermis is extensively innervated by neuropeptide-containing unmyelinated nerve fibers that extend vertically up from the dermis through the basement membrane to the stratum corneum. When the stratum corneum is irradiated with UVR, *trans*-UCA isomerizes to its more soluble *cis* form. In turn, *cis*-UCA can activate epidermal and dermal nerve fibers and stimulate the release of CGRP and substance P. Epidermal sensory nerve endings are also closely networked with keratinocytes, melanocytes, and Langerhans cells and as an integrated unit these cells can respond to the effects of UV irradiation. *cis*-UCA may initiate some of these responses, or alternatively the primary response to UVR may also be driven by UVR damage to membrane proteins, signaling molecules or nuclear DNA of epidermal cells.

As shown in Figs. 12.1 and 12.2, mediator traffic is two-way between epidermal cells and sensory nerves. UV-irradiated keratinocytes produce both NGF and α-MSH. As its name suggests, NGF can activate sensory nerves for neuropeptide release and may provide a more sustained signal than *cis*-UCA for neuropeptide release in UV-irradiated skin. However, NGF also contributes to the viability of nerves and mast cells of the skin. Similarly, α-MSH is anti-apoptotic and assists in repair of DNA of UVR-damaged epidermal cells including keratinocytes and melanocytes.

Dermal mast cells are key responder cells to the signals generated in the UV-irradiated epidermis, and in turn provide important mediators responsible for reduced immune responses to antigens applied to the irradiated site and to distant unirradiated sites. In particular, dermal mast cells degranulate in response to the neuropeptides, CGRP, substance P, and NGF. In turn, both local and systemic immune responses to antigens are modulated by the mast cell products, histamine, TNFα, and IL-10. Of course these are not the only pathways of UVR-induced immunomodulation; however, the use of neuropeptide-depleted and mast cell-deficient mice has shown that these cells and the neuroimmune system are important contributors to suppressed immune responses following UV irradiation of skin.

Summary for the Clinician

- UVR can regulate over-zealous immune diseases of the skin via stimulation of the neuroimmune system.
- Susceptibility to UVR-induced immunoregulation may vary from one individual to another according to dermal mast cell prevalence in skin and the plasticity of their peripheral sensory nervous system.
- The use of sunscreens filtering both UVA and UVB radiation is recommended until the relative contri butions of UVA and UVB wavelengths to UVR-immunomodulationis better understood.

References

1. Allanson M, Reeve VE (2005) Ultraviolet A (320–400 nm) modulation of ultraviolet B (290–320 nm) induced immune suppression is mediated by carbon monoxide. J Invest Dermatol 124:644–650
2. Amerio P, Toto P, Feliciani C, Suzuki H, Shivji G, Wang B, Sauder DN (2001) Rethinking the role of tumor necrosis factor-alpha in ultraviolet (UV) B-induced immunosuppression: altered immune response in UV-irradiated TNFR1R2 gene-targeted mutant mice. Br J Dermatol 144:952–957
3. Artuc M, Bohm M, Grutzkau A, Smorodchenko A, Zuberbier T, Lugen R, Henz BM (2006) Human mast cells in the neurohormonal network: expression of POMC, detection of precursor proteases, and evidence for IgE-dependent secretion of α-MSH. J Invest Dermatol 126:1976–1981
4. Asahina A, Hosoi J, Beissert S, Stratigos A, Granstein RD (1995) Inhibition of the induction of delayed-type and contact hypersensitivity by calcitonin gene-related peptide. J Immunol 154:3056–3061
5. Asahina A, Moro O, Hosoi J, Lerner EA, Xu S, Takashima A, Granstein RD (1995) Specific induction of cAMP in Langerhans cells by calcitonin gene-related peptide: relevance to functional studies. Proc Natl Acad Sci U S A 92:8328–8327
6. Baron ED, Fourtanier A, Compan D, Medaisko C, Cooper KD, Stevens SR (2003) High ultraviolet A protection affords greater immune protection confirming that ultraviolet A contributes to photoimmunosuppression in humans. J Invest Dermatol 121:869–875
7. Beissert S, Schwarz A, Schwarz T (2006) Regulatory T cells. J Invest Dermatol 126:15–24
8. Benrath J, Eschenfelder C, Zimmermann M, Gillardon F (1995) Calcitonin gene-relatd peptide, substance P and nitric oxide are involved in cutaneous inflammation following ultraviolet irradiation. Eur J Pharmacol 293:87–96
9. Bohm M, Wolff I, Scholzen TE, Robinson SJ, Healy E, Luger TA, Schwarz T, Schwarz A (2005) α-Melanocyte-stimulating hormone protects from ultraviolet radiation-induced apoptosis and DNA damage. J Biol Chem 280:5795802
10. Bohm M, Luger TA, Tobin DJ, Garcia-Borron JC (2006) Melanocortin receptor ligands: new horizons for skin biology and clinical dermatology. J Invest Dermatol 126:1966–1975
11. Botchkarev VA, Yaar M, Peters EMJ, Raychaudhuri SP, Botchkarev NV, Marconi A, Raychaudhuri SK, Paus R, Pincelli C (2006) Neurotrophins in skin biology and pathology. J Invest Dermatol 126:1719–1727
12. Cocchiara R, Bongiovanni A, Albegianni G, Azzolini A, Lampiasi N, Di Blasi F, Geraci D (1997) Inhibitory effect of neuraminidase on SP-induced histamine release and TNFα mRNA in rat mast cells: evidence of a receptor-independent mechanism. J Neuroimmunol 75:9–18
13. De Fabo EC, Noonan FP (1983) Mechanism of immune suppression by ultraviolet irradiation in vivo. I. Evidence for the existence of a unique photoreceptor in skin and its role in photoimmunology. J Exp Med 157:84–98
14. De Fabo EC, Noonan FP, Fears T, Merlino G (2004) Ultraviolet B but not Ultraviolet A radiation initiates melanoma. Cancer Res 64:6372–6376

15. Diffy BL (2002) Sources and measurement of ultraviolet radiation. Methods 28:4–13
16. Egan CL, Viglione-Schneck MJ, Walsh LJ, Green B, Trojanowski JQ, Whitaker-Menezes D, Murphy GF (1998) Characterization of unmyelinated axons uniting epidermal and dermal immune cells in primate and human skin. J Cutan Pathol 25:20–29
17. Eschenfelder CC, Benrath J, Zimmermann M, Gillardon F (1995) Involvement of substance P in ultraviolet irradiation-induced inflammation in rat skin. Eur J Neurosci 7:1520–1526
18. Garssen J, Buckley TL, van Loveren H (1998) A role for neuropeptides in UVB-induced systemic immunosuppression. Photochem Photobiol 68:205–210
19. Garssen J, De Gruijl F, Mol D, de Klerk A, Roholl P, van Loveren H (2001) UVA exposure affects UVB and *cis*-urocanic acid-induced systemic suppression of immune responses in Listeria monocytogenes-infected Balb/c mice. Photochem Photobiol 73:432–438
20. Gillardon F, Morano I, Zimmerman M (1991) Ultraviolet irradiation of the skin attenuates calcitonin gene-related peptide mRNA expression in rat dorsal root ganglion cells. Neurosci Lett 124:144–147
21. Gillardon F, Moll I, Michel S, Benrath J, Weihe E, Zimmerman M (1995) Calcitonin gene-related peptide and nitric oxide are involved in ultraviolet radiation-induced immunosuppression. Eur J Pharmacol 293:395–400
22. Gorman S, Tan JW, Thomas JA, Townley SL, Stumbles PA, Finlay-Jones JJ, Hart PH (2005) Primary defect in UVB-induced systemic immunomodulation does not relate to immature or functionally impaired APCs in regional lymph nodes. J Immunol 174:6677–6685
23. Grabbe S, Bhardwaj RS, Mahnke K, Simon MM, Schwarz T, Luger TA (1996) α-Melanocyte-stimulating hormone induces hapten-specific tolerance in mice. J Immunol 156:473–478
24. Grimbaldeston MA, Simpson A, Finlay-Jones JJ, Hart PH (2003) Determinants of dermal mast cell prevalence in human skin. Br J Dermatol 148:300–306
25. Grutzkau A, Henz BM, Kirchhof L, Luger T, Artuc M (2000) Alpha-melanocyte stimulating hormone acts as a selective inducer of secretory functions in human mast cells. Biochem Biophys Res Commun 278:14–19
26. Guhl S, Lee HH, Babina M, Henz BM, Zuberbier T (2005) Evidence for a restricted rather than generalized stimulatory response of skin-derived human mast cells to substance P. J Neuroimmunol 163:92–101
27. Hart PH, Jaksic A, Swift GJ, Norval M, El-Ghorr AA, Finlay-Jones JJ (1997) Histamine involvement in UVB- and *cis*-urocanic acid-induced systemic suppression of contact hypersensitivity responses. Immunology 91:601–608
28. Hart PH, Grimbaldeston MA, Swift GJ, Jaksic A, Noonan FP, Finlay-Jones JJ (1998) Dermal mast cells determine susceptibility to ultraviolet B-induced systemic suppression of contact hypersensitivity responses in mice. J Exp Med 187:2045–2053
29. Hart PH, Grimbaldeston MA, Swift GJ, Sedgwick JD, Korner H, Finlay-Jones JJ (1998) TNF modulates susceptibility to UVB-induced systemic immunomodulation in mice by effects on dermal mast cell prevalence. Eur J Immunol 28:2893–2901
30. Hart PH, Grimbaldeston MA, Swift GJ, Hosszu EK, Finlay-Jones JJ (1999) A critical role for dermal mast cells in *cis*-urocanic acid-induced systemic suppression of contact hypersensitivity responses in mice. Photochem Photobiol 70:807–812
31. Hosoi J, Murphy GF, Egan CL, Lerner EA, Grabbe S, Asahina A, Granstein RD (1993) Regulation of Langerhans cell function by nerves containing calcitonin gene-related peptide. Nature 363:159–162
32. Howes RA, Halliday GM, Barnetson RStC, Friedmann AC, Damian DL (2006) Topical capsaicin reduces ultraviolet radiation-induced suppression of Mantoux reactions in humans. J Dermatol Sci 44:113–115
33. Kahn HS, Tatham LM, Patel AV, Thun MJ, Heath CW (1998) Increased cancer mortality following a history of nonmelanoma skin cancer. JAMA 280:910–912
34. Kelly DA, Seed PT, Young AR, Walker SL (2003) A commercial sunscreen's protection against ultraviolet radiation-induced immunosuppression is more than 50% lower than protection against sunburn in humans. J Invest Dermatol 120:1–7
35. Khalil Z, Townley SL, Grimbaldeston MA, Finlay-Jones JJ, Hart PH (2001) *cis*-Urocanic acid stimulates neuropeptide release from peripheral sensory nerves. J Invest Dermatol 117:886–891
36. Kripke ML (1984) Immunological unresponsiveness induced by ultraviolet radiation. Immunol Rev 80:87–102
37. Kulms D, Zeise E, Poppelmann B, Schwarz T (2002) DNA damage, death receptor activation and reactive oxygen species contribute to ultraviolet radiation-induced apoptosis in an essential and independent way. Oncogene 21:5844–5851
38. Laihia JK, Attila M, Neuvonen K, Pasanen P, Tuomisto L, Jansen CT (1998) Urocanic acid binds to GABA but not to histamine (H1, H2 or H3) receptors. J Invest Dermatol 111:705–706
39. Lambert RW, Granstein RD (1998) Neuropeptides and Langerhans cells. Exp Dermatol 7:73–80
40. Lee E, Koo J, Berger T (2005) UVB phototherapy and skin cancer risk: a review of the literature. Int J Dermatol 44:355–360
41. Legat FJ, Wolf P (2006) Photodamage to the cutaneous sensory nerves: role in photoaging and carcinogenesis of the skin? Photochem Photobiol Sci 5:170–176
42. Legat FJ, Jaiani LT, Wolf P, Wang M, Lang R, Abraham T, Solomon AR, Armstrong CA, Glass JD, Ansel JC (2004) The role of calcitonin gene-related peptide in cutaneous immunosuppression induced by repeated subinflammatory ultraviolet irradiation exposure. Exp Dermatol 13:242–250
43. Marconi A, Vaschieri C, Zanoli S, Giannetti A, Pincelli C (1999) Nerve growth factor protects human keratinocytes from UVB-induced apoptosis. J Invest Dermatol 113:920–927
44. Mazzoni A, Young HA, Spitzer JH, Visintin A, Segal DM (2001) Histamine regulates cytokine production in maturing dendritic cells, resulting in altered T cell polarization. J Clin Invest 108:1865–1873
45. Meves A, Repacholi MH (2006) Does ultraviolet radiation affect vaccination efficacy? Int J Dermatol 45:1019–1024
46. Mouret S, Baudouin C, Charveron M, Favier A, Cadet J, Douki T (2006) Cyclobutane pyrimidine dimers are predominant DNA lesions in whole human skin exposed to UVA radiation. Proc Natl Acad Sci U S A 103:13765–13770

47. Naldi L, Fortina AB, Lovati S, Barba A, Gotti E, Tesari G, Schena S, Diociaiuti A, Nanni G, La Parola IL, Masini C, Piaserica S, Peserico A, Cainelli T, Remuzzi G (2000) Risk of nonmelanoma skin cancer in Italian organ transplant recipients: a registry-based study. Transplantation 70:1479–1484
48. Nghiem DX, Kazimi N, Clydesdale G, Ananthaswamy HV, Kripke ML. Ullrich SE (2001) Ultraviolet A radiation suppresses an established immune response: Implications for sunscreen design. J Invest Dermatol 117:1193–1199
49. Niizeki H, Alard P, Streilein JW (1997) Calcitonin gene-related peptide is necessary for UVB-impaired induction of contact hypersensitivity. J Immunol 159:5183–5186
50. Niizeki H, Kurimoto I, Streilein JW (1999) A substance P agonist acts as an adjuvant to promote hapten-specific skin immunity. J Invest Dermatol 112:437–442
51. Noonan FP, De Fabo EC (1992) Immunosuppression by ultraviolet B radiation: initiation by urocanic acid. Immunol Today 13:250–254
52. Peters EMJ, Ericson ME, Hosoi J, Seiffert K, Hordinsky MK, Ansel JC, Paus R, Scholzen TE (2006) Neuropeptide control mechanisms in cutaneous biology: physiological and clinical significance. J Invest Dermatol 126: 1937–1947
53. Redondo P, Garcia-Foncillas J, Okroujnov I, Bandres E (1998) Alpha-MSH regulates IL-10 expression by human keratinocytes. Arch Dermatol Res 290:425–428
54. Reeve VE, Domanski D, Slater M (2006) Radiation sources providing increased UVA/UVB ratios induce photoprotection dependent on the UVA dose in hairless mice. Photochem Photobiol 82:406–411
55. Sastry CM, Whitmore SE, Breysse PN, Morison WL, Strickland PT (2005) The effect of clinical UVA/B exposures on urinary urocanic acid isomer levels in individuals with Caucasian type (II/III) skin types. Dermatol Online J 11:11
56. Scholzen TE, Luger TA (2004) Neural endopeptidase and angiotensin-converting enzyme – key enzymes terminating the action of neuroendocrine mediators. Exp Dermatol 13(suppl 4):22–26
57. Scholzen TE, Brzoska T, Kalden D-H, O'Reilly F, Armstrong CA, Luger TA, Ansel JC (1999) Effect of ultraviolet light on the release of neuropeptides and neuroendocrine hormones in the skin: Mediators of photodermatitis and cutaneous inflammation. J Invest Dermatol Proc 4:55–60
58. Scholzen TE, Steinhoff M, Sindrilaru A, Schwarz A, Bunnett NW, Luger TA, Armstrong CA, Ansel JC (2004) Cutaneous allergic contact dermatitis responses are diminished in mice deficient in neurokinin 1 receptors and augmented by neurokinin 2 receptor blockage. FASEB J 18:1007–1009
59. Seiffert KS, Granstein RD (2002) Neuropeptides and neuroendocrine hormones in ultraviolet radiation-induced immunosuppression. Methods 28:97–103
60. Sleijffers A, Herreilers M, van Loveren H, Garssen J (2003) Ultraviolet B radiation induces upregulation of calcitonin gene-related peptide levels in human Finn chamber skin samples. J Biochem Photobiol B: Biol 69:149–152
61. Streilein JW, Alard P, Niizeki H (1999) Neural influences on induction of contact hypersensitivity. Ann NY Acad Sci 885:196–208
62. Townley SL, Grimbaldeston MA, Ferguson I, Rush RA, Zhang S-H, Zhou X-F, Conner JM, Finlay-Jones JJ, Hart PH (2002) Nerve growth factor, neuropeptides, and mast cells in ultraviolet-B-induced systemic suppression of contact hypersensitivity responses in mice. J Invest Dermatol 118:396–401
63. Tyrrell RM, Reeve VE (2006) Potential protection of skin by acute UVA irradiation – from cellular to animal models. Prog Biophys Mol Biol 92:86–91
64. Ullrich SE (2005) Mechanisms underlying UV-induced immune suppression. Mut Res 571:185–205
65. Van der Mei IA, Ponsonby AL, Dwyer T, Blizzard L, Simmons R, Taylor BV, Butzkueven H, Kilpatrick T (2003) Past exposure to sun, skin phenotype, and risk of multiple sclerosis: case control study. BMJ 327:316–321
66. Walterscheid JP, Ullrich SE, Nghiem DX (2002) Platelet-activating factor, a molecular sensor for cellular damage, activates systemic immune suppression. J Exp Med 195:171–179
67. Webber LJ. Whang E, De Fabo EC (1997) The effects of UVA-I (340–400 nm) and UVA-I+ II on the photoisomerization of urocanic acid in vivo. Photochem Photobiol 66:484–492
68. Woodward EA, Prele CM, Finlay-Jones JJ, Hart PH (2006) The receptor for *cis*-urocanic acid remains elusive. J Invest Dermatol 126:1191–1193

Section III

Neurobiology of Skin Appendages

Neurobiology of Hair

13

D.J. Tobin and E.M.J. Peters

Contents

Synonyms Box: Corticotropin-releasing Factor (CRF), Corticotropin-releasing Hormone; T cells, T lymphocytes; Immunocyte, Immune cell; Pigmentation, Melanogenesis; Nerve fibers, nerves; Pigment, Melanin; Hypothalamo-pituitary-adrenal axis (HPA), Hypothalamic-pituitary-adrenal axis (HPA); Cutaneous, Skin; CRF1, CRH-R1; CRF2, CRH-R2; Follicular papilla, Dermal papilla; Dopachrome tautomerase (DCT), Tyrosinase-related protein 2 (TRP-2); Arrector pili muscle, musculi arrectores pilorum, arrectores pilorum; Adrenalin/Noradrenalin, Epinephrine/ Norepinephrine

Key Features

- The hair follicle is one of a few body tissues with immune privilege status located in its transient inferior portion during the anagen growth phase, which is maintained by the production of potent immunosuppressive compounds.
- The hair follicle represents our body's only permanently regenerating organ, with two associated neural networks; one located around the hair follicle ostium, and the other around the isthmus/bulge region, and is associated with numerous mast cells.
- The hair follicle is both a source and a target of numerous cytokines, neuropeptides, neurohormones, neurotrophins, and neurotransmitters.
- A functionally organized hypothalamic-pituitary-adrenal axis (HPA) equivalent has been found in the skin and hair follicle with principal components CRF, POMC (and its cleavage peptides), and cortisol.
- Systemic neuro-immuno regulation of the hair follicle involves the sympathetic axis.
- Neuropeptides/ neurotrophins (e.g., Substance P, CGRP and NGF) regulate hair growth and perifollicular inflammation.

13.1 Brief Overview of Immunology of the Hair Follicle

Of the hair follicle's many surprises, perhaps one of the most fascinating revealed so far is its unique immunological status. This is further modulated as a function

R.D. Granstein and T.A. Luger (eds.), *Neuroimmunology of the Skin*,

of its life-long cycling. The latter is characterized by periods of active proliferation that drives elongation of a pigmented hair fiber during the anagen phase, followed by a massive apoptosis-driven loss of >60% of its proximal tissue mass during catagen, and subsequent entry into a relative resting phase during telogen before the telogen follicle commences the cycle again. Despite the flurry of exciting work in the 1970s that assigned the skin to so-called *first-level lymphoid organ* status [33], including the exquisite dissection of its cellular and humoral constituents at both the *innate* and adaptive immunity levels, researchers oddly neglected the skin's most prominent appendage – the hair follicle. Not only does our body's complement of hair follicles provide approximately 5 million potential ports of entry for microorganisms, the outward movement of the new hair shaft and displacement of the old club hair opens up a further potential access routes to the deeper and vascularized regions of the skin. Furthermore, the growing hair follicle is responsive to myriad immunomodulatory substances (e.g., cytokines, hormones, neuropeptides, drugs, etc), and the hair follicle undergoes dramatic alterations in the peri-follicular populations of both macrophages and mast cells during catagen [69].

Striking links between the hair follicle and the immune system can also be observed when things go wrong. This is dramatically seen in the presumptive autoimmune hair follicle disorder alopecia areata (AA), where the normally "immuno-silent" lower region of the growing hair follicle draws unwelcome attention from the immune system as it submits to a cyclic and reversible "autoimmune" attack of its transient structures [111]. This immunosilence is evidenced by several characteristics of the highly vascularized (but lymphatics-lacking) hair follicle, including the curious near-absence of MHC class I transplantation antigen expression in its proximal epithelium. Moreover, there is an active suppression of Natural Killer (NK) cell function in this tissue [49], which would usually be expected to respond to such low Class I-expressing cells. The total lack of MHC class II in the proximal hair follicle [70,119] is associated with a remarkable absence of functional Langerhans cells and T cells [20,26,40,68]. This immunologic peculiarity of the hair fiber-producing region of the hair follicle is very likely the product of intense selective pressure in mammalian evolution to retain an intact coat at all costs, that is, the avoidance of accidental or bystander immune damage. By contrast, Langerhans cells are present in the upper distal hair follicle, where they operate as sentinel cells responding to a range of noxious stimuli. Furthermore, the striking reduction in the incidence of dendritic T cells below the level of the hair follicle bulge correlates with reduced expression of T cell receptor stimulatory ligands there.

The maintenance of hair follicle immune privilege is facilitated by the production of potent immunosuppressive compounds (e.g., transforming growth factor beta1 [TGF-β1], α-melanocyte-stimulating hormone [α-MSH], adrenocorticotropic hormone [ACTH], etc.). These peptides may be particularly important for the hair follicle's cyclically active and melanin-forming melanocyte population. The latter are neural crest-derived cells, which engage in a biochemically hazardous synthesis of melanin characterized by the production of mutagenic intermediates [4], as well as reactive oxygen species via the oxidation of tyrosine and DOPA to melanin [41]. Despite this, melanogenesis must proceed without attracting a potentially catastrophic bystander immune recognition [69].

Emerging evidence is providing support for the involvement of the bone marrow-derived mast cell in hair follicle growth and cycling [63]. This immunocyte distributes strategically to the dense extracellular matrix-rich connective tissue sheath encapsulating the hair follicle [114,115], and more specifically close to the hair follicle's vasculature and innervation. Their numbers are greatest in hairy human skin. Skin mast cells share some functions with macrophages, including roles in anti-microbial/parasitic defense and immunomodulation [69], and are additionally a critical component of neurogenic inflammation (see later).

A significant stimulus for research into the hair follicle's immune privilege status comes from the interpretation that this may breakdown in immune-mediated hair follicle disorders (e.g., AA). Thus, the subsequent repair of the collapsed immune privilege may be of therapeutic benefit. It has been proposed by Paus and coworkers that in immunogenetically predisposed individuals, the hair follicle may be attacked by an immune system [37]. This may be stimulated by unknown trigger/response to a hair follicle autoantigen to induce interferon-γ (IFN-γ) production, especially during early stages of growth coincident with the reconstruction of its pigmentary unit [36]. IFN-γ upregulates the expression of MHC class I and II in the lower hair follicle. In this way, transiently expressed hair follicle anagen-associated antigens (now lacking the cover of their former immuno-privilege) may be presented to the immune system.

This may result in a full-blown anti-hair follicle immune response, which involves primarily CD8 T cells, and also CD4 T cells and possibly also hair follicle-specific IgG autoantibodies [35,112]. This lowering of the immune privilege *threshold* may be further exacerbated (and so increase still further the hair follicle's vulnerability to immune attack) if the production or function of locally sourced immunosuppressive cytokines and neurohormones (e.g., α-MSH) is deficient. Using organ culture of anagen hair follicles, Ito and colleagues reported that ectopic expression of MHC class I antigens can be normalized by treatment with α-MSH, IGF-1, or TGF-β1, all of which are generated locally in the hair follicle [49]. Thus, these agents may have therapeutic potential for the restoration of immune privilege and maintenance of MHC class I suppression in AA [48].

13.2 Brief Overview of Innervation of the Hair Follicle

Just as remarkable as the prominent position of the hair follicle in the design and function of the cutaneous immune system, the hair follicle has long been recognized as the most densely and complex innervated skin appendage [43]. Three tiers of nerve fiber bundles carry autonomic and peptidergic nerve fibers to their cutaneous targets. The epidermis, blood vessels, glands, and smooth muscles receive rather simple innervation structures. These structures mainly consist of individual nerve fibers, which terminate on or intermingle with their target cell populations. Their functional purposes are clear-cut, such as sensory perception, blood flow regulation, glandular secretion, or smooth muscle contraction. The occasional single nerve fiber, however, appears to terminate freely in the dermis or subcutis with a less clear biological role. It is conceivable that these nerve fibers contact individual mesenchymal and especially immune cells in these compartments since nearly every skin cell population including fibroblasts and mast cells was shown to entertain such contacts (c.f. [83]).

By comparison, nerve fibers that terminate on the hair follicle form two quite complex and distinct neuronal networks. With increasing size and function of the hair follicle these networks gain complexity and elaboration (c.f. [43]). However, the basic structure is common to virtually all hair follicles and species studied so far, ranging from the tiny and pigment-free vellus hair follicle to the impressive vibrissae follicle or from the mouse to the camel and is suggestive of a neuroimmune regulatory role in the control of hair cycling, on the one hand, and shaping the skin immune system, on the other.

In detail, one hair follicle neural network is located around the hair follicle ostium and innervates the hair follicle epithelium adjacent to the epidermis (FNA). Fibers constituting this network are fine and unmyelinated and contain a wide range of neuronal markers from neurotransmitters to neuropeptides and neurotrophins. They outnumber epidermal innervation and the wide range of nerve fiber subsets present suggests functions beyond the obvious afferent and efferent functions described above. Especially, their close neighborhood to immune cells in the hair follicle epithelium, such as dendritic cells which selectively populate this hair follicle compartment, clearly marks the hair follicle ostium as a high impact neuroimmune interaction site (Fig. 13.1).

Deeper into the skin the second hair follicle neural network organizes itself around the so-called isthmus and bulge region of the hair follicle. The precise location of this network is in between the isthmus of the sebaceous gland into the hair canal and the insertion of the arrector pili muscle into the hair follicle surrounding connective tissue sheath. This location, known as the isthmus and bulge region of the hair follicle, harbors hair follicle stem cells [27,29] responsible for hair cycling and hair pigmentation and also for wound healing and repopulation of the epidermis with pigment cells, for example, in vitiligo. Supplied by a nerve fiber bundle of the deep dermal plexus, the flattened longitudinal axons stretch upwards in a palisade manner along the basement membrane, which segregates the hair follicle epithelium from the surrounding mesenchyme. In the narrow space provided by the hooding sebaceous gland, these fibers terminate in a widening containing many neuro-mediator filled vesicles. Here they are free of their supporting Schwann cell lamellae and ready to shed their content upon stimulation. These nerve fibers have actually been observed to form intercellular junctions with processes of outer root sheath cells penetrating the basal lamina of the hair follicle [50], providing evidence for a direct impact of neuronal signaling molecules on the biology of cells in the stem cell region of the hair follicle.

Around them, c-fibers closely intermingled with stabilizing collagen fibrils circulate the hair follicle and the longitudinal fibers. Together they form a basket-like structure known variably as a lanceolate, palisade or Ruffini nerve ending, etc. The function of this neural network is only partially elucidated and comprises high

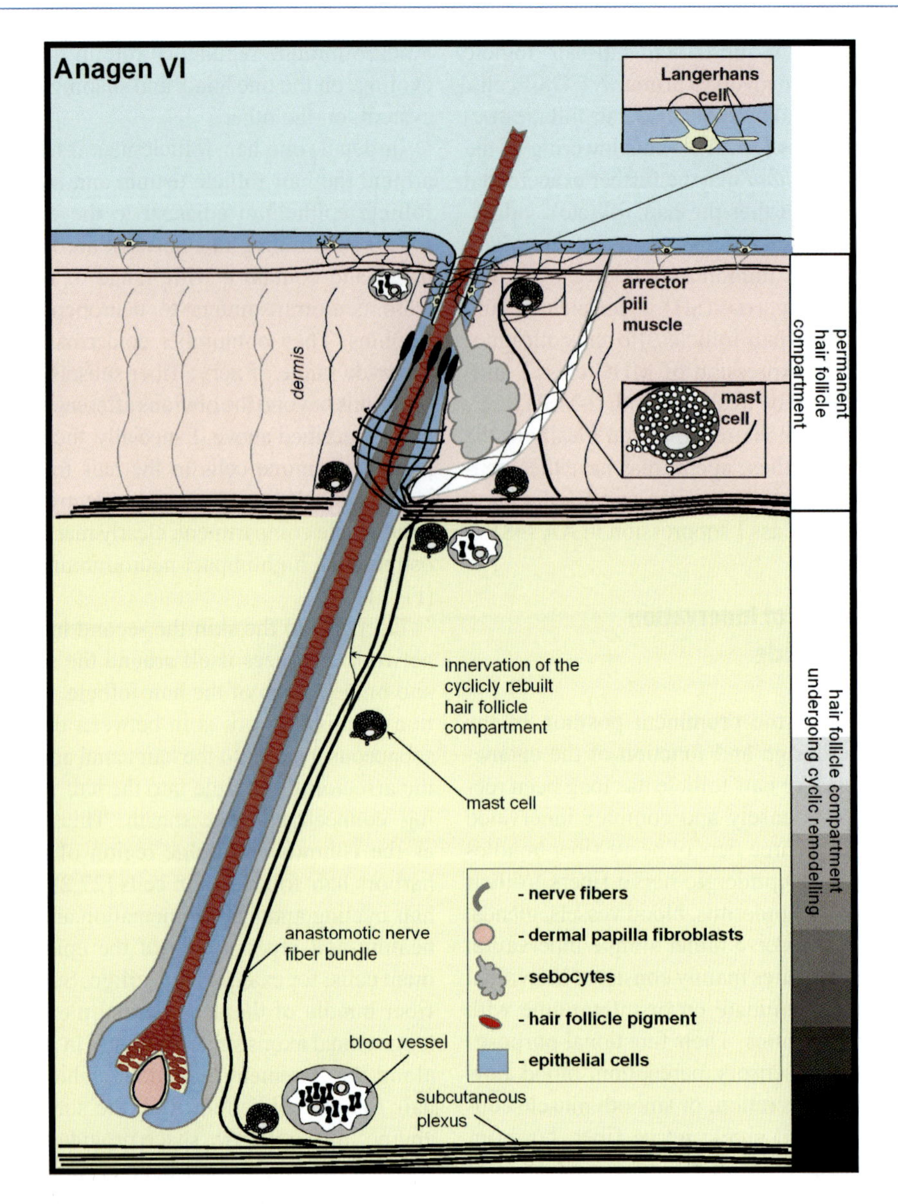

Fig. 13.1 Hair follicle neuroimmune network

threshold mechanical nociception detection, for example, velocity. Its density and complexity may house many more functions important to hair biology and skin immune responses alike. Depending on the researchers' main interest (anatomy, histochemistry, physiology, perception, etc.) the respective nomenclature has been used for the entire structure, including longitudinal and circular fibers or for only part of the structure that, and so has always been confusing. Accordingly, it was proposed to refer to this structure by the neutral name – follicular neural network B (FNB), which does not refer to either morphology or function [15].

In the neuroimmunological context (besides the possible direct impact on the isthmus and bulge cells) the nerve fiber bundle supplying fibers to the FNB is commonly accompanied by numerous mast cells. Passing through the dermis–subcutis border very close to the insertion of the arrector pili muscle into the connective tissue sheath, the FNB unites nerve fibers and innate immune cells to rapidly responding to environmental triggers close to the enigmatic hot-spot of hair follicle biology i.e. the isthmus and bulge region (Fig. 13.1). Acute, nonspecific neuroinflammatory responses to endogenous and exogenous environmental challenges, such as neurogenic inflammation, can therefore rapidly reach this change-sensitive control centre of the hair follicle and for example, contribute to premature termination of hair growth [8,75,92].

At the same time, the FNB is rich in calcitonin gene related peptide (CGRP) and other anti-inflammatory neuropeptides in almost all species studied [78]. This neuropeptide is known to suppress Langerhans cell activation [10,31,45]. Thus the FNB may contribute to the unique protection of this hair follicle compartment from specific inflammatory insult under physiological conditions, until the hyperactivation of neurogenic inflammation deleteriously overwhelms the hair follicle in case of damage (see later).

13.3 A Hypothalamo–Pituitary–Adrenal Axis equivalent in the Hair Follicle

The skin has enormous capacity to act as a peripheral neuroendocrine organ [94], and so it is expected to play a major role in maintaining a constant internal body environment or homeostasis [95,96]. Although such a view may have been considered heretical just a few years ago, with hindsight it should not be too surprising given the skin's strategic location between the external and internal environments. The hypothalamo–pituitary–adrenal axis (HPA) proper coordinates the body's response to systemic stress, and in the brain its most proximal component CRF (together with related urocortin I-III (URC I-II)-peptides) also regulates behavioral, autonomic, endocrine, reproductive, cardiovascular, gastrointestinal, metabolic, and immune systemic functions [49,95].

While the existence of an HPA equivalent in skin and its appendages is now increasingly well-established, we have only begun to explore its capability and functionality. An early query raised in the context of the existence of a cutaneous stress-system was its apparent lack of structural organization and compartmentalization, as compared to the HPA axis proper [60]. However, cutaneous production is not aimed at systemic effects and despite this difference, the HPA equivalent in skin is still functionally organized [49,94] and the same principal components are produced, including corticotropin releasing factor (CRF), the downstream proopiomelanocortin (POMC) with its multiple cleavage products (including most predominantly adrenocorticotropic hormone (ACTH), α-melanocyte stimulating hormone (α-MSH), β-endorphin, etc.) [53,89], as well as cortisol [49].

Thus, both the initiation and the termination of a classical stress response (i.e., via the cortisol-associated attenuation of CRF/POMC production) now appear to be present in the hair follicle and a role in immunological processes as well as hair cycling is conceivable. Remarkably, the human hair follicle expresses the genes and proteins for CRF and POMC, POMC-derived melanocortin peptides (α-MSH, ACTH, β-endorphin), pro-hormone convertases (PC-1 and PC-2), and the receptors for CRF (CRF1, CRF2) and melanocortins (e.g., MC-1R, μ-opiate receptor) [94,95,96,97,99,105]. In addition to HPA axis components, the skin and hair follicle are also invested with steroid, secosteroid, and a serotinergic/melatoninergic systems [49,57,100], and still others are likely to be found in this life-critical organ for most mammals.

13.3.1 Corticotropin-Releasing Factor (CRF) and the Hair Follicle

In skin cells, for example, melanocytes and dermal fibroblasts CRF initiates a cascade of events that is hierarchically ordered in a manner similar to that in the central HPA axis. In this way CRF activates its receptor CRF1 to induce cAMP accumulation followed by increased POMC gene expression and subsequent production of ACTH. In particular, melanocytes respond to CRF and ACTH with an enhanced production of cortisol/corticosterone [101,102]. We recently extended these studies and reported the expression of CRF (and the related urocortin) signaling systems in the human hair follicle (Fig. 13.2a) [54,104]. CRF and urocortin I peptides, as well as the receptors CRF1 and CRF2 were expressed in follicular keratinocytes, follicular papilla fibroblasts, and follicular melanocytes in vitro (Fig. 13.2a) [54,104]. The latter are also expressed in anatomically defined locations in human scalp anagen hair follicles, with the greatest expression of these peptides found in the differentiating keratinocytes of the pre-cortex and inner root sheath (Fig. 13.2b). Indeed, the expression of CRF peptide and of its receptors appears to correlate with keratinocyte differentiation status [54,104].

CRF/urocortin and their receptors are also expressed variably within follicular fibroblast subpopulations, critical for hair follicle growth and cycling [54,114,115]. Interestingly, the urocortin/CRF1 pairing was more highly expressed in fibroblasts of the follicular papilla than CRF/CRF2. Both CRF and urocortin signal via both CRF receptors; however, while

Hair Follicle Cell Populations *in vitro*

	FP cell	ORS Keratinocyte		Hair follicle Melanocyte	
		Undifferentiated	Differentiated	Undifferentiated	Differentiated
CRF	+	+	+	++	+++
Urocortin	++	–	–	++	+++
CRF1	++	++	+	++	++
CRF2	++	+	++	+	++

a

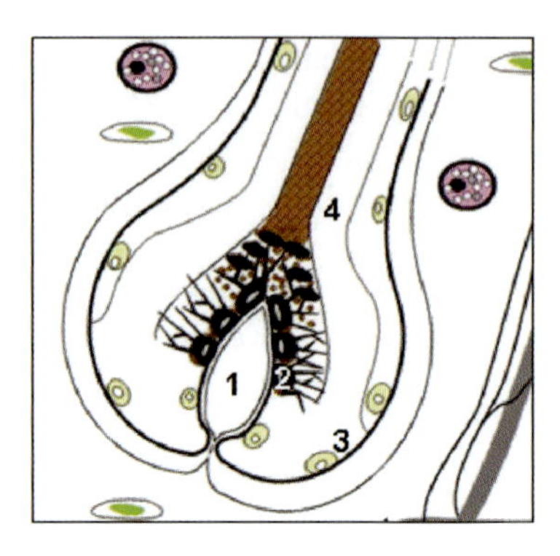

Hair Bulb Region *in situ*

	1	2	3	4	5
	FP	Hair bulb melanotic MCs	Peripheral matrix amelanotic MCs	Pre-cortex	ORS
CRF	+	–	+	++	+++
Urocortin	++	–	+	++	+++
CRF1	+	+/-	+	+	+
CRF2	+	–	+	+	+

b

Fig. 13.2 Expression of the CRF/Urocortin/CRF1,2 system in human scalp anagen hair follicles in situ and in isolated hair follicle cell populations in vitro. **a** *Primary hair follicle papilla fibroblasts* (FP). The pattern of expression of CRF, urocortin, CRF1, CRF2 was very heterogeneous in cultured follicle papilla cells. CRF expression was strongly nuclear, though also located in the cell cytoplasm. By contrast, urocortin expression was broadly cytoplasmic, where it exhibited a strong cytoskeletal pattern. The expression profiles of both CRF1/2 were broadly similar and were characterized by an intense granular cytoplasmic staining in a sub-population of cells; *Primary hair follicle keratinocytes* (ORS). CRF, urocortin, CRF1, and CRF2 were variably expressed in cultured hair follicle keratinocyte. CRF expression did not show any obvious correlation with cell differentiation and was characterized by both cytoplasmic and nuclear staining. By contrast, urocortin expression was weak/negative in matched follicular keratinocytes. CRF1 expression was detected in relatively high levels in all cells, where it exhibited a granular cytoplasmic distribution and also in cell nuclei. CRF2 expression was weak in the vast majority of cells; *Primary hair follicle melanocytes*. CRF, urocortin, CRF1, CRF2 were variably expressed in cultured hair follicle melanocytes. CRF expression correlated positively with differentiation status, with fine cytoplasmic staining and nuclear/nuclear membrane expression also found. By contrast, urocortin expression exhibited strong cell membrane localization. CRF1 expression was located both cytoplasmically and at the cell membrane, while CRF2 staining was intense at the cell membrane (adapted with permission from [53]). **b** The lower human hair follicle expresses CRF, urocortin, and associated receptors CRF1/2, which were variably detected in subpopulations of fibroblasts, keratinocytes, and melanocytes in the lower anagen hair follicle. For all four markers expression was broadly absent from melanotic melanocytes located above and around the follicular papilla (FP), and instead was detected in a minor sub-population of melanocytes located in the proximal/peripheral matrix region (adapted with permission from [54])

urocortin is equipotent for both receptors, CRH shows higher potency for CRF1 [118,121]. Thus, it would appear that signaling through CRF1 system predominates in this hair follicle cell population. Human dermal fibroblasts (but not epidermal keratinocytes) respond to CRF with stimulation of cAMP, induction of POMC gene and protein expression, and ACTH production and release [101]. Moreover, ex vivo stimulation of hair follicles with ACTH can upregulate cortisol in follicular papilla fibroblasts [48]. The greater expression of CRF message and protein in these hair growth inductive fibroblasts than in the noninductive interfollicular dermal fibroblasts is of particular interest and warrants further study. Given the close proximity of follicular papilla fibroblasts to hair bulb melanocytes and the observation that eumelanin is still produced in POMC knock-out C57BL/6 mice [123], it is possible that CRF (amongst other potential ligands) from the follicular papilla could stimulate melanogenesis in bulbar melanocytes.

A surprising finding of our immunolocalization studies to date is the striking apparent down-regulation in the expression of CRF, Urocortin, and CRF1, -2 receptors in the melanogenic zone of the anagen hair bulb, despite their being moderate expression close by in the most proximal and least differentiated hair bulb matrix. Moreover, the expression profiles of CRF1, -2 and their ligands do not fully overlap, as has also been observed in other tissues, including the brain [12]. However, these ligands did colocalize with their receptors in the differentiating keratinocytes of the keratogenous region of the growing hair follicle, and in this way concur with recent findings in human immortalized (HaCaT) and in normal epidermal keratinocytes where activation of CRF1 triggered both G0/1 arrest and early differentiation [84,95,124,125]. The functional significance of this expression in the pigmentary system is further emphasized by the pro-dentritogenic (Fig. 13.3a) and pro-melanogenic (Fig. 13.3b) effects of CRF1 agonists on follicular melanocytes in vitro [54]. Not only does this subpopulation of cutaneous melanocytes differentially express CRF, urocortin, CRF1, and CRF2 in situ, but CRF and modified CRF/urocortin peptides (selective for either CRF1 or for both CRF1 and CRF2) also modulate melanocyte phenotype in vitro by upregulating melanogenesis and dendricity. The phenotypic changes associated with CRF receptor agonists correlates with alterations in the expression and activity of melanogenic enzymes. Importantly, CRF is a potent inducer of tyrosinase protein, unlike urocortin (higher affinity for CRF2), which exerts little effect (Fig. 13.3c). Moreover, CRF is a potent stimulator of the *activity* of this rate-limiting enzyme for melanogenesis, whereas urocortin is not effective. CRF also increases the expression levels of the two other melanogenic enzymes, TRP-1 and DCT, whereas urocortin either reduces the expression of these enzymes or had no effect [54]. The relative absence of expression in the melanogenic zone of anagen hair follicle suggests that the CRF/CRF1 system may instead be functional during the early stages of follicular melanocyte differentiation, and rather becomes down-regulated in mature and fully melanogenic melanocytes. Consistent with this interpretation is the observed down-regulation of CRF1 during melanization of melanoma cells when compared with their amelanotic phenotype [97]. Thus, these findings suggest a role for CRF peptides in regulating human hair follicle melanocyte differentiation.

There is also evidence that CRF may act as a growth factor for melanocytes of the hair follicle in particular. CRF (10^{-7}–10^{-9} M) significantly stimulates melanocyte proliferation (Fig. 13.3d), while urocortin either has not effect or may even inhibit proliferation at 10^{-9} and 10^{-10} M [54]. These data suggests the involvement of both CRF1 and CRF2 receptors in the regulation of follicular melanocytes proliferation. Observed differences in the proliferative response of epidermal vs. follicular melanocytes both in situ and in vitro may be in part explained by the different pattern of CRF receptor expression; epidermal melanocytes express solely CRF1, while follicular melanocytes express both CRF1 and CRF2 [54,103]. Therefore, it is likely that several interdependent signaling pathways (i.e., Gsα/cAMP and Gq/11 /PLC, IP3, and PKC) are involved in mediating the effects of CRF and urocortin [38,98,120]. In addition, observations from other cell systems (e.g., endothelium) suggest that urocortin can signal via CRF2β to increase both cAMP and nitric oxide [64]. Both of these second messengers have been associated with melanogenesis in epidermal melanocytes [88]. Thus, our observation that a CRF2-selective urocortin analog fails to induce a melanogenic response suggests that hair follicle melanocytes may not express the relevant CRF2 isoform. In fact, it has been shown that CRF-related peptides exhibit a tenfold greater potency in stimulating CRF2β (in terms of adenylate cyclase activation) compared with either CRF2α or γ [122]. These differences in expression of CRF1 isoforms between these two distinct but related cutaneous melanocyte subpopulations may result in differential coupling to the activation of local POMC systems. Moreover, there may also be a requirement for significant cross-talk between the CRF1 and CRF2 signaling systems such that neither alone can provide optimal induction of melanocyte proliferation, dendricity, as well as melanogenesis. Although these tempting possibilities remain to be experimentally tested, if true

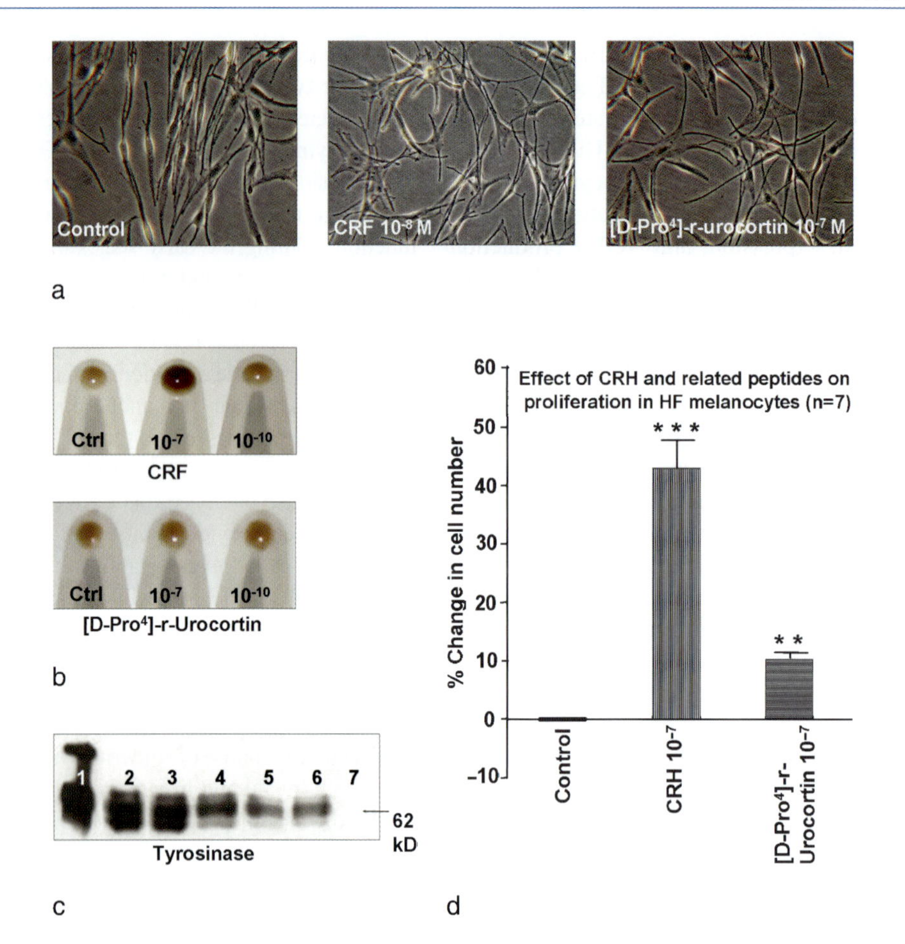

Fig. 13.3 CRF and Urocortin induce phenotypic change in human scalp hair follicle melanocytes in vitro. **a** Hair follicle melanocyte dendricity was stimulated by CRF, and less so by urocortin. **b** CRF, but not urocortin, stimulated melanogenesis in hair follicle melanocytes in culture. **c** CRF, but not urocortin, markedly upregulated tyrosinase protein expression in hair follicle melanocytes in culture. Lane 1: molecular weight marker; Lane 2: CRF 10^{-7} M; Lane 3: [D-Pro^5]-CRF 10^{-10} M; Lane 4: [D-Glu^{20}]-CRF 10^{-8} M; Lane 5: [D-Pro^4]-r-urocortin 10^{-10} M; Lane 6: unstimulated control; Lane 7: negative control. **d** CRF, but not urocortin, is a potent stimulator of proliferation in hair follicle melanocytes in culture. Results are expressed as a percentage increase in cell number over control unstimulated levels and are a mean ± SEM of seven cell lines (adapted with permission from [54])

it may be possible to down-regulate different aspects of melanocyte phenotype by alternately antagonizing signaling via either receptor type.

We have preliminary evidence that CRF/urocortin may also affect hair follicle fiber growth (Tobin et al, unpublished data). CRF significantly inhibits hair fiber elongation in organ cultured hair follicles [52], most likely due to the inhibition of keratinocyte proliferation in the anagen hair bulb and by inducing a corresponding and premature precipitation of the growing hair follicle into an apoptosis-driven catagen-like state. By contrast, urocortin did not inhibit hair fiber elongation in this assay and may even protect the hair follicle from entry in catagen. Thus, like in the epidermis, CRF signaling via CRF1 leads to inhibition of keratinocyte proliferation with corresponding induction of keratinocyte differentiation, while preferential activation with CRF2-selective urocortin analogs may invoke the opposite effect facilitating continuous hair fiber production. If a local stress response system is involved in regulation of the normal hair cycle, it may also have a role in hair follicle disorders like AA. It has been reported that the expression of CRF, ACTH, and α-MSH is raised in hair follicles from lesional AA skin [55].

13.3.2 Pro-Opiomelanocortin and its Melanocortin Peptides ACTH, α-MSH, and β-Endorphin and the Hair Follicle

Some skin and hair pigmentation phenotypes are linked to polymorphisms in the melanocortin 1 receptor

(MC-1R) gene [85]. The MC-1R receptor, a G protein-coupled receptor expressed at low levels on the surface of melanocytes, is activated by α-melanocyte stimulating hormone (α-MSH) and adrenocorticotropic hormone (ACTH) and to some extent by proopiomelanocortin itself [89]. Many lines of evidence indicate that melanocortin peptides are important regulators of skin and coat color [1]. Other persuasive evidence comes from the phenotypes that result from inactivating mutations in the POMC gene. For example, *POMC* null mutations in humans result in the red hair and fair skin phenotype (RHC phenotype) and this is thought to reflect a lack of ligands for the MC-1R [58,59]. In addition, *MC-1R* is highly polymorphic, and several of its alleles are associated with the red hair color (RHC) phenotype [85,86]. Although, the MC-1R/α-MSH, ACTH system occupies the dominant position in our current concept of the regulation of mammalian pigmentation [99], there is also accumulating evidence that non-MC-1R-dependent pathways can regulate pigmentation [90] in both epidermal and follicular melanocytes. For example, we have found that β-endorphin, operating via the μ-opiate receptor, is also able to modulate epidermal melanocyte biology in vitro and both the ligand and the high affinity receptor are expressed in situ [51,52].

Some cytokines and paracrine factors of the immune system, which are also produced in skin (e.g., interleukin [IL]-1α and IL-1β [34], can upregulate *MC-1R* mRNA in normal human melanocytes, while other cytokines suppress melanogenesis (tumor necrosis factor-α [TNF-α] [61] and TGF-β [62] by downregulating *MC-1R* expression in human melanocytes and mouse melanoma cells. α-MSH stimulates follicular melanogenesis in rodents by preferentially increasing the synthesis of eumelanin over pheomelanin (reviewed in [99,109]) in large part via the stimulation of tyrosinase activity at the transcriptional, translational, or post-translational levels [22,23].

In human epidermal melanocytes, α-MSH and ACTH stimulate melanogenesis, dendricity, and proliferation via action at the MC-1R [1,46,106,107,117], mediated predominately via the cAMP second messenger system [21,24]. We have recently also detected the expression of the α-MSH, ACTH/MC-1R system in human scalp hair follicles in situ (Fig. 13.4a) and in follicular cell sub-populations in vitro (Fig. 13.4b), including in melanocytes, keratinocytes, and fibroblasts [13,53]. POMC and its associated processing machinery and the MC-1R are expressed at the mRNA and protein level in cultured human follicular melanocytes, follicular keratinocytes, and follicular dermal papilla cells (Fig. 13.5). These findings are in agreement with the previous detection of *POMC* mRNA in these hair follicle cell sub-populations [51,52]. Importantly, α-MSH/ACTH peptides, POMC processing enzymes, and MC-1R are differentially expressed in hair follicle melanocytes in situ as a function of their anatomic location and melanogenic activity. Indeed, α-MSH/ACTH peptide expression is confined to a minor sub-population of melanocytes located in the outer root sheath and in melanocytes located in the most proximal and peripheral matrix region of the hair bulb. The role of this melanocyte subpopulation is unclear, but they may reflect a less differentiated pool of melanocytes involved in the reconstruction of the hair follicle pigmentary unit during the hair cycle [113].

Interestingly, α-MSH, ACTH, and POMC processing machinery were undetectable in the gp100-postive melanocytes located in the hair follicle melanogenic zone – the site of active melanin synthesis [53]. These findings suggest that POMC peptides exhibit a close relationship with melanocyte differentiation status and that this system may not be directly involved in the *maintenance* of melanogenesis during the growth or anagen phase of the hair cycle. This heterogeneous pattern of α-MSH and ACTH expression has also been reported for CRF and the CRF1 [48] – the chief regulator of pituitary POMC gene expression and the production and secretion of POMC peptides [2,99]. However, rare α-MSH and ACTH positive melanocytes can also be detected in the regressing epithelial strand of regressing catagen follicles and in the epithelial sac of telogen follicles. Some of these cells are likely to reflect apoptosis-resistant melanocytes of the previous proximal anagen bulb [28,110,113]. Therefore, on the basis of our current knowledge, expression of α-MSH and ACTH may be associated with the ability of some hair bulb melanocytes to survive the apoptosis-driven catagen process. Indeed, there is evidence in humans that ACTH may stimulate and/or prolong anagen, as overproduction of ACTH or therapeutic administration of ACTH causes acquired hypertrichosis associated with increased pigmentation [68].

Further evidence that POMC peptides have an important role to play in melanocyte differentiation is indicated by the observation that α-MSH, ACTH, and β-endorphin all increases dendricity (Fig. 13.6a), melanogenesis (Fig. 13.6b), and proliferation (Fig. 13.6c) in follicular melanocyte cultures [53,109]. ACTH 1–17 appears to be more effective at inducing melanogenesis in follicular melanocytes than either α-MSH

Hair bulb region *in situ*

	1	2	3	4
	FP	Hair bulb melanotic MCs	Peripheral matrix amelanotic MCs	Pre-cortex
α-MSH	+++	–	++	+
ACTH	+++	–	++	+
MC-1R	+++	–	++	–
PC1	+/–	–	++	+++
PC2	++	+/–	++	+
β-End	+	+++	+++	++
μ-OR	++	–	+++	++

a

Hair Follicle Cell Population

	FP cell	ORS Keratinocyte		Hair follicle Melanocyte	
		Undifferentiated	*Differentiated*	*Undifferentiated*	*Differentiated*
α-MSH	++	++	+	+++	–
ACTH	++	+	+	+++	+
MC-1R	+	+	+	++	+
PC1	+++	++	+	+++	–
PC2	+++	++	+	++	+/–
β-End	+	++	++	++	++
μ-OR	+	++	++	++	+

b

Fig. 13.4 Expression of POMC processing hormones, POMC peptides, and associated receptors in the human scalp anagen hair bulb in situ and in hair follicle cell populations in vitro. **a** α-MSH and ACTH peptides, PC1 and PC2 prohormone convertases, and the MC-1R and μ-opiate receptors were variably detected in subpopulations of keratinocytes, fibroblasts, and melanocytes in the lower anagen hair follicle in situ. For all markers, with the exception of β-endorphin, peptide and receptor expression was absent from melanogenic melanocytes located above and around the follicular papilla (FP), and instead was detected in a minor subpopulation of melanocytes located in the proximal/peripheral matrix region (adapted with permission from [53]). **b** α-MSH and ACTH peptides, PC1 and PC2 prohormone convertases, and the MC-1R and μ-opiate receptors were variably detected in cultured hair follicle-derived keratinocytes, fibroblasts, and melanocytes. *Primary hair follicle papilla fibroblasts* (FP)*:* The expression of the named markers was variable in cultured follicular papilla fibroblasts, with highest levels of expression confined to a sub-population of cells. Expression was strongest in the peri-nuclear region of the cells where it was distributed in a granular pattern. The expression of the MC-1R exhibited a distinctive granular distribution pattern that was marked in a sub-population of cells; *Primary hair follicle keratinocytes* (ORS):. The expression of named markers was variable in cultured hair follicle keratinocytes, with expression higher in keratinocytes with a basal cell phenotype compared with more differentiated keratinocytes. Expression of the POMC peptides α-MSH and ACTH, PC1/2, and MC-1R/μOR was higher in amelanotic melanocytes and was significantly downregulated in terminally differentiated pigmented bulbar melanocytes. A notable exception was β-endorphin, which was highly expressed in both melanocyte subtypes (adapted with permission from [53])

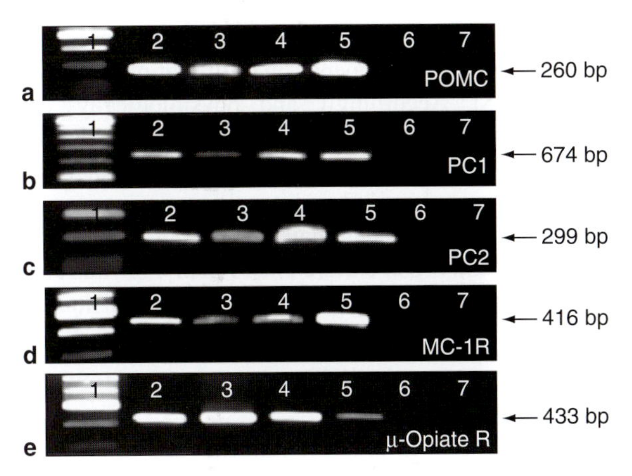

Fig. 13.5 Detection of POMC, PC1, PC2, and MC-1R and µOR-specific transcripts by RT-PCR in cultured follicular melanocytes, keratinocytes, and follicular papilla fibroblasts. **a** Detection of a 260 bp product specific for POMC; **b** Detection of 674 bp product specific for PC1, **c** Detection of 299 bp product specific for PC2; **d** Detection of a 416 bp product specific for MC-1R; **e** Detection of a 433 bp product specific for µ-opiate R.. Lane 1: DNA ladder, Lane 2: HFM, Lane 3: HFK, Lane 4: FPF; Lane 5 (**a**): plasmid containing POMC gene; Lane 5 (**b**–**d**): Epidermal melanocyte positive control; Lane 6 and 7: negative controls (adapted with permission from [53])

or β-endorphin, while all three are potent inducers of melanocyte dendricity. The melanogenic and dendritogenic effects of α-MSH and ACTH 1–17 in follicular melanocytes appear to correlate positively with increasing hair color, supporting earlier data showing that α-MSH binding sites maybe linked to hair color [65]. We have recently detected the secretion of intact POMC from hair follicle melanocytes in culture. This unexpected finding was associated with observed binding of POMC with the MC-1R (albeit at 10–100 times lower affinity than MSH or ACTH). Still, POMC induced a cAMP response, with corresponding induction of dendricity, melanogenesis, and proliferation in pigment cells [89]. These findings suggest that both constitutively-released POMC as well as its cleavage products can be potent biologic response modifiers produced by hair follicle cells.

13.3.3 Stress Has It: Actions of the Sympathetic Axis in Hair Follicle Neurobiology and Immunology

Besides the HPA, the sympathetic axis has long been recognized to play a role in skin and hair biology. Clinical observation has long reported the hair growth inhibitory side effects of treatment with β_2-adrenoreceptor blockers such as propranolol. Moreover, the hair follicle was also shown to express this receptor in the early growth phase in the enigmatic isthmus and bulge region, and to depend on noradrenalin signaling to conduct a fully functional hair cycle [16,73].

Apparently, in this context noradrenergic signaling primarily protects and promotes hair growth. Possibly this occurs not only by direct keratinocyte proliferation effects, but also via its neuroimmune interference. Derived from the innervation of the arrector pili muscle [16] or produced directly by keratinocytes and melanocytes [39,91], noradrenalin is both mast cell protective and in a wider sense also anti-inflammatory. This wider sense requires some explanation. Production of cytokines traditionally attributed as T-helper cell type 1 (TH1) cytokines such as IFN-γ or TNF-α promote cellular inflammatory responses and therefore count as proinflammatory, while so called TH2 cytokines (such as TGF-β) terminate cellular immune responses, conduct humoral immunity, and count as anti-inflammatory. The balance between these responses determines an effective host defense and is under tight regulation by neuroendocrine mechanisms (Fig. 13.7).

From systemic responses and direct neuroimmune interaction in lymphoid organs, we know that β-adrenergic signaling initiates a shift from TH1 to TH2 and suppresses the production of the potent hair growth terminator IFN-γ, while promoting the hair follicle immune privilege protective molecule TGF-β [32]. Moreover, it inhibits the local release of TNF-α by mast cells [11] and inhibits their degranulation induced by neuropeptides [87]. Thus systemic neuroimmune regulation through the sympathetic system may be mirrored in the skin and is complemented by regulation of neurogenic inflammation. However, the definite role of the sympathetic signaling system in the protection of the hair follicle immune privilege, inhibition of perifollicular cellular immune responses, and mast cell degranulation as well as its therapeutic potential in inflammatory hair growth disorders still awaits elucidation.

13.4 The Third Stress Axis in the Skin: Neuropeptides and Neurotrophins Regulate Hair Growth and Perifollicular Inflammation

13.4.1 Substance P and Inhibition of Hair Growth by Neurogenic Inflammation

A further neuroimmune link in the control of hair cycling as well as in the pathogenesis of common hair loss disorders such as telogen effluvium or AA is suggested by increased substance P (SP)-positive nerve fibers in early anagen, in stressed skin or in early AA lesions [7,43,75,76,77,92]. This neuropeptide has recently gained recognition as a prominent player in the cutaneous stress response and as a potent immune-modulator of both innate and specific immune responses [42,78].

Initial observations suggested that SP acts as a hair growth promoter, just like noradrenalin, since it enhanced keratinocytes proliferation in culture, and because hair growth depends on proliferating keratinocytes [108]. Accordingly and physiologically, the number of SP+ nerve fibers in the interfollicular dermis was found increased during the hair cycle in early anagen mouse skin [74] and it promoted anagen progression in early anagen hair follicles in murine full thickness skin organ culture [74]. Finally, SP was also able to promote anagen induction in telogen mouse

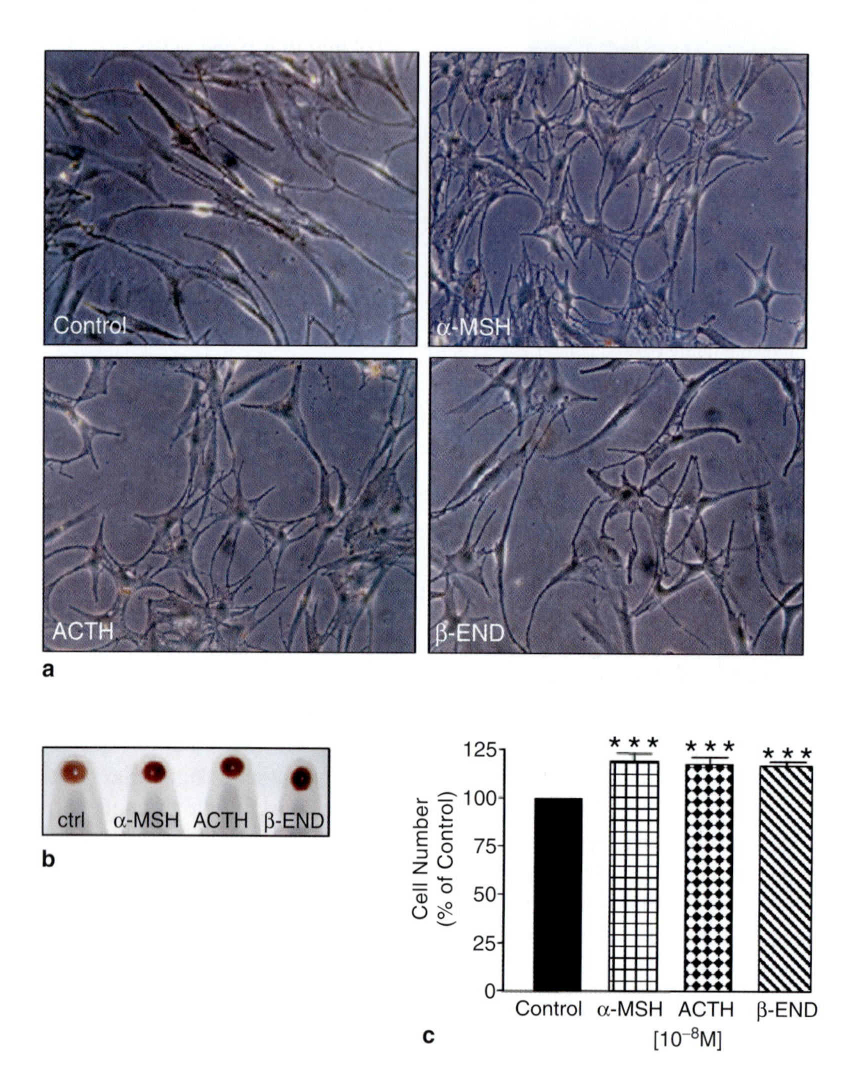

Fig. 13.6 POMC peptides induce phenotypic change in human scalp hair follicle melanocytes in vitro. **a** Hair follicle melanocyte dendricity was stimulated by α-MSH, ACTH 1–17, and β-endorphin. A marked (and broadly similar) increase in cell dendricity was seen 72h after Ac-α-MSH, ACTH 1–17, and β-endorphin stimulation (adapted with permission from [53]). **b** Melanogenesis and **c** proliferation in follicular melanocytes was stimulated Ac-α-MSH, ACTH 1–17, and β-endorphin. Cells with low basal melanin levels showed visible increases in melanogenesis after Ac-α-MSH, ACTH 1–17, and β-endorphin stimulation. Cell proliferation was assessed by determining cell counts before and after Ac-α-MSH, ACTH 1–17, and β-endorphin stimulation. Results are expressed as a percentage increase in cell number over control unstimulated levels and are a mean ± SEM of seven cell lines. Statistical significance was assessed by one way ANOVA *** = $p < 0.001$ (adapted with permission from [53])

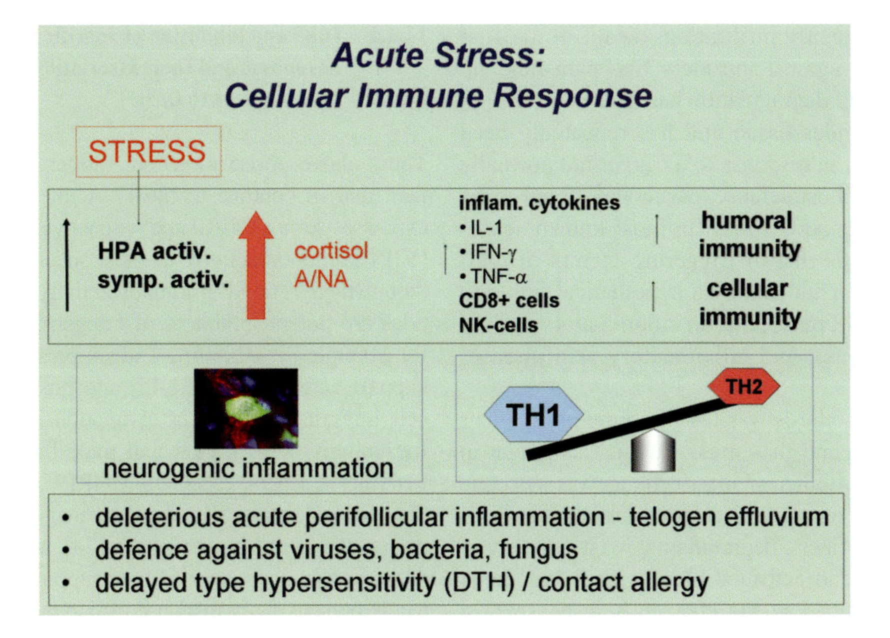

a

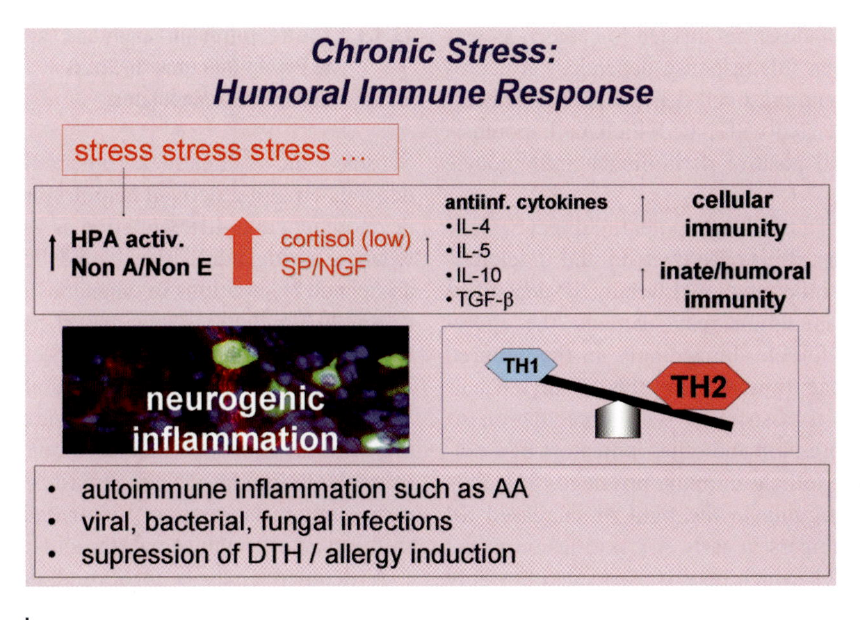

b

Fig. 13.7 Schematic representation of cutaneous stress responses and involved neuro-immune circuitry. **a** Summary of the discussed literature on acute stress responses and their implications for skin and hair follicle disease. **b** Summary of the discussed literature on chronic stress responses and their implications for skin and hair follicle disease

skin when released in low concentrations from slow release gel capsules [67].

However, SP holds a central position in the conduction of the so called neurogenic inflammation. In response to a wide variety of triggering challenges ranging from physical via chemical to biological SP degranulates mast cells via neurokinin receptors, releasing an array of pro-inflammatory mediators including histamine and TNF-α [5,78]. The result is a rapid reddening, swelling, and itching of the affected skin and subsequently infiltration, meant in the first place to defend against intruders like parasites and bacteria. Mast cell degranulation has deleterious lateral effects on bystander tissue and has repeatedly been reported to occur in response to a trigger not normally involved in local host defense, psychoemotional stress [6,93]. Thus, this adds stress (in most known senses of the term) to the list of triggering factors of mast cell degranulation and finally a hypothetical link was drawn between SP, mast cells, an inflammatory peripheral stress response, and inflammatory skin diseases [9,71,77,95].

When this hypothesis was tested on the murine hair follicle, it was found that stress was just as potent in degranulating perifollicular mast cells as was systemic SP [6]. A direct anatomical association between activated SP+ nerve fibers, degranulating mast cells, apoptotic skin cells, and activated blood vessel endothelia could be determined in the cutaneous response to a systemic psychoemotional stressor such as noise [76]. Moreover, stress was as effective a trigger for the transit of hair follicles from anagen to catagen as was SP itself. However, this response depended on neurokinin receptor 1 and mast cell-dependent mechanisms, which were associated with hair follicle cell apoptosis and MHC class II positive perifollicular immunocyte infiltration [7,8].

Thus, in case of a challenge, potential direct SP-hair growth promotion effects are overrun and a deleterious perifollicular inflammation driven by SP-dependent neurogenic inflammation may damage the growing anagen hair follicle. Intriguingly, in the cultured human anagen hair follicle assay this SP-driven hair follicle damage is associated with upregulation of MHC class I expression and other indicators of a collapse of the hair follicle immune privilege [81]. This is especially intriguing in the light of increased SP + nerve fiber numbers in early AA skin lesion in the murine AA model, which here is associated not only with activated mast cells but also with activated CD8+ T cells [92].

A new twist to neurogenic inflammation and neuroimmune communication is provided by the recent observation that longitudinal nerve fibers in the FNB express the H3 Histamine receptor, a receptor responsible for counter-regulatory downregulation of inflammatory responses and decreased nociception [25]. So even during a neurogenic inflammatory assault on the hair follicle, with the potential to terminate active hair growth, this hair follicle compartment may have protective neuroimmune defense mechanisms in place.

13.4.2 CGRP and Inhibition of Specific Immune Responses and Their Association with the Hair Growth Cycle

These above observations are further supported by the fact, that in contrast to the FNA, the FNB selectively expresses strong CGRP and vasoactive intestinal peptide (VIP) immunoreactivity in most species studies so far, though there is no SP-immunoreactivity. These neuropeptides are potent inhibitors of Langerhans cell function, for example by, inhibiting antigen presentation and NF-kappaB activation [10,31,45] and possibly Langerhans cell migration (Peters et al unpublished observation). Intriguingly, it was shown in mice that the density of FNA nerve fibers, especially CGRP+ ones, depends on the hair cycle stage. The densest innervation is observed during the early growth phase [74], while this precise phase is also characterized by a suppression of the contact hypersensitivity response [44] in the same model.

13.4.3 The Neurotrophin Family and Their Receptors as Potent Hair growth, Stress, and Immune Modulators

Sensory and sympathetic innervation in the skin depends on target derived neurotrophins both for their development and differentiation as well as their maintenance during adult live. In the light of hair cycle associated fluctuations in cutaneous innervation, neurotrophin fluctuations thus appear natural. Their role in the neuroimmune complex, however, is expanding rapidly, with new functions continuously identified for neurotrophin signaling. Besides neuronal plasticity, its implications now also include tissue remodeling, for example, during hair cycling, mast cell degranulation/ neurogenic inflammation, endothelial activation, TH2 shifting, and the cutaneous stress response [82].

A detailed analysis is beyond the scope of this chapter and has been provided in detail by others [17–19,82]. We therefore focus on the most recent findings described in neuroimmune hair follicle biology. Perhaps the most intriguing finding corresponds to the realization that the heretofore "low"-affinity pan-neurotrophin receptor p75NTR, a member of the death receptor family, is a high affinity receptor for the pro-neurotrophin of nerve growth factor (proNGF) [47]. In this combination it prominently induces premature termination of hair growth, catagen involution of the strongly p75NTR+ hair bulb, and marked apoptosis in the non-permanent part of the hair follicle [79,80].

NGF and SP appear to be closely linked in skin stress responses and inflammatory disease. Not only are they both elevated after stress and in cutaneous inflammatory disease [3,56,116], NGF and SP also act in unison in the stress-driven premature involution of the hair follicle [81,83]. Here NGF is not only responsible for SP+ innervation in the stressed skin [75], it can also be induced in keratinocytes by neuropeptides such as SP, pro-inflammatory cytokines [30] and the p75NTR [81]. Its effect on hair follicle biology [80,81] is thus twofold: direct hair follicle involution via p75NTR activation and perifollicular deleterious inflammation via mast cell degranulation [75] and altered neurotrophin receptor expression on the inflammatory infiltrate, for example, in AA [14,66].

13.5 Concluding Remarks

One of the fascinating implications of the above studies when taken together is that functional regulation of the hair follicle occurs by all known stress-response pathways implemented in the local neuroimmune network of the hair follicle. In case of the CRF/POMC system, for example, (including its regulatory principles) a symmetry or "self-similarity" of the melanocortin systems at both local (hair follicle) and systemic (CNS) levels is indicated [72]. In this way there appears to have been evolutionary selective pressures to reproduce preferred structural regulatory mechanisms (especially those involving important control and defense/stress responses) that have both stimulatory and an inhibitory feedback signals, and that these employ a universal biochemical language. In this context, the highly accessible hair follicle, and especially its continually remodeling hair follicle neuroimmune network, can provide a wonderful opportunity to dissect mechanisms underlying multiple basic cell biological phenomena including, neuroendocrinologic functioning.

Summary for the Clinician

Of all our phenotypic traits skin and hair communicate more immediate information to the observer than any other, and so has enormous psychosocial implications for us as individuals. Moreover, the skin occupies a very strategic location between the external and internal environments and so is critical in the preservation of the body homeostasis. There has been much excitement regarding the discovery of the huge capacity for the hair follicle to act as a peripheral neuro-immuno-endocrine organ that appears to be somewhat independent of central regulatory control. Striking links between the hair follicle and the immune system can also be observed clinically, e.g., during the break-down of immune privilege in the presumptive autoimmune hair follicle disorder alopecia areata, and so its repair may be of therapeutic benefit. Moreover, increased expression of CRF, ACTH and α-MSH has also been reported in hair follicles from alopecia areata lesions. Hair growth itself may be inhibited after β2-adrenoreceptor blocker use, though we await further evidence for a definitive role for sympathetic signalling before its therapeutic potential in inflammatory hair growth disorders can be realized. Evidence from murine models suggests that psychosocial stress may be as potent a perifollicular mast cells degranulator as systemic SP. While both SP and NGF act in stress-driven premature hair follicle involution, SP-associated damage of the hair follicle may contribute to loss of hair follicle immune privilege - a finding which concurs with increased SP+ nerve fibers in early alopecia areata lesions associated with activated mast cells and CD8+ T cells. In this rapidly developing field the involvement of the neuro-immuno-endocrine axis in hair follicle biology promises to provide new therapeutic tools to deal with disorders of the hair follicles.

References

1. Abdel-Malek Z, Swope VB, Suzuki I, Akcali C, Harriger MD, Boyce ST, et al (1995) Mitogenic and melanogenic stimulation of normal human melanocytes by melanotropic peptides. Proc Natl Acad Sci U S A 92:1789–1793
2. Aguilera G, Rabadan-Diehl C, Nikodemova M (2001) Regulation of pituitary corticotropin releasing hormone receptors. Peptides 22:769–774
3. Alleva E, Petruzzi S, Cirulli F, Aloe L (1996) NGF regulatory role in stress and coping of rodents and humans. Pharmacol Biochem Behav 54:65–72
4. Ames BN, Shigenaga MK, Hagen TM (1993) Oxidants, antioxidants, and the degenerative diseases of aging. Proc Natl Acad Sci U S A 90:7915–7922
5. Ansel JC, Brown JR, Payan DG, Brown MA (1993) Substance P selectively activates TNF-alpha gene expression in murine mast cells. J Immunol 150:4478–4485
6. Arck PC, Handjiski B, Hagen E, Joachim R, Klapp BF, Paus R(2001) Indications for a 'brain-hair follicle axis (BHA)': inhibition of keratinocyte proliferation and up-regulation of keratinocyte apoptosis in telogen hair follicles by stress and substance P. FASEB J 15:2536–2538
7. Arck PC, Handjiski B, Peters EM, Peter AS, Hagen E, Fischer A, et al (2003) Stress inhibits hair growth in mice by induction of premature catagen development and deleterious perifollicular inflammatory events via neuropeptide substance P-dependent pathways. Am J Pathol 162:803–814

8. Arck PC, Handjiski B, Kuhlmei A, Peters EM, Knackstedt M, Peter A, et al (2005) Mast cell deficient and neurokinin-1 receptor knockout mice are protected from stress-induced hair growth inhibition. J Mol Med (Berlin, Germany) 83:386–396
9. Arck PC, Slominski A, Theoharides TC, Peters EM, Paus R (2006) Neuroimmunology of stress: skin takes center stage. J Invest Dermatol 126:1697–1704
10. Asahina A, Moro O, Hosoi J, Lerner EA, Xu S, Takashima A, et al (1995) Specific induction of cAMP in Langerhans cells by calcitonin gene-related peptide: relevance to functional effects. Proc Natl Acad Sci U S A 92:8323–8327
11. Bissonnette EY, Befus AD (1997) Anti-inflammatory effect of beta 2-agonists: inhibition of TNF-alpha release from human mast cells. J Allergy Clin Immunol 100:825–831
12. Bittencourt JC, Vaughan J, Arias C, Rissman RA, Vale WW, Sawchenko PE (1999) Urocortin expression in rat brain: evidence against a pervasive relationship of urocortin-containing projections with targets bearing type 2 CRF receptors. J Comp Neurol 415:285–312
13. Bohm M, Eickelmann M, Li Z, Schneider SW, Oji V, Diederichs S, et al (2005) Detection of functionally active melanocortin receptors and evidence for an immunoregulatory activity of alpha-melanocyte-stimulating hormone in human dermal papilla cells. Endocrinology 146:4635–4646
14. Botchkarev VA (2003) Neurotrophins and their role in pathogenesis of alopecia areata. J Invest Dermatol Symp Proc 8:195–198
15. Botchkarev VA, Eichmuller S, Johansson O, Paus R (1997) Hair cycle-dependent plasticity of skin and hair follicle innervation in normal murine skin. J Comp Neurol 386:379–395
16. Botchkarev VA, Peters EM, Botchkareva NV, Maurer M, Paus R (1999) Hair cycle-dependent changes in adrenergic skin innervation, and hair growth modulation by adrenergic drugs. J Invest Dermatol 113:878–887
17. Botchkarev VA, Botchkareva NV, Albers KM, Chen LH, Welker P, Paus R (2000) A role for p75 neurotrophin receptor in the control of apoptosis-driven hair follicle regression. FASEB J 14:1931–1942
18. Botchkarev VA, Botchkareva NV, Peters EM, Paus R (2004) Epithelial growth control by neurotrophins: leads and lessons from the hair follicle. Prog Brain Res 146:493–513
19. Botchkarev VA, Yaar M, Peters EM, Raychaudhuri SP, Botchkareva NV, Marconi A, et al (2006) Neurotrophins in skin biology and pathology. J Invest Dermatol 126:1719–1727
20. Breathnach AS (1963) The distribution of Langerhans cells within the human hair follicle, and some observations on its staining properties with gold chloride. J Anat 97:73–80
21. Burbach JPH, Weigant VM (1999) Gene expression, biosynthesis and processing of pro-opiomelanocortin derived and vasopressin. In: De Weid D (ed) Neuropeptides: Basics and Perspectives, Elsevier, Amsterdam
22. Burchill SA, Thody AJ (1986) Melanocyte-stimulating hormone and the regulation of tyrosinase activity in hair follicular melanocytes of the mouse. J Endocrinol 111:225–232
23. Burchill SA, Virden R, Fuller BB, Thody AJ (1988) Regulation of tyrosinase synthesis by alpha-melanocytestimulating hormone in hair follicular melanocytes of the mouse. J Endocrinol 116:17–23
24. Busca R, Ballotti R (2000) Cyclic AMP a key messenger in the regulation of skin pigmentation. Pigment Cell Res 13:60–69
25. Cannon KE, Chazot PL, Hann V, Shenton F, Hough LB, Rice FL (2007) Immunohistochemical localization of histamine H3 receptors in rodent skin, dorsal root ganglia, superior cervical ganglia, and spinal cord: potential antinociceptive targets. Pain 129:76–92
26. Christoph T, Müller-Röver S, Audring H, Tobin D, Hermes B, Cotsarelis G, et al (2000) The human hair follicle immune system: cellular composition and immune privilege. Br J Dermatol 142(5):862–873
27. Chuong CM, Cotsarelis G, Stenn K (2007) Defining hair follicles in the age of stem cell bioengineering. J Invest Dermatol 127:2098–2100
28. Commo S, Bernard BA (2000) Melanocyte subpopulation turnover during the human hair cycle: an immunohistochemical study. Pigment Cell Res 13:253–259
29. Cotsarelis G (2006) Epithelial stem cells: a folliculocentric view. J Invest Dermatol 126:1459–1468
30. Dallos A, Kiss M, Polyanka H, Dobozy A, Kemeny L, Husz S (2006) Effects of the neuropeptides substance P, calcitonin gene-related peptide, vasoactive intestinal polypeptide and galanin on the production of nerve growth factor and inflammatory cytokines in cultured human keratinocytes. Neuropeptides 40:251–263
31. Ding W, Wagner JA, Granstein RD (2007) CGRP, PACAP, and VIP modulate Langerhans cell function by inhibiting NF-kappaB activation. J Invest Dermatol 127:2357–2367
32. Elenkov IJ, Wilder RL, Chrousos GP, Vizi ES (2000) The sympathetic nerve – an integrative interface between two supersystems: the brain and the immune system. Pharmacol Rev 52:595–638
33. Fichtelius KE, Groth O, Lidén S (1970) The skin, a first level lymphoid organ? Int Arch Allergy Appl Immunol 37:607–620
34. Funasaka Y, Chakraborty AK, Hayashi Y, Komoto M, Ohashi A, Nagahama M, et al (1998) Modulation of melanocyte-stimulating hormone receptor expression on normal human melanocytes: evidence for a regulatory role of ultraviolet B, interleukin-1alpha, interleukin-1beta, endothelin-1 and tumour necrosis factor-alpha. Br J Dermatol 139:216–224
35. Gilhar A, Ullmann Y, Berkutzki T, Assy B, Kalish RS (1998) Autoimmune hair loss (alopecia areata) transferred by T lymphocytes to human scalp explants on SCID mice. J Clin Invest 101:62–67
36. Gilhar A, Kam Y, Assy B, Kalish RS (2005) Alopecia areata induced in C3H/HeJ mice by interferon-gamma: evidence for loss of immune privilege. J Invest Dermatol 124:288–289
37. Gilhar A, Paus R, Kalish RS (2007) Lymphocytes, neuropeptides, and genes involved in alopecia areata. J Clin Invest 117:2019–2027
38. Grammatopoulos DK, Chrousos GP (2002) Functional characteristics of CRH receptors and potential clinical applications of CRH-receptor antagonists. TEM 13:436–444
39. Grando SA, Pittelkow MR, Schallreuter KU (2006) Adrenergic and cholinergic control in the biology of epidermis: physiological and clinical significance. J Invest Dermatol 126:1948–1965
40. Harrist TJ, Ruiter DJ, Mihm MC, Bhan AK (1983) Distribution of major histocompatibility antigens in normal skin. Br J Dermatol 109:623–633

41. Hegedus ZL (2000) The probable involvement of soluble and deposited melanins, their intermediates and the reactive oxygen side-products in human diseases and aging. Toxicology 145:85–101
42. Hendrix S, Peters EM (2007) Neuronal plasticity and neuroregeneration in the skin – the role of inflammation. J Neuroimmunol 184:113–126
43. Hendrix S, Peker B, Liezman C, Peters EMJ (2008) Skin and hair follicle innervation in experimental models: a review and guide for the exact and reproducible evaluation of neuronal plasticity. Exp Dermatol 17(3):214–227
44. Hoffman U, Tokura Y, Nishijima T, Takigawa M, Paus R (1996) Hair cycle-dependent changes in skin immune functions: anagen-associated depression of sensitization for contact hypersensitivity in mice. J Invest Dermatol 106:598–604
45. Hosoi J, Murphy GF, Egan CL, Lerner EA, Grabbe S, Asahina A, et al (1993) Regulation of Langerhans cell function by nerves containing calcitonin gene-related peptide. Nature 363:159–163
46. Hunt G, Donatien PD, Lunec J, Todd C, Kyne S, Thody AJ (1994) Cultured human melanocytes respond to MSH peptides and ACTH. Pigment Cell Res 7:217–221
47. Ibanez CF (2002) Jekyll-Hyde neurotrophins: the story of proNGF. Trends Neurosci 25:284–286
48. Ito N, Ito T, Betterman A, Paus R (2004) The human hair bulb is a source and target of CRH. J Invest Dermatol 122:235–237
49. Ito N, Ito T, Kromminga A, Bettermann A, Takigawa M, Kees F, et al (2005) Human hair follicles display a functional equivalent of the hypothalamic-pituitary-adrenal axis and synthesize cortisol. FASEB J 19:1332–1334
50. Kaidoh T, Inoue T (2000) Intercellular junctions between palisade nerve endings and outer root sheath cells of rat vellus hairs. J Comp Neurol 420:419–427
51. Kauser S, Schallreuter KU, Thody AJ, Gummer C, Tobin DJ (2003) Regulation of human epidermal melanocyte biology by beta-endorphin. J Invest Dermatol 120:1073–1080
52. Kauser S, Thody AJ, Schallreuter KU, Gummer CL, Tobin DJ (2004) beta-Endorphin as a regulator of human hair follicle melanocyte biology. J Invest Dermatol 123:184–195
53. Kauser S, Thody AJ, Schallreuter KU, Gummer CL, Tobin DJ (2005) A fully functional proopiomelanocortin/melanocortin-1 receptor system regulates the differentiation of human scalp hair follicle melanocytes. Endocrinology 146:532–543
54. Kauser S, Slominski A, Wei ET, Tobin DJ (2006) Modulation of the human hair follicle pigmentary unit by corticotropin-releasing hormone and urocortin peptides. FASEB J 20:882–895
55. Kim HS, Cho DH, Kim HJ, Lee JY, Cho BK, Park HJ (2006) Immunoreactivity of corticotropin-releasing hormone, adrenocorticotropic hormone and alpha-melanocyte-stimulating hormone in alopecia areata. Exp Dermatol 15:515–522
56. Kimata H (2003) Suckling reduces allergic skin responses and plasma levels of neuropeptide and neurotrophin in lactating women with atopic eczema/dermatitis syndrome. Int Arch Allergy Immunol 132:380–383
57. Kobayashi H, Kromminga A, Dunlop TW, Tychsen B, Conrad F, Suzuki N, et al (2005) A role of melatonin in neuroectodermal-mesodermal interactions: the hair follicle synthesizes melatonin and expresses functional melatonin receptors. FASEB J 19:1710–1712
58. Krude H, Gruters A (2000) Implications of proopiomelanocortin (POMC) mutations in humans: the POMC deficiency syndrome. TEM 11:15–22
59. Krude H, Biebermann H, Luck W, Horn R, Brabant G, Gruters A (1998) Severe early-onset obesity, adrenal insufficiency and red hair pigmentation caused by POMC mutations in humans. Nat Genetics 19:155–157
60. Lightman SL, Windle RJ, Julian MD, Harbuz MS, Shanks N, Wood SA, et al (2000) Significance of pulsatility in the HPA axis. Novartis Foundation symposium 227:244–257; discussion 257–260
61. Martinez-Esparza M, Jimenez-Cervantes C, Solano F, Lozano JA, Garcia-Borron JC (1998) Mechanisms of melanogenesis inhibition by tumor necrosis factor-alpha in B16/F10 mouse melanoma cells. FEBS 255:139–146
62. Martinez-Esparza M, Ferrer C, Castells MT, Garcia-Borron JC, Zuasti A (2001) Transforming growth factor beta1 mediates hypopigmentation of B16 mouse melanoma cells by inhibition of melanin formation and melanosome maturation. Int J Biochem Cell Biol 33:971–983
63. Maurer M, Fischer E, Handjiski B, von Stebut E, Algermissen B, Bavandi A, et al (1997) Activated skin mast cells are involved in murine hair follicle regression (catagen). Lab Iinvest 77:319–332
64. Miki I, Seya K, Motomura S, Furukawa K (2004) Role of corticotropin-releasing factor receptor type 2 beta in urocortin-induced vasodilation of rat aortas. J Pharmacol Sci 96:170–176
65. Nanninga PB, Ghanem GE, Lejeune FJ, Bos JD, Westerhof W (1991) Evidence for alpha-MSH binding sites on human scalp hair follicles: preliminary results. Pigment Cell Res 4:193–198
66. Palkina TN, Sharov AA, Sharov TY, Botchkarev VA (2005) Neurotrophins in autoimmune diseases: possible implications for alopecia areata. J Invest Dermatol Symp Proc 10:282
67. Paus R, Christoph T, Müller-Röver S (1999) Immunology of the hair follicle: a short journey into terra incognita. J Invest Dermatol Symp Proc 4:226–234
68. Paus R, Heinzelmann T, Schultz KD, Furkert J, Fechner K, Czarnetzki BM (1994) Hair growth induction by substance P. Lab Invest 71:134–140
69. Paus R, Nickoloff BJ, Ito T (2005) A 'hairy' privilege. Trends Immunol 26:32–40
70. Paus R, van der Veen C, Eichmuller S, Kopp T, Hagen E, Muller-Rover S, et al (1998) Generation and cyclic remodeling of the hair follicle immune system in mice. J Invest Dermatol 111:7–18
71. Pavlovic S, Daniltchenko M, Tobin DJ, Hagen E, Hunt SP, Klapp BF, et al (2008) Further exploring the brain-skin connection: stress worsens dermatitis via substance P-dependent neurogenic inflammation in mice. J Invest Dermatol 128:434–446
72. Peters A (2005) The self-similarity of the melanocortin system. Endocrinology 146:529–531
73. Peters EM, Maurer M, Botchkarev VA, Gordon DS, Paus R (1999) Hair growth-modulation by adrenergic drugs. Exp Dermatol 8:274–281
74. Peters EM, Botchkarev VA, Botchkareva NV, Tobin DJ, Paus R (2001) Hair-cycle-associated remodeling of the peptidergic

innervation of murine skin, and hair growth modulation by neuropeptides. J Invest Dermatol 116:236–245
75. Peters EM, Handjiski B, Kuhlmei A, Hagen E, Bielas H, Braun A, et al (2004) Neurogenic inflammation in stress-induced termination of murine hair growth is promoted by nerve growth factor. Am J Pathol 165:259–271
76. Peters EM, Kuhlmei A, Tobin DJ, Muller-Rover S, Klapp BF, Arck PC (2005) Stress exposure modulates peptidergic innervation and degranulates mast cells in murine skin. Brain Behav Immun 19:252–262
77. Peters EM, Arck PC, Paus R (2006a) Hair growth inhibition by psychoemotional stress: a mouse model for neural mechanisms in hair growth control. Exp Dermatol 15:1–13
78. Peters EM, Ericson ME, Hosoi J, Seiffert K, Hordinsky MK, Ansel JC, et al (2006b) Neuropeptide control mechanisms in cutaneous biology: physiological and clinical significance. J Invest Dermatol 126:1937–1947
79. Peters EM, Hendrix S, Golz G, Klapp BF, Arck PC, Paus R (2006c) Nerve growth factor and its precursor differentially regulate hair cycle progression in mice. J Histochem Cytochem 54:275–288
80. Peters EM, Stieglitz MG, Liezman C, Overall RW, Nakamura M, Hagen E, et al (2006d) p75 Neurotrophin receptor-mediated signaling promotes human hair follicle regression (Catagen). Am J Pathol 168:221–234
81. Peters EM, Liotiri S, Bodo E, Hagen E, Biro T, Arck PC, et al (2007a) Probing the effects of stress mediators on the human hair follicle: substance P holds central position. Am J Pathol 171:1872–1886
82. Peters EM, Raap U, Welker P, Tanaka A, Matsuda H, Pavlovic-Masnicosa S, et al (2007b) Neurotrophins act as neuroendocrine regulators of skin homeostasis in health and disease. Horm Metab Res 39:110–124
83. Peters EMJ, Ericson M, Hosoi J, Seiffert K, Hordinsky M, Ansel JC, et al (2006) Neuropeptidergic controls in cutaneous biology: Physiological and clinical significance. J Invest Dermatol 126(9):1937–47
84. Quevedo ME, Slominski A, Pinto W, Wei E, Wortsman J (2001) Pleiotropic effects of corticotrophin releasing hormone on normal human skin keratinocytes. In Vitro Cell Dev Biol Anim 37:50–54
85. Rees JL (2003) Genetics of hair and skin color. Annu Rev Genet 37:67–90
86. Rees JL, Healy E (1997) Melanocortin receptors, red hair, and skin cancer. J Invest Dermatol Symp Proc 2:94–98
87. Reynier-Rebuffel AM, Callebert J, Launay JM, Seylaz J, Aubineau P (1997) NE inhibits cerebrovascular mast cell exocytosis induced by cholinergic and peptidergic agonists. Am J Physiol 273:R845–R850
88. Romero-Graillet C, Aberdam E, Biagoli N, Massabni W, Ortonne JP, Ballotti R (1996) Ultraviolet B radiation acts through the nitric oxide and cGMP signal transduction pathway to stimulate melanogenesis in human melanocytes. J Biol Chem 271:28052–28056
89. Rousseau K, Kauser S, Pritchard LE, Warhurst A, Oliver RL, Slominski A, et al (2007) Proopiomelanocortin (POMC), the ACTH/melanocortin precursor, is secreted by human epidermal keratinocytes and melanocytes and stimulates melanogenesis. FASEB J 21:1844–1856
90. Schallreuter K, Slominski A, Pawelek JM, Jimbow K, Gilchrest BA (1998) What controls melanogenesis? Exp Dermatol 7:143–150
91. Schallreuter KU (1997) Epidermal adrenergic signal transduction as part of the neuronal network in the human epidermis. J Invest Dermatol Symp Proc 2:37–40
92. Siebenhaar F, Sharov AA, Peters EM, Sharova TY, Syska W, Mardaryev AN, et al (2007) Substance P as an immunomodulatory neuropeptide in a mouse model for autoimmune hair loss (alopecia areata). J Invest Dermatol 127:1489–1497
93. Singh LK, Pang X, Alexacos N, Letourneau R, Theoharides TC (1999) Acute immobilization stress triggers skin mast cell degranulation via corticotropin releasing hormone, neurotensin, and substance P: a link to neurogenic skin disorders. Brain Behav Immun 13:225–239
94. Slominski A (2005) Neuroendocrine system of the skin. Dermatology (Basel, Switzerland) 211:199–208
95. Slominski A, Wortsman J (2000) Neuroendocrinology of the skin. Endocr Rev 21:457–487
96. Slominski A, Ermak G, Hwang J, Chakraborty A, Mazurkiewicz JE, Mihm M (1995) Proopiomelanocortin, corticotropin releasing hormone and corticotropin releasing hormone receptor genes are expressed in human skin. FEBS Lett 374:113–116
97. Slominski A, Ermak G, Mazurkiewicz JE, Baker J, Wortsman J (1998) Characterization of corticotropin-releasing hormone (CRH) in human skin. J Clin Endocrinol Metab 83:1020–1024
98. Slominski A, Wortsman J, Pisarchik A, Zbytek B, Linton EA, Mazurkiewicz JE, et al (2001) Cutaneous expression of corticotropin-releasing hormone (CRH), urocortin, and CRH receptors. FASEB J 15:1678–1693
99. Slominski A, Tobin DJ, Shibahara S, Wortsman J (2004) Melanin pigmentation in mammalian skin and its hormonal regulation. Physiol Rev 84:1155–1228
100. Slominski A, Wortsman J, Tobin DJ (2005a) The cutaneous serotoninergic/melatoninergic system: securing a place under the sun. FASEB J 19:176–194
101. Slominski A, Zbytek B, Semak I, Sweatman T, Wortsman J (2005b) CRH stimulates POMC activity and corticosterone production in dermal fibroblasts. J Neuroimmunol 162:97–102
102. Slominski A, Zbytek B, Szczesniewski A, Semak I, Kaminski J, Sweatman T, et al (2005c) CRH stimulation of corticosteroids production in melanocytes is mediated by ACTH. Am J Physiol 288:E701–E706
103. Slominski A, Zbytek B, Pisarchik A, Slominski RM, Zmijewski MA, Wortsman J (2006a) CRH functions as a growth factor/cytokine in the skin. J Cell Physiol 206:780–791
104. Slominski A, Zbytek B, Zmijewski M, Slominski RM, Kauser S, Wortsman J, et al (2006b) Corticotropin releasing hormone and the skin. Front Biosci 11:2230–2248
105. Slominski AT, Botchkarev V, Choudhry M, Fazal N, Fechner K, Furkert J, et al (1999) Cutaneous expression of CRH and CRH-R. Is there a "skin stress response system?" Ann N Y Acad Sci 885:287–311
106. Sturm RA (2002) Skin colour and skin cancer – MC1R, the genetic link. Melanoma Res 12:405–416
107. Suzuki I, Cone RD, Im S, Nordlund J, Abdel-Malek ZA (1996) Binding of melanotropic hormones to the melanocortin receptor MC1R on human melanocytes stimulates proliferation and melanogenesis. Endocrinology 137: 1627–1633

108. Tanaka T, Danno K, Ikai K, Imamura S (1988) Effects of substance P and substance K on the growth of cultured keratinocytes. J Invest Dermatol 90:399–401
109. Tobin DJ, Kauser S (2006) Hair melanocytes as neuro-endocrine sensors – pigments for our imagination. Mol Cell Endocrinol 243:1–11
110. Tobin DJ, Paus R (2001) Graying: gerontobiology of the hair follicle pigmentary unit. Exp Gerontol 36:29–54
111. Tobin DJ, Fenton DA, Kendall MD (1991) Cell degeneration in alopecia areata. An ultrastructural study. Am J Dermatopathol 13:248–256
112. Tobin DJ, Orentreich N, Fenton DA, Bystryn JC (1994) Antibodies to hair follicles in alopecia areata. J Invest Dermatol 102:721–724
113. Tobin DJ, Slominski A, Botchkarev V, Paus R (1999) The fate of hair follicle melanocytes during the hair growth cycle. J Invest Dermatol Symp Proc 4:323–332
114. Tobin DJ, Gunin A, Magerl M, Handijski B, Paus R (2003a) Plasticity and cytokinetic dynamics of the hair follicle mesenchyme: implications for hair growth control. J Invest Dermatol 120:895–904
115. Tobin DJ, Gunin A, Magerl M, Paus R (2003b) Plasticity and cytokinetic dynamics of the hair follicle mesenchyme during the hair growth cycle: implications for growth control and hair follicle transformations. J Invest Dermatol Symp Proc 8:80–86
116. Toyoda M, Nakamura M, Makino T, Hino T, Kagoura M, Morohashi M (2002) Nerve growth factor and substance P are useful plasma markers of disease activity in atopic dermatitis. Br J Dermatol 147:71–79
117. Tsatmali M, Ancans J, Thody AJ (2002) Melanocyte function and its control by melanocortin peptides. J Histochem Cytochem 50:125–133
118. Wei ET, Thomas HA, Christian HC, Buckingham JC, Kishimoto T (1998) D-amino acid-substituted analogs of corticotropin-releasing hormone (CRH) and urocortin with selective agonist activity at CRH1 and CRH2beta receptors. Peptides 19:1183–1190
119. Westgate GE, Craggs RI, Gibson WT (1991) Immune privilege in hair growth. J Invest Dermatol 97:417–420
120. Wiesner B, Roloff B, Fechner K, Slominski A (2003) Intracellular calcium measurements of single human skin cells after stimulation with corticotropin-releasing factor and urocortin using confocal laser scanning microscopy. J Cell Sci 116:1261–1268
121. Wille S, Sydow S, Palchaudhuri MR, Spiess J, Dautzenberg FM (1999a) Identification of amino acids in the N-terminal domain of corticotropin-releasing factor receptor 1 that are important determinants of high-affinity ligand binding. J Neurochem 72:388–395
122. Wille S, Sydow S, Palchaudhuri MR, Spiess J, Dautzenberg FM (1999b) Identification of amino acids in the N-terminal domain of corticotropin-releasing factor receptor 1 that are important determinants of high-affinity ligand binding. J Neurochem 72:388–395
123. Yaswen L, Diehl N, Brennan MB, Hochgeschwender U (1999) Obesity in the mouse model of pro-opiomelanocortin deficiency responds to peripheral melanocortin. Nat Med 5:1066–1070
124. Zbytek B, Slominski AT (2005) Corticotropin-releasing hormone induces keratinocyte differentiation in the adult human epidermis. J Cell Physiol 203:118–126
125. Zbytek B, Pikula M, Slominski RM, Mysliwski A, Wei E, Wortsman J, et al (2005) Corticotropin-releasing hormone triggers differentiation in HaCaT keratinocytes. Br J Dermatol 152:474–480

Neurobiology of Sebaceous Glands

14

M. Böhm and T.A. Luger

Contents

Key Features

- The human sebaceous gland is both a target organ and a source for various neuromediators.
- Neuromediators appear to act in an autocrine, a paracrine and/or a classical endocrine fashion to modulate sebaceous gland function.
- The so far identified neuromediators have been shown to regulate sebocyte differentiation, proliferation, and inflammatory responses.
- The presence of virtually all components of the proopiomelanocortin system indicates expression of a local hypothalamic–pituitary–adrenal axis within the sebaceous gland.
- Selected neuromediators and their receptors are also present in periglandular nerve fibers, suggesting bilateral communication between the sebaceous gland and the nervous system.

Synonyms Box: ACTH Adrenocorticotropin, CRH Corticotropin-releasing hormone, CRH-R Corticotropin-releasing hormone receptor, CGRP calcitonin gene-related peptide, DOR δ-Opioid receptor, ED Endorphin, HPA Hypothalamic-pituitary-adrenal, IL Interleukin, MC-R Melanocortin receptor, MOR μ-Opioid receptor, NEP Neutral endopeptidase, NDP-α-MSH [Nle4, D-Phe7]-α-MSH, NGF Nerve growth factor, NPY Neuropeptide Y, POMC Proopiomelanocortin, SP Substance P, Trk Tyrosine kinase, VIP Vasointestinal peptide, VPAC VIP receptor

14.1 Introduction

Our present view that the sebaceous gland is an integral part of the cutaneous neuroendocrine and neuroimmune system is based on both basic science and clinical experience. Accordingly, research over the last years has shown that sebocytes share many features with epidermal keratinocytes with regard to expression of neuropeptide receptors and generation of neuropeptides [28]. In this context, the close anatomical connection between the sebaceous gland and the hair follicle, the latter an established immune privileged organ within the skin and both a prominent source of a plethora of neuroendocrine mediators and target site for neuropeptides, should also be noticed [4]. The concept of the sebaceous gland as a stress-sensitive organ is also supported by clinical wisdom and observations [2,20], which recently could be recapitulated by a study linking the stress experienced by university students with the course of acne vulgaris [9].

R.D. Granstein and T.A. Luger (eds.), *Neuroimmunology of the Skin*,

This review summarizes our current basic knowledge of "neuromediators" and their receptors in sebaceous gland biology. The term "neuromediator" has been chosen deliberately to include not only the *bona fide* neuropeptides, but also neurotransmitters, some which are chemically not peptides but amino acid derivatives. According to the aim of this book, the emphasis of the following chapter is on the human sebaceous gland but where necessary related interesting findings from animal models as well as from modified sebaceous glands of man are mentioned. The chapter is structured into two principal sections: the first section describes the *state of the art* of neuromediators and their receptors expressed exclusively by sebocytes; the second section summarizes our current knowledge of neuromediators and neuromediator receptors expressed by nerve fibers around the human sebaceous acini.

14.2 The Sebaceous Gland: An Integral Part of the Neuroimmune System

14.2.1 Neuromediators and Neuromediator Receptors Expressed by Sebocytes

14.2.1.1 Melanocortins

One of the earliest findings that established a functional connection between the neuroendocrine system and the sebaceous gland was the demonstration of a regulatory effect of α-melanocyte-stimulating hormone (MSH) and adrenocorticotropin (ACTH) on sebum production by the rat preputial gland [11,27,38], a modified sebaceous gland of rodents. Both ACTH and α-MSH were originally characterized as neurohormones released from the rat pituitary gland and proposed to act directly or indirectly on the preputial gland to increase sebum secretion. Removal of the neurointermediate lobe decreased sebum secretion to a level comparable with that following total hypophysectomy [37]. Since the pars intermedia of the rodent pituitary gland is a source for α-MSH, this melanocortin was coined a "sebotrophin" specifically increasing the biosynthesis of wax esters and sterols in the rat [39]. However, in vivo studies in the rodent model could not substantiate a direct effect of α-MSH on the sebum rate in adrenalectomized animals, indicating that both α-MSH and ACTH act via androgen induction in the adrenal gland [36,38].

Recent studies on human sebocytes, however, clearly demonstrate that these cells are direct target cells for melanocortin peptides. By using the immortalized sebocyte cell line SZ95, created by transfecting primary human facial sebocytes with the coding region of the Simian virus-40 large T antigen, we showed a direct immunomodulatory effect of α-MSH on sebocytes [5]. The SZ95 cell line was previously reported to exhibit all morphologic, phenotypic, and functional characteristics of normal human sebocytes, therefore, underscoring the usefulness of this in vitro model for human sebocyte studies [47]. Accordingly, secretion of the chemokine intrerleukin (IL)-8 was dose-dependently inhibited by α-MSH in SZ95 sebocytes both in the absence or presence of the proinflammatory mediator IL-1β [5]. An effective concentration of α-MSH in the nanomolar range suggested the presence of specific α-MSH binding sites in SZ95 sebocytes. Autoradiography studies could furthermore confirm the presence α-MSH binding sites, which were upregulated by incubating the cells with IL-1β or tumor necrosis factor-α (Böhm and Schiöth, unpublished findings). These findings may suggest a role for α-MSH as a negative regulator of inflammation in the sebaceous gland. Subsequent studies addressing the effect of ACTH as well as the superpotent MSH analogue [Nle4, D-Phe7]-α-MSH (NDP-α-MSH) in a primary human sebocyte culture system derived from facial skin revealed that both melanocortin peptides increased cytoplasmic lipid droplet formation at a level similar to stimulation with bovine pituitary extract. Moreover, NDP-α-MSH increased squalene synthesis in a dose-dependent manner with peak synthesis of squalene at 10 nM [44].

The biological effects of melanocortins on sebocytes are mediated via melanocortin receptors (MC-Rs), which are small heptahelical surface receptors and which belong to the superfamily of G protein-coupled receptors. Activation of all five so-far identified and cloned MC-Rs leads to activation of adenylate cyclase and an increase in intracellular cAMP [21]. In man, there is increasing evidence that both MC-1Rs and MC-5Rs are expressed within the sebaceous glands as shown by immunohistochemistry [5,15,35,44,45]. In particular, MC-5R has been suggested as a marker of sebocyte differentiation since, in human skin, MC-5R immunoreactivity was most pronounced in terminally differentiated cells while MC-1R was mostly detected in the peripheral nondifferentiated epithelia of the sebaceous acini. Moreover, in cultured primary human sebocytes, RNA expression of MC-5R was detected only at the onset of differentiation and in fully differentiated cells containing lipid granules [45]. The latter finding is in accordance with the detection of MC-1R but not MC-5R in the human sebocyte cell line SZ95 which, under routine culture conditions, rapidly proliferates but shows few signs of terminal differentiation [5]. It is therefore possible that the MC-1R expressed by human sebocytes is more involved in immunoregulation, as pointed out above, whereas MC-5R is more concerned with sebogenesis. Indeed, transgenic mice with targeted disruption of MC-5R displayed reduced sebum secretion, lack of NDP-MSH radiolabeling of the preputial glands and loss of NDP-MSH-induced cAMP increase in membrane fractions from these glands [8]. In situ expression of MC-5R could be demonstrated by in situ hybridization of secretory epithelia in both the preputial gland as well as sebaceous glands in the skin of normal mice. However, we do not know whether mice with targeted disruption or signal deficiency of MC-1R have defective sebum production. Of note, the observed reduction in sebum production in mice with targeted disruption of MC-5R was mild [8]. Preliminary evidence exists that MC-1R is also expressed in the sebaceous glands of murine skin (Böhm et al., unpublished).

Given the capacity of the skin to express MC-Rs in numerous cell types as well as to generate proopiomelanocortin (POMC)-derived peptides upon proinflammatory stressors, it is also of some interest to know whether the sebaceous gland is stimulated to produce melanocortin peptides in an autocrine or a paracrine fashion. Preliminary evidence exists that POMC expression can be induced in human primary sebocytes by the artificial cAMP inducers forskolin and choleratoxin [17]. On the other hand, POMC RNA expression was undetectable in SZ95 sebocytes even after stimulation with natural prototypical POMC-inducers including proinflammatory cytokines, α-MSH, and corticotropin-releasing hormone (Böhm et al., unpublished findings). In vivo, POMC and POMC peptide expression may be induced in the sebaceous gland by natural, yet to be identified factors. Accordingly, expression of POMC RNA has been detected by RT-PCR in laser capture microdissected human sebaceous glands [19]. More recently, immunostaining for both α-MSH and ACTH was reported in hair follicle epithelia and in secretory epithelia of the sebaceous gland in patients with alopecia areata and control patients. In situ staining of α-MSH in the aforementioned cells was more accentuated in lesional skin of patients with alopecia areata [18], suggesting induction of the cutaneous hypothalamic pituitary adrenal (HPA) axis by disease stress.

In summary, these findings demonstrate that sebocytes, via expressing specific receptors, are target cells for the action of melanocortin peptides delivered by the classical endocrine route or via the cutaneous melanocortin system [7]. Melanocortin receptors expressed by human sebocytes mediate, on one side, immunoregulatory effects (such as chemokine secretion) and, on the other side, cell-differentiating effects (such as lipid production) of (NDP)-α-MSH or ACTH. The detection of POMC and melanocortin peptides in the sebaceous gland itself, moreover, suggests that a "seboglandular" HPA axis is operational, which could serve as a regulator of local homeostasis during cutaneous inflammatory and immune responses.

14.2.1.2 Endogenous Opioids

The endogenous opioid β-endorphin (β-ED) is another POMC-derived peptide and, like α-MSH and ACTH, is considered a prototypical stress-responsive neurohormone. Until now, β-ED immunoreactivity could only be verified in sebaceous glands of rodents. Immunoreactivity for β-ED was reported in sebaceous glands of mice, but only in the telogen phase of the hair cycle. β-ED immunoreactivity increased in the pilosebaceous glands with increasing time after induction of anagen [29]. β-ED immunostaining was also detectable in guinea pig sebaceous glands of normal skin with an increase during immune inflammation, that is, after experimental induction of a passive Arthus reaction, the latter suggesting that sebaceous glands participate in the regulation of the cutaneous immune system [12]. In contrast, immunohistochemical studies on human skin failed to detect β-ED immunostaining [30]. In light of the detected POMC RNA and the immunoreactivity for α-MSH and ACTH in sebocytes in situ as outlined above [18,19] more studies are needed to definitely rule out β-ED production by the human sebaceous gland.

Regarding the functional role of β-ED in the sebaceous gland, there is preliminary evidence that β-ED, in addition to its potential immunoregulatory role, may act as another lipidogenic hormone [6]. In vitro treatment of SZ95 sebocytes with β-ED in chemically defined medium induced cytoplasmic lipid droplet formation and suppressed cell proliferation induced by epidermal growth factor. Most interestingly, β-ED stimulated lipogenesis and specifically increased the amount of C16:0, C16:1, C18:0, C18:1, and C18:2 fatty acids to an extent similar to linoleic acid. To elucidate the mechanism of action of β-ED in human sebocytes, expression studies on opioid receptors (ORs), that is, the μ-opioid receptor (MOR) and the δ-OR (DOR) were performed. RT-PCR analysis, Western immunoblotting, and immunocytochemical studies disclosed the presence of MORs in both SZ95 sebocytes as well as in the human sebaceous gland in situ [6]. In contrast, expression of DORs was not detectable. The detected MOR immunoreactivity in human sebocytes supports previous findings describing the expression of MORs in human epidermal keratinocytes in vitro as well as in epidermal and follicular keratinocytes and various glandular epithelia in normal human skin in situ [3].

These data extend our present knowledge about the seboglandular POMC system towards endogenous opioids (such as β-ED) and ORs (such as the MOR). Recent preliminary findings demonstrate that the MOR is expressed by human sebocytes in vitro and in situ. β-ED appears to regulate cell proliferation, lipidogenesis, and possibly immune and inflammatory responses within the sebaceous gland.

14.2.1.3 Corticotropin-Releasing Hormone

There is evidence that corticotropin-releasing hormone (CRH), its binding protein (CRH-BP), and corticotropin receptors (CRH-Rs) are also expressed in human sebocytes [19,48]. These findings are in accordance with the now established role of CRH as a master player in the cutaneous stress response and support the idea of a cutaneous analogon of the HPA axis [31]. By using laser-capture microdissection with RT-PCR, CRH, CRH-R, and POMC were detected in sebaceous cells [19]. The presence of CRH in the human sebaceous gland was recently supported by immunohistochemical studies of skin samples from patients with alopecia areata, in which sebaceous glands in lesional skin appear to express more CRH immunoreactity than sebaceous glands in normal skin [18]. In vitro, expression of CRH, CRH-BP, and CRH-R1 and 2 was also detected at the mRNA and protein levels in SZ95 sebocytes. CRH-R1 was the predominant type of CRH-R expressed by human sebocytes. Functionally, CRH is likely to act as an autocrine or a paracrine factor with pro-differentiating effects [48]. CRH induced lipidogenesis and enhanced mRNA expression of 3β-hydroxysteroid dehydrogenase/Δ^{5-4}-isomerase, an enzyme that converts dehydroepiandrosterone to testosterone. In contrast to α-MSH, CRH did not modulate IL-1β-induced IL-8 release by SZ95 sebocytes in vitro. Testosterone suppressed CRH-R1 and CRH-R2 mRNA expression in SZ95 sebocytes, while growth hormone switched CRH-R1 mRNA expression to CRH-R2 [48].

These findings highlight a role for CRH as a lipogenic regulator of human sebocytes in concert with other hormones (e.g., testosterone or growth hormone). Concomitant expression of CRH with POMC-derived peptides, such as α-MSH within human sebaceous glands, furthermore points towards an integral role of CRH and its receptor in the neuroimmune response of this glandular structure.

14.2.1.4 Miscellaneous

In contrast to the increasing body of data emphasizing the role of the POMC system in sebaceous gland activity and neuroimmunomodulation, comparatively little is known about the expression of other neuromediators and their receptors by sebocytes. It should also be noted that the majority of the following studies rely on immunohistochemical approaches, warranting a closer examination by biochemical and molecular biology methods as well as in sebocyte culture systems.

Within the human system, immunoreactivity for nerve growth factor (NGF) and its high affinity receptor, tyrosine kinase A (TrkA), has been detected in the sebaceous gland especially in peripheral cells [1]. This contradicts earlier findings in which NGF and TrkA were only detectable in sebaceous glands of acne patients [40]. Interestingly, secretions from the Meibomian gland, a modified sebaceous eyelid gland crucially involved in production of the outermost layer of the tear film, contained the TrkC, the receptor for the NGF-related neuropeptide neurotrophin-3 [42]. At the functional level, it was reported that IL-6 can induce nerve growth factor expression by sebocytes in human facial organ cultures [40]. The authors suggested that during inflammatory stress, proinflammatory mediators such as IL-6 may induce the NGF system in the sebaceous gland, leading to increased periglandular innervation, possibly followed by neurogenic inflammatory stimuli.

In another immunohistochemical study, serotonin immunoreactivity was detected within the cytoplasm of the sebaceous glands of patients with chronic eczema [16]. Of note, skin cells have the full enzymatic capacity to synthesize the neurotransmitter/neuromediator serotonin [32]. In light of the well known proinflammatory actions of serotonin and the increasing identification of its receptors on various cell types, some of which already have been identified in various skin cell types [32], efforts should be fruitful to define the relevant receptor for serotonin in human sebocytes and to establish their function.

Recent data suggest that the sebaceous gland is also a target site for somatostatin [14], a neuroendocrine peptide expressed by the nervous system but also by endocrine cells in the periphery (pancreas, gut, thyroid, and others). In human skin, somatostatin immunoreactive nerve fibers were described some time ago. Somatostatin consists of two bioactive forms, a 14 amino acid peptide, and a congener of somatostatin-14 extended at the N-terminal domain [22]. Using immunohistochemistry, it was found that sebaceous glands express all five subtypes of the somatostatin receptors [14]. In light of the pleiotropic effects of somatostatin, including regulation of other neuromediators, for example, CRH, immunomodulation, proliferation, and apoptosis, exploring the serotonin system in the sebaceous should be highly interesting.

In another study it was reported that the vanilloid receptor subtype 1 (VR1/TRPV1) is expressed not only in cutaneous sensory nerve fibers, but also in various resident skin cells, including the sebaceous gland [33]. VR1 is a nonselective cation channel that binds vanilloids, for example, the alkaloid capsaicin but also endogenous cannabinoids [34]. In the sebaceous gland, VR1 immunoreactivity was most prominent in differentiated sebocytes, whereas the undifferentiated cells were largely negative [33].

Preliminary evidence exists that human sebocytes express vasoactive intestinal polypeptide (VIP) receptors (VPAC) with VPAC2 exhibiting the most pronounced expression in cells surrounded by VIP-immunoreactive nerve fibers. Sebocytes also were found to express neuropeptide Y (NPY) and calcitonin gene-related peptide (CGRP) receptors [26].

14.3 Neuromediators and Neuromediator Receptors Expressed by Periglandular Nerve Fibers

The density of nerve fibers around the human sebaceous glands in normal skin is generally sparse and, accordingly, this appendage structure is classically considered not be innervated. On the other hand, a few reports describe the presence of neuromediator and neuromediator receptor-positive nerve fibers around sebaceous glands and around Meibomian glands in man and in other species.

Expression of the DOR has been studied in the rat skin. DOR-immunoreactive nerve fibers were found to be associated with hair follicle sebaceous glands, including the ducts [43]. All DOR immunoreactive nerve fibers around the glandular structures also contained CGRP. In sebaceous glands of guinea pig skin, CGRP- and NPY-immunoreactive nerve fibers were also detected [13]. Light and electron microscopy studies of rat skin furthermore revealed the presence of SP immunoreactive nerve fibers along the outer border of sebaceous glands [23].

The above findings have been extended by studies on the modified sebaceous Meibomian gland of the eyelid in humans and in several other species. Accordingly, nerve fibers with colocalization of CGRP were detected around the outside of the rat Meibomian gland acini [43]. The presence of CGRP as well as VIP, NPY, and substance P (SP) was also found in nerve fibers near the Meibomian gland acini in guinea pigs and monkeys (*Macacca fascicularis* and *Macacca mulatta*) [10,24]. The human Meibomian gland was likewise reported to contain SP- and CGRP-positive nerve fibers around the acini [24]. In another report, VIP-positive nerve-like structures were identified in the stroma of the human Meibomian gland [25].

Regarding the human sebaceous gland and the role of periglandular nerve fibers, it was reported that the number of SP immunoreactive nerve fibers around sebaceous glands were more numerous in lesional skin of patients with acne than in skin of healthy subjects [41]. Expression of the SP-inactivating cell surface protein neutral endopeptidase (NEP) was observed within sebaceous germinative cells of acne patients but not of healthy volunteers. Exogenous SP treatment increased expression of NEP within sebaceous germinative cells and resulted in an increase in the size of the sebaceous glands as well as in the number of lipid granules in a skin organ model [41]. These data suggest that SP could act as a neurogenic inflammatory mediator involved in the pathogenesis of acne vulgaris.

14.4 Summary

The data presented in this review demonstrate that the sebaceous gland is both a pleiotropic target structure and source for various neuromediators (Table 14.1). Multiple neuromediators, either delivered via the classical endocrine route (e.g., during activation of the

Table 14.1 Neuromediators and neuromediator receptors in the sebaceous gland[a]

	Expression in sebocytes	Expression in periglandular Nerve fibers
Neuromediator	CRH POMC α-MSH ACTH NGF Serotonin Somatostatin NPY CGRP	SP CGRP NPY VIP
Neuromediator Receptor	MC-1R, MC-5R CRH-R1, CRH-R2 MOR TrkA, TrkC VPAC1/2 VR1	DOR

[a] Including the Meibomian gland

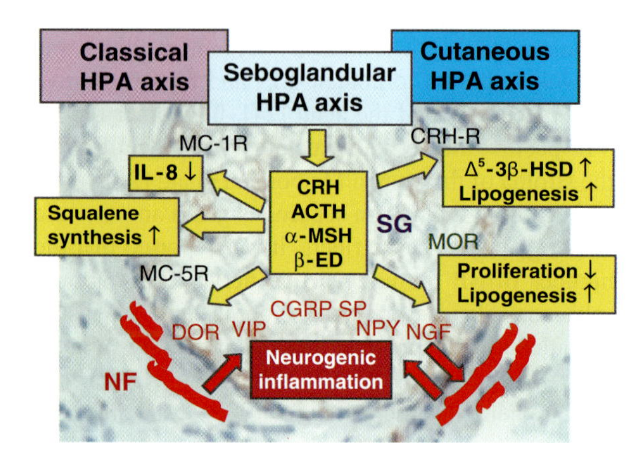

Fig. 14.1 The various neuromediators and neuromediators receptors expressed by the sebaceous gland and periglandular nerve fibers. *SG* sebaceous gland, *NF* nerve fibers. All other abbreviations are explained already in the text

classical HPA axis), by paracrine or autocrine pathways (e.g., by activation of the cutaneous or seboglandular HPA axis), or by periglandular nerve fibers (i.e., during neurogenic inflammation), can act as regulators of exocrine, immune, and inflammatory effector functions of sebocytes (Fig. 14.1). This concept [46] supports the view of a "psychotropic" nature of the sebaceous gland, which not only helps to understand diseases of sebocyte dysfunction, for example, acne vulgaris as a stress-sensitive disease, but also points towards novel and fascinating future treatment options for such diseases. Given the possibility of neuropharmacological intervention against many of the herein described neuromediators (e.g., via melanocortin peptides and their derivatives, their antagonists, opiate antagonists, SP inhibitors, serotonin antagonists, etc.), further exploration of the neurobiology of the sebaceous gland is a promising research field for the basic scientist and dermatologist.

Summary for the Clinician

- The human sebaceous gland is a prominent target for a variety of neuromediators including classical neuropeptides and neurotransmitters
- Via expression of specific receptors for such neuromediators, the sebaceous gland is susceptible to stress
- Understanding this important connection between stress and the sebaceous gland will optimize the relation between the clinical dermatologist and the patient affected by disorders of the sebaceous gland, e. g. acne.
- Molecular targeting of neuromediator receptors expressed by sebocytes may be expected to become a future therapy for inflammatory disorders of the sebaceous gland.

References

1. Adly MA, Assaf HA, Nada EA, Soliman M, Hussein M (2006) Expression of nerve growth factor and its high-affinity receptor, tyrosine kinase A proteins, in the human scalp skin. J Cutan Pathol 33:559–568
2. Baldwin HE (2002) The interaction between acne vulgaris and the psyche. Cutis 70:133–139
3. Bigliardi PL, Bigliardi-Qi M, Buechner S, Rufli T (1998) Expression of mu-opiate receptor in human epidermis and keratinocytes. J Invest Dermatol 111:297–301
4. Böhm M, Luger TA (1998) The pilosebaceous unit is part of the immune system. Dermatatology 196:75–79
5. Böhm M, Schiller M, Ständer S, Seltmann H, Li Z, Brzoska T, Metze D, Schiöth HB, Skottner A, Seiffert K, Zouboulis CC, Luger TA (2002) Evidence for expression of melanocortin-1 receptor in human sebocytes in vitro and in situ. J Invest Dermatol 118:533–535
6. Böhm M, Li Z, Ottavini M, Picardo M, Zouboulis CC, Ständer S, Luger TA (2004) Beta-endorphin modulates lipogenesis in human sebocytes. Invest Dermatol 123:A10 (Abstract)
7. Böhm M, Luger TA, Tobin DJ, Garcia-Borron JC (2006) Melanocortin receptor ligands: new horizons for skin biology and clinical dermatology. J Invest Dermatol 126:1966–1975
8. Chen W, Kelly MA, Opitz-Araya X, Thomas RE, Low MJ, Cone RD (1997) Exocrine gland dysfunction in MC5-R-deficient mice: evidence for coordinated regulation of

exocrine gland function by melanocortin peptides. Cell 91:789–798
9. Chiu A, Chon SY, Kimball AB (2003) The response of skin disease to stress: changes in the severity of acne vulgaris as affected by examination stress. Arch Dermatol 139:897–900
10. Chung CW, Tigges M, Stone RA (1996) Peptidergic innervation of the primate meibomian gland. Invest Ophthalmol Vis Sci 37:238–245
11. Ebling FJ (1977) Hormonal control and methods of measuring sebaceous gland activity. J Invest Dermatol 62:161–171
12. Gendek-Kubiak H (1998) Immunohistochemical demonstration of beta-endorphin in guinea pig sebaceous glands of normal skin and during immune inflammation. Folia Histochem Cytobiol 36:15–18
13. Gendek-Kubiak H, Kmiec BL (2004) Immunolocalization of CGRP, NPY and PGP 9.5 in guinea pig skin. Folia Morphol (Warsz) 63:115–117
14. Hagströmer L, Emtestam L, Stridsberg M, Talme T (2006) Expression pattern of somatostatin receptor subtypes 1–5 in human skin: an immunohistochemical study of healthy subjects and patients with psoriasis or atopic dermatitis. Exp Dermatol 15:950–957
15. Hatta N, Dixon C, Ray AJ, Phillips SR, Cunliffe WJ, Dale M, Todd C, Meggit S, Birch-MacHin MA, Rees JL (2001) Expression, candidate gene, and population studies of the melanocortin 5 receptor. J Invest Dermatol 2001:116: 564–570
16. Huang J, Li G, Yin D, Chi R (2004) Immunohistochemical study of serotonin in lesions of chronic eczema. Int Dermatol 43:723–726
17. Huang Q, Anthonavage M, Li W, Eisinger M (2002) The role of α-MSH in human sebaceous lipogenesis. J Invest Dermatol 119:275 (Abstract)
18. Kim HS, Cho DH, Kim HJ, Lee JY, Cho BK, Park HJ (2006) Immunoreactivity of corticotropin-releasing hormone, adrenocorticotropin and a-melanocyte-stimulating hormone in alopecia areata. Exp Dermatol 15:515–522
19. Kono M, Nagata H, Umemura S, Kawana S, Yoshiyuki R (2001) In situ expression of corticotropin-releasing hormone (CRH) and proopiomelanocortin (POMC) genes in human skin. FASEB J 15:2297–2299
20. Koo JY, Smith LL (1991) Psychologic aspects of acne. Pediatr Dermatol 8:185–188.
21. Mountjoy KG, Robbins LS, Mortrud MT, Cone RD (1992) The cloning of a family of genes that encode the melanocortin receptors. Science 257:1248–125
22. Patel YC (1999) Somatostatin and its receptor family. Front Neuroendocrinol 20:157–198
23. Ruocco I, Cuello AC, Shigemoto R, Ribeiro-da-Silva A (2001) Light and electron microscopic study of the distribution of substance P-immunoreactive fibers and neurokinin-1 receptors in the skin of the rat lower lip. J Comp Neurol 432:466–480
24. Seifert P, Spitznas M (1996) Immunocytochemical and ultrastructural evaluation of the distribution of nervous tissue and neuropeptides in the meibomian gland. Graefes Arch Clin Exp Ophthalmol 234:648–656
25. Seifert P, Spitznas M (1999) Vasoactive intestinal polypeptide (VIP) innervation of the human eyelid glands. Exp Eye Res 68:685–692
26. Seiffert K, Zouboulis CC, Seltmann H, Granstein RD (2000) Expression of neuropeptide receptors by human sebocytes and stimulatory effect of their agonists on cytokine production. Horm Res 53:102 (Abstract)
27. Shuster S, Thody AJ (1974) The control and measurement of sebum secretion. J Invest Dermatol 61:172–190
28. Slominski A, Wortsman J (2000) Neuroendocrinology of the skin. Endocr Rev 21:457–487
29. Slominski A, Paus R, Mazurkiewicz J (1992) Proopiomelanocortin expression in the skin during hair growth in mice. Experientia 48:50–54
30. Slominski A, Wortsman J, Mazurkiewicz JE, Matsuoka L, Dietrich J, Lawrence K, Gorbani A, Paus R (1993) Detection of proopiomelanocortin-derived antigens in normal and pathologic human skin. J Lab Clin Med 122:627–628
31. Slominski A, Wortman J, Luger T, Paus R, Solomon S (2000) Corticotropin releasing hormone and proopiomelanocortin involvement in the cutaneous response to stress. Physiol Rev 80:979–1020
32. Slominski A, Wortsman J, Tobin DJ (2005) The cutaneous serotoninergic/melatoninergic system: securing a place under the sun. FASEB J 19:176–194
33. Ständer S, Moormann C, Schumacher M, Buddenkotte J, Artuc M, Shpacovitch V, Brzoska T, Lippert U, Henz BM, Luger TA, Metze D, Steinhoff M (2004) Expression of vanilloid receptor subtype 1 in cutaneous sensory nerve fibers, mast cells, and epithelial cells of appendage structures. Exp Dermatol 13:129–139
34. Szolcsanyi J (2000) Are cannabinoids endogenous ligands for the VR1 capsaicin receptor? Trends Pharmacol Sci 21:2003–2004
35. Thiboutot D, Sivarajah A, Gilliland K, Cong Z, Clawson G (2001) The melanocortin 5 receptor is expressed in human sebaceous glands and rat preputial cells. J Invest Dermatol 115:614–619
36. Thody AJ, Shuster S (1971) Effect of adrenalectomy and adrenocorticotrophic hormone on sebum secretion in the rat. J Endocrinol 49:325–238
37. Thody AJ, Shuster S (1972) Control of sebum secretion by the posterior pituitary. Nature 237:346–347
38. Thody AJ, Shuster S (1989) Control and function of sebaceous glands. Physiol Rev 69:383–416
39. Thody AJ, Cooper MF, Bowden PE, Meddis D, Shuster S (1976) Effect of α-melanocyte-stimulating hormone and testosterone on cutaneous and modified sebaceous glands in the rat. J Endocrinol 71:279–288
40. Toyoda M, Nakamura M, Morohashi M (2002) Neuropeptides and sebaceous glands Eur J Dermatol 12:422–427
41. Toyoda M, Nakamura M, Makino T, Kagoura M, Morohashi M (2002) Sebaceous glands in acne patients express high levels of neutral endopeptidase. Exp Dermatol 11:241–247
42. Tsai PS, Evans JE, Green KM, Sullivan RM, Schaumberg DA, Richards SM, Dana MR, Sullivan DA (2006) Proteomic analysis of human meibomian gland secretions. Br J Ophthalmol 90:372–377
43. Wenk HN, Honda CN (1999) Immunohistochemical localization of delta opioid receptors in peripheral tissues. J Comp Neurol 408:567–567
44. Zhang L, Anthonavage M, Huang Q, Li WH, Eisinger M (2003) Proopiomelanocortin peptides and sebogenesis. Ann N Y Acad Sci 994:154–161

45. Zhang L, Li W-H, Anthonavage M, Eisinger M (2006) Melanocortin-5 receptor: a marker of human sebocyte differentiation. Peptides 27:413–420
46. Zouboulis CC, Böhm M (2004) Neuroendocrine regulation of sebocytes – a pathogenetic link between stress and acne. Exp Dermatol 13(Suppl 4):31–35
47. Zouboulis CC, Seltmann H, Neitzel H, Orfanos CE (1999) Establishment and characterization of an immortalized human sebaceous gland cell line (SZ95). J Invest Dermatol 113:1011–1020
48. Zouboulis CC, Seltmann H, Hiroi N, Chen WC, Young M, Oeff M, Scherbaum WA, Orfanos CE, McCann SM, Bornstein SR (2002) Corticotropin releasing hormone: an autocrine hormone that promotes lipogenesis in human sebocytes. Proc Natl Acad Sci U S A 99: 7148–7153

Neurobiology of Skin Appendages: Eccrine, Apocrine, and Apoeccrine Sweat Glands

15

K. Wilke, A. Martin, L. Terstegen, and S.S. Biel

Contents

Synonyms Box: Acetylcholine, neurotransmitter in both the peripheral nervous system and central nervous system, acts as main activator of the eccrine sweat gland, being released from the periglandular nerve endings; Acrosyringium, intra-epidermal part of the sweat gland; Apocrine sweat gland, mainly present in the armpits and around the genital area, secretes a fluid that is rich in proteins and lipids; Botulinum toxin (BTX), a neurotoxin produced by the bacterium *Clostridium botulinum,* inhibits the fusion of acetylcholine vesicles with pre-synaptic membranes and thereby inhibits sweat secretion effectively; Catecholamines, hormones that are released by the adrenal glands in situations of stress ("fight-or-flight" hormones), namely *Epinephrine* (Adrenaline)

Key Features

- *Eccrine sweat gland innervation is mostly cholinergic*: The major neurotransmitter released from the periglandular nerve endings is acetylcholine, whose concentration determines the sweat rate in humans. Eccrine sweat glands express various muscarinic acetylcholine receptor subtypes and eccrine sweat secretion can be blocked effectively by anti-muscarinic substances.
- *Eccrine sweat secretion can be blocked effectively by botulinum toxin (BTX)*: Botulinum neurotoxins are metalloproteases that cleave SNARE (soluble n-ethylmaleimide sensitive factor attachment protein receptor) proteins, thereby inhibiting the trafficking process of acetylcholine vesicles, their fusion with pre-synaptic membranes and finally the release of acetylcholine.
- *Dermcidin (DCD)*: Patients suffering from atopic dermatitis have reduced DCD concentrations in eccrine sweat. Therefore, in contrast to healthy patients, sweating does not lead to a reduction of bacteria on the skin.
- *Frey's syndrome*: This pathological state of gustatory sweating which affects the area of the cheek can occur after parotid surgery due to disruption of parasympathetic secretomotor fibers. These fibers anastomose with sudomotor sympathetic fibers of the skin, thereby gaining control of sweat gland activity.
- *Apocrine sweat gland innervation is believed to be adrenergic*: The apocrine secretion is controlled via epinephrine and norepinephrine but it is still unclear whether activation of apocrine glands takes place via sympathetic innervation or via circulating catecholamines.
- *Cystic fibrosis*: Patients suffering from cystic fibrosis show increased sweat chloride levels due to a defect in CFTR, which plays an important role in chloride reabsorption from sweat.

R.D. Granstein and T.A. Luger (eds.), *Neuroimmunology of the Skin*,

and *Norepinephrine* (Noradrenaline); Cystic fibrosis transmembrane conductance regulator (CFTR), a protein involved in the transport of chloride ions across cell membranes, plays an important role in the reabsorption of ions from primary sweat; Dermcidin (DCD), an antimicrobial peptide being expressed in eccrine sweat glands and is believed to play a role in innate host defense mechanisms; Eccrine sweat gland, plays an important role in body temperature regulation, secretes a fluid mainly composed of water and various ions; Frey's syndrome, a pathological state of gustatory sweating, which affects the area of the cheek, and may occur after parotid surgery; Major histocompatibility complex (MHC), a chromosomal segment that codes for cell-surface histocompatibility antigens and is the principal determinant of tissue type and transplant compatibility; Muscarinic acetylcholine receptors, acetylcholine receptors that are more sensitive to muscarine than to nicotine, play a role in signal transduction processes of eccrine sweat secretion

15.1 Introduction

Sweat glands are cutaneous appendages that, besides other functions, enable mammals to regulate their body temperature. According to their morphology and their mode of secretion, sweat glands can be classified as eccrine or apocrine. This distinction was introduced by Schiefferdecker in 1922 [58]. Moreover, in 1987, Sato described glands that showed the characteristics of both eccrine and apocrine glands and therefore were named apoeccrine [55,56].

15.1.1 Eccrine Sweat Glands

Eccrine sweat glands already exist at birth and are widely distributed over the whole skin surface with only two exceptions: lips and glans penis [57]. Depending on individual variation, there are 1.6–5 million sweat glands found across the human body with an average density of 200 sweat glands/cm^2 ranging from 64 sweat glands/cm^2 on the back to 700 sweat glands/cm^2 on the palms and soles [15,18,57]. An eccrine gland consists of a single tubule ranging from about 4–8 mm in total length. The intra-epidermal part of the sweat gland is called the acrosyringium followed by the straight duct, a coiled duct, and a coiled, secretory portion which is found in the deep dermis. Glands appear as coils in their natural three dimensional context, where one coil may consist of more than one gland tubule [75].

The acrosyringium consists of epithelial cells with no clear distinction or border to the epidermis. The epithelial cells differentiate towards the lumen; therefore, cornified cells can be found inside the lumen. The luminal diameter is between 20 and 60 µm, whereas the dermal duct, whether straight or coiled, has a smaller inner diameter of only 10–20 µm. The coiled duct is described as an epithelium of two to three layers of epithelial cells [24,37]. These are connected at numerous sites by desmosomes and intercellular junctions. A functional barrier is thus created between the luminal and extracellular compartments [22]. The inner cells are rich in tonofilaments, yielding a thick terminal web or cortical zone called cuticula. The outer or basal cells are surrounded by a continuous basal membrane, which in turn is enclosed by a collagenous and fibrocyte sheath [53]. The basic function of the sweat duct is the reabsorption of ions from the primary sweat fluid. In this process, besides others, the chloride channel cystic fibrosis transmembrane conductance regulator (CFTR), being abundantly expressed in the apical membranes of the duct, plays an important role in chloride reabsorption [49]. This is why patients suffering from cystic fibrosis due to a defect in CFTR show increased sweat chloride levels [34].

The secretory portion of the sweat gland consists of a tubule with an outer diameter of 60–120 µm and an inner diameter of 30–40 µm. The overall diameter of the coil is around 500–700 µm. Three different cell types can be distinguished within the secretory portion: clear cells, dark cells, and myoepithelial cells [24]. Clear cells contain many mitochondria. Between adjacent cells, intercellular canaliculi can be found, which are specific for eccrine secretory cells [9]. Their role in secretion together with convoluted cell membranes of adjacent cells is not fully understood [21]. Dark cells, in contrast to clear cells, can be intensely stained with eosin, toluidine blue, and methylene blue [23]. Moreover, they are osmiophillic, appear denser, and are also called granular cells because of their abundant granules, whereas clear cells are also called agranular. Dark cells form the luminal cell layer and clear cells the basal cell layer, while the function of the myoepithelial cells remains unclear.

Eccrine sweat can vary in composition, depending on hydration, exercise, state of health, and region of the body [36,38,46]. Besides water, that accounts for 99% of eccrine sweat, further components are sodium, chloride, potassium, calcium, magnesium, lactate, ammonia, amino acids, urea, and bicarbonate [20,36,61]. In addition, several proteins and peptides, for example, cysteine proteinases [79], DNAse I [78], lysozyme, Zn-α2-glycoprotein [40], CRISP-3 [68], and dermcidin (DCD) [59], have been identified in eccrine sweat. Some of these, like DCD, an antimicrobial peptide, that is expressed constitutively in eccrine glands, are believed to play a role in innate host defense mechanisms [50,59]. Recently, it could be shown that sweating leads to a reduction of bacteria on the skin of healthy subjects, but not in patients with atopic dermatitis having reduced DCD concentrations in eccrine sweat [51].

15.1.2 Apocrine Sweat Glands

Apocrine glands already exist at birth but do not become active until puberty. They are restricted to hairy body areas as they open and secrete into the hair canal and can only be found in the axillary, mammary, perineal, and genital region. The density of apocrine glands in the axillary region have been investigated in a study performed by Sato [56], in which he found 8–43 clear apocrine and up to 54 apoeccrine glands/cm^2. With an outer diameter of about 800 μm, an apocrine gland coil clearly exceeds that of an eccrine gland. The outer diameter of the gland tubule is about 120–200 μm, while the inner diameter is around 80–100 μm. It is hardly possible to differentiate between a duct and a secretory coil as the apocrine duct is very short and can be found in close vicinity to hair [2]. The secretory coil consists of two different types of cells: secretory and myoepithelial cells. Although secretory cells are typically tubular shaped, they can vary in shape due to their secretory activity. Their nucleus is located in the basal cell region and the cytoplasm is filled with mitochondria and different granules. The cell membranes are convoluted; microvilli are present towards the lumen. Pinching-off as the mode of secretion can be observed at the ultrastructural level. The function of the surrounding myoepithelial cells remains to be elucidated.

The fluid secreted by the apocrine sweat gland is an oily, odorless substance, containing proteins, lipids, and steroids [32]. However, it can not be excluded that apocrine secretions are mixed with sebum, as both apocrine and sebaceous glands open into the hair follicle. Recently, it was shown that two apocrine proteins, referred to as apocrine secretion odor-binding proteins 1 and 2 (ASOB1 and ASOB2) function as carrier proteins for volatile odor molecules, for example, 3-methyl-2-hexenoic acid, that are subsequently released by bacterial enzymes. ASOB2 was shown to be identical to the lipocalin apolipoprotein D (apoD) [26,80]. ASOB1 shares homology to the α-chain of apolipoprotein J (apoJ) [63]. As in other species lipocalins serve as carrier proteins for pheromones, an analogous function has been suggested for ASOB1/2.

Recent investigations suggest that genes encoding the major histocompatibility complex (MHC) influence human body odor as well as body odor preferences. MHC proteins are cell-surface proteins that bind short polypeptides and present them to T lymphocytes. Doing so, they play a vital role in the complex immunological dialogue that occurs between T cells and other cells, allowing the body to differentiate from oneself. Recent studies show that women score male body odors more pleasant when they differ from these men regarding their MHC genotype and that these preferences depend on the women's hormonal status. The function of the MHC-dependent mating preferences is not yet clear. Yet, it is speculated that an increased heterozygosity in MHC genes may lead to a higher resistance to infectious diseases in offspring and avoids inbreeding [72–74].

15.1.3 Apoeccrine Sweat Glands

The apoeccrine gland is a mixed type gland which was first described by Sato in 1987 [56]. Apoeccrine glands are believed to develop during puberty from eccrine glands as the proportion of eccrine glands decreases with age. Like apocrine glands, apoeccrine glands can also be found in the anogenital region [70]. It is very likely that the apoeccrine gland is also restricted to hairy body regions.

A typical morphological characteristic is the irregular shape of the tubule and of the cells. As the apoeccrine gland is composed of myoepithelial cells, eccrine secretory cells, and apocrine secretory cells, the identification is possible with specific markers corresponding to the particular cell types, that is, Phalloidin, S-100, or CD15 [76,77].

Apoeccrine glands secrete an eccrine-like watery fluid. As they share multiple similarities with eccrine glands concerning morphology, marker proteins, and mode of secretion [55,56,77], it is conceivable that the composition of apoeccrine sweat resembles that of eccrine sweat. Sato and Sato [55] determined sodium and potassium concentrations in the duct of isolated apoeccrine glands and found similar values as for eccrine sweat. However, until now the composition of apoeccrine sweat has not been elucidated, as it is not feasible to discriminate between eccrine and apoeccrine sweat during sweat collection in the axilla.

15.2 Innervation of Sweat Glands

15.2.1 Eccrine Sweat Glands

Eccrine sweat glands are innervated by postganglionic sympathetic fibers. Spinal cord segments from T2 to T8 provide innervation to the skin of the upper limbs, from T1 to T4 to the skin of the face, from T4 to T12 to the skin of the trunk, and from T10 to L2 to the skin of the lower limbs [20]. Normally, norepinephrine is the peripheral neurotransmitter of sympathetic innervation. The eccrine sweat gland displays an exception of this general rule, as the major neurotransmitter released from the periglandular nerve endings is acetylcholine, whose concentration determines the sweat rate in humans [Fig. 15.1]. Eccrine sweat glands express various muscarinic acetylcholine receptor subtypes in myoepithelial cells (m2-m5 AChR) as well as in acinar cells (m1, m3, m4-AChR) [31]. It is assumed that this receptor type is responsible for primary eccrine sweat secretion as the secretion can be blocked effectively by anti-muscarinic substances. Besides the muscarinic receptors, ion transporters as well as aquaporins seem to be involved in the secretory process [42,43]. Additionally, it could be shown that eccrine sweating can be induced locally by adrenergic stimulation [54,71], although it is not yet clear whether this response has physiological relevance. Furthermore, recent investigations demonstrate that several other receptors such as β2-adrenoceptor [65], VIP-receptor [30], EGF-receptor [52], vanilloid receptor-1 [64], and nicotinic acetylcholine receptors [31] are abundant in sweat glands. However, the complete secretory mechanism is not yet fully understood.

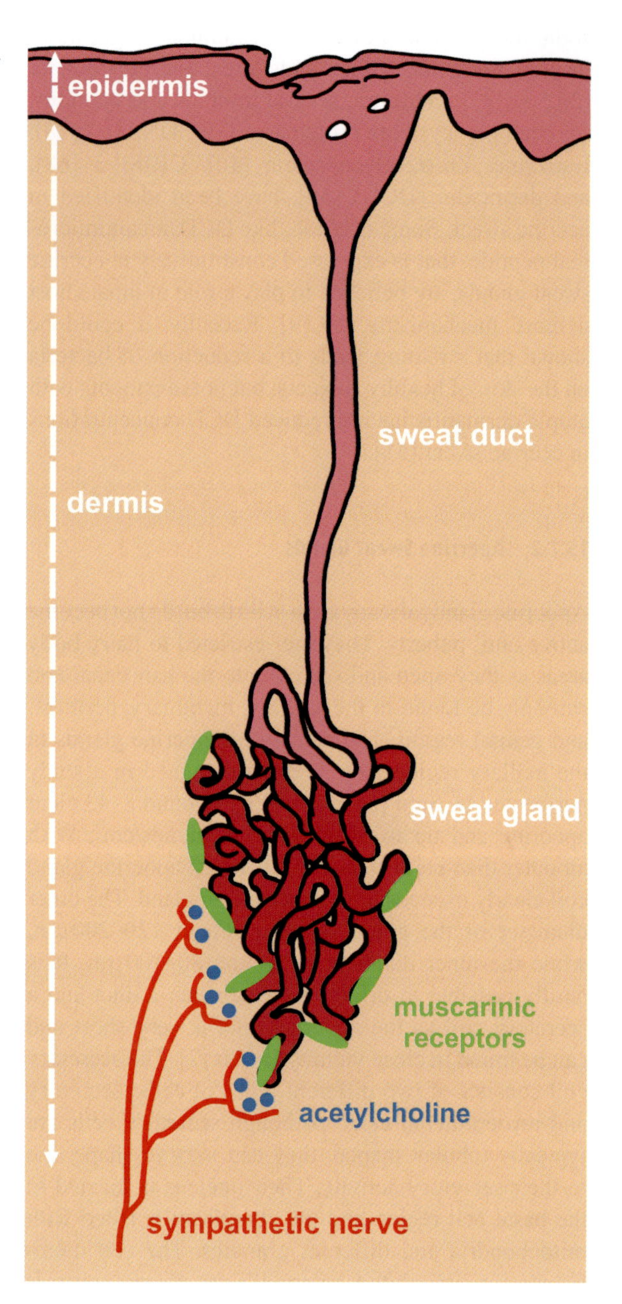

Fig. 15.1 Innervation of eccrine sweat glands. Acetylcholine is released from the periglandular nerve endings of sympathetic cholinergic fibers and subsequently binds to the corresponding muscarinic acetylcholine receptors of the sweat gland

15.2.2 Apocrine Sweat Glands

Apocrine sweat glands respond to emotional stimuli such as anxiety, pain, or sexual arousal. The apocrine secretion is controlled via epinephrine and norepinephrine [5,35,41]. However, it remains unclear whether activation of apocrine glands takes place via sympathetic innervation or via circulating catecholamines. Interestingly, until now no adrenergic receptors have been identified in human apocrine sweat glands.

15.2.3 Apoeccrine Sweat Glands

In vitro, apoeccrine glands show a very high cholinergic sensitivity but can also be stimulated with β-adrenergic and α-adrenergic agonists. Their responsiveness to cholinergic as well as adrenergic stimuli is greater than that of eccrine glands to these stimuli. This enhanced responsiveness might be the reason for the higher overall sweat rate of apoeccrine sweat glands compared to other sweat gland types, suggesting that apoeccrine sweat glands contribute strongly to axillary sweating [55,56].

15.3 Perspiration Control

Sweat gland function is a key element in the regulation of skin and body temperature, which is crucial for survival; constant body core temperatures above 40°C result in protein denaturation and cell death, which finally lead to multiple organ failure. Consequently, the decrease of body core temperature under conditions of high environmental temperature or under physiological stress is the most important role of perspiration. Thermal energy is released by the evaporation of sweat from the skin surface and as a result, skin and body core temperature are lowered. Perspiration can also be stimulated due to emotional stress or consumption of spicy food. However, these stimuli are less well understood.

15.3.1 Thermal Sweating

Being distributed over the whole body surface, eccrine sweat glands contribute strongly to the thermoregulatory process. Along with vasodilatation in the skin, thermoregulatory sweating decreases body temperature under heat stress conditions. Failure of this mechanism can lead to hyperthermia and death. The center for thermoregulation is the hypothalamus. Therefore, eccrine sweat gland activity is directly controlled by the central nervous system. Besides changes in core body temperature, this center also responds to hormones, endogenous pyrogens, physical activity, and emotions [25]. The sum of internal body temperature and mean skin temperature is the main drive for thermal sweating, whereas the influence of the first exceeds the one of the second by a factor of 10 [27,28]. In addition, sweat rate is affected by local cutaneous thermal conditions, leading to an increased sweat rate due to augmentation of local skin temperature. The mechanism by which this reaction is controlled remains unclear. However, possible explanations are a greater release of neurotransmitter as well as an increased sensitivity of sweat glands to a specific agonists during conditions of higher local skin temperature [11,39,45]. Furthermore, thermoregulatory sweating is influenced by many other internal factors such as gender, menstrual cycle and circadian rhythm as well as external factors like humidity [7,16,66].

15.3.2 Emotional Sweating

Emotional sweating, as the name implies, occurs in response to excitement, fear, anxiety, pain, and a multitude of other disturbed states. Emotional sweating arises independently from ambient temperature and can occur over the whole body surface but is most evident on palms, soles, and in the axillary region [4,8,14]. Emotional sweating is often mentioned in the course of the so-called fight-or-flight response as a primitive acute stress response that is induced by the hypothalamus and leads to the release of norepinephrine and epinephrine from the adrenal medulla [Fig. 15.2]. Finally, this reaction induces a series of physical responses, for example, increase of heart rate and rise of blood pressure leading to an immediate provision of energy.

Emotional sweating of the axillary region does not occur until the age of puberty. This is why apocrine

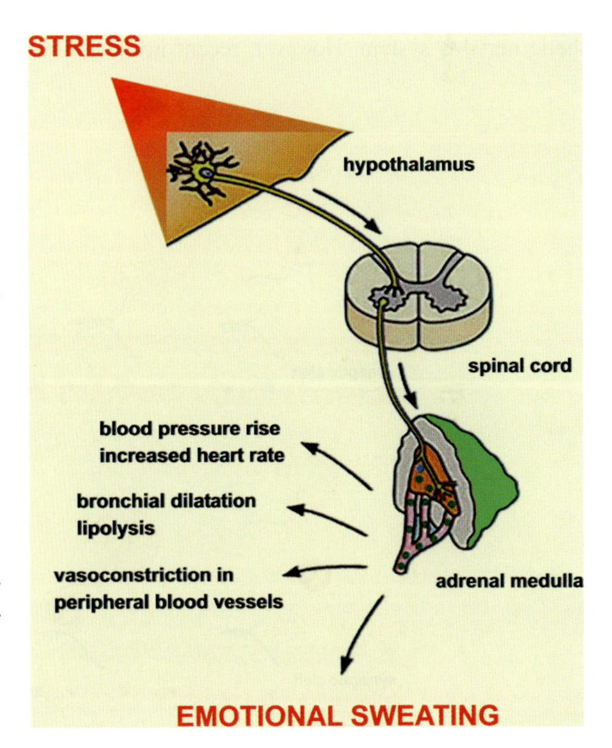

Fig. 15.2 Potential activation mechanism of emotional sweating during fight-or-flight response. Under stressful conditions, the human body reacts with the fight-or-flight response, which is induced by the hypothalamus. In the course of this response, emotional sweating might be activated directly via circulating catecholamines from the adrenal medulla

and apoeccrine glands are believed to play a major role in emotional axillary sweating because they are known to become active during this developmental stage [58,76]. Apocrine sweat glands respond vigorously to emotive stimuli and are physiologically activated by adrenergic stimulation [19,41,60]. However, it is still unclear whether this stimulation takes place via innervation or circulating catecholamines from the adrenal medulla. The function of apocrine sweat secretion has not been completely elucidated until now. One hypothesis deals with apocrine odors exhibiting a pheromone-like effect [1,80]. As apoeccrine glands show strong cholinergic and adrenergic sensitivity, their involvement in emotional sweating is very likely [55].

Emotional sweating of palms and soles can already occur in babies [67]. It has evolved as an improved fleeing reaction in different mammals during evolution as palmoplantar sweating increases friction and thus prevents slipping during running or climbing in stressful situations [3]. Moreover, sweating of palms and soles is mainly induced by emotive stimuli, not by high ambient temperature [29]. Emotional sweating of palms and soles only involves eccrine sweat glands that are typically activated by cholinergic fibers of the sympathetic nervous system. However, recent investigations on subjects suffering from anhidrosis due to deficits in cholinergic transmission indicate that adrenergic stimulation is also present in palms and soles [41].

15.3.3 Gustatory Sweating

Under certain conditions sweat secretion can be induced by ingestion. The exact mechanism is still unknown. Nevertheless, two different mechanisms are conceivable inducers of gustatory sweating. First, ingestion causes an increase in metabolism, which leads to elevated body temperature followed by thermal sweating as a direct effect. Second, a mild form of gustatory sweating is induced by hot and spicy food as an indirect effect that is confined to the face, the scalp, and the neck. This reaction is believed to be driven by the substance capsaicin, which binds to warm sensors in the oral cavity, leading to a thermoregulatory response [6,33,48].

Frey's syndrome, a pathological state of gustatory sweating, which affects the area of the cheek, can occur after parotid surgery. Presumably this unilateral form of gustatory sweating results from disruption of parasympathetic secretomotor fibers. These fibers

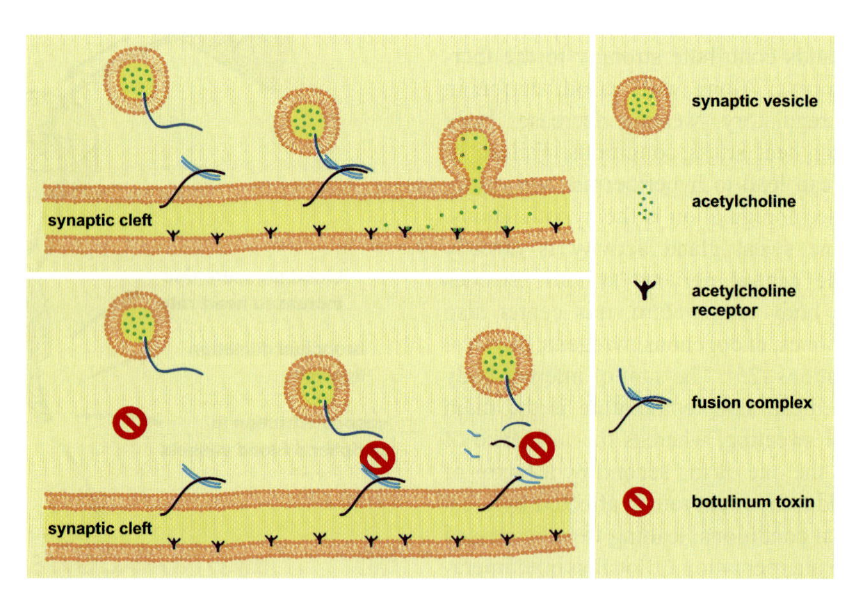

Fig. 15.3 Inhibition of acetylcholine release by botulinum toxin. *Top*: Docking of synaptic acetylcholine vesicles to the pre-synaptic membrane and subsequent membrane fusion is mediated by the SNARE protein fusion complex. Acetylcholine is released into the synaptic cleft and binds to acetylcholine receptors on the postsynaptic membrane. *Bottom:* After entering the nerve endings, botulinum toxin cleaves SNARE proteins and thus inhibits membrane fusion of synaptic vesicles and acetylcholine release

anastomose with sudomotor sympathetic fibers of the skin, thereby gaining control of sweat gland activity [13,44]. Besides Frey's syndrome, gustatory sweating is a rare complication in diabetic patients with autonomic neuropathy [69].

15.3.4 Inhibition of Sweat Secretion by Botulinum Toxin

Eccrine sweat secretion can be blocked effectively by botulinum toxin (BTX), a neurotoxin produced by the bacterium *Clostridium botulinum*, which acts on autonomic cholinergic nerve terminals. Botulinum neurotoxins are metalloproteases that cleave SNARE (soluble n-ethylmaleimide sensitive factor attachment protein receptor) proteins, thereby inhibiting the trafficking process of acetylcholine vesicles, their fusion with presynaptic membranes, and finally the release of acetylcholine. There are seven serologically distinguishable BTX types (A-G) acting on the proteins synaptobrevin, syntaxin, and SNAP-25 (25-kDa synaptosome-associated protein) that form the ternary SNARE complex [Fig. 15.3]. As an example, BTX A, the pharmaceutically most relevant toxin, cleaves SNAP-25. BTX consists of ~50 kDa light chain and ~100 kDa heavy chain, which are linked by a disulfide bond and leads to blockade of acetylcholine release in a three-step process. First, the toxin binds to the presynaptic membrane and is subsequently internalized by endocytotic vesicles. Second, the vesicular lumen is acidified by the action of an ATPase proton pump. The pH-change induces a conformational change of the BTX protein leading to its insertion into the vesicular membrane. Finally, the light chain is released into the cytosol displaying its proteolytic activity. Because of its local action, long-lasting effect, and high efficacy, BTX injections are used to treat different forms of hyperhidrosis [10,12,17,47,62].

Summary for the Clinician

Sweat glands, mostly eccrine sweat glands which are distributed over the whole body surface, play a key role in the regulation of skin and body temperature. The center for thermoregulation is the hypothalamus. Therefore, eccrine sweat gland activity is directly controlled by the central nervous system. Normally, norepinephrine is the peripheral neurotransmitter of sympathetic innervation. The eccrine sweat gland displays an exception of this general rule, as the major neurotransmitter released from the periglandular nerve endings is acetylcholine, whose concentration determines the sweat rate in humans. Eccrine sweat secretion can be blocked effectively by botulinum toxin (BTX), which acts on autonomic cholinergic nerve terminals. One important eccrine sweat component is Dermcidin (DCD), an antimicrobial peptide, which plays a role in innate host defense mechanisms. Patients with atopic dermatitis have reduced DCD concentrations in eccrine sweat.

Emotional sweating, often mentioned in the course of the so-called fight-or-flight response, is a primitive acute stress response that is induced by the hypothalamus and leads to the release of norepinephrine and epinephrine from the adrenal medulla. Emotional sweating involves eccrine sweat glands when occurring palmoplantar, and apocrine sweat glands when occurring axillar. Apocrine glands already exist at birth but do not become active until puberty. They are restricted to hairy body areas as they open and secrete into the hair canal and can only be found in the axillary, mammary, perineal, and genital region. The function of apocrine sweat secretion has not been completely elucidated until now. One hypothesis deals with apocrine odors exhibiting a pheromone-like effect. Moreover, recent investigations suggest that genes encoding the major histocompatibility complex (MHC) influence human body odor as well as body odor preferences. MHC proteins play a vital role in the complex immunological dialogue that occurs between T cells and other cells, allowing the body to differentiate from oneself. Yet, it is speculated that an increased heterozygosity in MHC genes may lead to a higher resistance to infectious diseases in offspring and avoids inbreeding.

Eccrine and apocrine sweat glands as well as apoeccrine sweat gland, a mixed type, are very complex structures whose functions are still not fully understood but play important roles in terms of thermoregulation, regulation of skin flora as well as olfactory communication between subjects.

References

1. Ackerl K, Atzmueller M, Grammer K (2002) The scent of fear. Neuro Endocrinol Lett 23(2):79–84
2. Ackermann AB (1997) Embryologic, histologic, and anatomic aspects, In: Stamathis (ed) Histologic Diagnosis of Inflammatory Skin Diseases, Wiiliams & Wilkins, Baltimore, pp 3–56
3. Adelman S, Taylor CR, Heglund NC (1975) Sweating on paws and palms: what is its function? Am J Physiol 229(title5):1400–1402
4. Allen JA, Jenkinson DJ, Roddie IC (1973) The effect of -adrenoceptor blockade on human sweating. Br J Pharmacol 47(3):487–497
5. Aumuller G, Wilhelm B, Seitz J (1999) Apocrine secretion–fact or artifact? Anat Anz. 181(title5): 437–446.

6. Bronshvag MM (1978) Spectrum of gustatory sweating, with especial reference to its presence in diabetics with autonomic neuropathy. Am J Clin Nutr 31(2):307–309
7. Candas V, Libert JP, Vogt JJ (1983) Sweating and sweat decline of resting men in hot humid environments. Eur J Appl Physiol Occup Physiol 50(2):223–234
8. Chalmers TM, Keele CA (1952) The nervous and chemical control of sweating. Br J Dermatol 64(2):43–54
9. Charles A (1960) An electron microscope study of the eccrine sweat gland. J Invest Dermatol 34:81–88
10. Delgado MR (2003) Botulinum neurotoxin type A. J Am Acad Orthop Surg 11(5):291–294
11. DiPasquale DM, Buono MJ, Kolkhorst FW (2003) Effect of skin temperature on the cholinergic sensitivity of the human eccrine sweat gland. Jpn J Physiol 53(6):427–430
12. Dolly O (2003) Synaptic transmission: inhibition of neurotransmitter release by botulinum toxins. Headache 43 Suppl 1:S16–S24
13. Dunbar EM, et al (2002) Understanding gustatory sweating. What have we learned from Lucja Frey and her predecessors? Clin Auton Res 12(3):179–184
14. Eisenach JH, Atkinson JL, Fealey RD (2005) Hyperhidrosis: evolving therapies for a well-established phenomenon. Mayo Clin Proc 80(5):657–666
15. Fiedler HP (1968) Der Schweiß, 2 ed. Weiler im Allgäu, Buchdruckerei Holzer
16. Garcia AM, et al (2006) Luteal phase of the menstrual cycle increases sweating rate during exercise. Braz J Med Biol Res 39(9):1255–1261
17. Glogau RG (2004) Hyperhidrosis and botulinum toxin A: patient selection and techniques. Clin Dermatol 22(1):45–52
18. Goldsmith LA (1998) Biology of eccrine and apocrine sweat glands, In: Fitzpatrick TB (ed) Dermatology in General Medicine, Mc Graw-Hill, New York, pp 155–164
19. Goodall M (1970) Innervation and inhibition of eccrine and apocrine sweating in man. J Clin Pharmacol J New Drugs 10(4):235–246
20. Groscurth P (2002) Anatomy of sweat glands. Curr Probl Dermatol 30:1–9
21. Hashimoto K (1971) Demonstration of the intercellular spaces of the human eccrine sweat gland by lanthanum I. The secretory coil. J Ultrastruct Res 36:249–262
22. Hashimoto K (1971) Demonstration of the intercellular spaces of the human eccrine sweat gland by lanthanum II. The duct. J Ultrastruct Res 37:504–520
23. Hashimoto K, Hori K, Aso M (1986) Sweat glands, In: Bereiter-Hahn J, Matoltsy AG, Richards KS (ed) Biology of the Integument 2 Vertebrates, Springer-Verlag, Berlin, pp 339–356
24. Hibbs RG (1958) The fine structure of human eccrine sweat glands. Am J Anat 103(2):201–217
25. Holzle E (2002) Pathophysiology of sweating. Curr Probl Dermatol 30:10–22
26. Jacoby RB, et al (2004) Detection and quantification of apocrine secreted odor-binding protein on intact human axillary skin. Int J Cosmetic Sci 26:37–46
27. Jessen C (2001) Temperature Regulation in Humans and Other Mammals. Springer–Verlag, Berlin Heidelberg, pp 193
28. Johnson JM, Proppe DW (1996) Cardiovascular adjustments to heat stress, In: Fregly MJ, Blatteis CM (ed) Handbook of Physiology. Section 4: Environmental Physiology, Oxford University Press, Oxford, pp 215–243
29. Kerassidis S (1994) Is palmar and plantar sweating thermoregulatory? Acta Physiol Scand 152(3):259–263
30. Kummer W, Herbst WM, Heym C (1990) Vasoactive intestinal polypeptide receptor-like immunoreactivity in human sweat glands. Neurosci Lett 110(3):239–243
31. Kurzen H, et al (2004) Phenotypical and molecular profiling of the extraneuronal cholinergic system of the skin. J Invest Dermatol 123(5):937–949
32. Labows JN, et al (1979) Steroid analysis of human apocrine secretion. Steroids 34(3):249–258
33. Lee TS (1954) Physiological gustatory sweating in a warm climate. J Physiol 124(3):528–542
34. LeGrys VA (2001) Assessment of sweat-testing practices for the diagnosis of cystic fibrosis. Arch Pathol Lab Med 125(11):1420–1424
35. Lonsdale-Eccles LN, La LC (2003) Axillary hyperhidrosis: eccrine or apocrine? PG-2–7. Clin Exp Dermatol 28(1):2–7
36. Mitsubayashi K (1994) Analysis of metabolites in sweat as a measure of physical condition. Analytika Chimica Acta 289:27–34
37. Montgomery I, et al (1985) The effects of thermal stimulation on the ultrastructure of the human atrichial sweat gland. II. The duct The effects of thermal stimulation on the ultrastructure of the human atrichial sweat gland. I. The fundus. Br J Dermatol 112(2):165–177
38. Morgan RM, Patterson MJ, Nimmo MA (2004) Acute effects of dehydration on sweat composition in men during prolonged exercise in the heat. Acta Physiol Scand 182(1):37–43
39. Nadel ER, et al (1971) Peripheral modifications to the central drive for sweating. J Appl Physiol 31(6):828–833
40. Nakayashiki N (1990) Sweat protein components tested by SDS-polyacrylamide gel electrophoresis followed by immunoblotting. Tohoku J Exp Med 161(1):25–31
41. Nakazato Y, et al (2004) Idiopathic pure sudomotor failure: anhidrosis due to deficits in cholinergic transmission. Neurology 63(8):1476–1480
42. Nejsum LN, et al (2002) Functional requirement of aquaporin-5 in plasma membranes of sweat glands. Proc Natl Acad Sci U S A 2:2
43. Nejsum LN, Praetorius J, Nielsen S (2005) NKCC1 and NHE1 are abundantly expressed in the basolateral plasma membrane of secretory coil cells in rat, mouse, and human sweat glands. Am J Physiol Cell Physiol 20:20
44. Nolte D, et al (2004) [Botulinum toxin for treatment of gustatory sweating. A prospective randomized study]. Mund Kiefer Gesichtschir 8(6):369–375
45. Ogawa T, Sugenoya J (1993) Pulsatile sweating and sympathetic sudomotor activity. Jpn J Physiol 43(3):275–289
46. Patterson MJ, Galloway SD, Nimmo MA (2000) Variations in regional sweat composition in normal human males. Exp Physiol 85(6):869–875
47. Penna P (2002) Botulinum neurotoxin therapy: overview of serotypes A and B. Pharmacy and Therapeutics, September 2002
48. Pierau FK (1996) Peripheral thermosensors. In: Fregly MJ, Blatteis CM (ed) Handbook of Physiology. Section 4:

Environmental Physiology, Oxford University Press, Oxford, pp 85–104
49. Reddy MM, Quinton PM (1994) Rapid regulation of electrolyte absorption in sweat duct. J Membr Biol 140(1):57–67
50. Rieg S, et al (2004) Dermcidin is constitutively produced by eccrine sweat glands and is not induced in epidermal cells under inflammatory skin conditions. Br J Dermatol 151(3):534–539
51. Rieg S, et al (2005) Deficiency of dermcidin-derived antimicrobial peptides in sweat of patients with atopic dermatitis correlates with an impaired innate defense of human skin in vivo. J Immunol 174(12):8003–8010
52. Saga K, Jimbow K (2001) Immunohistochemical localization of activated EGF receptor in human eccrine and apocrine sweat glands. J Histochem Cytochem 49(5):597–602
53. Sato K, (1998) Biology of the eccrine sweat gland, In: Fitzpatrick TB (ed) Dermatology in General Medicine, pp 221–241
54. Sato K, Sato F (1981) Pharmacologic responsiveness of isolated single eccrine sweat glands. Am J Physiol 240(title1): R44–R51
55. Sato K, Sato F (1987) Sweat secretion by human axillary apoeccrine sweat gland in vitro. Am J Physiol 252: R181–R187
56. Sato K, Leidal R, Sato F (1987) Morphology and development of an apoeccrine sweat gland in human axillae. Am J Physiol 252:R166–R180
57. Sato K, et al (1989) Biology of sweat glands and their disorders. I. Normal sweat gland function. J Am Acad Dermatol 20(4):537–563
58. Schiefferdecker P (1922) Die Hautdrüsen des Menschen und der Säugetiere, ihre biologische und rassenanatomische Bedeutung, sowie die Muscularis sexualis. Zoologica 27(72)
59. Schittek B, et al (2001) Dermcidin: a novel human antibiotic peptide secreted by sweat glands. Nat Immunol 2(12):1133–1137
60. Shelley WB, Hurley HJ Jr (1953) The physiology of the human axillary apocrine sweat gland. J Invest Dermatol 20(4):285–297
61. Shirreffs SM, Maughan RJ (1997) Whole body sweat collection in humans: an improved method with preliminary data on electrolyte content. J Appl Physiol 82(1):336–341
62. Singh BR (2000) Intimate details of the most poisonous poison. Nat Struct Biol 7(8):617–619
63. Spielman AI, et al (1998) Identification and immunohistochemical localization of protein precursors to human axillary odors in apocrine glands and secretions. Arch Dermatol 134(title7):813–818
64. Stander S, et al (2004) Expression of vanilloid receptor subtype 1 in cutaneous sensory nerve fibers, mast cells, and epithelial cells of appendage structures. Exp Dermatol 13(3):129–139
65. Steinkraus V, et al (1996) Autoradiographic mapping of beta-adrenoceptors in human skin. Arch Dermatol Res 288(9):549–553
66. Stephenson LA, Kolka MA (1988) Effect of gender, circadian period and sleep loss on thermal responses during exercise. In: Gonzalez RR (ed) Human Performance Physiology and Environmental Medicine at Terrestrial Extremes, Cooper, Carmel, IN, pp 267–304
67. Storm H (2001) Development of emotional sweating in preterms measured by skin conductance changes. Early Hum Dev 62(2):149–158
68. Udby L, et al (2002) An ELISA for SGP28/CRISP-3, a cysteine-rich secretory protein in human neutrophils, plasma, and exocrine secretions. J Immunol Methods 263(1–2):43–55
69. Urman JD, Bobrove AM (1999) Diabetic gustatory sweating successfully treated with topical glycopyrrolate: report of a case and review of the literature. Arch Intern Med 159(8):877–878
70. van der Putte SCJ (1991) Anogenital "sweat" glands. Am J Dermatopathol 13(6):557–567
71. Warndorff JA, Neefs J (1971) A quantitative measurement of sweat production after local injection of adrenalin. J Invest Dermatol 56(5):384–386
72. Wedekind C, Furi S (1997) Body odour preferences in men and women: do they aim for specific MHC combinations or simply heterozygosity? Proc Biol Sci 264(1387):1471–1479
73. Wedekind C, Penn D (2000) MHC genes, body odours, and odour preferences. Nephrol Dial Transplant 15(9):1269–1271
74. Wedekind C, et al (1995) MHC-dependent mate preferences in humans. Proc Biol Sci 260(1359):245–249
75. Wilke K (2005) Struktur- und Funktionsaufklärung von Schweißdrüsen und ihre Interaktion mit Antitranspirantien. Technische Universität Hamburg-Harburg, Hamburg, p 182
76. Wilke K, et al (2004) Immunolabelling is essential for the differentiation of human axillary apoeccrine glands. J Invest Dermatol 123(5):A93
77. Wilke K, et al (2005) Are sweat glands an alternate penetration pathway? Understanding the morphological complexity of the axillary sweat gland apparatus. Skin Pharmacol Physiol 19(1):38–49
78. Yasuda T, et al (1996) A new individualization marker of sweat: deoxyribonuclease I (DNase I) polymorphism. J Forensic Sci 41(5):862–864
79. Yokozeki H, Hibino T, Sato K (1987) Partial purification and characterization of cysteine proteinases in eccrine sweat. Am J Physiol 252(6 Pt 2):R1119–R1129
80. Zeng C, et al (1996) A human axillary odorant is carried by apolipoprotein D. Proc Natl Acad Sci U S A 93(13): 6626–6630

Section IV

The Nervous System and the Pathophysiology of Skin Disorders

Neurophysiology of Itch

16

G. Yosipovitch and Y. Ishiuji

Contents

Key Features

- A subset of nociceptive histamine sensitive C neurons has been demonstrated to be dedicated for transmission of itch.
- Histamine independent nocicpetive C nerves that are mechanosensitive are capable of transmitting itch.
- Central pathways are involved in itch transmission both spinally and in the brain.
- Chronic itch shares similar patterns to chronic pain.
- Both central and peripheral mediators are important in pruritus.
- Neurotrophic factors, particularly NGF, have a role in chronic itch.

Synonyms Box: Itch, pruritus; alloknesis, when touch or brush-evoked itch occurs around an itching site; allodynia, when inflamed skin reacts to gentle mechanical stimuli by giving rise to a perception of pain

16.1 Introduction

Itch (Latin pruritus) is been defined as unpleasant sensation that elicits the desire to scratch. It is a dominant symptom of skin disease, almost all inflammatory skin diseases can itch [64,69]. In most cases, itch results from interaction of the brain–skin axis [27,39]. Itch has many similarities to pain; both are unpleasant sensations which consist of multidimensional phenomena including sensory discriminative, cognitive, evaluative, and motivational components. A significant difference between both sensations is the behavioral response patterns – while pain elicits a reflex withdrawal, itch leads to a scratch reflex [44]. Nevertheless, both can lead to serious impairment of quality of life. Another unique feature of itch is that it is restricted to the skin and some adjoining mucosa such as conjunctiva. Itch sensations emanate from activity of nerve fibers in the epidermis and upper layers of dermis; nerve fibers located in deeper layers of the reticular dermis and subcutaneous fat do not seem to transmit itch.

R.D. Granstein and T.A. Luger (eds.), *Neuroimmunology of the Skin*,

16.2 The Neural Basis of Itch

The neurophysiological basis for itch was unclear for decades. Historically three theories were proposed to explain the neuronal mechanism of itch [34].

Specificity theory suggested that there is a group of primary sensory neurons that respond to pruritogenic stimuli and no other. The existence of labeled lines for itch sensation has been supported by the findings [45] of specific C nerve fibers that transmit itch peripherally in humans and further by findings of histamine sensitive neurons in the spinothalamic tract [2]. However, there may be other nerve fibers that transmit itch as well, especially in patients suffering from chronic itch.

The *Intensity theory* was proposed [61] suggesting that itch is a subluminal form of pain and low activity in nerve fibers induces itch that will turn into pain by increased stimulation. However, this theory has been abandoned since direct testing of noxious stimuli (thermal and mechanical) in threshold doses do not illicit itch. Microneurography has also helped to disprove this historic concept that pruritus and pain are simply responses of the same neurons to mild vs. intense stimuli, respectively. Moreover, many treatments that inhibit pain do not inhibit itch and vice versa. For example, the effect of μ-opiods that inhibit pain and may actually aggravate itch.

The *Selective theory* recognizes the absence of specific populations of sensory neurons dedicated to signaling of itch. It suggests instead that a subset of afferent nociceptors that respond to pruritogenic stimuli have different central connectivity's and activate separate central neurons. This theory has recent support from studies using noxious stimuli, including thermal, mechanical, and chemical using bradykinin [22,25] that induced itch rather than pain in patients with chronic itch.

16.2.1 Histamine Sensitive C Nerve Fibers

Histamine was found to directly stimulate histamine type 1 receptors on C pruritoceptors [45]. These neurons are sensitive to pruritogenic and thermal stimuli, but not mechanical stimuli. The response pattern of these fibers matched the time course of itch sensation reported by participants (see Fig. 16.1). They have exceptionally slow conduction velocity, unusually wide innervation territories, and represent no more than 5% of total C fibers. These C fibers were shown to have spontaneous activity in a recent microneurographic study of a patient with chronic itch [46]. In contrast, the vast majority of C fibres are sensitive to mechanical and heat stimuli and are entirely insensitive to histamine [47].

The co-responsiveness of this subset of C neurons to temperature change as well as pruritic stimuli is of interest because raising the temperature of skin lowers the threshold of receptors to pruritic stimuli [14] and most pruritic patients complain of aggravation of pruritus in a warm environment.

Another important question is whether other histamine receptors are activated during itch induction; recent studies suggest that other receptors such as histamine receptor 4 [4] induce itch in mice. However, their role in human itch remains to be elucidated.

16.2.2 Histamine Independent Itch Fibers

Mechanically induced itch, for example, itch associated with exposure to wool, without an accompanying flare reaction, cannot be explained by activation of histamine fibers. Moreover, chronic itch associated with most inflammatory and systemic diseases does not respond to antihistamines, suggesting that there are other pathways for itch transmission, which are histamine independent.

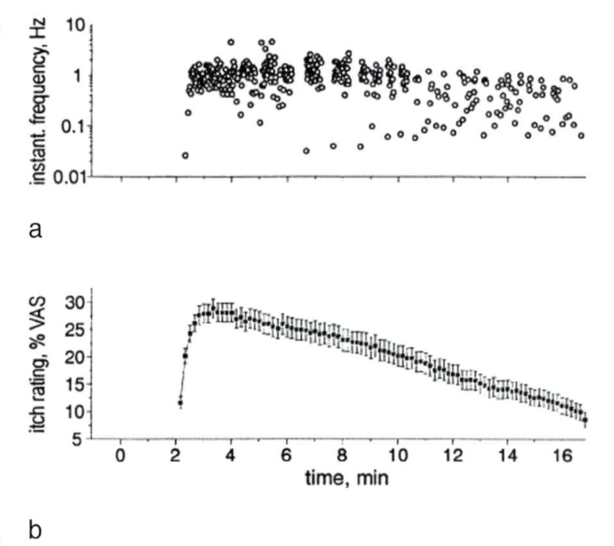

Fig 16.1 **a** Electrophysiological recordings of discharge frequency of histamine sensitive C-fibers in the peroneal nerve after histamine iontophoresis. The unit was not spontaneously active before histamine application. **b** Average itch magnitude ratings of the healthy volunteers after an identical histamine stimulus. Ratings were at 10 s intervals on a VAS with the end points no itch. Error bars indicate SE. This figure is modified, with permission, from [45]

Itch can be generated without flare reaction using electrical stimuli [26] as well as with spicules of Cowhage [49] via mechano-sensitive C fibers [27]. Peripheral activation of itch can involve several classes of C-fibers, both histamine mechano-insensitive as well as mechano-sensitive.

16.2.3 Central Pathways

In the spinal cord, nociceptor C neurons synapse with secondary transmission neurons in the gray matter of the dorsal horn. These neurons then cross over and ascend in the lateral spinothalamic tract to the thalamus. Recent studies using microneurography have identified a subclass of lamina I spinothalamic tract neurons specifically and selectively excited by iontophoretically administered histamine [2]. Thus, pruritus is transmitted by dedicated neurons not only peripherally, but also centrally (Fig. 16.2).

16.2.3.1 Higher Centers Involved in Itch Processing

Supra-spinal processing of itch and its corresponding scratch response has not been studied as extensively as pain brain imaging. Most of the studies examined itch response with histamine stimulation in healthy volunteers using positron emission tomography (PET) and functional MRI with the BOLD response [33,36,63]. Induction of itch by intradermal injection of histamine and histamine skin-prick elicits coactivation of the anterior cingulate cortex, insular cortex and premotor and supplementary motor area, inferior parietal lobe, with a left-hemisphere predominance as well as the cerebellum [7,11,23]. The substantial coactivation of the motor area supports the clinical observation that itch is inherently linked to a desire to scratch. In these studies, the "intention to scratch" was mirrored by functional increases in blood flow in the cortical motor areas and cerebellum. Activation of multiple brain

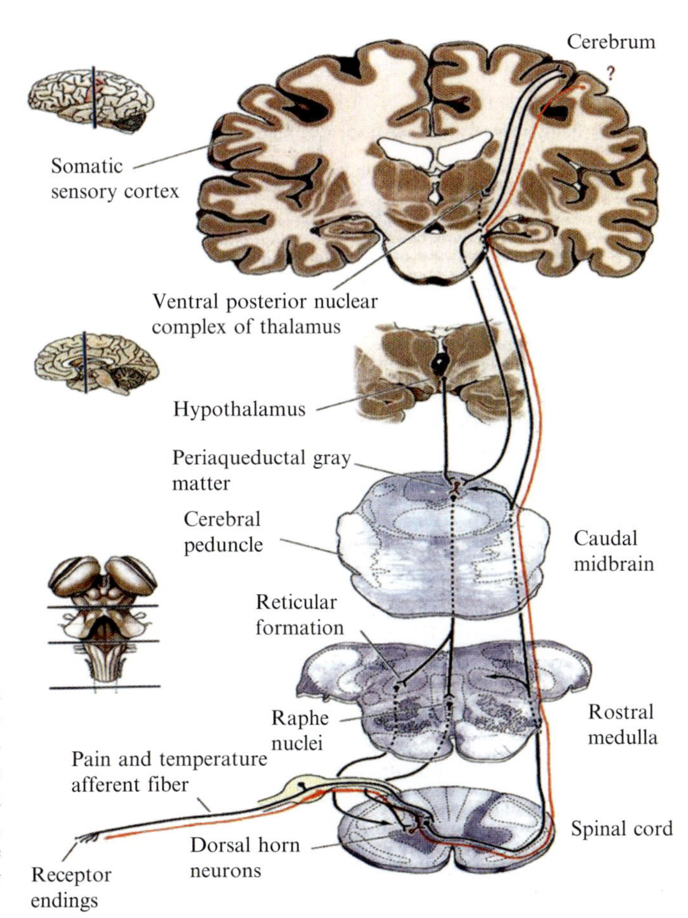

Fig. 16.2 Neurophysiology of itch. Information on itch and pain transmission is conveyed through unmyelinated C nerve fibers that both use the lateral spinothalamic tract. Studies have demonstrated specific C nerve fibers that transmit itch. Itch representation in the brain activates several regions similar to pain, such as the premotor cortex and anterior cingulate cortex but does not seem to activate somatosensory cortex I and II

areas suggests that there is not a sole itch centre, and shows the multidimensionality of the itch sensation. Pain demonstrates a similar pattern of brain activation involving many of the same cortical regions [11]. However, there are subtle differences in the activation pattern between itch and pain; for example, in itch there is no detectable activation of the somatosensory cortex II and the activation of the somatosensory cortex I is minimal, while both are significantly activated in pain. A possible explanation is that there is weaker activation of these primary and secondary somatosensory cortices. Thus, differences between pain and itch processing may not result from distinct brain centers, but rather, reflect different activation patterns. The anterior cingulate cortex has a significant role in aversiveness to sensory stimuli and is also associated with motivation for rewarding events. This could explain patients' desire for repetitive scratching. A recent study has examined the effect of an allergen-induced itch in patients with a history of atopy as well as histamine prick tests in healthy volunteers. This study demonstrated activation of multiple regions, which were not documented simultaneously previously, notably the insula, the anterior cingulate cortex, and the orbitofrontal cortex. These results correlated to itch intensity. The activation of the orbitofrontal cortex as well as the limbic system may reflect the subjects desire to relieve itch by scratching [31].

16.2.4 Chronic Itch and Peripheral and Central Sensitization

Chronic itch can be related to different types of itch such as pruritoceptive itch originating from skin disease, neuropathic itch due to pathology in the nervous system, as well as itch due to systemic and psychiatric causes [5,58,65]. Chronic itch shares many similarities with chronic pain, both have similar peripheral sensitization as well as central nervous system sensitization.

16.2.4.1 Peripheral Sensitization in Chronic Itch

Patients with chronic itch have increased skin innervation density [60]. Moreover, increased nerve growth factor (NGF) expression has been found in patients with atopic dermatitis in the skin and serum [15,57]. These same neurotrophins are elevated in chronic localized pain and are known to sensitize primary afferent fibers [35].

16.2.4.2 Central Sensitization in Chronic Itch

Chronic itch leads to sensitization of second order neurons in the dorsal horn, thereby leading to increased sensitivity to itch. There are two forms of increased sensitivity to itch. First, alloknesis (itchy skin) [44,50,65] can be observed when touch- or brush-evoked itch occurs around an itching site [51]. This phenomenon is analogous to the better known allodynia, in which inflamed skin reacts to gentle mechanical stimuli by giving rise to a perception of pain. Like allodynia, it requires ongoing activity in primary afferents and is elicited by low threshold mechanoreceptor A beta fibers. Alloknesis is common; for example, alloknesis is a prominent feature of atopic dermatitis [20] and explains patients' complaints of severe pruritus associated with sweating, sudden changes in temperature, dressing, and undressing. The second type of increased sensitivity to itch is termed punctuate hyperknesis, in which a prick induces intense itch sensation in the area surrounding histamine induction, similar to the phenomena in chronic pain coined punctuate hyperalgesia [44].

Noxious pain and repetitive scratching have been known to inhibit itch for decades. These stimuli have been shown to inhibit histamine induced itch in healthy controls using psychophysical methods [70]; however; studies in patients with chronic itch demonstrate that painful stimuli, such as electrical stimuli and heat pain, may be perceived as itch [22,25,26,29]. An analogous phenomenon has been noted in chronic pain patients who suffer from post-herpetic neuralgia (PHN), in which histamine iontophoresis is perceived as painful [3]. These findings indicate that pain-induced inhibition of pruritus may be compromised in patients with chronic itch. This may also explain why scratching aggravates itch and induces a vicious circle of scratching inducing itch.

16.2.5 Pruritus Receptor Unit

Removal of the epidermis abolishes perception of pruritus, suggesting that pruritus receptor units are located predominantly within this layer. Light microscope and ultrastructural studies of human skin have shown the existence of intraepidermal nerve fibers with 'free' nonspecialized nerve endings extending to the stratum granulosum [21]. To date, it has not been possible to prove that these include free nerve endings serving the sensation of pruritus, but demonstration that many of these fibers stain positively for neuropeptides suggests that this is indeed the case, and that pruritus is transmitted in the epidermis by C fibers. C fibers are primary afferent nociceptors. Moreover, keratinocytes express a variety of neural mediators and receptors, all of which appear to be involved in the itch sensation [9]. These include opioids, proteases, substance P, nerve growth factor (NGF), neurotrophin 4, and their respective receptors, including μ- and κ-opioid receptors, PAR-2, vanilloid receptors, TRKA, TRPV ion channels, and cannabinoid receptors 1 and 2. Keratinocytes also have voltage-gated ATP channels and adenosine similar to C nerve fibers (see Chap. 2). Since these channels have a role in pain [28], these findings suggest that keratinocytes may act as itch receptors in specific inflammatory conditions.

16.3 Neuromediators of Pruritus

There are many mediators that cause pruritus in inflammatory skin diseases. Both central and peripheral mediators are important in pruritus. Most such mediators also cause other signs of inflammation (pain, erythema due to vasodilation, increased vascular permeability). Several of these cause pruritus indirectly by evoking release of histamine and other mediators from mast cells (e.g., substance P and several opioid peptides) or by potentiating the actions of other mediators (e.g., prostaglandin E_1). This chapter will cover in brief recent findings related to neuromediators associated with pruritus. For a detailed description of neuromediators see Chap. 2.

16.3.1 Proteinases

Recent work has revealed the mechanism whereby mast cell-derived tryptase contributes to the neurogenic inflammatory response [53]. Studies using dermal microdialysis have shown that levels of tryptase and its receptor (PAR-2) are elevated fourfold in atopic dermatitis [54]. PAR-2 is also highly expressed in the epidermis of atopic dermatitis patients. PAR2 agonists induce itch in patients with atopic eczema, suggesting that itch in atopic eczema is associated with activation of these receptors. Administration of a PAR2 inhibitor inhibited itch in a mouse model. Currently, there are no published human studies using PAR2 antagonists in treatment of itch. The role of PAR1 and PAR4, recently discovered in nerve fibers, in itch is unknown. It is noteworthy that proteinase activity can also be found in common allergens [16] and staphylococcal skin infections, both of which are known to aggravate atopic dermatitis and itch. The importance of epidermal serine proteases in eliciting itch has been further demonstrated in a mouse model, which showed that overexpression of a serine protease caused severe itch and scratching [19].

16.3.2 Opioid Peptides

It is well known that the perception of pruritus is modified by endogenous opiates via central opioid receptors [43]. The concept of central pruritus and possible involvement of pruritic mediators located in the central nervous system is becoming increasingly recognized as important in both cutaneous and systemic diseases, with implications for treatment. Opioids appear to induce itch via two possible mechanisms. First, by degranulation of cutaneous mast cells [12]; Second, via a direct central and peripheral pruritogenic effect by activating μ-opioid receptors [6,12,13,18]. It has recently been suggested that generalized pruritus is induced by an imbalance between the μ- and κ-opioid systems [30,59]. Activation of μ-opioid receptors stimulates itch perception, whereas κ-opioid receptor stimulation inhibits μ-receptor effects both centrally and peripherally [38,56,59]. Administration of butorphanol a μ-antagonist and a κ-agonist demonstrated a rapid and marked improvement in patients with intractable pruritus [8]. Nociceptin, the endogenous peptide ligand for opioid receptor like-1 (ORL1) receptor, has also been implicated in cutaneous inflammation, pain, and pruritus. Recent data using a mouse model suggests that nociceptin acts on ORL1 receptor expressed on keratinocytes to produce leukotriene B, which induces scratching [1]. Nociceptin-induced scratching was significantly inhibited by treatment with systemic naloxone.

16.3.3 Neurotrophins

Neurotrophins are factors that regulate growth and function of nerve cells (see Chap. 9). The prototypical neurotrophic factor is nerve growth factor (NGF). Increased levels of epidermal NGF correlate with the proliferation of terminal cutaneous nerves and upregulation of neuropeptides. NGF is known to induce sprouting of nerve fibers, sensitization of nerve endings, axonal transport in spinal ganglia (DRG cells), and increased expression of neuropeptides.

Keratinocytes express high levels of NGF, which is required not only for survival and regeneration of sensory neurons but also controls the responsiveness of such neurons to external stimuli [37,41]. There is a significant correlation between plasma levels of NGF with disease activity in atopic dermatitis [57]. The most sensitive marker of pruritus in atopic dermatitis was NGF receptor in urine (TRKA). Increased cutaneous gene expression of NGF was found in mast cells, keratinocytes, and fibroblasts of atopic dermatitis patients and was highly associated with plasma levels of NGF, thus providing further data that NGF may contribute significantly to itch in atopic dermatitis [17]. Other neurotrophins, such as neurotrophin 4, have been recently found to be upregulated in keratinocytes of atopic dermatitis patients [15].

16.4 Other Peripheral Mediators of Itch

A number of other peripheral mediators appear to be involved in itch or attenuating itch sensation, such as neuromediators that activate ion channels that are part of the transient receptor potential (TRP) channel family. TRPV1 is a vanilloid receptor located on both C nerve fibers and keratinocytes [52]. This receptor is activated by capsaicin as well as endogenous substances such as cannabinoids, prostaglandins, and various neurotrophins. Cannabinoid receptors CB1 and CB2 are located in the epidermis; recently, a cannabinoid agonist has been shown to inhibit histamine-induced itch [42]. Another TRP that may have a role in itch attenuation is TRPM8, which is expressed in C nerve fibers. TRPM8 is thought to be a thermosensor for coolness and is activated by menthol and induces analgesia in chronic neuropathic pain [40].

16.5 Immune Cells as Itch Mediators

There is a significant cross-talk between the cutaneous nervous system and cutaneous immune system (see Chap. 4) [48,55]. Neuropeptides released by cutaneous sensory nerves can directly modulate skin and immune cell functions. Neuropeptides activate transcription factors and regulate the expression of adhesion molecules and proinflammatory cytokines, thereby modulating immune and inflammatory reactions [32]. This interaction is bidirectional: cytokines and chemokines have been shown to regulate primary nerve afferents via receptor activation. Immune diseases such as psoriasis, atopic dermatitis, and lichen planus are associated with significant itch and involve cutaneous nerve fibers and neuropeptides.

Interleukin 2 (IL-2) causes pruritus on intradermal injection [62]. High doses of IL-2 administered to cancer patients causes intense generalized pruritus. Supernatants of mitogen stimulated leukocytes in patients with atopic eczema contain large amounts of IL-2 and IL-6. Moreover, treatments with topical immunomodulators such as tacrolimus, pimecrolimus, which inhibit the production of Il-2, are known to inhibit itch. As yet it is not clear whether this is a direct receptor-mediated effect or an indirect effect via mast cells or endothelial cells.

TNF alpha is known to sensitize nociceptive nerve endings via its effect on TNF alpha receptors [69]; however, its role in itch is unclear. Interleukin 31 is a newly discovered cytokine produced by T helper 2 cells. In a recent study using a transgenic mouse model, overexpression of this cytokine led to severe scratching and dermatitis [10]. Keratinocytes express the IL-31 receptor; however, the mechanism by which IL-31 induces itch is unclear.

16.6 The Effect of Psychological Stress on Itch

Patients with different types of itch often report that emotional stress aggravates their itch [66–68]. Atopic eczema and psoriasis are two inflammatory diseases where stress clearly can induce or aggravate the itch and the disease [66,68]. One of the most intriguing aspects of studies of itch will be understanding the influence of the neuroendocrine system in the complex interaction of the hypothalamus–pituitary–adrenal (HPA) axis. Mediators such as cortisol, adrenocorticotropin-releasing hormone, and noradrenaline may have a role in itch transmission. Drugs that can inhibit or reduce stress responses or mediators in the brain have been shown to alter itch intensity [24,71].

16.7 Conclusions

Our understanding of the neurophysiology and neurochemistry of itch has grown tremendously in recent years. Numerous neurotransmitter systems and receptor mechanisms have been identified. Continued focus on neuronal factors involved in itch transmission will lead to better understanding of this complex symptom and offer promising treatments.

Summary for the Clinician

Itch is the dominant symptom of skin disease; almost all inflammatory skin diseases can itch. Without adequate treatment, pruritus can significantly impair patients' quality of life. In most cases, itch results from interaction of the brain–skin axis. Chronic itch has many similarities to chronic pain. Both conditions have hypersensitivity of peripheral nerve fibers and the central nervous system. Targeting the neural system, the common pathway for itch transmission by inhibition, and reduction of itch intensity will hopefully translate into effective new treatments for pruritus.

References

1. Andoh T, Yageta Y, Takeshima H, et al (2004) Intradermal nociceptin elicits itch-associated responses through leukotriene B(4) in mice. J Invest Dermatol 123:196–201
2. Andrew D, Craig AD (2001) Spinothalamic lamina I neurons selectively sensitive to histamine: a central neural pathway for itch. Nat Neurosci 4:72–77
3. Baron R, Schwarz K, Kleinert A, et al (2001) Histamine-induced itch converts into pain in neuropathic hyperalgesia. Neuroreport 12:3475–3478
4. Bell JK, McQueen DS, Rees JL (2004) Involvement of histamine H4 and H1 receptors in scratching induced by histamine receptor agonists in Balb C mice. Br J Pharmacol 142:374–380
5. Bernhard JD (1994) Neurogenic pruritus and strange sensations. In: Bernhard JD (eds) Itch Mechanisms and Management of Pruritus. New York, McGraw-Hill, pp 165–202
6. Bernstein JE, Swift R (1979) Relief of intractable pruritus with naloxone. Arch Dermatol 115:1366–1367
7. Darsow U, Drzezga A, Frisch M, et al (2000) Processing of histamine-induced itch in the human cerebral cortex: a correlation analysis with dermal reactions. J Invest Dermatol 115:1029–1033
8. Dawn AG, Yosipovitch G (2006) Butorphanol for treatment of intractable pruritus. J Am Acad Dermatol 54(3):527–531
9. Denda M (2002) New strategies to improve skin barrier homeostasis. Adv Drug Deliv Rev 54(Suppl 1):S123–S130
10. Dillon SR, Sprecher C, Hammond A, et al (2004) Interleukin 31, a cytokine produced by activated T cells, induces dermatitis in mice. Nat Immunol 5:752–760
11. Drzezga A, Darsow U, Treede RD, et al (2001) Central activation by histamine-induced itch: analogies to pain processing: a correlational analysis of O-15 H2O positron emission tomography studies. Pain 92:295–305
12. Fjellner B, Hagermark O (1982) Potentiation of histamine-induced itch and flare responses in human skin by the enkephalin analogue FK-33–824, beta-endorphin and morphine. Arch Dermatol Res 274:29–37
13. Fjellner B, Hagermark O (1984) The influence of the opiate antagonist naloxone on experimental pruritus. Acta Derm Venereol 64:73–75
14. Fruhstorfer H, Hermanns M, Latzke L (1986) The effects of thermal stimulation on clinical and experimental itch. Pain 24:259–269
15. Grewe M, Vogelsang K, Ruzicka T, et al (2000) Neurotrophin-4 production by human epidermal keratinocytes: increased expression in atopic dermatitis. J Invest Dermatol 114:1108–1112
16. Grobe K, Poppelmann M, Becker WM, et al (2002) Properties of group I allergens from grass pollen and their relation to cathepsin B, a member of the C1 family of cysteine proteinases. Eur J Biochem 269:2083–2092
17. Groneberg DA, Serowka F, Peckenschneider N, et al (2005) Gene expression and regulation of nerve growth factor in atopic dermatitis mast cells and the human mast cell line-1. J Neuroimmunol 161:87–92
18. Hagermark O (1992) Peripheral and central mediators of itch. Skin Pharmacol 5:1–8
19. Hansson L, Backman A, Ny A, et al (2002) Epidermal overexpression of stratum corneum chymotryptic enzyme in mice: a model for chronic itchy dermatitis. J Invest Dermatol 118:444–449
20. Heyer G, Ulmer FJ, Schmitz J, et al (1995) Histamine-induced itch and alloknesis (itchy skin) in atopic eczema patients and controls. Acta Derm Venereol 75:348–352
21. Hilliges M, Wang L, Johansson O (1995) Ultrastructural evidence for nerve fibers within all vital layers of the human epidermis. J Invest Dermatol 104:134–137
22. Hosogi M, Schmelz M, Miyachi Y, et al (2006) Bradykinin is a potent pruritogen in atopic dermatitis: a switch from pain to itch. Pain 126(1–3):16–23
23. Hsieh JC, Hagermark O, Stahle-Backdahl M, et al (1994) Urge to scratch represented in the human cerebral cortex during itch. J Neurophysiol 72:3004–3008
24. Hundley JL, Yosipovitch G (2004) Mirtazapine for reducing nocturnal itch in patients with chronic pruritus: a pilot study. J Am Acad Dermatol 50:889–891
25. Ikoma A, Fartasch M, Heyer G, et al (2004) Painful stimuli evoke itch in patients with chronic pruritus: central sensitization for itch. Neurology 62:212–217
26. Ikoma A, Handwerker H, Miyachi Y, et al (2005) Electrically evoked itch in humans. Pain 113:148–154
27. Ikoma A, Steinhoff M, Stander S, et al (2006) The neurobiology of itch. Nat Rev Neurosci 7:535–547
28. Inoue K, Koizumi S, Fuziwara S, et al (2002) Functional vanilloid receptors in cultured normal human epidermal keratinocytes. Biochem Biophys Res Commun 291:124–129
29. Ishiuji Y, Dawn, A, Fountain J, et al (2006) Scratching and noxious heat aggravate itch perception in atopic dermatitis. J Invest Dermatol 126(4 Suppl):5 (abstract)
30. Kumagai H, Maruyama S, Gejyo F, et al (2003) Role of mu-and kappa-opioid systems in systemic and peripheral itch, and effects of a novel kappa-agonist, TRK-820. 2nd International Workshop for the Study of Itch
31. Leknes SG, Bantick S, Willis CM, et al (2006) Itch and motivation to scratch: an investigation of the central and peripheral correlates of allergen- and histamine-induced itch in humans. J Neurophysiol 97:415–422
32. Luger TA (2002) Neuromediators – a crucial component of the skin immune system. J Dermatol Sci 30:87–93

33. McGlone F, Rukwied R, Howard M, et al (2004) Histamine-induced discriminative and affective responses revealed by functional MRI. In: Yosipovitch G, Greaves MW, Fleischer Jr AB, McGlone F (eds) Itch: Basic Mechanisms and Therapy, 1st edn. New York, Marcel Dekker, pp 51–61
34. McMahon SB, Koltzenburg M (1992) Itching for an explanation. Trends Neurosci 15:497–501
35. Mendell LM, Albers KM, Davis BM (1999) Neurotrophins, nociceptors, and pain. Microsc Res Tech 45:252–261
36. Mochizuki H, Tashiro M, Kano M, et al (2003) Imaging of central itch modulation in the human brain using positron emission tomography. Pain 105:339–346
37. Nakamura M, Toyoda M, Morohashi M (2003) Pruritogenic mediators in psoriasis vulgaris: comparative evaluation of itch-associated cutaneous factors. Br J Dermatol 149:718–730
38. Pan ZZ (1998) mu-Opposing actions of the kappa-opioid receptor. Trends Pharmacol Sci 19:94–98
39. Paus R, Schmelz M, Biro T, et al (2006) Frontiers in pruritus research: scratching the brain for more effective itch therapy. J Clin Invest 116:1174–1186
40. Proudfoot CJ, Garry EM, Cottrell DF, et al (2006) Analgesia mediated by the TRPM8 cold receptor in chronic neuropathic pain. Curr Biol 16:1591–1605
41. Raychaudhuri SP, Raychaudhuri SK (2004) Role of NGF and neurogenic inflammation in the pathogenesis of psoriasis. Prog Brain Res 146:433–437
42. Rukweid R, Dvorak M, Watkinson A, et al (2004) Putative role of cannabinoids in experimentally induced itch and inflammation in human skin. In: Yosipovitch G, Greaves MW, Fleischer Jr AB, McGlone F (eds) Itch: Basic Mechanisms and Therapy, 1st edn. New York, Marcel Dekker, pp 115–130
43. Schmelz M (2002) Itch – mediators and mechanisms. J Dermatol Sci 28:91–96
44. Schmelz M (2005) Itch and pain. Dermatol Ther 18: 304–307
45. Schmelz M, Schmidt R, Bickel A, et al (1997) Specific C-receptors for itch in human skin. J Neurosci 17:8003–8008
46. Schmelz M, Hilliges M, Schmidt R (2003) Active "itch fibers" in chronic pruritus. Neurology 61:564–566
47. Schmelz M, Schmidt R, Weidner C, et al (2003) Chemical response pattern of different classes of C-nociceptors to pruritogens and algogens. J Neurophysiol 89:2441–2448
48. Scholzen T, Armstrong CA, Bunnett NW, et al (1998) Neuropeptides in the skin: interactions between the neuroendocrine and the skin immune systems. Exp Dermatol 7:81–96
49. Shelley WB, Arthur RP (1955) Studies on cowhage (Mucuna pruriens) and its pruritogenic proteinase, mucunain. AMA Arch Derm 72:399–406
50. Simone DA, Alreja M, Lamotte RH (1991) Psychophysical studies of the itch sensation and itchy skin ("alloknesis") produced by intracutaneous injection of histamine. Somatosens Mot Res 8:271–279
51. Simone DA, Nolano M, Johnson T, et al (1998) Intradermal injection of capsaicin in humans produces degeneration and subsequent reinnervation of epidermal nerve fibers: correlation with sensory function. J Neurosci 18:8947–8959
52. Stander S, Moormann C, Schumacher M, et al (2004) Expression of vanilloid receptor subtype 1 in cutaneous sensory nerve fibers, mast cells, and epithelial cells of appendage structures. Exp Dermatol 13:129–139
53. Steinhoff M, Vergnolle N, Young SH, et al (2000) Agonists of proteinase-activated receptor 2 induce inflammation by a neurogenic mechanism. Nat Med 6:151–158
54. Steinhoff M, Neisius U, Ikoma A, et al (2003) Proteinase-activated receptor-2 mediates itch: a novel pathway for pruritus in human skin. J Neurosci 23:6176–6180
55. Steinhoff M, Stander S, Seeliger S, et al (2003) Modern aspects of cutaneous neurogenic inflammation. Arch Dermatol 139:1479–1488
56. Togashi Y, Umeuchi H, Okano K, et al (2002) Antipruritic activity of the kappa-opioid receptor agonist, TRK-820. Eur J Pharmacol 435:259–264
57. Toyoda M, Nakamura M, Makino T, et al (2002) Nerve growth factor and substance P are useful plasma markers of disease activity in atopic dermatitis. Br J Dermatol 147:71–79
58. Twycross R, Greaves MW, Handwerker H, et al (2003) Itch: scratching more than the surface. QJM 96:7–26
59. Umeuchi H, Togashi Y, Honda T, et al (2003) Involvement of central mu-opioid system in the scratching behavior in mice, and the suppression of it by the activation of kappa-opioid system. Eur J Pharmacol 477:29–35
60. Urashima R, Mihara M (1998) Cutaneous nerves in atopic dermatitis. A histological, immunohistochemical and electron microscopic study. Virchows Arch 432:363–370
61. von Frey M (1922) Zur Physiologie der Juckempfindung. Arch Neerland Physiol 7:142–145
62. Wahlgren CF, Tengvall LM, Hagermark O, et al (1995) Itch and inflammation induced by intradermally injected interleukin-2 in atopic dermatitis patients and healthy subjects. Arch Dermatol Res 287:572–580
63. Walter B, Sadlo MN, Kupfer J, et al (2005) Brain activation by histamine prick test-induced itch. J Invest Dermatol 125:380–382
64. Yosipovitch G (2003) Pruritus: an update. Curr Prob Dermatol 15:137–164
65. Yosipovitch G, Greaves MW (2004) Definitions of itch. In: Yosipovitch G, Greaves MW, Fleischer Jr AB, McGlone F (eds) Itch: Basic Mechanisms and Therapy, 1st edn. New York, Marcel Dekker, pp 1–4
66. Yosipovitch G, Goon A, Wee J, et al (2000) The prevalence and clinical characteristics of pruritus among patients with extensive psoriasis. Br J Dermatol 143:969–973
67. Yosipovitch G, Ansari N, Goon A, et al (2002) Clinical characteristics of pruritus in chronic idiopathic urticaria. Br J Dermatol 147:32–36
68. Yosipovitch G, Goon AT, Wee J, et al (2002) Itch characteristics in Chinese patients with atopic dermatitis using a new questionnaire for the assessment of pruritus. Int J Dermatol 41:212–216
69. Yosipovitch G, Greaves MW, Schmelz M (2003) Itch. Lancet 361:690–694
70. Yosipovitch G, Fast K, Bernhard JD (2005) Noxious heat and scratching decrease histamine-induced itch and skin blood flow. J Invest Dermatol 125:1268–1272
71. Zylicz Z, Krajnik M, Sorge AA, et al (2003): Paroxetine in the treatment of severe non-dermatological pruritus: a randomized, controlled trial. J Pain Symptom Manage 26:1105–1112

Neuroimmunologic Cascades in the Pathogenesis of Psoriasis and Psoriatic Arthritis

17

S.P. Raychaudhuri and S.K. Raychaudhuri

Contents

Synonyms Box:

Abbreviations: *CGRP* Calcitonin gene-related protein, *NGF* Nerve growth factor, *NGF-R* Nerve growth factor receptor, *p75^{NTR}* p75 neurotrophin receptor, *RANTES* regulated upon activation, normal T cell expressed and secreted, *SCID* Severe combined immunodeficient, *SP* Substance P, *TrkA* Receptor tyrosine kinase A, *VIP* Vasoactive intestinal peptide

Key Features

- Psoriasis and its associated systemic inflammatory arthritis is a chronic inflammatory disease of unknown etiology. Significant progress has been made in elucidating the pathogenesis of psoriasis; still the molecular basis of the inflammatory and proliferative processes of psoriasis is largely unknown. The role of neurogenic inflammation has provided a new dimension in understanding the pathogenesis of various cutaneous and systemic inflammatory diseases such as atopic dermatitis, urticaria, rheumatoid arthritis, ulcerative colitis and bronchial asthma.
- In this article we have addressed certain key events with respect to the role of neurogenic inflammation in the development of a psoriatic lesion. Significant are the proliferation of nerves, upregulation of neuropeptides and increased levels of nerve growth factor (NGF).
- In immunoperoxidase studies, we found that keratinocytes in lesional and non-lesional psoriatic tissue express high levels of NGF in the terminal cutaneous nerves of psoriatic lesions and that there is a marked upregulation of the NGF receptors: p75 neurotrophin receptor (p75^{NTR}) and tyrosine kinase A (TrkA). Keratinocytes of psoriatic plaques express increased levels of NGF and it is likely that murine nerves will promptly grow into the transplanted plaques on a severe combined immunodeficient (SCID) mouse. Indeed, we have noted marked proliferation of nerve fibers in transplanted psoriatic plaques on a SCID mouse compared with the few nerves seen in transplanted normal human skin. These observations, as well as a recent report suggesting therapeutic efficacy following manipulation of TrkA induced signal in the SCID-psoriasis xenograft model, further substantiate a

R.D. Granstein and T.A. Luger (eds.), *Neuroimmunology of the Skin*,

contributing role of NGF and its receptor system in the pathogenesis of psoriasis.

› As psoriasis and psoriatic arthritis are within the spectrum of the same disease process, currently we are exploring the role of NGF/NGF-R in the pathophysiology of psoriatic arthritis.

› A new discipline has emerged in clinical pharmacology focusing on development of drugs targeting the neuropeptides (NP), NP receptors and the NGF/NGF-R system We are in the process of developing novel therapeutic agents for psoriasis and psoriatic arthritis by manipulating TrkA induced signal transductions.

17.1 Introduction

Psoriasis is a chronic inflammatory dermatologic disease. It affects nearly 2% of the world population and both males and females are equally prone to psoriasis [16]. This common skin disease can appear at any age [48]. The lesions of psoriasis are characterized by erythema, scaling, and infiltration. Elbows, knees, and scalp are common sites of involvement, although no part of the skin is resistant to psoriasis. Psoriasis is a non-fatal life long disease, but on occasions psoriasis can be a source of significant morbidity. Erythroderma, extensive pustular lesions, and an associated systemic inflammatory polyarthritis are severe forms of psoriasis.

As of now there is no cure for psoriasis. The pathogenesis of psoriasis is incompletely understood. There is substantial evidence that activated T lymphocytes play a key role in the pathogenesis of psoriasis. Cytokines, chemokines, growth factors, adhesion molecules, neuropeptides, and T cell receptors act in integrated ways to evolve in unique inflammatory and proliferative processes typical of psoriasis. The concept of neuroimmunology as it relates to psoriasis is relatively new. In last 20 years, independent studies from several psoriasis research centers have proposed a possible role of neurogenic inflammation in the pathogenesis of psoriasis.

In this chapter we discuss interactions of the neuroimmune system with respect to T cell functions, endothelial cell biology, chemokine expression, and cell trafficking. We also provide evidence from in vivo and in vitro studies about the role of neuroendocrine factors in chronic inflammatory human diseases such as psoriasis and psoriatic arthritis.

17.2 Neurogenic Inflammation in Psoriasis

The process of antidromic stimulation of dorsal roots resulting in vasodilatation, exudation of plasma, and migration of leukocytes is referred to as neurogenic inflammation. Neurogenic inflammation results due to release of neuropeptides from unmyelinated sensory nerve endings. Our research group has a special interest in neurogenic inflammation. Correlating the clinical observation that stress exacerbates psoriasis and psoriasis is symmetrically distributed, Farber et al. proposed a role for cutaneous nerves and neuropeptides in the pathogenesis of psoriasis [18]. Subsequently many investigators, including us, have reported an upregulation of neuropeptides such as SP, VIP, and CGRP along with marked proliferation of terminal cutaneous nerves in psoriatic lesions [12,41,42,68]. Neuropeptides can play significant role in the inflammatory and proliferative processes of psoriasis. SP is chemotactic to neutrophils [61], activates T cells [10], VIP is mitogenic to keratinocytes [24], CGRP acts synergistically with SP to stimulate keratinocyte proliferation [73], and both VIP and CGRP are potent mitogens for endothelial cells [25].

In several case reports it has been reported that in areas of anesthesia active plaques of psoriasis resolves [47]. Here we provide brief information on three patients. A 68-year-old Caucasian male had chronic plaque psoriasis involving the elbows, forearms, knees, and legs. The patient underwent a reconstructive surgery on the left knee for osteoarthritis. By 6–8 weeks following surgery, a large plaque on the lateral surface of the left leg resolved. On examination the skin at the resolved site was found to be anesthetic, probably due to nerve damage following surgery. A comparable plaque on the contralateral leg remained active. In an elderly lady, psoriasis plaques on left forearm resolved at sites of anesthesia following a radical mastectomy for breast cancer, whereas in the contralateral forearm psoriasis remained active. In another patient it was observed that psoriasis resolved at the anesthetic area over knee and, with the return of sensation, psoriasis reappeared at the same site.

These clinical and histopathological observations that psoriasis resolves at sites of anesthesia, neuropeptides are upregulated in lesions, and there is a marked proliferation of terminal cutaneous nerves in psoriatic plaques provided us with convincing cause to study the cellular and molecular mechanisms for these unique events. As nerve growth factor (NGF) plays a role in regulating innervation [76] and upregulating neuropeptides [34,60],

Table 17.1 Expression Of NGF in the keratinocytes and NGF-R within papillary dermal nerves in lesional/nonlesional psoriatic skin, inflammatory dermatoses including lichen planus, and normal skin

Type of skin	NGF⁺ KC/mm²	NGF-R⁺/3 mm
Psoriasis		
A. Lesional	84.68 ± 46.35 (n = 8)	34.0 ± 23.0 (n = 26)
B.Nonlesional	44.80 ± 29.96 (n = 8)	28.1 ± 4. 5 (n = 8)
Normal skin	18.88 ± 11. 76 (n = 5)	18.88 ± 11.76 (n = 8)
Lichen plaus	7.54 ± 16. 86 (n = 5)	
Inflammatory dermatoses		12.75 ± 22.0 (n = 12)

All values displayed as mean number of positively stained keratinocytes/mm² (KC/mm²) in epidermis and papillary dermal nerves/3 mm biopsy ± standard deviation
NGF nerve growth factor, *NGF-R* nerve growth factor-receptor

Table 17.2 Inflammogenic properties Of NGF

1. Degranulates mast cells
2. Upregulates expression of SP, CGRP
3. Activates T cells
4. Upregulates expression of RANTES in keratinocytes
5. Recruits inflammatory cellular infiltrates
6. Induces expression of ICAM-1 on endothelial cells
7. Promotes angiogenesis

we decided to investigate the expression of NGF/NGF-R in the lesional and nonlesional psoriatic skin, normal skin, and other inflammatory skin diseases.

In an immunohistochemical study (Table 17.1), we found that keratinocytes in lesional and nonlesional psoriatic tissue express high levels of NGF compared to the controls [50]. Fantini et al. have observed similar results in tissue extracts; it was found that levels of NGF are higher in psoriatic lesions [15].

Several functions of NGF are relevant to the inflammatory and proliferative processes of psoriasis (Table 17.2). Nerve growth factor promotes keratinocyte proliferation and protects keratinocytes from apoptosis [74,44]. NGF degranulates mast cells and induces migration of these cells; both are early events in a developing lesion of psoriasis [2,43]. In addition, NGF activates T lymphocytes and recruits inflammatory cell infiltrates [64,6,31]. In one of our recent studies, we have observed NGF induces expression of the potent chemokine RANTES in keratinocytes. RANTES is chemotactic for resting CD4+ memory T cells, and activated naive and memory T cells [58]. It is possible that in a developing psoriatic lesion, upregulation of NGF induces the influx of mast cells and lymphocytes, which in turn initiates an inflammatory reaction contributing to the pathogenesis of psoriasis.

17.3 Koebner Phenomenon: Role of NGF/NGF-R System in the Evolution of an Isomorphic Response

A wound induces a reaction characterized by proliferation of keratinocytes, fibroblasts, vascular elements, nerves, and an accumulation of inflammatory cells. In non-psoriatics, healing stops after a finite time depending on the nature of the wound. In patients with psoriasis a wound frequently results in papulosquamous lesions.

Recent reports suggest that NGF produced by the keratinocytes plays a role in wound healing. NGF promotes axonal regeneration and reinnervation of terminal cutaneous nerves. Upregulation of NGF in injured skin has been confirmed [39]. In developing lesions of psoriasis, we have observed is marked expression of NGF in the basal keratinocytes following 24 h of trauma induced by tape stripping, whereas with the same procedure in patients without psoriasis there was no induction of NGF in the epidermis.

These findings suggest that the increased expression of NGF in the keratinocytes of lesional and nonlesional psoriatic tissue may be an early event in the pathogenesis of psoriasis. As in the primed keratinocytes of nonlesional psoriatic skin, the basal level of NGF is upregulated (Table 17.1); proliferation of keratinocytes induced by a wound will result in significantly higher levels of NGF in lesion-free skin of a psoriasis patient compared to a healthy individual. Elevated levels of NGF will induce an inflammatory response (Table 17.2), proliferation of nerves, and upregulation of neuropeptides such as SP, CGRP. Increased levels of neuropeptides and NGF, in addition to their pro-inflammatory effects, will induce keratinocyte proliferation [24,44,73,74], which in turn will result in increased expression of NGF. Thus, a vicious cycle of a proliferative and inflammatory process is established (Fig. 17.1) in one who is genetically psoriatic. In subjects without psoriasis, the expression of NGF is 3–4 times less per square millimeter of epidermis, compared to nonlesional psoriatic skin (Table 17.1). The healing events, therefore, do not generate the critical levels of NGF and neuropeptides to initiate or maintain cascades essential for a chronic inflammatory reaction.

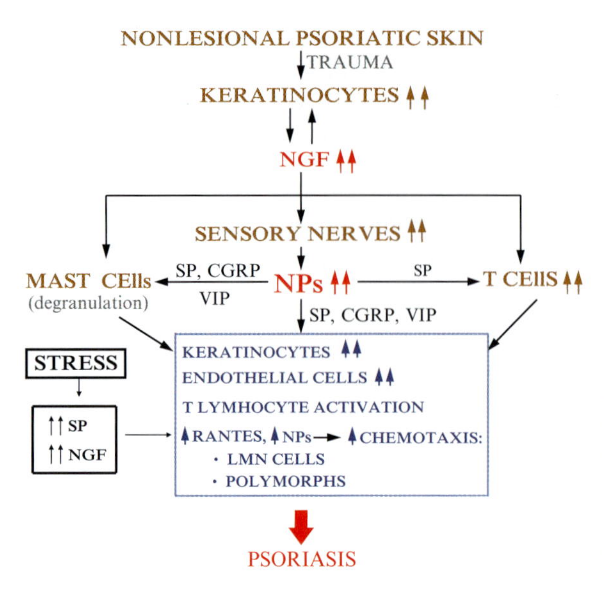

Fig. 17.1 Role of neurogenic inflammation in the pathogenesis of psoriasis (↑/↑↑ = increase in amount/number, *LMN* lymphomononuclear cells, *NPs* neuropeptides)

Stressful events can alter substance P levels in the CNS and in the periphery. In an animal model it has been reported that stress can increase levels of substance P in the adrenal glands by activating the descending autonomic fibers [66]. Some of the descending autonomic fibers innervate opioid interneurons in the dorsal horn and as interneurons exists in the spinal cord for the substance P containing nerves, and it is conceivable that descending autonomic paths can cause release of cutaneous neuropeptides [20]. Studies have reported that psychosocial stressful events result in increased levels of NGF in blood and the NGF mRNA synthesis in the hypothalamus [35,3]. Thus, it is likely that a similar cascade of events as mentioned in the preceding paragraph may occur in "distressed" psoriatic patients.

17.4 Is Psoriasis a Neuroimmunologic Disease?

To designate psoriasis purely as a T cell disease is too simplistic. Some investigators consider psoriasis to be an autoimmune disease induced by an antigen [8,65,13]. Up to now the alleged role of an antigen in psoriasis is hypothetical; no antigen has yet been discovered for psoriasis. An antigen-induced T cell activation process alone fails to clarify various salient features of psoriasis. It does not explain the Koebner phenomenon, the symmetrical distribution of psoriasis lesions, proliferation of cutaneous nerves, and the upregulation of neuropeptides in psoriatic tissue [18,12,41,1,69]. It does not have an answer either for the striking clinical observation that psoriasis resolves at sites of anesthesia [47].

In the last two decades, extensive work has been done to explore the immunological mechanisms involved in psoriasis. An active role of T cells is strongly substantiated by the following observations: (1) Immunotherapy targeted specifically against CD4+ T cells clears active plaques of psoriasis [22]. (2) In SCID mice, transplanted nonlesional psoriatic skin converts to a psoriatic plaque subsequent to intradermal administration of T cells activated with an antigen cocktail [75]. However, it is equally true that psoriasis treated with agents such as calcipotriol and etretinate, which affect the differentiation process of keratinocytes, are very effective in psoriasis. Neither calcipotriol nor etretinate are effective in other T cell-mediated cutaneous diseases such as atopic dermatitis or contact dermatitis.

Though psoriasis has been claimed to be an autoimmune disease, the antigen or specific endogenous factors responsible for activation of T cells in psoriasis is still unknown. Regarding induction of psoriasis in transplanted nonlesional skin in the SCID mouse model, the T cells were activated with an antigen cocktail [75]. As such artificial antigen cocktail does not exist in a lesional or nonlesional psoriatic skin, and it is possible that local epidermal and dermal factors like NGF and SP may be responsible for lesional T lymphocyte activation. Recently we have identified increased levels of RANTES in psoriatic keratinocytes [52]. Increased levels of RANTES induced by NGF may also be a contributing factor for the activation of the lesional T cells [53].

Nickoloff and his group reported that they could induce psoriasis by injecting lymphomononuclear cells activated with substance P in the SCID mouse model (personal communication). In a double-blinded, placebo controlled study, we addressed the role of NGF/NGF-R in psoriasis in an in vivo system using the SCID mouse model of psoriasis [54] and in human by tape stripping method [56]. We have observed a novel observation that in the SCID mouse model of psoriasis, autologous immunocytes activated with NGF can convert transplanted nonlesional skin to a psoriatic plaque in 3 weeks.

17.5 Significance of NGF and Its Receptor System in Psoriatic Arthritis

Psoriatic arthritis is a systemic inflammatory disease mainly involving the skin and the joints. It is reasonable to postulate that the skin and joint involvement will share common pathophysiologic processes. Both psoriasis and psoriatic arthritis share several similar key biologic events in respect to HLA phenotyping, cell trafficking, nature of T cell phenotypes, cytokine profiles, and angiogenesis. Methotrexate, retinoids, cyclosporine, and TNF antagonists are effective for both psoriasis and psoriatic arthritis.

Coexistence of psoriasis and psoriatic arthritis, sharing of critical pathologic principles including tropism of CD8 lymphocytes/neutrophils in epidermis/synovium, and clinical efficacy of different groups of pharmacological agents provide evidence that psoriatic skin disease and psoriatic arthritis are manifestation of a multiorgan disease that likely share a common pathophysiologic origins. NGF influences several essential biological events of an inflammatory process such as induction of adhesion molecules and upregulation of chemokines. NGF receptors have been identified in both B and T lymphocytes. NGF is a growth factor, and its antiapoptotic function in different types of cells including keratinocytes, Schwann cells, and lymphocytes is well established. NGF is a key regulatory molecule for nerve proliferation and upregulation of inflammogenic neuropeptides such as substance P and CGRP. NGF is mitogenic to endothelial cells and promotes angiogenesis. All of these functions of NGF/NGF-R are potentially relevant in maintenance or initiation of the inflammatory cascades contributing to the pathogenesis of psoriasis and psoriatic arthritis.

The role of NGF/NGF-R system in inflammation and inflammatory diseases is an active field of research. There is no animal model for psoriatic arthritis. Transgenic mice expressing the human TNF gene (Tg197) are reported to have upregulation of NGF in the synovium of inflamed joints; and subcutaneous injection of NGF antibody was noted to be therapeutically beneficial in these mice [4]. It has been reported that human synovial fibroblasts produce and release NGF, and TrkA NGF receptors are expressed in normal human synovial fibroblasts. Also, both TNF-alpha and IL-1 can promote synthesis of NGF in cultured synovial fibroblasts [36]. In patients with psoriatic arthritis and other forms of inflammatory arthritis, no comprehensive work has been carried out to address the role of NGF and its receptor system. Very limited information is available about the role of neurogenic inflammation in psoriatic arthritis; only a few case reports. Nonetheless, these reports indicate that substance P release from the synovial membrane into joint fluid is blocked by nerve damage, and digital denervation prevented development of arthritis in the interphalangeal joints. There are reports that NGF and substance P levels in the synovial fluid of psoriatic arthritis patients may be elevated [26]. In a very recent report, modulation of intra-articular innervation and neuropeptides has been suggested as a potential new and unused therapeutic approach in ameliorating joint inflammation in psoriatic arthritis [29]. Thus we believe, NGF and its receptor system may play an important role in the inflammatory cascades of psoriatic arthritis. The results of our initial investigations suggest that NGF levels are increased in the synovial fluid of posoriatic arthritis patients compared to patients with osteoarthritis. Also we have noticed marked expression of TrkA in endothelial cells of the synovium of psoriatic arthritis patients.

17.6 Knowledge of Psychoneuroimmunology Provides a New Dimension for the Treatment of Psoriasis

The role of neuroimmunologic cascades in the pathogenesis of psoriasis has provided a new approach for the treatment of psoriasis. Understanding the factors that increase the morbidity of a disease is essential for providing effective clinical care. A treatment regimen without controlling the exogenous and endogenous factors responsible for the increased severity of psoriasis is only partially effective. There is unequivocal evidence that stress is a triggering factor for the appearance or exacerbation of psoriasis. This indicates that, in addition to the standard therapies available to dermatologists, it is advisable to consider stress relaxation measures. Psychological evaluation is a sine qua non for the management of psoriasis. It is essential to understand whether a patient has a primary emotional disorder or whether emotional instability is secondary to the psychosocial impact of psoriasis. In selected patients, antidepressants or anxiolytics may be appropriate. The other aspect of psychological evaluation is to find out the underlying physical and emotional stresses, the patient's personality, and the way the patient react to a stressful situation. Analysis of these factors will indicate the desirability of various stress relaxation modalities such as hypnosis, biofeedback, meditation, visual imagery, and cognitive organization. A change in lifestyle or attitude can have significant effect on the course of psoriasis. Studies have demonstrated that psychologic interventions such as hypnosis [70] and biofeedback [27,37] are helpful in the treatment of psoriasis. At the Psoriasis Research Institute, we initiated a total care program [17]. By total care of psoriasis, we mean exemplary skin care with the available pharmacological agents and, in addition, control of the associated factors that affect the morbidity of psoriasis [49]. Total care encompasses the following: complete physical evaluation, evaluation of the type and extent of psoriasis, a personal wellness questionnaire analysis to measure patient's current health status, psychological examination, teaching stress reduction techniques, and patient education (self-help/mutual aid group). We expect patients adhering to a total care program will require less medication and have longer periods of remission.

A more direct approach would be to design drugs to counter the inflammatory and proliferative cascades induced by NGF and neuropeptides. To date, relatively few clinical studies have used this principal to treat inflammatory diseases. Neuropeptide receptor analogues have been reported to block the inflammatory effects of neuropeptides, such as plasma extravasation [77], nociceptor flexor reflexes [72], and erythematous responses [68]. In vitro effects of neuropeptides on lymphocytes [23], keratinocytes [45], and mast cells [30] can be modulated by neuropeptide receptor antagonists as well. These results indicate that neuropeptide analogues can be applied for inhibiting the inflammatory and vascular changes associated with psoriasis.

Peptide T, a synthetic octapeptide, is a protease resistant analog of VIP [68]. The first report of the efficacy of peptide T in psoriasis arose from an anecdotal case report where psoriasis in an AIDS patient significantly improved following the intravenous infusion of peptide T [57]. Subsequently, intravenous use of peptide T has been reported to improve psoriasis [71,38]. Farber et al. evaluated the efficacy of peptide T by a direct administration into psoriatic lesion by mini-osmotic pump [63]. In a double blind, placebo controlled study the authors reported that infusion of peptide T in nanogram amounts improved psoriatic lesions both clinically and histopathologically. The mechanism of action peptide T in psoriasis is not clear; possibilities suggested are antagonizing the action of VIP, up-regulation of somatostatin in the psoriatic lesions, and immunomodulation [19,28,51].

A somatostatin analog (Sandostatin) is the other neuropeptide analog reported to be efficacious in psoriasis [11]. However, a high frequency of gall stones was noted among these patients. Somatostatin is a well known SP inhibitor [32]. Another neuropeptide modulating agent used as a therapeutic agent in psoriasis is capsaicin (*trans*-8methyl-*N*-vanillyl-G-nonenamide), the extract of the hot pepper, which depletes substance P from the sensory C nerve fibers [21]. Topical use of capsaicin has been reported to be effective in psoriasis [9] but it is unsuitable because it causes significant burning of the skin.

Substance P antagonists may also be useful therapeutic agent for psoriasis. Spantide, a structural analogue of substance P, has been reported to cause inhibition of delayed-type cutaneous hypersensitivity reactions in healthy human volunteers [67]. Spantide also can inhibit substance P-induced keratinocyte proliferation in vitro [45]. Peptide antagonists are metabolically unstable and can cause hypersensitivity reaction. The discovery of CP-96,345, a synthetic nonpeptide substance P receptor (NK-1 receptor) antagonist, has opened new avenues to evaluate the effects of NK-1 receptor antagonism in man [62]. In animal models, CP-96,345 has been found to inhibit plasma exudation induced by substance P and can block nociceptor responses to noxious cutaneous stimuli [33,46]. Spantide II is a peptide with a molecular weight of 1670.2, which binds to neurokinin-1 receptor (NKR-1) and blocks proinflammatory activities associated with substance P. Spantide II can effectively be delivered to epidermis and dermis to exert a significant anti-inflammatory activity on the reduction of inflammation in a mouse model of allergic contact dermatitis [5]. These results indicate the applicability of substance P antagonists in various inflammatory conditions, including psoriasis. Currently various synthetic substance P antagonists are being evaluated in psoriasis and other inflammatory diseases.

Several peptidases including neutral endopeptidase (NEP), angiotensin-converting enzyme (ACE), and dipeptidyl amino peptidase IV (DAP IV) are responsible for degradation of neuropeptides [14]. Glucocorticoids can induce NEP [7]. This provides another explanation for the efficacy of glucocorticoids in the treatment of psoriasis. Neurogenic inflammation induced by substance P can be suppressed with exogenous human recombinant NEP [40]. The effector phase of allergic contact dermatitis (ACD) responses was examined in NEP(−/−) knockout and NEP(+/+) wild-type mice and compared with the irritant contact dermatitis response in these animals. NEP was found to be normally immunolocalized in epidermal keratinocytes and dermal blood vessels. The ACD ear swelling response was 2.5-fold higher in animals lacking NEP and was accompanied by a significant increase in plasma extravasation and infiltration of inflammatory leukocytes [59]. The augmented ACD response in NEP(−/−) animals could be abrogated by either administration of a neurokinin receptor 1 antagonist or by repeated pretreatment with topical capsaicin. This suggests that neurogenic inflammation induced by substance P can be suppressed and provides the logistics of using exogenous human recombinant NEP for therapy of inflammatory diseases [59].

Although clinical and laboratory studies suggest a critical role for NGF and its receptor (NGF-R) in the inflammatory process of psoriasis, direct evidence has been lacking. To determine the significance of the NGF/NGF-R system in the inflammatory process of psoriasis, we evaluated the effects of K252a, a high affinity NGF receptor inhibitor [55]. In this double-blinded, placebo-controlled study, we addressed the role of NGF/NGF-R in psoriasis in an in vivo system using the severe combined immunodeficient (SCID) mouse–human skin model of psoriasis. Transplanted psoriatic plaques on the SCID mice (n = 12) were treated with K252a, a high affinity NGF receptor blocker. Psoriasis significantly improved following 2 weeks of therapy. The length of the rete pegs changed from 308.57 ± 138.72 to 164.64 ± 64.78 µm ($p < 0.01$, Student's *t*-test). The control group, treated with normal saline, did not improve. A similar improvement of psoriasis was observed by directly antagonizing NGF with a NGF neutralizing antibody.

Elucidation of the molecular and cellular mechanisms responsible for the pathogenesis of psoriasis had been significantly handicapped due to lack of an ideal animal model. Recent establishment of the SCID-human skin chimeras with transplanted psoriasis plaques has opened new vistas to study the molecular complexities involved in psoriasis [54]. Histologic and immunologic features of psoriasis can be maintained in the transplanted plaques for more than 6 months. Using this model we have established that modulation of NGF and its high affinity receptor results in striking histological and clinical improvement of psoriasis. We are in the process of designing safe and effective therapeutic agents that can provide a prolonged remission of psoriasis.

In this article we have addressed certain key clinicopathological events, which directly or indirectly regulate the development of a psoriatic lesion. Significant are the proliferation of nerves, upregulation of neuropeptides, and increased levels of NGF. Clearance of psoriatic lesions at sites of anesthesia following nerve injury suggests an indisputable role for neurogenic inflammation in the pathogenesis of psoriasis. In serial biopsies from induced Koebner lesions, we noticed marked upregulation of NGF in Koebner-positive lesions following 24 h of cutaneous trauma. Synthesis of NGF reached its peck levels in the second week and remained persistently high thereafter. This suggests that the NGF/NGF-R system is functionally active at a very early phase of the inflammatory and proliferative processes of psoriasis. In SCID mice with xenograft transplants, we observed a marked proliferation of NGF-R (p75) positive nerve fibers in the transplanted psoriatic plaques compared to the transplanted normal human skin. This observation substantiates the in vivo effect of NGF produced by keratinocytes in a psoriatic plaque. Using this model we have established that modulation of NGF and its high affinity receptor (TrkA) results in striking histological and clinical improvement of psoriasis.

Our focus in this article is psoriasis because the role of neurogenic inflammation, neuropeptides, NGF/NGF-R system, and neuroimmunologic cascades has been studied most extensively in this disease. Currently, we are working to elucidate the role of neuropeptides and NGF/NGF-R system in the pathogenesis of psoriatic arthritis. Neuropeptides, and especially substance P, has been studied in several human diseases [59]. Various exogenous and endogenous stimuli can induce or upregulate neuropeptide release from sensory nerves in the lung, including allergen, histamine, prostaglandins, and leukotrienes. Patients with asthma are hyper-responsive to substance P and NK-1R expression is increased in their bronchi. Neurogenic inflammation also participates in virus-associated respiratory infection, nonproductive cough, allergic rhinitis, and sarcoidosis. Substance P regulates smooth muscle contractility, epithelial ion transport, vascular permeability, and immune function in the gastrointestinal tract. Elevated levels of Substance P and upregulated NK-1R expression have been reported in the rectum and colon of patients with inflammatory bowel disease (IBD) and correlate with disease activity. Increased levels of SP are found in the synovial fluid and serum of patients with rheumatoid arthritis (RA) and NK-1R mRNA is upregulated in RA synoviocytes.

A new discipline is emerging in clinical pharmacology focusing on development of drugs targeting the neuropeptides (NP), the NP receptors, and the NGF/NGF-R system. Currently, we are evaluating antagonists to selected neuropeptides, NGF, and NGF-R (p75/TrkA), with the expectation of developing novel therapies for psoriasis and psoriatic arthritis. In this article we have discussed our observations and reviewed the current literature regarding the role of neurogenic inflammation in inflammatory diseases. It is likely that in the near future unique neuropharmacologic drugs will be used for the treatment of a wide variety of inflammatory disorders.

Summary for the Clinician

We have demonstrated that following cutaneous trauma in a developing psoriasis lesion keratinocyte proliferation and upregulation of NGF in basal keratinocytes are earlier events that precede epidermotropism of T lymphocytes. Further, we have demonstrated that NGF secreted by the psoriatic keratinocytes is functionally active and keratinocytes of psoriatic patients produce higher levels of NGF compared to normal individuals. NGF influences all key pathologic events of psoriasis; NGF is mitogenic to keratinocytes, promotes angiogenesis, and activates T cells. This compelling evidence persuaded us to develop therapeutic strategies for psoriasis by manipulating NGF/TrkA interaction and its down-stream signal transduction events. To substantiate the significance of NGF/NGF-R system in the pathogenesis of psoriasis, we evaluated the therapeutic efficacy of K252a, a high-affinity NGF receptor inhibitor, and NGF-neutralizing antibody. Transplanted psoriatic plaques in the severe combined immunodeficient (SCID) mouse–human skin model were treated with intralesional injections of K252a and NGF-neutralizing antibody. In this study, we have demonstrated that K252a, an inhibitor of signal transductions induced by NGF/NGF-R interaction, is therapeutically effective in psoriasis. Efficacy was evidenced by decreased thickness of the rete pegs, reduced infiltrates, and normalization of the stratum corneum, whereas the control group treated with normal saline did not improve. A role for NGF in the pathogenesis of psoriasis is further substantiated by our observation that a similar improvement could be reproduced by directly inhibiting NGF with a NGF-neutralizing antibody. Results of this study provided direct evidence that NGF/NGF-R contributes to a pathologic process of a human disease, and therapeutic manipulation of NGF/TrkA interaction is plausible for the treatment psoriasis.

References

1. Al'Abadie MSK, Senior HJ, Bleehen SS, et al (1992) Neurogenic changes in psoriasis. An immunohistochemical study. J Invest Dermatol 98:535
2. Aloe L, Levi-Mantalcini R (1977) Mast cells increase in tissues of neonatal rats injected with the nerve growth factor. Brain Res 133:358–366
3. Aloe L, Alleva E, De Simone R (1990) Changes of NGF level in mouse hypothalamus following intermale aggressive behavior: biological and immunohistochemical evidence. Behav Brain Res 39:53–61
4. Aloe L, Probert L, Kollias G, et al (1993) The synovium of transgenic arthritic mice expressing human tumor necrosis factor contains a high level of nerve growthfactor. Growth Factors 9:149–155
5. Babu RJ, Kikwai L, Jaiani LT, et al (2004) Percutaneous absorption and anti-inflammatory effect of a substance P receptor antagonist: spantide II. Pharm Res 21:108–113
6. Bischoff SC, Dahinden CA (1992) Effect of nerve growth factor on the release of inflammatory mediators by mature human basophils. Blood 79:2662–2669
7. Borson DB, Gruenert DC (1991) Glucocorticoids induce neutral endopeptidase in transformed human tracheal epithelial cells. Am J Physiol 260:L83–L89
8. Bos JD, Hulsebosch HJ, Krieg SR, et al. (1983) Immunocompetent cells in psoriasis: in situ immunophenotyping with monoclonal antibodies. Arch Dermatol Res 275:181–189
9. Bernstein JE, Parish LC, Rapaport M, et al (1986) Effects of topically applied capsaicin on moderate and severe psoriasis vulgaris. J Am Acad Dermatol 15:504–507
10. Calvo CF, Chavanel G, Senik A (1992) Substance P enhances interleukin-2 expression in activated human T cells. J Immunol 148:3498–3504
11. Camisa C, O'Dorisio TM, Maceyko RF, et al (1990) Treatment of psoriasis with chronic subcutaneous administration of somatostatin analog 201–295 (sandostatin). An open lable pilot study. Clev Clin J Med 57:71–76
12. Chan J, Smoller BR, Raychaudhuri SP, et al (1997) Intraepidermal nerve fiber expression of calcitonin gene-related peptide, vasoactive intestinal peptide and substance P in psoriasis. Arch Dermatol Res 289:611–616
13. Chang JC, Smith LR, Froning KJ, et al (1997) Persistance of T-cell clones in psoriatic lesions. Arch Dermatol 133:703–708
14. Erdos EG, Skidgel RA (1989) Neutral endopeptidase and related regulators of peptide hormones. FASEB J 3:145–151
15. Fantini F, Magnoni C, Brauci-Laudeis L, Pincelli C (1995) Nerve growth factor is increased in psoriatic skin. J Invest Dermatol 105:854–855
16. Farber EM, Peterson JB (1961) Variations in the natural history of psoriasis. Calif Med 95:6–11
17. Farber EM, Raychaudhuri SP (1997) Concept of total care: a third dimension in the treatment of psoriasis. Cutis 59:35–39
18. Farber EM, Nickoloff BJ, Recht B, et al (1986) Stress, symmetry, and psoriasis: possible role of neuropeptides. J Am Acad Dermatol 14:305–311
19. Farber EM, Cohen EN, Trozak DJ, et al (1991) Peptide T improves psoriasis when infused into lesions in nanogram amounts. J Am Acad Dermatol. 25:658–664
20. Farber EM, Rein G, Lanigan SW (1991) Stress and psoriasis – psychoneuroimmunologic mechanisms. Int J Dermatol 30:8–12
21. Fitzgerald M (1983) Capsaicin and sensory neurons: a review. Pain 15:109–130
22. Gottlieb AB, Lebwohl M, Shirin S, et al (2000) Anti-CD4 monoclonal antibody treatment of moderate to severe psoriasis vulgaris: results of a pilot, multicenter, multiple-dose, placebo-controlled study. J Am Acad Dermatol 43:595–604

23. Gozes Y, Brenneman DE, Fridkin M, et al (1991) A VIP antagonist distinguishes spinal cord receptors on spinal cord cells and lymphocytes. Brain Res 540:319–321
24. Haegerstrand A, Jonzon B, Dalsgaard CJ, et al (1989) Vasoactive intestinal polypeptide stimulates cell proliferation and adenylate cyclase activity of cultured human keratinocytes. Proc Natl Acad Sci U S A 86:5993–5996
25. Hagerstrand A, Dalsgaard CJ, Jonzon B, et al (1990) Calcitonin gene-related peptide stimulates proliferation of human endothelial cells. Proc Natl Acad Sci U S A 87:3299–3303
26. Halliday DA, Zettler C, Rush RA, et al (1998) Elevated nerve growth factor levels in the synovial fluid of patients with inflammatory joint disease. Neurochem Res 23:919–922
27. Hughes HH, England R, Goldsmith DA (1981) Biofeedback and psychotherapeutic treatment of psoriasis: a brief report. Psycho Rep 48:99–102
28. Johansson O, Hilliges M, Talme T, et al (1994) Somatostatin immunoreactive cells in lesional psoriatic human skin during Peptide T treatment. Acta Derm Venereol (Stockh) 74:106–109
29. Kane D, Lockhart JC, Balint PV, et al (2005). Protective effect of sensory denervation in inflammatory arthritis (evidence of regulatory neuroimmune pathways in the arthritic joint). Ann Rheum Dis 64:325–327
30. Krumins SA, Broomfield C (1992) Evidence of NK 1 and NK 2 tachykinin receptors and their involvement in histamine release in a murine mast cell line. Neuropeptides 21:65–72
31. Lambiase A, Bracci-Laudiero L, Bonini S, et al (1997) Human CD4 + T cell clones produce and release nerve growth factor and express high-affinity nerve growth factor receptors. J Allergy Clin Immunol 100:408–414
32. Leeman SE, Krause JE, Lembeck F (eds) (1991) Substance P and related peptides: cellular and molecular physiology. 18–21 July 1990, Worcester, Massachusetts, Proceedings; Ann N Y Acad Sci 632:1–58, 263–271
33. Lei YH, Barnes PJ, Rogers DF (1992) Inhibition of neurogenic plasma exudation in guinea-pig airways by CP-96, 345, a new nonpeptide NK1 receptor antagonist. Br J Pharmacol 105:261–262
34. Lindsay RM, Harmar AJ (1989) Nerve growth factor regulates expression of neuropeptides genes in adult sensory neurons. Nature 337:362–364
35. Luppi P, Levi-Montalcini R, Bracci-Laudiero L, et al (1993) NGF is released into plasma during human pregnancy: an oxytocin-mediated response? Neuroreport 4:1063–1065
36. Manni L, Lundeberg T, Fiorito S, et al (2003) Nerve growth factor release by human synovial fibroblasts prior to and following exposure to tumor necrosis factor-alpha, interleukin-1 beta and cholecystokinin-8: the possible role of NGF in the inflammatory response. Clin Exp Rheumatol 21:6176–6124
37. Marcer D (1986) Biofeedback and Related Therapies in Clinical Practice. Aspen Publishers, Rockville, MD
38. Marcusson JA, Lazega D, Pert CB, et al (1989) Peptide T and psoriasis. Acta Derm Venereol Suppl (Stockh) 146:117–121
39. Matsuda H, Koyama H, Sato H, et al (1998) Role of nerve growth factor in cutaneous wound healing: accelerating effects in normal and healing-impaired diabetic mice. J Exp Med 187:297–306
40. Nadel JA (1991) Neutral endopeptide modulates neurogenic inflammation. Eur Respir J 4:745–754
41. Naukkarinen A, Nickoloff BJ, Farber EM (1989) Quantification of cutaneous sensory nerves and their substance P content in psoriasis. J Invest Dermatol 92:126–129
42. O'Connor TM, O'Connell J, O'Brien DI, et al (2004) The role of substance P in inflammatory diseases. J Cell Physiol 201:167–180
43. Pearce FL, Thompson HL (1986) Some characteristics of histamine secretion from rat peritoneal mast cells stimulated with nerve growth factor. J Physiol 372:379–393
44. Pincelli C, Haake AR, Benassi L, et al (1997) Autocrine nerve growth factor protects human keratinocytes from apoptosis through its high affinity receptor (TRK): a role for BCL-2. J Invest Dermatol 109:757–764
45. Rabier M, Wilkinson DI (1991) Neuropeptides modulate Leukotrine B4 Mitogenicity toward cultured keratinocytes. Clin Res 39:536a
46. Radhakrishna V, Henry JL (1991) Novel substance P antagonist, CP-96, 345, blocks responses of cat spinal dorsal horn neurons to noxious cutaneous stimulation and substance P. Neurosci Lett 132:39–43
47. Raychaudhuri SP, Farber EM (1993) Are sensory nerves essential for the development of psoriasis lesions? J Am Acad Dermatol 28:488–489
48. Raychaudhuri SP, Gross J (2000) A comparative study of pediatric onset psoriasis with adult onset psoriasis. Pediatr Dermatol 17:174–178
49. Raychaudhuri SP, Gross J (2000) Psoriasis risk factors: role of lifestyle practices. Cutis 66:348–352
50. Raychaudhuri SP, Jiang W-Y, Farber EM (1998) Psoriatic keratinocytes express high levels of nerve growth factor. Acta Derm Venereol 78:84–86
51. Raychaudhuri SP, Farber EM, Raychaudhuri SK (1999) Immunomodulatory effects of peptide T on Th 1/Th 2 cytokines. Int J Immunopharmacol 21:609–615
52. Raychaudhuri SP, Jiang WY, Farber EM, et al (1999) Upregulation of RANTES in psoriatic keratinocytes: a possible pathogenic mechanism for psoriasis. Acta Derm Venereol 79:9–11
53. Raychaudhuri SP, Farber EM, Raychaudhuri SK (2000) Role of nerve growth factor in RANTES expression by keratinocytes. Acta Derm Venereol 80:247–250
54. Raychaudhuri SP, Dutt S, Raychaudhuri SK, et al (2001) Severe combined immunodeficiency mouse-human skin chimeras: a unique animal model for the study of psoriasis and cutaneous inflammation. Br J Dermatol 144:931–939
55. Raychaudhuri SP, Sanyal M, Weltman H, et al (2004) K252a, a high-affinity nerve growth factor receptor blocker, improves psoriasis: an in vivo study using the severe combined immunodeficient mouse-human skin model. J Invest Dermatol 122:812–819
56. Raychaudhuri SP, Jiang WY, Raychaudhuri SK (2008) Revisiting the Koebner phenomenon: role of NGF and its receptor system in the pathogenesis of psoriasis. Am J Pathol 172:961–971
57. Ruff MR (1989) Peptide T. Drugs Future 14:1049–1051
58. Schall TJ (1991) Biology of the RANTES/SIS cytokine family. Cytokine 3:165–183
59. Scholzen TE, Steinhoff M, Bonaccorsi P, et al (2001) Neutral endopeptidase terminates substance P-induced inflammation in allergic contact dermatitis. J Immunol 166:1285–1291

60. Schwartz J, Pearson J, Johnson E (1982) Effect of exposure to anti-NGF on sensory neurons of adult rats and guinea pigs. Brain Res 244:378–381
61. Smith HC, Barker JN, Morris RW, et al (1993) Neuropeptides induce rapid expression of endothelial cell adhesion molecules and elicit granulocytic infiltration in human skin. J Immunol 151:3274–3282
62. Snider RM, Constantine JW, Lowe JA, et al (1991) A potent nonpeptide antagonist of the substance P (NK1) receptor. Science 251:435–437
63. Talme T, Lund-Rosell B, Sundquist KG, et al (1994) Peptide T: a new treatment for psoriasis? Acta Derm Venereol (Stockh) 186:76–78
64. Thorpe LW, Werrbach-Perez K, Perez-Polo JR (1987) Effects of nerve growth factor on the expression of IL-2 receptors on cultured human lymphocytes. Ann N Y Acad Sci 496:310–311
65. Valdimarsson H, Baker BS, Jonsdottir I, et al (1986) Psoriasis: a disease of abnormal keratinocyte proliferation induced by T lymphocytes. Immunol Today 7:256–259
66. Vaupel R, Jarry H, Schlomer HT, et al (1988) Differential response of substance P containing subtypes of adrenomedullary cells to different stressors. Endocrinology 123: 2140–2145
67. Wallengren J (1991) Substance P antagonist inhibits immediate and delayed type cutaneous hypersensitivity reactions. Br J Dermatol 124:324–328
68. Wallengren J, Moller H (1988) Some neuropeptides as modulators of experimental contact allergy. Contact Dermatitis 19:351–354
69. Wallengren J, Ekman R, Sunder F (1987) Occurrence and distribution of neuropeptides in human skin. An immunocytochemical and immunohistochemical study on normal skin and blister fluid from inflamed skin. Acta Derm Venereol (Stockh) 67:185–192
70. Waxman D (1973) Behavior therapy of psoriasis – a hypnoanalytic and counter conditioning technique. Postgrad Med J 49:591–595
71. Wetterberg L, Alexius B, Saaf J, et al (1987) Peptide T in treatment of AIDS (letter). Lancet 1:159
72. Wiesenfeld-Hallin Z, Xu Xj, Hakanson R, et al (1990) The specific antagonistic effect of intrathecal spantide II on substance P stimulation-induced facilitation of the nociceptive flexor reflex in rat. Brain Res 526:284–290
73. Wilkinson DI (1989) Mitogenic effect of substance P and CGRP on Keratinocytes. J Cell Biol 107:509a
74. Wilkinson DI, Theeuwes MI, Farber EM (1994) Nerve growth factor increases the mitogenicity of certain growth factors for cultured human keratinocytes: A comparison with epidermal growth factor. Exp Dermatol 3:239–245
75. Wrone-Smith T, Nickoloff BJ (1996) Dermal injection of immunocytes induces psoriasis. J Clin Invest 98:1878–1887
76. Wyatt S, Shooeter EM, Davies AM (1990) Expression of the NGF receptor gene in sensory neurons and their cutaneous targets prior to and during innervation. Neuron 2:421–427
77. Xu Xj, Hao JX, Wiesenfeld-Hallin Z, et al (1991) Spantide II, a novel tachykinin antagonist inhibits plasma extravasation induced by antidromic C-fiber stimulation in rat hind paw. Neuroscience 42:731–737

Neuroimmunology of Atopic Dermatitis

18

A. Steinhoff and M. Steinhoff

Contents

Key Features

- Sensory and autonomic nerves are capable of modulating vascular responses (white dermographism, inflammatory vasodilatation, and edema), immunomodulation of T-cells and mast cells as well as keratinocyte proliferation (eczema).
- Sensory nerves can receive signals from keratinocytes or endothelial cells after stimulation of the latter by inflammatory stimuli. This may result in activating or inhibitory signaling, depending on the stimulus.
- There is no doubt that the nervous system of the skin has an impact in the pathophysiology of AD.

Synonyms Box: cytokine: mediators mainly released by immune cells, neurotrophin: growth factors modulating nerve function, neuropeptide: peptides released by sensory neurons (rarely autonomic nerves), Interleukin: cytokines regulating immune responses, neurotransmitter: classical mediators of autonomic nerves (adrenaline, acetylcholine)

Abbreviations: *ACh* Acetylcholine, *AD* Atopic dermatitis, *BDNF* Brain-derived neurotrophic factor, *CB* Cannabinoid, *CGRP* Calcitonin gene related peptide, *CNS* Central nervous system, *Derf Dermatophagoides farinae*, *FK506* Immunosuppressant that prolongs allograft survival, *H4R* Histamine receptor-4, *H3R* Histamine receptor-3, *HMC-1* Human mast cell line, *HU210* Synthetic cannabinoid agonist, *IFN-ϕ* Interferon-γ, *IL* Interleukin, *MIF* Macrophage migration inhibitory factor, *MSH* Melanocyte-stimulating hormone, *NGF* Nerve growth factor, *NK1R* Neurokinin-1-receptor, *NT4/NT3* Neurotrophin-4/neurotrophin-3, *PACAP* Pituitary adenylate cyclase activating polypeptide, *PAR* Proteinase-activated receptor, *POMC* Pro-opio-melano-corticotropin, *RA* Rheumatoid arthritis, *RT PCR* Reverse transcription polymerase chain reaction, *SMS* Somatostatin-derived peptide, *SP* Substance P, *SST* Somatostatin, *TH* T-helper cell, *trk* Tropomyosin-related kinase receptors, *TRPV1* Transient receptor potential vanilloid type-1, *VIP* Vasoactive intestinal peptide

18.1 Introduction

Sensory and autonomic nerves play a role in the regulation of skin homeostasis and disease. A role for the nervous system in the pathophysiology of atopic dermatitis (AD) is also reflected by the fact that the term

R.D. Granstein and T.A. Luger (eds.), *Neuroimmunology of the Skin*,

"neurodermatitis" has a long and well-accepted history for this disease. A dysregulation of neuropeptides and neurotrophins has been described in different stages of AD. In AD, neurocutaneous interactions influence a variety of functions such as cell growth, immune defense, inflammation, and itching. This interaction is mediated by primary afferent sensory fibers as well as autonomic nerves releasing neuromediators, which activate specific receptors on many target cells during the inflammatory and pruritic process. A variety of receptors for neuropeptides, neurotrophins, and neurotransmitters as well as ion channels or cytokines are involved in the pathophysiology of AD. This neuroimmune interaction is controlled by endopeptidases, which are able to terminate neuropeptide-induced inflammatory or immune responses. This chapter discusses recent advances in our understanding of the role of nerves in the pathophysiology of AD and suggests that the peripheral as well as the central nervous system represents a crucial component of the inflammatory and pruritic amplification cycle in this disease. This knowledge may help to develop new strategies for the treatment of AD and sensitive skin with barrier dysfunction.

18.2 Neuroimmune Aspects of Atopic Dermatitis

In acute as well as chronic lesions of AD, an increased staining of cutaneous nerves or altered concentrations of neuropeptides have been observed [17,19,30,36,47, 51,78,82,85]. Furthermore, characteristics of the triple response (erythema, wheal, flare) of "neurogenic inflammation" as well as pruritus have been observed after injection of neuropeptides into human skin [23,24].

In patients with AD, tachykinin receptors have been detected on blood vessels and keratinocytes by autoradiography [43,72]. Additionally, NK1R expression on endothelial cells was diminished after UVA irradiation, whereas NK1R expression on keratinocytes was unchanged, indicating a differential regulation of this receptor in different target cells by UV light and during cutaneous inflammation. Interestingly, an altered expression pattern of neurokinin receptors was observed after UV-A irradiation in patients with AD. UVA irradiation seems to reduce skin inflammation through the modulation of NK 1R expression on endothelial cells [72].

The tachykinin substance P (SP) may have a modulatory effect on proliferation and cytokine mRNA expression in monocytes in response to dust mites (*Dermatophagoides farinae* Der f) in AD patients. Moreover, SP promoted Der f-induced proliferation and upregulated interleukin (IL)-10 mRNA expression while downregulating IL-5 mRNA expression. Of note, the proliferation in high responders was associated with the upregulation of IL-2 mRNA expression and the induction of IL-5 mRNA expression, indicating that SP modulated the immune responses of T cells towards Der f by promoting the proliferation and altering the cytokine profile in these patients. The authors concluded that these changes may control clinical manifestations of AD [93], especially when considering the elevated levels of SP in AD lesions [30,36].

Neuropeptides can exert different effects during inflammation. For example, SP and vasoactive intestinal peptide (VIP) have opposing effects on the release of TH_1- and TH_2-related cytokines in AD. Although SP increased both the TH_2 cytokine IL-4 and the TH_1 cytokine interferon (IFN)-γ, the release of both cytokines was inhibited by VIP in vitro. These data suggest that a clear understanding about the pathophysiological impact of the detected neuropeptides and neuropeptide receptors as modulators of the TH_1/TH_2 balance in atopic dermatitis is still missing.

VIP also influences the induction of pruritus. Intracutaneous injection of VIP led to dose-dependent pruritus as well as wheal and flare in normal and atopic skin. Furthermore, an increase in blood flow was measured after combined VIP and acetylcholine (ACh) administration in AD patients suffering from acute AD, whereas flare area and plasma extravasation were significantly reduced after single VIP and combined VIP and ACh injections, respectively [56,57]. Recent in vivo studies in human skin suggest that active vasodilatation depends solely on functional cholinergic fibres, but not on ACh itself [6]. Downregulation of VIP-receptor expression and elevated VIP serum levels in AD patients strongly support a role for this receptor in this disease [19,84].

The role of POMC-derived neuropeptides in the pathogenesis of AD is suggested by the observation that α-MSH modulates IgE production and the finding of increased levels of POMC-peptides in the skin of AD patients [59,60]. Recently, the role of β-endorphin in the pathophysiology of AD was investigated. It was found that the immunoreactivity for β-endorphin in keratinocytes and unmyelinated sensory nerve fibers was changed as compared to normal controls. β-endorphin-positive keratinocytes were clustered around μ-opioid receptor-positive nerves [8]. Moreover, the μ-opioid receptor expression was diminished in skin biopsies from patients with AD. This was due to that fact that the receptor was internalized after activation in keratinocytes of AD patients [9]. Because morphine and endorphins are known to be involved in the transmission of itch in sensory nerves, one may speculate that the endorphin/

μ-opioid receptor system may be implicated in inflammatory and pruritic processes of AD [70]. However, these types of preliminary studies support, but do not prove, a direct role for neuropeptides in AD. This has to await further clinical studies using neuropeptide inhibitors.

Recent studies suggest that neurotrophins may participate in the pathophysiology of AD. Nerve-growth factor (NGF) and its high- and low-affinity receptors tropomyosin-related kinase receptor (trk)A and trkB, respectively, are upregulated in the skin of AD patients (Fig. 18.1). Recent evidence indicates that NGF supports nerve-sprouting and may thereby modulate itch perception in inflamed skin as well as neurogenic inflammation.

In AD, an enhanced expression and release of NGF has been described in mast cells and keratinocytes, less so in fibroblasts [20]. Plasma levels of NGF were also increased in those patients. Interestingly, NGF induced the release of histamine and tryptase from a mast cell line (HMC-I). This finding, together with the known effects on keratinocytes, may lead one to speculate that NGF may regulate mast cell–nerve and keratinocyte–nerve interactions in the skin during AD.

In addition to their neurotrophic effects, neurotrophins can also be regarded as inflammatory cytokines. In particular, they exert a networking effect on neurons, classical cutaneous cells (keratinocytes, fibroblasts, Langerhans cells) as well as invading immune cells, which simultaneously operate as a source of, as well as a target for, neurotrophins during inflammation. Neurotrophins have also been verified to exert angiogenic and microvascular remodeling activities during allergic airway inflammation [45]. In human keratinocytes, neurotrophin (NT)-4-production was induced by IFN-γ in vitro and the expression was increased in AD [18]. Of note, the NT-4 protein accumulated in the epidermis while NT-3 staining was exclusively found in the dermal compartment, suggesting a differential role for specific neurotrophins in the pathophysiology of AD. However, NT-4 but not NT-3 expression was markedly increased in the IFN-γ-injected skin, suggesting an impact of NT-4 in the chronic phase of AD. In addition, prurigo nodularis lesions of AD patients also stained intensively for NT-4, indicating the involvement of this neurotrophin in AD and in prurigo.

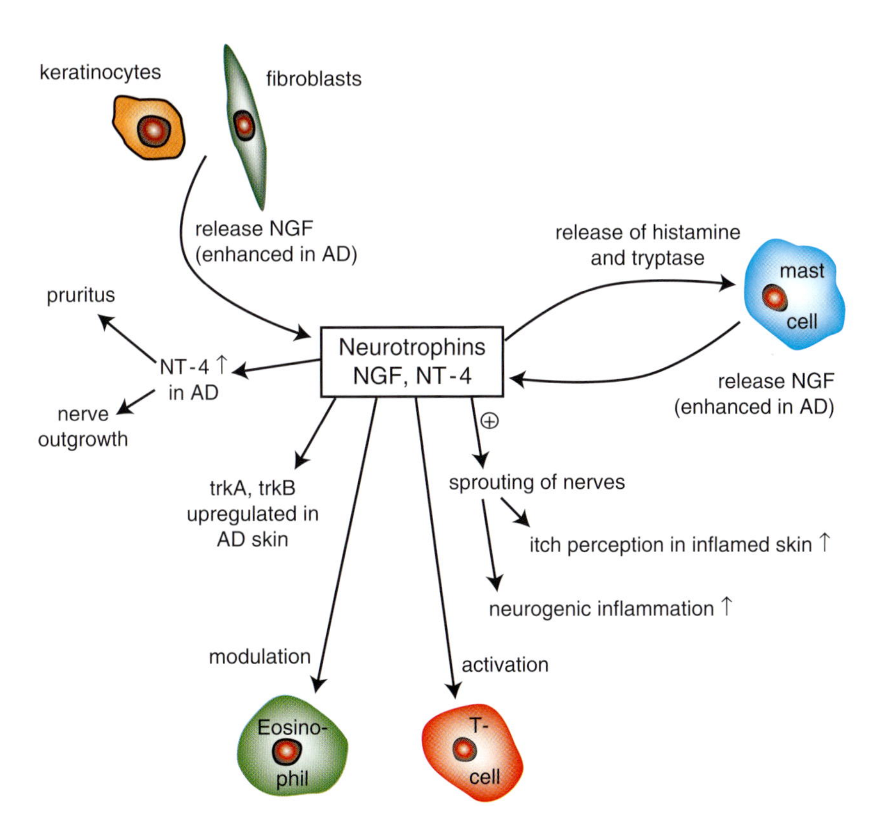

Fig. 18.1 Neurotrophins exert a networking effect on neurons, classical cutaneous cells (keratinocytes, fibroblasts, Langerhans cells) as well as invading immune cells which simultaneously operate as a source of, as well as a target for, neurotrophins during inflammation

18.3 Neurovascular Aspects of Atopic Dermatitis

The skin of AD patients is characterized by atypical vascular responses characterized by vasoconstriction instead of vasodilatation. This may be triggered by mechanical (white dermographism) or chemical (e.g., acetylcholine) stimuli. These patients show a pronounced and prolonged vasoconstriction after exposure to cold and a lower temperature of the body appendages. The underlying pathophysiological mechanisms of these phenomena, however, are still unclear. Sensory as well as autonomic nerves may be involved. In autonomic nerves, in addition, it is still under debate whether neuropeptides such as neurotensins or only classical neurotransmitters such as the beta-adrenergic receptor system are implicated in vascular events during AD [62].

Because of the close anatomical association between nerves and the dermal vasculature, a modulatory role of nerves on the recruitment of lymphocytes to the site of inflammation has been suggested. This may be supported by the finding that the relation of TH_1- to TH_2 cells in skin lesions of AD patients can be influenced by stress, for example, by neuronal manipulation of cytokines, chemokines, or cell adhesion molecules. In particular, mechanisms such as oxidative stress pathways, nerve–mast cell interactions as well as modulation of the synthesis and release of chemokines, amines, reactive oxygen products, glucocorticoids, macrophage migration-inhibitory factor (MIF), proteases, or neuropeptides delineate the complex mechanisms linking AD to "stress" molecules and pathways (Fig. 18.2). In sum, the vascular endothelium exerts a fundamental position for the control of acute and chronic inflammatory responses during AD.

The dermal vasculature is highly innervated and neuropeptides are differently regulated in nerves around vascular cells during AD (reviewed in [73]). This influences inflammation as well as pruritus, for example, by inducing the release of mediators from mast cells [IL-4, tumor necrosis factor (TNF)-α, histamine, tryptase]. As said, neurotrophins such as NGF or brain-derived neurotrophic factor (BDNF) regulate the function of mast cells as well as eosinophils as effector cells in AD [52]. Thus, a dysregulated interaction between the skin vasculature and sensory nerves are ultimately involved in pathogenic events during AD (Fig. 18.2).

Furthermore, AD patients suffer from exaggerated vasodilator responses to emotional stress with consequent pruritus and scratching. Of note, during vasodilatation the threshold for itch is decreased, the duration is prolonged, and the frequency of night-time scratching movements is dramatically enhanced, suggesting an interaction of nerves and vasculature in the pathophysiology of AD.

The neuropeptide galanin inhibits plasma extravasation. Galanin and galanin-binding sites, which are located around blood vessels [38] in rat skin, are upregulated and indicate the role of the galanin system in the cutaneous microvasculature [65].

On the other hand, the neuropeptide pituitary adenylate cyclase activating polypeptide (PACAP) is a potent vasodilatator [77], Of note, immunohistochemical as well as reverse transcription polymerase chain reaction (RT PCR) studies demonstrated a close association of PACAP-immunoreactive fibers with mast cells as well as dermal blood vessels in AD skin, supporting the idea of PACAP as a regulator of human vasculature in vivo.

18.4 Role of Nerves in the Pathophysiology of Pruritus in Atopic Dermatitis

Pruritus is the cardinal symptom of AD [88]. Anatomically, it is localized to the skin or mucosa and has a punctate, phasic quality. Although pruritus is experienced as a sensation arising in the skin, strictly speaking, it is also an extracutaneous event, because the central nervous system (CNS) is ultimately involved [48]. The intensity of itch we feel, for example, after an insect bite, represents a neuronal projection of a centrally formed sensation into defined regions of the integument (localized pruritus), or into large territories of our body surface (generalized pruritus) [48,89].

Clinically, pruritus has a high impact on the quality of life in AD. Interestingly, removal of the epidermis abolishes pruritus. However, the selective block of myelinated nerve fibers does not prevent itch sensation and, C-fibers, and probably certain subtypes of Aδ-fibers mediate chemical, thermal, and probably osmotic or electrical stimuli to the CNS. Accordingly, both the peripheral cutaneous and central nervous systems coordinate the sensation of itch, which results in the autonomic reflex of scratching, which is a chronic symptom of AD.

Neurophysiological data revealed that the excitation of neuropeptide-containing C-fibers transmits pruritus [63,64]. Various studies have shown that the number of sensory and neuropeptide-containing nerve fibers is increased in skin lesional of AD [69,85].

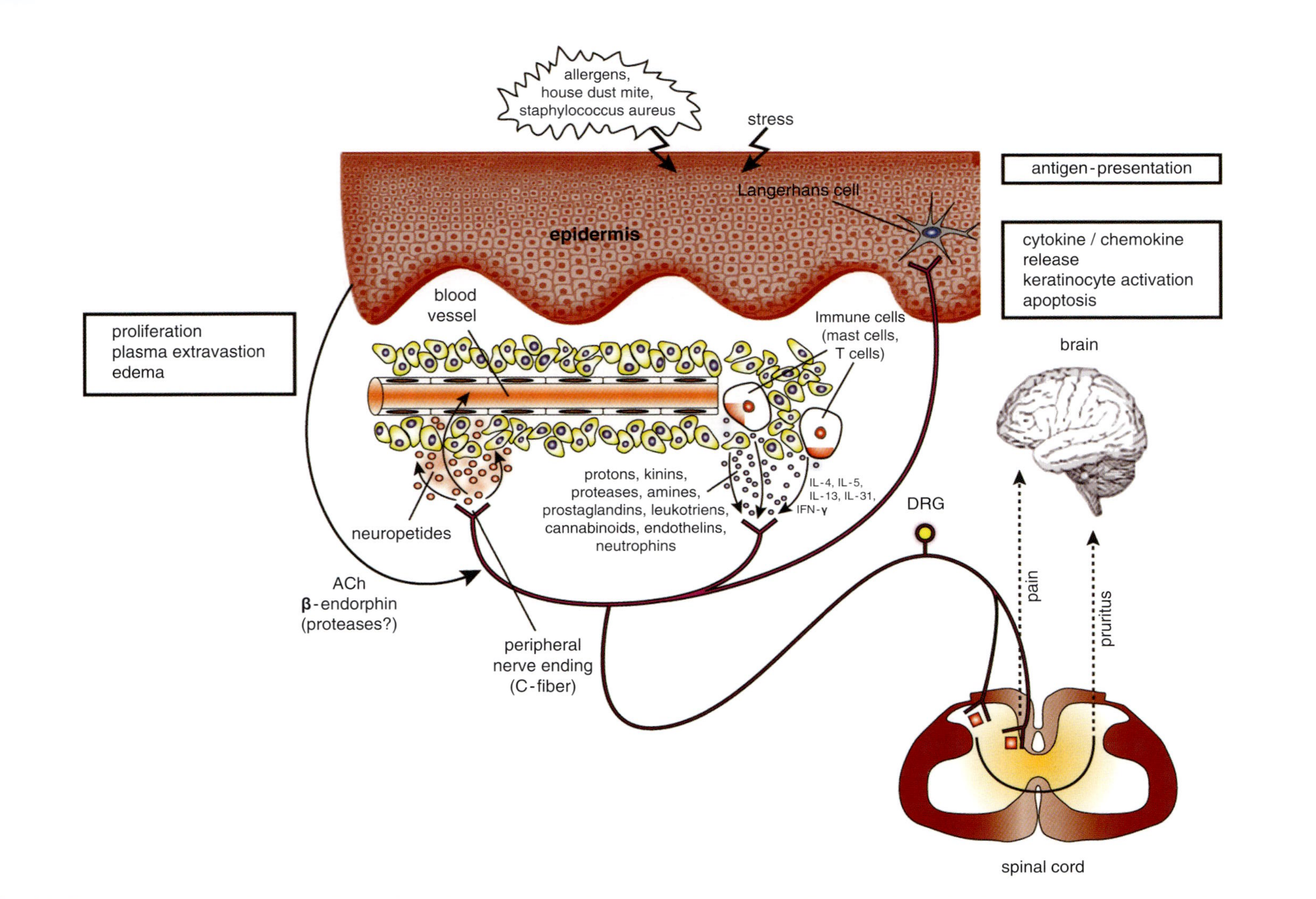

Fig. 18.2 Triggered by exogenous factors like allergens, house dust mites, or *staphylococcus aureus* as well as endogenous factors such as stress mediators are released by keratinocytes (ß-endorphin), endothelial cells, mast cells (tryptase, histamine, TNF-α, leukotriens), and T-lymphocytes (IL-31). These mediators bind to specific receptors (e.g., for histamine, prostaglandins, cannabinoids, proteases, pH-changes, heat, cold) on peripheral ends of primary afferent neurons. The first bodies of the afferent neurons are located in dorsal root ganglia before the neurites have their first synapses in the dorsal horn of the spinal cord and are connected to neurons of the same or collateral side, which transmit chemical or electrical signals to the CNS. To study the gene expression status or the release of neuropeptides from primary peripheral nerves, the cell bodies located in dorsal root ganglia are experimentally examined

A wide range of *peripheral* itch-inducing stimuli that are generated within or administered to the skin can trigger pruritus. An armada of mediators (Fig. 18.2) with amines (histamine, serotonin in rodents), prostanoids (prostaglandins, leukotrienes), kinins, kallikreins, proteases (tryptase), cytokines, protons, and others suffice to produce itching and/or edema and erythema upon stimulation. The crucial mediators involved in the pathophysiology of AD, however, are still unknown.

The most extensive substance studied so far in AD is histamine, which has a well established role in mediating pruritogenic effects in urticaria. Its role in AD is still uncertain. Recently, additional histamine receptors were cloned and characterized such as the histamine-receptor-4 (H4R). H4R is associated with the induction of itch in mice [5]. In addition, H3R is involved in scratching behavior of mice [26]. When positron emission tomography was used after histamine application, certain areas of the CNS were found to indeed participate in itch perception by humans [27]. Furthermore, patients with AD appear to have an altered histamine response and a decreased ability of sensory nerves to signal itching to the CNS [21]. The fact that antihistamines fail to eliminate itch in many other skin diseases (e.g., AD) suggests that other mediators and mechanisms are involved in this process. Intradermal injection of neuropeptides such as SP into human skin provoke itch, along with the characteristics of neurogenic inflammation, such as wheal and flare. These responses were inhibited by antihistamines and the mast cell stabilizer compound 48/80, indicating an involvement of mast cell mediators in this process. This is also the case for VIP, somatostatin, secretin, and neurotensin [22,56,87,88]. Interestingly, the potent vasodilatator CGRP does not stimulate histamine release from mast cells and does not mediate pruritogenic effects in humans [14]. However, there is evidence that SP-induced itch responses can be mediated by NK1R activation in mice [2], supporting a direct effect of SP in mediating pruritus in vivo.

Intracutaneous injection of both VIP and acetylcholine (ACh) induce wheal, flare, and a dose-dependent pruritus in healthy skin and in patients with AD. However, the subjective pruritus score did not differ between combined injections of VIP and ACh from ACh-injections alone [56]. This suggests a predominant role of ACh over VIP involved in the pathophysiology of pruritus in patients with AD. Moreover, a higher density of sensory nerve fibers with a larger diameter was observed by electron microscopy in lichenified lesions of AD [85]. In notalgia paresthetica, a neuropathia characterized by pruritus, pain, and hyperalgesia, immunostaining for several neuropeptides revealed that affected areas had a significant increase in intradermal nerve fibers as well as epidermal dendritic cells, suggesting that sensory nerve fibers are involved in the pathogenesis of this disease. Agents such as capsaicin, which deplete neuropeptides from sensory neurons, have been shown to have a therapeutic effect in diseases associated with pruritus and pain.

Intradermally injected morphine induced an itch response and low doses of β-endorphin or enkephalins intensify histamine-induced pruritus, although single doses of opioids at the same concentrations did not provoke pruritus [14,15]. Opioid μ-receptors seem to play an important role in the central neural mechanisms of itch sensation, because, in addition to analgesia, morphine provokes pruritus when applied intrathecally and the opioid antagonist naloxone is effective in abolishing or diminishing itch in mice, monkeys, and humans [7,80,81]. Thus, endogenous opioids may participate in the transmission or modulation of pruritic stimuli to the cortex, although the precise mechanisms and the central role of opioids for itch responses are still unknown. Opioids may be involved in the pathophysiology of cholestatic pruritus [33,34]. Importantly, effective treatment of pruritus varies amongst different itchy disorders. For example, in contrast with skin-derived pruritus, cholestatic pruritus can be significantly reduced by application of an antagonist to 5-hydroxytryptamine, although serotonin is synthesized in platelets and probably human melanocytes, not in human mast cells or nerve fibers [31].

Like opioids, cannabinoids have been the focus of pruritus research [37,49,76], partly because cannabinoid receptor-1 (CB1) and transient receptor potential vanilloid type-1 (TRPV1) are highly colocalized in small-diameter primary afferent neurons [37]. Moreover, CB1 agonists effectively suppressed histamine-induced pruritus in humans [13], suggesting involvement of CB1 and cannabinoids in mast-cell-dependent itching. Furthermore, under inflammatory conditions [67], endogenous cannabinoids such as anandamide are capable of activating and sensitizing TRPV1, thereby switching their neuronal effect from inhibition [1] to excitation and sensitization [86]. Finally, cannabinoid receptors are also constitutively expressed by human non-neuronal skin cells such as keratinocytes [41,71], and induce release of β-endorphin from murine keratinocytes [29]. Thus, cannabinoids may be involved in the neuronal–non-neuronal cellular network of pruritogenic and painful stimuli arising in or from skin. Consequently, coadministration of a TRPV1 agonist with a CB1 agonist would lead to an anti-pruritic response, and may prevent acute burning sensations induced by capsaicin stimulation, because CB agonists (e.g., anandamide, HU210) would prevent the excitation induced by capsaicin [54,58].

The idea that proteinases, for example, those from plants or bacteria, are involved in the induction of pruritus is rather old [66]. Papain and trypsins, as well as tryptase, are capable of inducing itch responses in humans. This may at least in part be mediated by the activation of PARs. Mast cells store large amounts of proteases including tryptase and chymase in their cytoplasm. Recently, a new mechanism was demonstrated where activated dermal mast cells localized in the proximity of afferent unmyelinated C-fibers directly induce pruritoceptive itch [74]. Via the activation of the G protein-coupled receptor proteinase-activated receptor-2 (PAR-2), tryptase transmits itch perception and in parallel mediates neuropeptide release, which in turn may activate mast cells through neurokinin recep tors [74]. Interestingly, PAR-2 was shown to be involved in the pathophysiology of itching in AD patients. Accordingly, the concentration of the ligand, tryptase, was also enhanced, suggesting a role for the tryptase-PAR-2 system in this context. Interestingly, the concentration of histamine was not enhanced in lesional skin of the patients [74]. These results were recently confirmed in experimentally induced murine itch models, using either PAR-2 agonists in mice or the trypase-inhibitor nafomastate [83].

Very recently, a new itch pathway was identified in AD that linked the inflammatory and pruritic response via the IL-31 pathway, because mice overexpressing IL-31 developed skin lesions and pruritus that were similar to this disease [12]. IL-31 expression is not only increased in patients with AD but also in those with ACD, 2 pruritic skin disorders. In both types of eczema, expression of IL-31 is associated with the expression of the TH2 cytokines IL-4 and IL-13. [44]. Moreover, the receptor for IL-31 and IL-31 RA was found to be highly expressed in a mouse model of AD [68]. This finding sheds a new light on the interaction between cytokines and peripheral nerves in pruritus, although the role of this interaction in humans is not completely understood as yet. Taken together, these findings show that IL-31 may represent a novel target for antipruritic drug development.

Unfortunately, most knowledge gained from animal experiments so far is not directly transferable to humans, as people may have different expression patterns of the cytokine in question or its receptor compared to animals. This is currently under investigation.

In summary, future studies with specific agonists and antagonists to neuropeptides and their receptors, endopeptidases, proteases, neurotrophins, or cytokines, as well as the pharmacological modulation of peripheral, spinal, or central neuronal mechanisms will be helpful to determine the role of sensory nerves and develop better treatments for pruritus in AD.

18.5 Therapeutic Consequences of Neuroimmune Interactions for Atopic Dermatitis

Since our understanding of the interactions of the skin and nervous system continues to expand, novel therapies will likely be developed to treat the inflammatory skin diseases that are mediated through neuroinflammation. Specific pharmacologic targets for the development of new agents will include the neuropeptides released in the skin, neuropeptide receptors expressed on target cells in the skin, proteases that degrade neuropeptides, agents that modify the function of vanilloid receptors, and growth factors that influence cutaneous innervation. These approaches promise the development of more specific therapies for a wide range of chronic, debilitating skin diseases.

Most clinical experience so far has been developed with the topical agent capsaicin. The usefulness of capsaicin has been demonstrated in several inflammatory diseases, including intestinal and airway inflammation or arthritis. Capsaicin was effective for the treatment of painful diabetic neuropathy [55], cold urticaria [28], AD [91,92], herpes infection, tumor pain, and different forms of pruritus [40]. Thus capsaicin and its analogs appear to be useful in reducing pain, pruritus, or neurogenic inflammation. Capsaicin was effective in some patients with ophthalmicus neuralgia due to herpes zoster [16,35,42]. Pretreatment of patients with IgE-mediated responses to capsaicin, for example, significantly reduced the flare response after histamine- or SP-injection but enhanced erythema responses after UV irradiation, tuberculin reaction, and contact dermatitis [90]. These contradictory skin reactions are probably due to more complex mechanisms of neuropeptide regulation and receptor stimulation in various diseases not yet completely understood. However, antagonists to the neurokinin-1 receptor (NK1R) may be beneficial for the treatment of AD because of their anti-inflammatory and antipruritic capacity. These findings also point out the need for further studies regarding the role of capsaicin as well as more specific and potent antagonists to TRPV1 in human skin diseases. This is important because the side effects of topical capsaicin treatment (such as burning sensations and hyperesthesia) currently limit its use. Furthermore, capsaicin treatment was ineffective in some patients [16]. One alternative may be resiniferatoxin, an ultrapotent analog of capsaicin (reviewed in [11]). Further, pharmacological and clinical studies and the development of new capsaicin analogs and synthetic capsaicin receptor ligands are indispensable to clarify the effectiveness of capsaicin as a therapeutic agent for skin diseases [4,10,32,39,79].

Other currently more experimental treatments for cutaneous neuroinflammation include the use of UV-irradiation. UVA irradiation, for instance, modulated the expression of tachykinin receptors in AD [72]. Topical treatment with the tricyclic antidepressant doxepin significantly inhibited histamine or SP-induced wheal and/or flare responses in patients with AD [61], suggesting a new therapeutic approach to mast cell- and neuropeptide-associated diseases. A major drawback of this therapy, however, is the sensitizing capacity of this drug, leading to allergic contact dermatitis. It also evoked sedative effects in some patients.

Cannabinoids have been shown to reduce hyperalgesia and neurogenic inflammation via interaction with cannabinoid-1 receptors and inhibition of neurosecretion (CGRP) from peripheral terminals of nociceptive primary afferent nerve fibers in rat hind paw [53,54]. Moreover, cannabinoid-2 (CB-2) receptor inhibition may be beneficial for the treatment of pain and itch [29].

Somatostatin (SST) and its receptors seem to play a regulatory, largely inhibitory, role in immune responses. A SST analogue peptide (SMS 201–995) enhanced the immunosuppressive effect of FK506 (tacrolimus) in rat spleen cells in vitro. Moreover, combined therapy with the SST analogue and FK506 at very low doses led to effective immunosuppression without any undesirable side effects, indicating a therapeutic synergistic effect of this neuropeptide analogue by decreasing the toxicity of other immunosuppressives [50]. This finding is in good agreement with the finding that neuronal SST, after stimulation by capsaicin, exerts anti-inflammatory effects in a murine model of experimentally induced contact dermatitis [3].

Recent results suggest the potential of antagonists to ligands of protease-activated receptors (PARs) as targets for anti-inflammatory therapy by downregulating inflammation as well as pruritus [25,46,75].

The finding that IL-31 is a cytokine that links the mechanism of inflammation and pruritus in AD opens up a new field for specifically targeting inflammatory mediators that are associated with the neuroimmune network in the skin. Thus, antagonists for IL-31 R may be beneficial for the treatment of inflammation and pruritus in AD patients.

In summary, in view of a key role in the skin for interactions between the nervous and immune systems through various mediators, the possibility of producing anti-inflammatory agents through activities against ligands or receptors will be an important task for research in dermatological therapy.

Acknowledgements

Supported by grants from the DFG (STE 1014/2–1), IZKF Münster (STEI2/027/06), SFB 293 (A14), SFB 492 (B13), IMF Münster, C.E.R.I.E.S., Paris; Serono, Germany; Rosacea foundation, Galderma, France, and the Novartis foundation (to M.S.).

Summary for the Clinician

There is no doubt that the nervous system of the skin has an impact in the pathophysiology of AD. Sensory as well as autonomic nerves are capable of modulating vascular responses (white dermographism, inflammatory vasodilatation, and edema), immunomodulation of T-cells and mast cells as well as the control of keratinocyte proliferation (eczema). Endogenous as well as exogenous trigger factors for AD activate epidermal as well as dermal nerves in the skin, thereby inducing and amplifying the itch response. AD can be induced, for example, by direct activation of TRPV1 receptors on sensory nerves by pH changes or peptides. Secondarily, sensory nerves can receive signals from keratinocytes or endothelial cells after stimulation of the latter by inflammatory stimuli. This may result in activating or inhibitory signaling, depending on the stimulus.

When stimulated, nerve fibers release neuromediators of different chemical origin, predominantly peptides, which target skin cells expressing specific neuropeptide receptors. Homeostasis is accomplished by peptidases, which degrade neuropeptides, and neurotrophins that influence innervation and receptor expression in the ganglia of primary afferent neurons. In addition, anti-inflammatory and/or analgetic (antipruritic) neuropeptides such as cannabinoids are able to reconstitute or establish homeostasis. Therefore, certain neuropeptides and transmitters can also have therapeutic effects in inflammatory, pruritic diseases such as AD. However, the precise ability of the cutaneous nervous system to regulate pro- and anti-inflammatory events as well as host defense in AD remains to be determined.

The use of modern techniques (molecular biology, nanotechnology, genomics, and proteomics) along with pharmacological approaches in animals and humans will lead to a deeper understanding of the neuroimmune system in the pathophysiology of AD and, probably, treatment.

References

1. Akerman S, Kaube H, Goadsby PJ (2004) Anandamide acts as a vasodilator of dural blood vessels in vivo by activating TRPV1 receptors. Br J Pharmacol 142:1354–1360
2. Andoh T, Nagasawa T, Satoh M, Kuraishi Y (1998) Substance P induction of itch-associated response mediated by cutaneous NK1 tachykinin receptors in mice. J Pharmacol Exp Ther 286:1140–1145
3. Banvolgyi A, Palinkas L, Berki T, Clark N, Grant AD, Helyes Z, Pozsgai G, Szolcsanyi J, Brain SD, Pinter E (2005) Evidence for a novel protective role of the vanilloid TRPV1 receptor in a cutaneous contact allergic dermatitis model. J Neuroimmunol 169:86–96
4. Barak LS, Oakley RH, Laporte SA, Caron MG (2001) Constitutive arrestin-mediated desensitization of a human vasopressin receptor mutant associated with nephrogenic diabetes insipidus. Proc Natl Acad Sci U S A 98:93–98
5. Bell JK, McQueen DS, Rees JL (2004) Involvement of histamine H4 and H1 receptors in scratching induced by histamine receptor agonists in Balb C mice. Br J Pharmacol 142:374–380
6. Bennett LA, Johnson JM, Stephens DP, Saad AR, Kellogg DL Jr. (2003) Evidence for a role for vasoactive intestinal peptide in active vasodilatation in the cutaneous vasculature of humans. J Physiol 552:223–232
7. Bergasa NV, Talbot TL, Alling DW, Schmitt JM, Walker EC, Baker BL, Korenman JC, Park Y, Hoofnagle JH, Jones EA (1992) A controlled trial of naloxone infusions for the pruritus of chronic cholestasis. Gastroenterology 102:544–549
8. Bigliardi-Qi M, Sumanovski LT, Buchner S, Rufli T, Bigliardi PL (2004) Mu-opiate receptor and Beta-endorphin expression in nerve endings and keratinocytes in human skin. Dermatology 209:183–189
9. Bigliardi-Qi M, Lipp B, Sumanovski LT, Buechner SA, Bigliardi PL (2005) Changes of epidermal mu-opiate receptor expression and nerve endings in chronic atopic dermatitis. Dermatology 210:91–99
10. Biro JC, Benyo B, Sansom C, Szlavecz A, Fordos G, Micsik T, Benyo Z (2003) A common periodic table of codons and amino acids. Biochem Biophys Res Commun 306:408–415
11. Biro T, Acs G, Acs P, Modarres S, Blumberg PM (1997) Recent advances in understanding of vanilloid receptors: a therapeutic target for treatment of pain and inflammation in skin. J Investig Dermatol Symp Proc 2:56–60
12. Dillon SR, Sprecher C, Hammond A, Bilsborough J, Rosenfeld-Franklin M, Presnell SR, Haugen HS, Maurer M, Harder B, Johnston J, Bort S, Mudri S, Kuijper JL, Bukowski T, Shea P, Dong DL, Dasovich M, Grant FJ, Lockwood L, Levin SD, LeCiel C, Waggie K, Day H, Topouzis S, Kramer J, Kuestner R, Chen Z, Foster D, Parrish-Novak J, Gross JA (2004) Interleukin 31, a cytokine produced by activated T cells, induces dermatitis in mice. Nat Immunol 5:752–760
13. Dvorak M, Watkinson A, McGlone F, Rukwied R (2003) Histamine induced responses are attenuated by a cannabinoid receptor agonist in human skin. Inflamm Res 52:238–245
14. Fjellner B, Hagermark O (1981) Studies on pruritogenic and histamine-releasing effects of some putative peptide neurotransmitters. Acta Derm Venereol 61:245–250
15. Fjellner B, Hagermark O (1982) Potentiation of histamine-induced itch and flare responses in human skin by the enkephalin analogue FK-33–824, beta-endorphin and morphine. Arch Dermatol Res 274:29–37
16. Frucht-Pery J, Feldman ST, Brown SI (1997) The use of capsaicin in herpes zoster ophthalmicus neuralgia. Acta Ophthalmol Scand 75:311–313
17. Glinski W, Brodecka H, Glinska-Ferenz M, Kowalski D (1994) Increased concentration of beta-endorphin in sera of patients with psoriasis and other inflammatory dermatoses. Br J Dermatol 131:260–264
18. Grewe M, Vogelsang K, Ruzicka T, Stege H, Krutmann J (2000) Neurotrophin-4 production by human epidermal keratinocytes: increased expression in atopic dermatitis. J Invest Dermatol 114:1108–1112
19. Groneberg DA, Welker P, Fischer TC, Dinh QT, Grutzkau A, Peiser C, Wahn U, Henz BM, Fischer A (2003) Down-regulation of vasoactive intestinal polypeptide receptor expression in atopic dermatitis. J Allergy Clin Immunol 111:1099–1105
20. Groneberg DA, Bester C, Grutzkau A, Serowka F, Fischer A, Henz BM, Welker P (2005) Mast cells and vasculature in atopic dermatitis – potential stimulus of neoangiogenesis. Allergy 60:90–97
21. Heyer G, Hornstein OP, Handwerker HO (1989) Skin reactions and itch sensation induced by epicutaneous histamine application in atopic dermatitis and controls. J Invest Dermatol 93:492–496
22. Heyer G, Ulmer FJ, Schmitz J, Handwerker HO (1995) Histamine-induced itch and alloknesis (itchy skin) in atopic eczema patients and controls. Acta Derm Venereol 75:348–352
23. Heyer G, Koppert W, Martus P, Handwerker HO (1998) Histamine and cutaneous nociception: histamine-induced responses in patients with atopic eczema, psoriasis and urticaria. Acta Derm Venereol 78:123–126
24. Heyer GR, Hornstein OP (1999) Recent studies of cutaneous nociception in atopic and non-atopic subjects. J Dermatol 26:77–86
25. Hollenberg MD, Compton SJ (2002) International Union of Pharmacology. XXVIII. Proteinase-activated receptors. Pharmacol Rev 54:203–217
26. Hossen MA, Sugimoto Y, Kayasuga R, Kamei C (2003) Involvement of histamine H3 receptors in scratching behaviour in mast cell-deficient mice. Br J Dermatol 149:17–22
27. Hsieh JC, Hagermark O, Stahle-Backdahl M, Ericson K, Eriksson L, Stone-Elander S, Ingvar M (1994) Urge to scratch represented in the human cerebral cortex during itch. J Neurophysiol 72:3004–3008

28. Husz S, Toth-Kasa I, Kiss M, Dobozy A (1994) Treatment of cold urticaria. Int J Dermatol 33:210–213
29. Ibrahim MM, Porreca F, Lai J, Albrecht PJ, Rice FL, Khodorova A, Davar G, Makriyannis A, Vanderah TW, Mata HP, Malan TP Jr. (2005) CB2 cannabinoid receptor activation produces antinociception by stimulating peripheral release of endogenous opioids. Proc Natl Acad Sci U S A 102:3093–3098
30. Jarvikallio A, Harvima IT, Naukkarinen A (2003) Mast cells, nerves and neuropeptides in atopic dermatitis and nummular eczema. Arch Dermatol Res 295:2–7
31. Johansson O, Liu PY, Bondesson L, Nordlind K, Olsson MJ, Lontz W, Verhofstad A, Liang Y, Gangi S (1998) A serotonin-like immunoreactivity is present in human cutaneous melanocytes. J Invest Dermatol 111:1010–1014
32. Johnson GD, Stevenson T, Ahn K (1999) Hydrolysis of peptide hormones by endothelin-converting enzyme-1. A comparison with neprilysin. J Biol Chem 274:4053–4058
33. Jones EA, Bergasa NV (1992) The pruritus of cholestasis and the opioid system [clinical conference]. JAMA 268: 3359–3362
34. Jones EA, Bergasa NV (1999) The pruritus of cholestasis [In Process Citation]. Hepatology 29:1003–1006
35. Kaminska R, Naukkarinen A, Horsmanheimo M, Harvima IT (1999) Suction blister formation in skin after acute and repeated mast cell degranulation. Acta Derm Venereol 79:191–194
36. Katsuno M, Aihara M, Kojima M, Osuna H, Hosoi J, Nakamura M, Toyoda M, Matsuda H, Ikezawa Z (2003) Neuropeptides concentrations in the skin of a murine (NC/Nga mice) model of atopic dermatitis. J Dermatol Sci 33:55–65
37. Klein TW (2005) Cannabinoid-based drugs as anti-inflammatory therapeutics. Nat Rev Immunol 5:400–411
38. Kofler B, Berger A, Santic R, Moritz K, Almer D, Tuechler C, Lang R, Emberger M, Klausegger A, Sperl W, Bauer JW (2004) Expression of neuropeptide galanin and galanin receptors in human skin. J Invest Dermatol 122:1050–1053
39. Lazar J, Szabo T, Kovacs L, Blumberg PM, Biro T (2003) Distinct features of recombinant rat vanilloid receptor-1 expressed in various expression systems. Cell Mol Life Sci 60:2228–2240
40. Lotti T, Teofoli P, Tsampau D (1994) Treatment of aquagenic pruritus with topical capsaicin cream. J Am Acad Dermatol 30:232–235
41. Maccarrone M, Di Rienzo M, Battista N, Gasperi V, Guerrieri P, Rossi A, Finazzi-Agro A (2003) The endocannabinoid system in human keratinocytes. Evidence that anandamide inhibits epidermal differentiation through CB1 receptor-dependent inhibition of protein kinase C, activation protein-1, and transglutaminase. J Biol Chem 278:33896–33903
42. Menke JJ, Heins JR (1999) Treatment of postherpetic neuralgia. J Am Pharm Assoc (Wash) 39:217–221
43. Misery L (1998) Langerhans cells in the neuro-immuno-cutaneous system. J Neuroimmunol 89:83–87
44. Neis MM, Peters B, Dreuw A, Wenzel J, Bieber T, Mauch C, Krieg T, Stanzel S, Heinrich PC, Merk HF, Bosio A, Baron JM, Hermanns HM (2006) Enhanced expression levels of IL-31 correlate with IL-4 and IL-13 in atopic and allergic contact dermatitis. J Allergy Clin Immunol 118:930–937
45. Nockher WA, Renz H (2006) Neurotrophins in allergic diseases: from neuronal growth factors to intercellular signaling molecules. J Allergy Clin Immunol 117:583–589
46. Ossovskaya VS, Bunnett NW (2004) Protease-activated receptors: contribution to physiology and disease. Physiol Rev 84:579–621
47. Ostlere LS, Cowen T, Rustin MH (1995) Neuropeptides in the skin of patients with atopic dermatitis. Clin Exp Dermatol 20:462–467
48. Paus R, Schmelz M, Biro T, Steinhoff M (2006) Frontiers in pruritus research: scratching the brain for more effective itch therapy. J Clin Invest 116:1174–1186
49. Paus R, Schmelz M, Biro T, Steinhoff M (2006) Scratching the brain for more effective "itch" therapy – Frontiers in pruritus research. J Clin Invest 116:1174–1186
50. Perego C, Lattuada D, Casnici C, Gatti S, Orsenigo R, Panagiotis S, Franco P, Marelli O (1998) Study of the immunosuppressive effect of SMS 201–995 and its synergic action with FK 506. Transplant Proc 30:2182–2184
51. Pincelli C, Fantini F, Massimi P, Girolomoni G, Seidenari S, Giannetti A (1990) Neuropeptides in skin from patients with atopic dermatitis: an immunohistochemical study. Br J Dermatol 122:745–750
52. Raap U, Kapp A (2005) Neuroimmunological findings in allergic skin diseases. Curr Opin Allergy Clin Immunol 5:419–424
53. Richardson JD, Aanonsen L, Hargreaves KM (1998) Antihyperalgesic effects of spinal cannabinoids. Eur J Pharmacol 345:145–153
54. Richardson JD, Kilo S, Hargreaves KM (1998) Cannabinoids reduce hyperalgesia and inflammation via interaction with peripheral CB1 receptors. Pain 75:111–119
55. Ross DR, Varipapa RJ (1989) Treatment of painful diabetic neuropathy with topical capsaicin [letter] [see comments]. N Engl J Med 321:474–475
56. Rukwied R, Heyer G (1998) Cutaneous reactions and sensations after intracutaneous injection of vasoactive intestinal polypeptide and acetylcholine in atopic eczema patients and healthy controls. Arch Dermatol Res 290:198–204
57. Rukwied R, Heyer G (1999) Administration of acetylcholine and vasoactive intestinal polypeptide to atopic eczema patients. Exp Dermatol 8:39–45
58. Rukwied R, Watkinson A, McGlone F, Dvorak M (2003) Cannabinoid agonists attenuate capsaicin-induced responses in human skin. Pain 102:283–288
59. Rupprecht M, Hornstein OP, Schluter D, Schafers HJ, Koch HU, Beck G, Rupprecht R (1995) Cortisol, corticotropin, and beta-endorphin responses to corticotropin-releasing hormone in patients with atopic eczema. Psychoneuroendocrinology 20:543–551
60. Rupprecht M, Salzer B, Raum B, Hornstein OP, Koch HU, Riederer P, Sofic E, Rupprecht R (1997) Physical stress-induced secretion of adrenal and pituitary hormones in patients with atopic eczema compared with normal controls. Exp Clin Endocrinol Diabetes 105:39–45
61. Sabroe RA, Kennedy CT, Archer CB (1997) The effects of topical doxepin on responses to histamine, substance P and prostaglandin E2 in human skin. Br J Dermatol 137:386–390
62. Schallreuter KU, Pittelkow MR, Swanson NN, Beazley WD, Korner C, Ehrke C, Buttner G (1997) Altered catecholamine

synthesis and degradation in the epidermis of patients with atopic eczema. Arch Dermatol Res 289:663–666
63. Schmelz M (2001) A neural pathway for itch. Nat Neurosci 4:9–10
64. Schmelz M, Schmidt R, Bickel A, Handwerker HO, Torebjork HE (1997) Specific C-receptors for itch in human skin. J Neurosci 17:8003–8008
65. Schmidhuber SM, Santic R, Tam CW, Bauer JW, Kofler B, Brain SD (2007) Galanin-like peptides exert potent vasoactive functions in vivo. J Invest Dermatol 127:716–721
66. Shelley WB, Arthur RP (1957) The neurohistology and neurophysiology of the itch sensation in man. AMA Arch Derm 76:296–323
67. Singh ME, McGregor IS, Mallet PE (2005) Repeated exposure to Delta(9)-tetrahydrocannabinol alters heroin-induced locomotor sensitisation and Fos-immunoreactivity. Neuropharmacology 49:1189–1200
68. Sonkoly E, Muller A, Lauerma AI, Pivarcsi A, Soto H, Kemeny L, Alenius H, Dieu-Nosjean MC, Meller S, Rieker J, Steinhoff M, Hoffmann TK, Ruzicka T, Zlotnik A, Homey B (2006) IL-31: a new link between T cells and pruritus in atopic skin inflammation. J Allergy Clin Immunol 117:411–417
69. Stander S, Steinhoff M (2002) Pathophysiology of pruritus in atopic dermatitis: an overview. Exp Dermatol 11:12–24
70. Stander S, Gunzer M, Metze D, Luger T, Steinhoff M (2002) Localization of micro-opioid receptor 1A on sensory nerve fibers in human skin. Regul Pept 110:75–83
71. Stander S, Schmelz M, Metze D, Luger T, Rukwied R (2005) Distribution of cannabinoid receptor 1 (CB1) and 2 (CB2) on sensory nerve fibers and adnexal structures in human skin. J Dermatol Sci 38:177–188
72. Staniek V, Liebich C, Vocks E, Odia SG, Doutremepuich JD, Ring J, Claudy A, Schmitt D, Misery L (1998) Modulation of cutaneous SP receptors in atopic dermatitis after UVA irradiation. Acta Derm Venereol 78:92–94
73. Steinhoff M, Stander S, Seeliger S, Ansel JC, Schmelz M, Luger T (2003) Modern aspects of cutaneous neurogenic inflammation. Arch Dermatol 139:1479–1488
74. Steinhoff M, Neisius U, Ikoma A, Fartasch M, Heyer G, Skov PS, Luger TA, Schmelz M (2003) Proteinase-activated receptor-2 mediates itch: a novel pathway for pruritus in human skin. J Neurosci 23:6176–6180
75. Steinhoff M, Buddenkotte J, Shpacovitch V, Rattenholl A, Moormann C, Vergnolle N, Luger TA, Hollenberg MD (2005) Proteinase-activated receptors: transducers of proteinase-mediated signaling in inflammation and immune response. Endocr Rev 26:1–43
76. Steinhoff M, Bienenstock J, Schmelz M, Maurer M, Wei E, Biro T (2006) Neurophysiological, neuroimmunological and neuroendocrine basis of pruritus. J Invest Dermatol 1268:1705–1718
77. Steinhoff M, Steinhoff A, Homey B, Luger TA, Schneider SW (2006) Role of vasculature in atopic dermatitis. J Allergy Clin Immunol 118:190–197
78. Sugiura H, Omoto M, Hirota Y, Danno K, Uehara M (1997) Density and fine structure of peripheral nerves in various skin lesions of atopic dermatitis. Arch Dermatol Res 289:125–131
79. Szolcsanyi J, Sandor Z, Petho G, Varga A, Bolcskei K, Almasi R, Riedl Z, Hajos G, Czeh G (2004) Direct evidence for activation and desensitization of the capsaicin receptor by *N*-oleoyldopamine on TRPV1-transfected cell, line in gene deleted mice and in the rat. Neurosci Lett 361:155–158
80. Thomas DA, Williams GM, Iwata K, Kenshalo DR Jr., Dubner R (1992) Effects of central administration of opioids on facial scratching in monkeys. Brain Res 585:315–317
81. Thomas DA, Oliveras JL, Maixner W, Dubner R (1992) Systemic morphine administration attenuates the perceived intensity of noxious heat in the monkey. Pain 49:129–135
82. Toyoda M, Morohashi M (1998) Morphological assessment of the effects of cyclosporin A on mast cell – nerve relationship in atopic dermatitis. Acta Derm Venereol 78:321–325
83. Ui H, Andoh T, Lee JB, Nojima H, Kuraishi Y (2006) Potent pruritogenic action of tryptase mediated by PAR-2 receptor and its involvement in anti-pruritic effect of nafamostat mesilate in mice. Eur J Pharmacol 530:172–178
84. Umemoto N, Kakurai M, Okazaki H, Kiyosawa T, Demitsu T, Nakagawa H (2003) Serum levels of vasoactive intestinal peptide are elevated in patients with atopic dermatitis. J Dermatol Sci 31:161–164
85. Urashima R, Mihara M (1998) Cutaneous nerves in atopic dermatitis. A histological, immunohistochemical and electron microscopic study. Virchows Arch 432:363–370
86. van der Stelt M, Trevisani M, Vellani V, De Petrocellis L, Schiano Moriello A, Campi B, McNaughton P, Geppetti P, Di Marzo V (2005) Anandamide acts as an intracellular messenger amplifying Ca2 + influx via TRPV1 channels. EMBO J 24:3026–3037
87. Wahlgren CF (1992) Pathophysiology of itching in urticaria and atopic dermatitis. Allergy 47:65–75
88. Wahlgren CF (1995) Measurement of itch. Semin Dermatol 14:277–284
89. Wallengren J (2005) Neuroanatomy and neurophysiology of itch. Dermatol Ther 18:292–303
90. Wallengren J, Moller H (1986) The effect of capsaicin on some experimental inflammations in human skin. Acta Derm Venereol 66:375–380
91. Weisshaar E, Forster C, Dotzer M, Heyer G (1997) Experimentally induced pruritus and cutaneous reactions with topical antihistamine and local analgesics in atopic eczema. Skin Pharmacol 10:183–190
92. Weisshaar E, Heyer G, Forster C, Handwerker HO (1998) Effect of topical capsaicin on the cutaneous reactions and itching to histamine in atopic eczema compared to healthy skin. Arch Dermatol Res 290:306–311
93. Yokote R, Yagi H, Furukawa F, Takigawa M (1998) Regulation of peripheral blood mononuclear cell responses to Dermatophagoides farinae by substance P in patients with atopic dermatitis. Arch Dermatol Res 290:191–197

Stress and Urticaria

19

M.A. Gupta

Contents

Key Features

- Underlying etiology of urticaria is not identifiable in about 70% of cases.
- Psychogenic factors are important among approximately 50% of urticaria patients.
- In the presence of a stressful situation, the patient may develop an urticarial rash as a result of an allergic response; the same allergen may not produce the rash when the stressor is removed.
- Childhood abuse and posttraumatic stress disorder may be associated with urticaria and/or angioedema when the patient is exposed to abuse–reminiscent stimuli.
- The effect of stress on uritcaria is most likely mediated by corticotropin releasing factor (CRH), which can increase mast cell degranulation. CRH is elevated during stress and in psychiatric disorders such as major depression and PTSD.

Synonyms Box:

Angioedema, similar to hives but involves swelling below the dermis (hives involve swelling in the superficial portion of the dermis); '*Body memories*'— body sensations or symptoms that may symbolically or literally capture some aspect of the trauma. They may be experienced during a traumatic flashback and may represent a conversion reaction. *Conditioned response*— a learned or acquired response to a previously neutral stimulus. For example, exposure to a certain food item (initially a neutral stimulus) in the presence of the unconditioned stimulus (eg., severe stress) that triggers an unconditioned response (eg. hives), may, at a later date, result in the development of hives when the patient is exposed to the same food item (now a conditioned stimulus).

19.1 Definition

Urticaria or hives has a spectrum of clinical presentations, ranging from occasional localized wheals to widespread recurrent whealing and angioedema [12]. Fifteen to twenty percent of individuals in the general population usually develop hives at some point in their lives. Urticaria is characterized by short-lived swelling of the skin, mouth, and/or genitalia due to transient leakage of plasma from small blood vessels into the surrounding connective tissue. Superficial

R.D. Granstein and T.A. Luger (eds.), *Neuroimmunology of the Skin*,

swellings of the dermis present as wheals; deeper swelling of the dermis and subcutaneous and/or submucosal tissue is called angioedema. The wheals are usually pruritic and often pale in the center because of intense edema, before maturing into pink superficial plaques that resolve over 2–24h, without leaving a mark on the skin unless the skin has been damaged by excoriation. Angioedema, in contrast to wheals, may be more painful than itchy, and generally takes longer time to resolve. In urticaria, the wheals and angioedema often coexist, but each may occur alone. Any pattern of recurrent urticaria occurring at least twice a week for 6 weeks is called chronic urticaria [12]. Therefore, by definition, all chronic urticaria goes through an acute (<6 weeks duration) phase. The diagnosis of urticaria is primarily clinical and additional screening tests often add little information to the diagnosis or underlying cause [12]. The importance of histamine as a mediator of urticaria has been recognized for many years; mast cell degranulation with release of histamine is central to the development of the wheals and angioedema [12]. However, the specific reason for this enhanced release of histamine and the underlying cause of the urticaria is often not clear. The underlying etiology of urticaria was not identifiable in 70% [13] to 79% [5] of cases and psychogenic factors were important in 48% [13] to 35.5% [5] of cases, respectively. It has been observed [5], however, that 'the role of psychological factors is extremely difficult to assess'. Czubalski and Rudzki [9] have examined psychosomatic factors in 48 patients with specific types of urticaria and reported exacerbation by emotional factors occurred in 77% of cholinergic urticaria and 82% with dermographism. About 81.8% of the 18 patients with dermographism had "abnormal" electroencephalograms [9]. Emotional factors did not exacerbate cold urticaria [9]. *Adrenergic urticaria* is a rare entity that is largely stress induced [22,48]; symptoms can be reproduced by intradermal injection of epinephrine and norepinephrin and treated with the beta-blocker propranolol [22]. Adrenergic urticaria may coexist with cholinergic urticaria [35]. There have been case reports of a relationship between angioedema and neurological symptoms [57], suggesting both focal and generalized central nervous system involvement.

19.2 Role of Psychosocial Stress

A large body of the literature on stress and urticaria is more than 25-years-old. Wittkower and Russell (1953) [65] classified urticaria under "Skin diseases in which psychosomatic factors are of particular importance"; Obermayer (1955) [37] under "Dermatoses in which emotional factors usually constitute an important component"; and Whitlock (1976) [63] has observed that "Urticaria is one of two skin conditions – atopic dermatitis being the other – which most frequently has been ascribed to emotional disturbances." Rook and Wilkinson (1979) [45] have categorized urticaria under "Dermatoses in which emotional predisposing, precipitating or perpetuating factors are often of importance." Medansky and Handler (1981) [32] classified urticaria under "the collaborative group" where organic causes and emotional disorders collaborate in different degrees to cause the skin disorder, and note that 'most experienced dermatologists are convinced that emotional factors play an important role in 25–50% of chronic urticaria cases'. Koblenzer (1983) [26] classified urticaria under "a condition in which strong psychogenic factors are imputed"; and Panconesi (1984) [40] has referred to urticaria as an "emotional allergy." He has described two forms of psychogenic urticaria [41]: (i) the *acute emotional form* that follows specific events with a clear cause–effect relationship; and (ii) the *chronic recurrent form* that "seems to be structured deeply on a psychodynamic basis as a psychosomatic disease". Panconesi further notes [41] that in specific immunologic tests, IgE-dependent induction is evident in only a minority (<20%) of cases. Some patients with chronic urticaria have been show to have lower serum levels of dehydroepiandrosterone sulfate (DHEA-S) during the active period of the disease [25]; patients with low serum DHEA-S reported greater psychological distress as exemplified by a lower sense of coherence and higher anxiety and depression scores [4]. DHEA-S and the DHEA-S/Cortisol ratio are important in an individual's response to stress and may play a significant role in modulating the vulnerability of the organism to the negative effects of stress [36].

19.2.1 Types of Stress

A wide range of psychological and stress ratings have been described in the literature [17], which largely reflects the most prevalent view of psychosomatic disorders at the time of the respective studies. Measures of psychological comorbidity range from assessment of emotional states and personality variables to psychodynamic formulations and diagnoses of comorbid psychiatric disorders. Various studies refer to *emotional* or *psychological* factors. Several studies have examined the association of stressful or catastrophic major life events and other stressful events with strong emotional valence for the patient;

some studies have examined traumatic life events such as abuse and/or neglect during childhood and later life. Finally, there is increasing recognition of the impact of urticaria upon the quality of life of the patient [2,15,53] – this disease related stress may further impact the course of urticaria. Psychiatric comorbidity can significantly increase quality of life impairment in chronic urticaria [53].

19.2.2 Personality Variables and Psychodynamic Formulations

Whitlock [63] has quoted some of the early work of Erasmus Wilson (1863), who mentioned "mental excitement, anxiety, and nervous debility" as a cause of urticaria, and Hillier (1865), who also felt that "nervous excitement was associated with the onset of urticaria." Warin and Champion [62] have quoted the work of Heinz Auspitz (1835–1886) who, influenced by the work of contemporary neurologists, placed urticaria with the *angioneuroses* to describe a local disturbance of the tonicity of blood vessels; and Ernst Schwimmer of Budapest (1837–1898) who called urticaria a "vasomotor sensory neurosis." Up to the 1930s, urticaria was described as a *vasomotor neurosis* [62]. There is extensive literature from prior to the 1960s, which has been discussed in some of the major reviews of psychosomatic dermatology [37,63,65].

A wide range of personality traits have been described in urticaria; however, overall, there is little evidence to substantiate a specific "hives personality" [63]. Stokes et al. [56] in a study involving 100 urticaria patients described "the urticariogenic personality" as "high tension, competitive personality, keyed to high pitch and perpetually intent on destination...." They reported "psychoneurogenous elements" in the backgrounds of 83.5% of urticaria patients vs. 24% of controls with psoriasis, acne, and impetigo [56]. Rees [44] studied 100 patients (76 urticaria only, 16 angioedema, and 8 both urticaria and angioedema) and described the "most common emotional precipitants" as "anxiety and tension" followed by "feelings of resentment and hostility," and "humiliation and embarrassment," while "tension arising from arousal of sex desires without satisfaction" was a precipitant in some women. These patients [44] had "more neurotic symptoms such as tension, anxiety, and depression than controls". In contrast, Wittkower [64] described the personality characteristics of 35 urticaria patients and observed that "these patients have in common an inability to tolerate the denial of love and a tendency to react angrily to any situations of this kind". He [64] observed that "situations which objectively or subjectively intensified either their need for affection or their sense of frustration precipitated the onset or recurrence of urticaria". Wittkower [64] further notes that "their angry reactions may be openly displayed, completely repressed, or expressed in the disguised form of urticarial lesions". Graham and Wolf [11] studied 30 cases of urticaria and reported that the patients had "a passive attitude towards punishment and failed to feel or express hostility." They concluded that "the vascular reactions in the disease are the same as those that occur after actual trauma to the skin" [11]. The above are examples of the wide-ranging and sometimes apparently discrepant personality types described by various authors. In view of this, Rees [44] comments that "the importance of personality type is not so much a matter of specificity but rather to the extent that it determines reaction to stresses, production of emotional tension, and the degree to which emotions are suppressed and inadequately expressed."

There are wide-ranging psychodynamic formulations, which are discussed in the earlier reviews [37,63,65]. Wittkower [64] refers to the role of "repressed aggressiveness"; "masochism" where, "instead of hurting, scratching, and attacking their love denying objects, these individuals scratch themselves and develop attacks of urticaria"; "repressed exhibitionism"; "skin erotism"; and "secondary gain". Shipman et al. [49] reached similar conclusions and observed that "patients were overtly submissive and pleading or very aggressive, and they all seemed to have strong underlying hostility, particularly masochistic trends". Shoemaker [50] described "the crucial affective state in an attack of urticaria" and observed "at the time the urticarial reaction occurs the patient appears to be hung dead center between the extremes of passivity and aggression, immobilized in his conflict betwixt abject dependence and destructive rage". A recent study [8] of 41 chronic idiopathic urticaria patients reported a correlation between difficulties with anger regulation and pruritus in urticaria.

19.2.3 Psychological Factors

Graham and Wolf (1950) [11] studied the stress-mediated pathogenesis of urticaria in 30 patients with recurrent urticaria of at least 3 weeks duration. Controls were hospital personnel without skin diseases. A detailed life history was obtained with particular reference to the setting in which the first and subsequent urticarial attacks occurred. In the experimental setting these issues were brought up with the "aim to arouse in the patient the same emotional state as that which had been associated with the event when they actually occurred" [11]. The skin temperature and the reactive hyperemia threshold were employed as measures of cutaneous vascular function during the mental interviews, "since they reflect the activity of the arterioles and the minute vessels, respectively". When examining the relationship between life situations and symptoms, Graham and Wolf [11] observed that the central theme in stressful situations that precipitated hives was a feeling of being wronged or injured, usually by someone in a close family relationship. The patients became "intensely resentful and the urticaria developed". Their assessment of vascular changes during stressful interviews revealed that "all patients were seen to flush... and 5 subjects had urticarial lesions while discussing their problems." The skin temperature and response to mechanical stimuli increased and the reactive hyperemia threshold decreased [11]. These changes indicate that "the essential difficulty is an increased tendency of both arterioles and minute vessels to dilatation". Graham and Wolf further comment that "the vessels behave as they would have if the person had actually been receiving blows." This dilating tendency was further exemplified by the increased sensitivity of the skin to locally applied histamine (23/24 patients vs. 6/25 controls) and pilocarpine (23/24 patients vs. 3/25 controls), with development of a wheal during emotionally neutral situations [11].

In a study involving 100 patients, Rees (1957) [44] observed that the concept of stress as an external force or stimulus is meaningful only if taken in conjunction with individual susceptibility and reaction. He noted that "the same force may be stressful to some and not to others and may be stressful at one time to the individual and not at another". Overall, psychological factors were considered to play a causal role among 68% of his patients – among 39% psychological factors played a "dominant" role while they played a "subsidiary role" among 29%. In 13% psychological factors were considered subsidiary to allergic and infective causes.

Green et al. (1965) [13] reviewed 236 cases of chronic urticaria at the Mayo Clinic (152 patients with urticaria; 14 with angioedema; 70 with both urticaria and angiodema at some time) and reported that an allergic or physical factor was determined among only 20 patients or about 8% of the total sample. The primary etiological factor was classified as "indeterminate" for 165/ 236 or 70% of the patients [13] and "psychogenic" for the remaining 22%. A further evaluation of other possible secondary contributory factors revealed psychogenic factors to be important in an additional 26% of the "indeterminate" group. Therefore, in total, psychogenic factors were assessed as being important among 48% or almost half of the patients with chronic urticaria. It is important to note that 113/236 patients were assessed as having an "atopic constitution" or a family and/or personal history of allergy; however, specific allergic factors were not identified as the primary etiology for their urticaria [13].

Champion et al. (1969) [5] examined 554 consecutive patients with urticaria and reported that in 79% the urticaria was of "unknown etiology." Psychological factors were considered to be the "main cause" among 11.5% and a "significant contributory cause" in a further 24% of cases [5], with a total of 35.5% of cases where psychological factors were considered to play an important role. Psychological factors were less frequently observed in children and men over 45 years; however, they were implicated in 44.5% of women over 15 years of age. In severe cases of urticaria, psychological factors were less frequent with a prevalence of 4%. When psychological factors were involved the total duration of disease was significantly longer by about 1 year [5].

Juhlin (1981) [23] evaluated 330 patients with recurrent urticaria using a questionnaire. Fifty men and 50 women had only urticaria and the rest had both urticaria and angioedema. All patients had recurrent attacks for more than 3 months. They [23] reported that "severe depression or psychiatric problems were mentioned by 16%," mainly women, and 7% reported that "stress and nervousness made their urticaria worse".

Steinhardt [54] studied two series of patients between 1942 and 1947 and between 1948 and 1953, and reported that psychogenic factors were responsible for chronic urticaria in 15% of the earlier series and 39.5% of the later series. He attributed this difference to the increase in awareness among physicians of the importance of psychological factors in urticaria. Badoux et al [1] showed that 74 patients with chronic idiopathic urticaria had psychopathologic ratings that were higher than healthy controls, similar to asthmatic patients, and lower than psychologically distressed patients. Depression scores were lower among 34 patients with chronic urticaria than among 34 patients with idiopathic generalized pruritus [47]. A study of 19 patients with chronic idiopathic urticaria [52] showed higher psychopathological scores for somatization, obsessive-compulsiveness, interpersonal sensitivity, depression, and anxiety compared to controls.

In a study of 296 dermatology patients with pruritic dermatoses, including 32 patients with chronic idiopathic urticaria, the severity of pruritus increased with the severity of depressive symptoms [19]. A recent study [38] of 84 patients with chronic idiopathic urticaria and 75 controls reported that 60% of the urticaria patients had a psychiatric diagnosis, the most frequent (40%) diagnosis being depressive disorders. Another study [60] of 89 patients with chronic idiopathic urticaria who were evaluated using structured clinical interviews, 49.4% had at least one major psychiatric disorder and 44.9% were diagnosed with a personality disorder. Among patients with a major psychiatric disorder 43.3% had an anxiety disorder and 13.5% had a major depressive disorder with 20.2% meeting the criteria for any mood disorder. The most commonly diagnosed personality disorders were obsessive –compulsive (30.3%) and avoidant (18%) personality. There are case reports of association of cold urticaria with hypomania during winter [21] and recurrent idiopathic urticaria with panic disorder [16].

19.2.4 Stress from Life Events

Rees [44] observed that the onset of urticaria was associated with stressful life situations in 51% in contrast to the control group of hernia and appendicitis, which were associated with stress in 8% of cases. In urticaria "the stress was sudden and severe in 11% and prolonged in 40%". In the case of "sudden stress," the onset of illness followed "bereavement of a loved person in 5 patients and illness or accident to a loved person in 3 patients". In the case of "prolonged stress" there were a variety of situations and stresses associated with family and marital problems, work and financial problems, "sex problems", "engagement difficulties" with periods of "humiliation and anxiety arising from objections by relatives", "humiliating experiences", and "situations evoking marked jealousy reaction". Lyketsos et al. [30] reported that a sample of 28 urticaria patients scored significantly higher in stress, as measured by the Social Readjustment Rating Scale, experienced in the year preceding the onset or exacerbation of their illness, than 38 inpatients with other skin diseases. Fava et al. [10] administered Paykel's interview for stressful life events to 20 patients with chronic urticaria and 20 patients with fungal infections who served as controls, and reported that 90% of the urticaria patients reported at least one life event before illness vs. 50% of controls; the total number of life events in the urticaria group was almost 3 times the number of events in the control group. In a study of 43 patients, Michaelsson [34] reported that "increased mental tension and fatigue" were spontaneously reported as the main exacerbating or precipitating factor by 77% of the sample. In another study of 53 patients, Teshima et al. [59] observed that a stressful event such as a change in school and residence, change in type of work, promotion, marriage, birth of a child, or emotional shock had occurred before the onset of urticaria in 30% of cases. Griesemer and Nadelson [14] reported an emotional trigger among 68% of urticaria patients. In aTaiwanese study of 75 patients with chronic idiopathic urticaria vs. 133 controls with tinea pedis, Yang et al. [66] reported that in the 6 months preceding disease onset, patients with urticaria had experienced significantly more life events, higher subjective weighting of impacts from life events, and more somatic symptoms such as insomnia. Insomnia is typically associated with stress and psychiatric pathology, and the high comorbidity of insomnia with urticaria is most likely an index of psychological distress in these patients. In a Turkish study [38] involving 84 patients with chronic idiopathic urticaria, 81% of patients believed that their disease was due to stress.

19.2.5 Traumatic and Catastrophic Life Events

Traumatic life events represent a more severe variety of stress where the human capacity to cope is overwhelmed [20]. There have been case studies of development of acute urticaria immediately after major traumatic events [33,43,46] such an earthquake [55], patient's favorite football (soccer) team performing poorly at a World Cup game [33], or acute realization by the patient that he had been the victim of business fraud [43]. Other case studies suggest that such fear-related psychophysiological responses may also manifest as acute angioedema [6] or urticaria, which may be mistakenly attributed to latex allergy [29] in the patient with a phobia of dental procedures. Teshima et al. [58] have shown increased plasmin activity during emotional stress in 2 patients with hereditary angioedema, which flared with emotional stress. In larger scale studies of catastrophic events such as earthquakes [3,39], it has been noted during the first 3 months after the event, most dermatologic complaints are infections and injuries related to the immediate damage to the infrastructure, while the stress-mediated dermatoses such as urticaria, eczema, etc. had a higher incidence 4–6 months post-earthquake.

19.2.6 Childhood Abuse and Developmental Trauma

Shoemaker (1963) [50] examined 40 urticaria patients and observed "some form of parental rejection" in their histories. He describes a very traumatic home environment such as "shifting placement in foster homes," "drab endurance of institutional anonymity," and "history of violently embattled parents matched numerically the incidence of actual separation from a parent by death, divorce, or desertion." Shoemaker [50] further observed that these patients, who were products of an embattled childhood, tended to be caught up in an equally embattled life situation as adults and their urticaria occurred within the context of these traumatic settings. The role of severe childhood trauma often associated with severe emotional, sexual, and physical abuse during early development is possibly under-recognized as an underlying factor in idiopathic urticaria. Individuals with childhood abuse are more vulnerable psychologically to the development of trauma-related syndromes such as post-traumatic stress disorder (PTSD) as adults. These individuals may therefore also be more predisposed to react somatically to traumatic stress. The following are three illustrative cases describing psychogenic urticaria in patients with histories of severe childhood abuse:

Case 1. This 36-year-old woman was referred for psychiatric evaluation because of recurrent angioedema involving her tongue and floor of her mouth that had necessitated several visits to the emergency department. Extensive dermatologic studies including immunologic workup were all negative. Psychiatric history revealed that the patient had been recurrently orally sexually abused by her stepfather as a child. The patient was diagnosed with PTSD. This stepfather had moved back to the patient's city recently and this had coincided with the development of angioedema. As the patient worked through her trauma issues and her dysregulated emotional state was managed with pharmacotherapy and psychotherapy, her angioedema was no longer a problem.

Case 2. This 45-year-old woman with a history of severe childhood physical and emotional abuse and neglect by her mother and sexual abuse by the father was diagnosed with PTSD at the age of 40 years after she reacted by impulsively hitting and injuring her husband. The patient described that the husband had made an advance that triggered memories of abuse by her father. The patient had a history of recurrent urticaria of recent onset for which she had been put on a tapering dose of oral corticosteroids on at least four occasions. Extensive dermatological and immunological workups were all negative. The patient was managed with mood stabilizing agents and psychotherapy and there has been no recurrence of her urticaria for over 1 year.

Case 3. This 38-year-old married woman with a recent diagnosis of PTSD secondary to severe childhood trauma, where she may have been physically beaten and sexually abused in a cult, presented with recurring linear welts on her left arm as she was undergoing therapy for her complex PTSD. The patient recalled that the linear welts had occurred in the same location where she had experienced trauma after being tied down and abused as a child in the cult. The urticarial reaction was diagnosed as a "body memory" or conversion reaction. After extensive therapy including eye movement desensitization and reprocessing (EMDR) [18], the patient no longer experienced the urticarial reaction.

Childhood maltreatment can result in classically conditioned associations between abusive stimuli and negative emotions [20,61]. Such traumatic memories are not remembered per se, and are evoked by events that are similar to the original abuse situation. Therefore, later in life, exposure to abuse-reminiscent stimuli and memories may produce strong memories and negative affects or the memories may be re-experienced as a somatic level as 'body memories'.

19.3 Urticaria as a Conditioned Response

There are cases where, in the presence of a stressful situation, the patient may develop an urticarial rash as a result of an allergic response to a food or drug, and the same food or drug may not be associated with an urticarial rash when the psychosocial stressor is removed. In such cases, "the patient's state of emotional tension may constitute an associated, if not dominant causative factor, and may raise or lower the threshold for an allergic response" [37]. An individual may develop stress induced hives as part of an unconditioned response to stress. Exposure to the stressful situation while eating an item of food (initially a neutral stimulus) may subsequently result in hives when the patient eats the same food item, as part of a conditioned response.

Rudzki et al. [46] have observed a significant placebo response in urticaria in nonhypnotized patients among whom 94% showed improvement in chronic urticaria with placebo instead of an antihistamine. There is extensive earlier literature on the effect of hypnotic suggestion on urticaria and immediate and delayed type hypersensitivity reactions. Obermayer [37] reviewed the case of a patient who was allergic to quinine. Urticaria could be produced when the suggestion of taking quinine was made under hypnosis; while actual administration of quinine during a hypnotic period, when it was suggested that the drug was not quinine, failed to produce urticaria [37]. However, controlled studies of hypnotic suggestion on the histamine-induced wheal in patients with urticaria and controls showed no significant effect of hypnosis [27], and more recent studies [28] have failed to demonstrate the effect of hypnosis on delayed-type hypersensitivity reactions.

19.4 Neuroimmunology

The possible neuroimmunological mechanisms underlying the effect of stress on urticaria have been discussed in depth in earlier chapters. The skin appears to have the equivalent of a hypothalamic–pituitary–adrenal (HPA) axis, including local expression of corticotropin-releasing hormone (CRH) and its receptors (CRH-R) [42]. Acute stress and intradermal administration of CRH stimulate skin mast cells and increase vascular permeability through CRH-R1 activation. Acute stress increases skin CRH that can trigger mast cell-dependent vascular permeability [31]. Furthermore, psychiatric disorders such as major depression and PTSD have been associated with increased levels of CRH [7,24,51]. The effect of CRH on mast cell degranulation is one of the important mechanisms underlying stress-mediated symptom exacerbation in urticaria.

Summary for the Clinician

There is extensive literature dating back to the 1800s supporting the role of stress and psychological factors in urticaria, and most authors report that stress plays an important role in 25–50% of cases. The placebo response in urticaria has been reported to be over 90% in one controlled study. The impact of urticaria upon the quality of life can also be a source of significant stress for the patient. Urticaria or angioedema may develop as an allergic response to a food or drug in the presence of a stressful situation, and the same allergen may not produce urticaria once the stress is removed. Recent neuroimmunologic studies indicate that stress and psychiatric disorders, such as major depression and post-traumatic stress disorder are associated with increased levels of CRH, which can trigger mast cell degranulation and increased vascular permeability, culminating in urticaria. These findings suggest that psychological factors should be assessed in all cases of urticaria, even in the minority of patients where a specific cause has been identified, as even in these cases stress can modulate the course of the disease.

References

1. Badoux A, Levy DA (1994) Psychologic symptoms in asthma and chronic urticaria. Ann Allergy 72:229–234
2. Baiardini I, Giardini A, Pasquali M, et al (2003) Quality of life and patients' satisfaction in chronic urticaria and respiratory allergy. Allergy 58:621–623
3. Bayramgurler D, Bilen N, Namli S, et al (2002) The effects of 17 August Marmara earthquake on patient admittances to our dermatology department. J Eur Acad Dermatol Venereol 16:249–252
4. Brzoza Z, Kasperska-Zajac A, Badura-Brzoza K, et al (2008) Decline in dehydroepiandrosterone sulfate observed in chronic urticaria is associated with psychological distress. Psychosomatic Medicine 70(6): 723–728
5. Champion RH, Roberts SOB, Carpenter RG et al (1969) Urticaria and angio-oedema. A review of 554 patients. Br J Dermatol 81:588–597
6. Chue PWY (1976) Acute angioneurotic edema of the lips and tongue due to emotional stress. Oral Surg 41:734–738
7. Claes SJ (2004) Corticotropin-releasing hormone (CRH) in psychiatry: from stress to psychopathology. Ann Med. 36:50–61
8. Conrad R, Gieser F, Haidl G, et al (2008) Relationship between anger and pruritus perception in patients with chronic idiopathic urticaria and psoriasis. J Eur Acad Derm Venereol [March 18, 2008 Epub ahead of print]
9. Czubalski K, Rudzki E (1977) Neuropsychic factors in physical urticaria. Dermatologica 154:1–4
10. Fava GA, Perini GI, Santonastaso P, et al (1980) Life events and psychological distress in dermatologic disorders: psoriasis, chronic urticaria and fungal infections. Br J Med Psychol 53:277–282
11. Graham DT, Wolf S (1950) Pathogenesis of urticaria: experimental study of life situations, emotions and cutaneous vascular reactions. JAMA 143:1396–1402
12. Grattan CEH, Sabroe RA, Greaves MW (2002) Chronic urticaria. J Am Acad Dermatol 46:645–657
13. Green GR, Koelsche GA, Kierland RR (1965) Etiology and pathogenesis of chronic urticaria. Ann Allergy 23:30–36
14. Griesemer RD, Nadelson T (1979) Emotional aspects of cutaneous disease. In: Fitzpatrick IM, Eisen AZ, Wolff K, et al (eds) Dermatology in General Medicine. McGraw-Hill, New York
15. Grob JJ, Revuz J, Ortonne JP, et al (2005) Comparative study of the impact of chronic urticaria, psoriasis and atopic dermatitis on the quality of life. Br J Dermatol 152:289–295
16. Gupta MA, Gupta AK (1995) Severe recurrent urticaria associated with panic disorder: a syndrome responsive to

selective serotonin reuptake inhibitor (SSRI) antidepressants? Cutis 56:53–54
17. Gupta MA, Gupta AK (1996) Psychodermatology: an update. J Am Acad Dermatol 34:1030–1046
18. Gupta MA, Gupta AK (2002) Use of eye movement desensitization and reprocessing (EMDR) in the treatment of dermatologic disorders. J Cut Med Surg 6:415–421
19. Gupta MA, Gupta AK, Schork NJ, et al (1994) Depression modulates pruritus perception: a study of pruritus in psoriasis, atopic dermatitis and chronic idiopathic urticaria. Psychosom Med 56:36–40
20. Gupta MA, Lanius RA, Van der Kolk BA (2005) Psychologic trauma, posttraumatic stress disorder and dermatology. Dermatol Clin 23:649–656
21. Harvey NS, Mikhail WI (1986) Seasonal hypomania in a patient with cold urticaria. Br J Psychiatry 149:238–241
22. Haustein UF (1990) Adrenergic urticaria and adrenergic pruritus. Arch Derm Venereol (Stockh)70:82–84
23. Juhlin L (1981) Recurrent urticaria: clinical investigations of 330 patients. Br J Dermatol 104:369–381
24. Kasckow JW, Baker D, Geracioti TD Jr. (2001) Corticotropin-releasing hormone in depression and posttraumatic stress disorder. Peptides 22:845–851
25. Kasperska- Zajac A, Brzoza Z, Rogala B (2008) Lower serum dehydroepiandroterone sulfate concentration in chronic idiopathic urticaria: a secondary transient phenomenon? Br J Dermatol June 27 [Epub ahead of print]
26. Koblenzer CS (1983) Psychosomatic concepts in dermatology. A dermatologist- psychoanalyst's viewpoint. Arch Dermatol 119:501–512
27. Levine MI, Geer JH, Kost PF (1966) Hypnotic suggestion and the histamine wheal. J Allergy 37:246–250
28. Locke SE, Ransil BJ, Zachariae R, et al (1994). Effect of hypnotic suggestion on the delayed-type hypersensitivity response. JAMA 272:47–52
29. Longley AJ, Fiset L, Getz T, et al (1994). Fear can mimic latex allergy in patients with dental phobia. General Dentistry 42:236–240
30. Lyketsos GC, Stratigos GC, Tawil G, et al (1985) Hostile personality characteristics, dysthymic states and neurotic symptoms in urticaria, psoriasis and alopecia. Psychother Psychosom 44:122–131
31. Lytinas M, Kempuraj D, Huang M, et al. (2003) Acute stress results in skin corticotropin-releasing hormone secretion, mast cell activation and vascular permeability, an effect mimicked by intradermal corticotropin-releasing hormone and inhibited by histamine-1 receptor antagonists. Int Arch Allergy Immunol 130:224–231
32. MedanskyRH,HandlerRM(1981)Dermatopsychosomatics: classification, physiology and therapeutic approaches. J Am Acad Dermatol 5:125–136
33. Merry P (1987) World Cup urticaria. J Royal Soc Med 80:779
34. Michaelsson G (1969) Chronic urticaria. Acta Derm Venereol 49:404–416
35. Mihara S, Hide M (2008) Adrenergic urticaria in a patient with cholinergic urticaria. Br J Dermatol 158: 629–630
36. Morgan III CA, Southwick S, Hazlett G et al (2004) Relationships among plasma dehydroepiandrosterone sulfate and cortisol levels, symptoms of dissociation, and objective performance in humans exposed to acute stress. Arch Gen Psych 61: 819–825
37. Obermayer ME (1955) Psychocutaneous Medicine. Charles C Thomas Publishers, Springfield, IL, pp 266–278
38. Ozkan M, Oflaz SB, Kocaman N et al (2007) Psychiatric morbidity and quality of life in patients with chronic idiopathic urticaria. Ann Allergy Asthma Immunol 99: 29–33
39. Oztas MO, Onder M, Oztas P, et al (2000) Early skin problems after Duzce earthquake. Int J Dermatol 39: 952–958
40. Panconesi E (1984) Psychosomatic dermatology. Clin Dermatol 2:94–179
41. Panconesi E, Hautmann G (1996) Psychophysiology of stress in dermatology. Dermatol Clin 14:399–421
42. Papadopoulou N, Kalogeromitros D, Staurianeas NG, et al (2005) Corticotropin-releasing hormone receptor-1 and histidine decarboxylase expression in chronic urticaria. J Invest Dermatol. 125(5):952–955
43. Pistiner M, Pitlik S, Rosenfeld J (1979) Psychogenic urticaria. Lancet 2:1383
44. Rees 1 (1957) An aetiological study of urticaria and angioneurotic oedema. J Psychosom Res 2:172–189
45. Rook A, Wilkinson DS (1979) Psychocutaneous disorders. In: Rook A, Wilkinson DS, Ebling FJG(eds) Textbook of Dermatology, Vol. III, 3rd ed., Blackwell Scientific Publications, Oxford, pp 2023–2035
46. Rudzki E, Borkowski W, Czubalski K (1970) The suggestive effect of placebo on the intensity of chronic urticaria. Acta Allergol 25:70–73
47. Sheehan-Dare RA, Henderson MJ, Cotterill JA (1990) Anxiety and depression patients with chronic urticaria and generalized pruritus. 123:769–774
48. Shelley WB, Shelley ED (1985). Adrenergic urticaria: a new form of stress induced hives. Lancet 2:1031–1033
49. Shipman WG, Shoemaker RJ, Levine MI, et al (1959) The problem of psychosomatic specificity in chronic urticaria: a case report. J Clin Psychol (1959) 15(2):133–136
50. Shoemaker RJ (1963) A search for the affective determinants of chronic urticaria. Psychosomatics 4:125–132
51. Smith MA, Davidson J, Ritchie JC, et al (1989) The corticotropin-releasing hormone test in patients with posttraumatic stress disorder. Biol Psychiatr 26(4):349–355
52 Sperber J, Shaw J, Bruce S (1989) Psychological components and the role of adjunct interventions in chronic idiopathic urticaria. Psychother Psychsom 51:135–141
53. Staubach P, Eckhardt- Henn A, Dechene M (2006). Quality of life in patients with chronic urticaria is differentially impaired and determined by psychiatric comorbidity. Br J Dermatol 154:294–298
54. Steinhardt MJ (1954) Urticaria and angioedema: statistical survey of five hundred cases. Ann Allergy 12(6):659–670
55. Stewart JH, Goodman MM (1989) Earthquake urticaria. Cutis 43:340
56. Stokes JH, Kulchar GV, Pillsbury DM (1935) Effect on the skin of emotional and nervous states: etiological background of urticaria with special reference to the psychoneurogenous factor. Arch Dermatol Syphilol 31:470–480
57. Sunder TR, Balsam MJ, Vengrow MI (1982). Neurological manifestations of angioedema. JAMA 247:2005–2007
58. Teshima H, Inoue S, Ago Y, et al (1974) Plasminic activity and emotional stress. Psychother Psychosom 23:218–228

59. Teshima H, Kubo C, Kihara H, et al (1982) Psychosomatic aspects of skin disease from the standpoint of immunology. Psychother Psychosom 37:165–175
60. Uguz F, Engin B, Yilmaz E (2008) Axis I and Axis II diagnoses in patients with chronic idiopathic urticaria. J Psychosom Res 64: 225–229
61. Van der Kolk BA, Fisler R (1995) Dissociation and the fragmentary nature of traumatic memories: overview and exploratory study. J Traumatic Stress 8:505–525
62. Warin RP, Champion RH, Urticaria (1974) In: Rook A (ed)Major Problems in Dermatology, Vol. 1, W. B. Saunders, London
63. Whitlock FA (1976) Psychophysiological aspects of skin disease. W B Saunders, London, pp154–164
64. Wittkower ED (1953) Studies of the personality of patients suffering from urticaria. Psychosom Med 15(2): 116–126
65. Wittkower E, Russell B (1953) Emotional factors in skin disease. A Psychosomatic Medicine Monograph. Paul B. Hoeber Inc, New York, pp 133–150
66. Yang HY, Sun CC, Wu YC, et al (2005) Stress, insomnia and chronic idiopathic urticaria. J Formos Med Assoc 104:254–263

Acne Vulgaris and Rosacea

20

C.C. Zouboulis

Contents

Key Features

- There is increasing evidence that the nervous system (for example emotional stress) can influence the course of acne.
- Neuropeptides are expressed in the skin, in which they exhibit a number of immunomodulatory influences on cellular differentiation.
- Substance P, ectopeptidases, the CRH system, α-MSH and endogenous opioids exhibit activity on the pilosebaceous unit.
- Flushing or transient erythema in rosacea is controlled by humoral substances and neural stimuli.
- The major neuropeptides probably involved in rosacea include substance P, VIP and CRH.
- Ultraviolet radiation is an initial stress factor for the development of inflammatory lesions in rosacea.

20.1 Acne Vulgaris

Synonyms Box: ACME (= ακμή)

20.1.1 Introduction

The possibility of a causative influence of emotional stress, especially of stressful life events, on the course of acne has long been postulated [47,56]. Clinical wisdom and experience [8], as well as many anecdotal observations [4,30,45] and uncontrolled case series [71], support this opinion [118]. Using a self-administered questionnaire sent to 4,000 adult women aged 25–40 years, Poli et al. reported that stress was recorded as the cause of acne in 50% of their patients [71]. However, the role of stressful events in the triggering or exacerbation of acne has not been explored in detail [67]. Only one current study met the acceptable methodological standards for stress measurement: Chiu et al., in a prospective study of 22 university students, found that patients with acne may experience a worsening of their disease during examinations [18]. Changes in acne severity correlated highly with increasing stress, suggesting that emotional stress from external sources may have a significant influence on acne.

R.D. Granstein and T.A. Luger (eds.), *Neuroimmunology of the Skin*,

20.1.2 Neuropeptides and Acne

Acne is a chronic inflammatory, exclusively human disease of the pilosebaceous unit, mostly affecting the sebaceous gland follicles – usually referred to as sebaceous follicles – located on the face, chest, shoulders and back, where they are most common. The aetiology of acne is not yet fully clarified but it is widely accepted that its pathogenesis is multi-factorial, with abnormal follicular differentiation and increased cornification, enhanced sebaceous gland activity and hyperseborrhea, bacterial hypercolonization as well as inflammation and immunological host reaction being the major contributors.

Ongoing research is modifying the classical view of acne pathogenesis. The beginning of micro-comedone formation is associated with aberrant differentiation of the follicular epithelium [50,64], vascular endothelial cell activation and inflammatory events [42], which supports the hypothesis that acne may represent a genuine inflammatory disease [116,122]. The identification of upstream mechanisms leading to the characteristic clinical lesions, the comedones, which also seem to represent a sign of follicular inflammation, supports this hypothesis. A hereditary background, androgens, skin lipids, inflammatory signalling and regulatory neuropeptides seem to be involved in this multi-factorial process [122]. As reported elsewhere, communication and reciprocal regulation between nervous, endocrine and immune systems are essential for biological stability and responses to external and internal challenges [15]. In particular, neuropeptides, hormones and cytokines act as signalling molecules that mediate communication between the three interacting systems. Analogous to central responses to stress, which involve predominantly the hypothalamic–pituitary–adrenal (HPA) axis, it has been proposed that the skin may share similar mediators [2,57,90]. Neuropeptides, originally described in central nervous tissue [19], are also expressed in the skin, in which they exhibit a number of immunomodulatory influences on cellular differentiation [11,89,91,115,117]. Associations among the central nervous system, the autonomous nervous system, the endocrine glands and the peripheral organs (especially the sebaceous glands, the sweat glands, hair and vessels) were proposed as early as in 1964 by Stüttgen [92]. Since then, receptors for the following neuropeptides have been reported in human sebaceous glands: corticotropin-releasing hormone [26,121], α-melanocyte stimulating hormone [13,27,96], β-endorphin [55], vasoactive intestinal peptide (VIP), neuropeptide Y and calcitonin gene-related peptide (CGRP) [84].

20.1.3 Role of Neuropeptides for Regulation of Clinical Inflammation in Acne

There is current evidence that regulatory neuropeptides with hormonal and non-hormonal activity may control the development of clinical inflammation in acne. Numerous substance P immunoreactive nerve fibres were detected in close apposition to the sebaceous glands and expression of the substance P-inactivating enzyme neutral endopeptidase was observed within sebaceous germinative cells of acne patients [102]. In vitro experiments using an organ culture system demonstrated that substance P induced expression of neutral endopeptidase in sebaceous glands in a dose-dependent manner. Neutral endopeptidase belongs to the group of ectopeptidases, which are expressed in the sebaceous glands and may be involved in the development and/or resolution of acne lesions [98]. On the other hand, treatment of sebocytes with interleukin (IL)-1β, which resulted in marked increase of IL-8 release [121], was partially blocked by co-incubation of the cells with α-melanocyte-stimulating hormone in a dose-dependent manner [13]. IL-8 and IL-6 are involved in the development of acne lesions [3] and their release by sebocytes can also be induced by corticotropin-releasing hormone by an IL-1β-independent pathway [48]. The latter neuropeptide also induces the synthesis of sebaceous lipids in vitro [48,121]. Adrenocorticotropic hormone evokes adrenal dehydroepiandrosterone to regulate skin inflammation [2]. These current findings indicate that central [91] or topical 'stress' [2,117] (Fig. 20.1) may, indeed, influence the feedback regulation; thus inducing the development of clinical inflammation in early acne lesions.

20.1.4 Substance P and the Ectopeptidases

The ectopeptidase neutral endopeptidase (CD10) has been reported to be significantly upregulated within sebaceous glands of acne patients, but is not detectable in those of healthy subjects [102]. Ectopeptidases are a group of enzymes which are ubiquitously expressed and have pleiotropic functions [5]. Interaction with agonistic antibodies or inhibitors revealed that ecto-enzymes, beyond their proteolytic activity, influence fundamental biological processes such as growth,

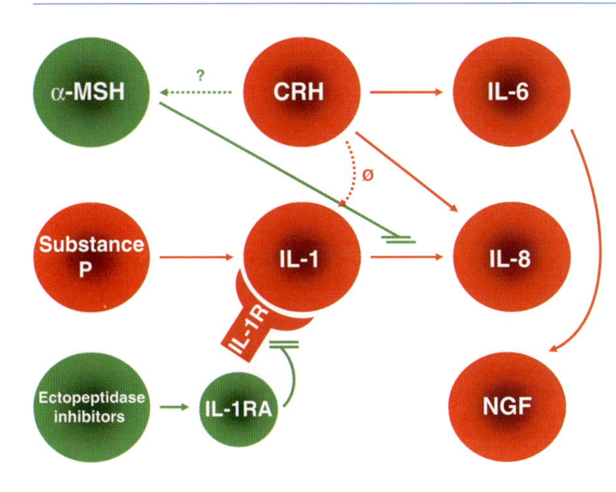

Fig. 20.1 Neuropeptide – cytokine/chemokine signalling in human sebaceous glands and human sebocytes. *α-MSH* α-melanocyte stimulating hormone, *CRH* corticotropin-releasing hormone, *IL-6* interleukin-6, *IL-1* interleukin 1, *IL-8* interleukin-8, *IL-1R* interleukin-1 receptor, *IL-1RA* interleukin-1 receptor antagonist, *NGF* neural growth factor, ø no influence, *question mark* unknown regulation in sebocytes, *equal to* inhibition, red = proinflammatory activity, green = anti inflammatory activity

apoptosis, differentiation, adhesion, motility, invasion, cell–cell interaction, angiogenesis and transformation [21,40,80]. This implies a possible role for these enzymes as targets to influence pathophysiologic conditions. The skin represents an interesting new target organ of ectopeptidase inhibitors, not only because skin diseases involve the activation of immune cells, but also because ectopeptidases are expressed on human keratinocytes and sebocytes in vivo and are upregulated in hyperproliferative skin diseases, for example, psoriasis and acne [37,65,78].

Neutral endopeptidase inactivates substance P, a neuropeptide belonging to the tachykinin family, which can induce neurogenic inflammation [81]. Substance P immunoreactive nerve fibers have been detected in close apposition to sebaceous glands of acne patients but not to those of healthy volunteers [100,102] (Fig. 20.2). In vitro experiments using an organ culture system demonstrated that substance P induced the expression of neutral endopeptidase within sebaceous germinative cells in a dose-dependent manner [102]. Moreover, a significant increase was detected in the size of the sebaceous glands and in the number of sebum vacuoles in sebocytes on treatment with substance P [100]. Substance P stimulates IL-1 expression in human sebocytes in vitro [51].

Neutral endopeptidase is often co-expressed in vivo with other ectopeptidases, such as dipeptidyl peptidase IV (DPIV) and aminopeptidase N (APN) [80] (Fig. 20.3).The pharmacological inhibition of DPIV (CD26) as well as APN (CD13) affects growth, cytokine production and typical functions of human peripheral T cells both in vitro and in vivo [54,77,79]. Inhibitors of DPIV have potent immunosuppressive and anti-inflammatory effects in various disease models such as murine experimental autoimmune encephalomyelitis, collagen- and alkyldiamone-induced arthritis and rat cardiac transplantation, and are presently undergoing clinical phase II and III trials for type II diabetes [98]. APN inhibitors have shown therapeutic efficacy in analgesia models and tumour neoangiogenesis, and bestatin is clinically used as an immunomodulator in cancer patients.

Moreover, inhibitors of DPIV and APN suppress keratinocyte and sebocyte proliferation in vitro [25,97,98], partially restore keratinocyte differentiation in vivo [97] and induce terminal differentiation of sebocytes in vitro [98]. In addition, they increase the expression of the anti-inflammatory cytokine IL-1 receptor antagonist (IL-1RA) in sebocytes and keratinocytes in vitro.

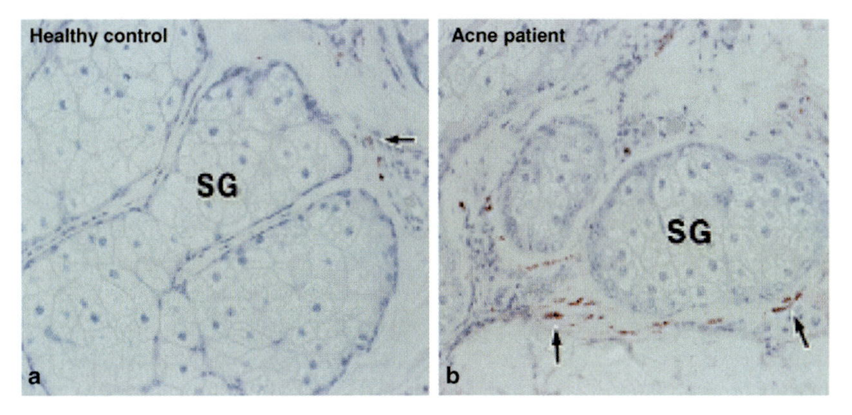

Fig. 20.2 Immunohistochemical staining with substance P (*arrows*) around sebaceous glands (SG) in a control subject (**a**) and in a patient with acne (**b**) (from [102])

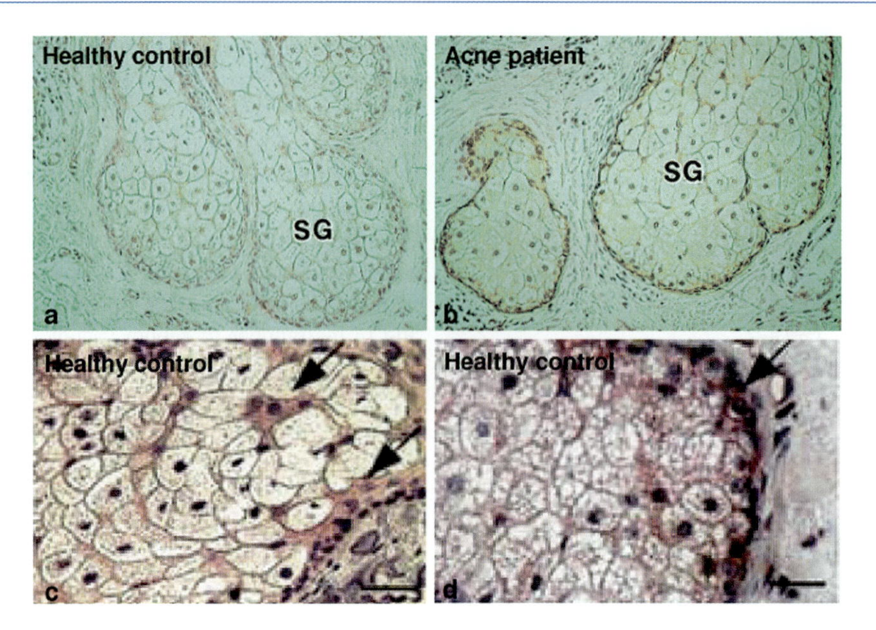

Fig. 20.3 Expression of neutral endopeptidase within SG in a control subject (**a**) and in a patient with acne (**b**). Neutral endopeptidase staining is restricted to the sebaceous germinative cells (from [102]). Detection of dipeptidyl peptidase IV (**c**) and aminopeptidase N (**d**) within the sebaceous glands of control subjects (from [98]). Arrows indicate positive labelling of the sebaceous germinative cells

The proinflammatory cytokine IL-1α represents a hallmark in early acne pathogenesis. Increased levels can be found in uninvolved follicles [6], micro-comedones [1,42] and abundantly in extracted mature comedones [38]. Its functional relevance was studied in an in vitro organ model, where it induced a follicular hyperkeratosis in isolated sebaceous infundibula [33]. This process could be blocked or reversed by the application of IL-1RA, indicating that the ratio of IL-1α/IL-1RA is involved in regulating growth and differentiation of follicular keratinocytes. It is likely that infundibular keratinocytes are a major source of the increased IL-1 levels around acne follicles. On the other hand, sebocytes "stressed" in vitro produce IL-1α on the mRNA and protein level [119] and in vivo investigations detected presence of IL-1α immunoreactivity and mRNA in sebaceous glands [6,10]. Treatment of sebocytes with IL-1β results in marked increase of IL-8 release [121], whereas IL-8 is probably involved in the development of acne lesions [3].

In addition to substance P, antigens of stationary, but not exponential, phase live bacteria seem to induce IL-1α in follicular keratinocytes and sebocytes [31,63,69], but there is no correlation between levels of any cytokine, in particular IL-1α, and numbers of follicular micro-organisms [39].

20.1.5 Corticotropin-Releasing Hormone and Its Analogues

Corticotropin-releasing hormone (CRH), the most proximal module of the cutaneous hypothalamus–pituitary–adrenal-like axis, its binding protein (CRHBP) and corticotropin receptors (CRH-Rs) act as a central regulatory system of the HPA axis [89,91]. Pro-CRH processing into CRH appears to be similar at the central and peripheral levels, including the skin. Current studies have confirmed the presence of a complete CRH/CRHBP/CRH-R system in human sebocytes. CRH and CRH-Rs have been detected in human sebaceous glands at the mRNA and protein levels in tissue [26,44] (Fig. 20.4). CRH, CRHBP, CRH-R1 and CRH-R2 are also expressed in human sebocytes in vitro at the mRNA and protein levels [48,121]. CRH is likely to serve as an important autocrine hormone in sebocytes with a homeostatic pro-differentiation activity. It directly inhibits proliferation, induces lipid synthesis and enhances mRNA expression of Δ5–3β-hydroxysteroid dehydrogenase in human sebocytes in vitro, the enzyme that converts dehydroepiandrosterone to testosterone [24,121]. Testosterone and growth hormone, which also enhance sebaceous lipid synthesis, antagonize CRH in human sebocytes in vitro by down-regulating or

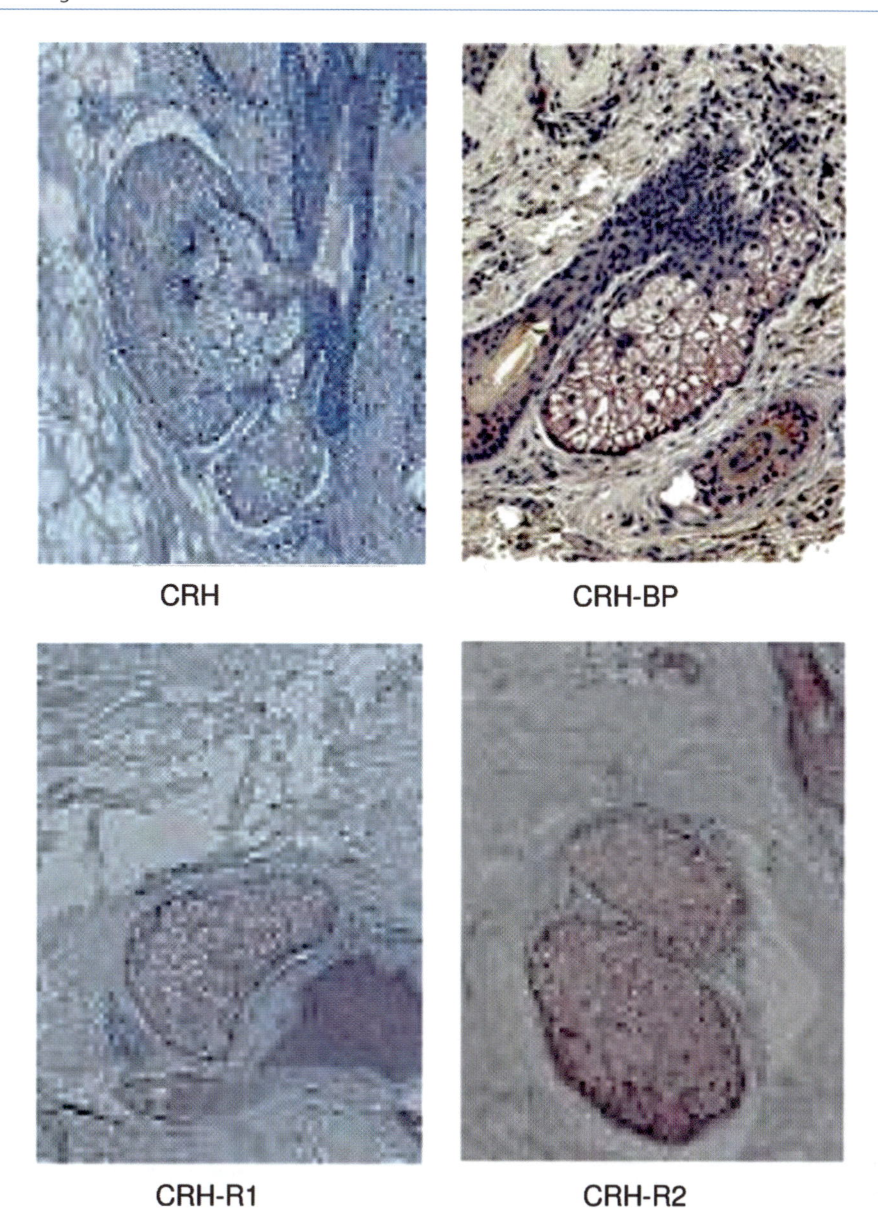

Fig. 20.4 Expression of corticotropin-releasing hormone (CRH), CRH-binding protein (CRH-BP), CRH receptor 1 (CRH-R1) and CRH receptor 2 (CRH-R2) in sebaceous glands of control subjects. CRH barely is expressed, whereas strong expression of CRH-BP and CRH-R2 and weaker of CRH-R1 are detected (from [110])

modifying CRH-R expression, respectively [121]. The induction of sebaceous lipids by CRH is CRH-R1-mediated. CRH also enhances the release of IL-6 and IL-8 in human sebocytes in vitro by an IL-1β-independent pathway [48]. The CRH analogue urocortin also inhibits sebocyte proliferation in vitro, while urotensin and sauvagine seem to be inactive.

On the other hand, CRH enhances keratinocyte immunoactivity by upregulating the interferon-γ-stimulated expression of homing-associated cell adhesion molecules and intercellular adhesion molecule-1 and of the HLA-DR antigen [74]. In addition, it enhances IL-6 and inhibits IL-1β production in human keratinocytes [112].

These findings implicate a major involvement of CRH in the clinical development of seborrhoea and acne, as well as in other skin disorders and diseases associated with alterations in the formation of sebaceous lipids.

20.1.6 Melanocortins

The role of α-melanocyte-stimulating hormone (α-MSH) as a "sebotropin" has been postulated decades ago based on its effect on the rat preputial gland [22,86,99], but only recently it was shown that human sebocytes express melanocortin receptors. Cloning of the melanocortin receptors [61], application of novel research tools and the generation of the SZ95 immortalized human sebaceous gland cell line [120] have paved the way for novel and ongoing studies aimed at ultimately delineating the action of α-MSH in sebocytes and its role in acne.

First, melanocortin (MC)-5 receptor knock-out mice were shown to exhibit reduced exocrine gland function [17]. The transgenic animals show reduced sebum secretion, lack of [NIe (4), D-Phe (7)] α-MSH radiolabelling of the preputial glands and a loss of [NIe (4), D-Phe (7)] α-MSH-induced cyclic adenosine monophosphate increase in membrane fractions from these glands. MC-5 receptor expression has also been demonstrated in microdissected human facial sebaceous glands, as well as in facial skin biopsy specimens, by immunohistochemistry [35,96]. On the other hand, in situ expression of the MC-1 receptor was also detected in skin specimens from patients with acne vulgaris and healthy individuals [13,27].

The MC-1 receptor is also expressed in hair follicle epithelium, while interfollicular epidermis is largely non-reactive [12].

Interestingly, MC-1 receptor immunostaining is more accentuated in the sebaceous glands of involved and non-involved skin from acne patients than in sebaceous glands from healthy individuals [27] (Fig. 20.5). Moreover, the reactivity of the MC-1 receptor is most prominent in basal and differentiating peripheral sebocytes, which are the biologically most active cells of the sebaceous gland. MC-1 receptor expression has been shown to be upregulated by proinflammatory signals [9,34], whereas IL-1 increases MC-1 receptor expression and the number of detectable α-MSH binding sites in human sebocytes [27]. The latter generates a negative feedback mechanism for α-MSH, which exerts direct anti-inflammatory actions, that is, inhibition of IL-1-mediated IL-8 secretion [13], indicating an α-MSH-mediated cytoprotection from harmful cytotoxic stimuli released during inflammation.

Under situations of increased systemic or cutaneous stress, as in patients with acne vulgaris, aberrant α-MSH

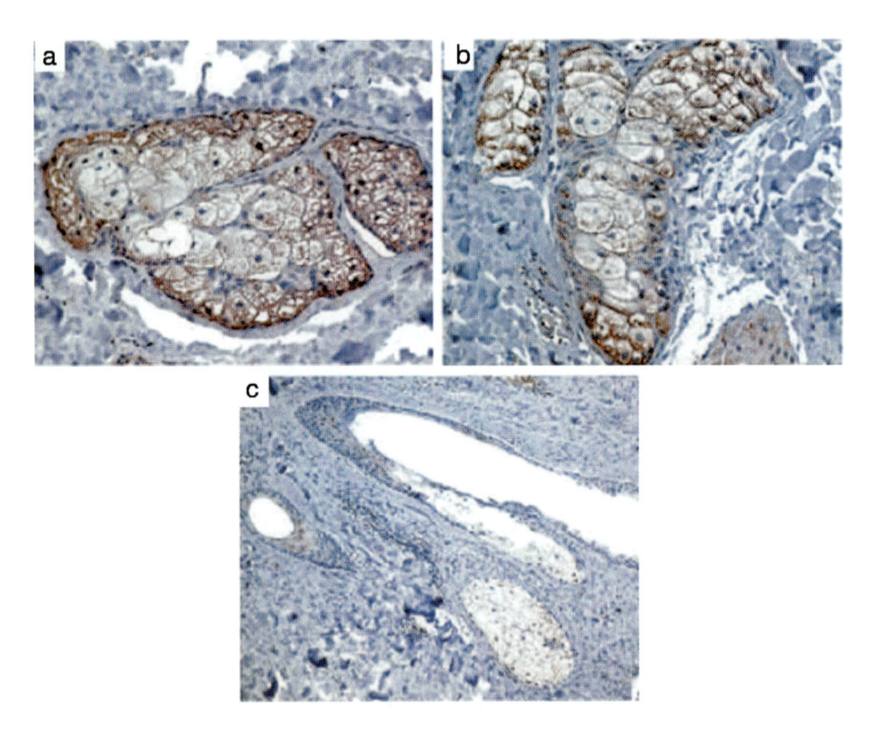

Fig. 20.5 Localization of melanocortin-1 receptor in sebaceous glands of involved (**a**) and uninvolved skin (**b**) of acne patients and in healthy controls (**c**). Very strong and strong immunostaining of melanocortin-1 receptor in almost all sebocytes of involved and uninvolved skin is observed; reactivity is most prominent in basal and differentiating peripheral sebocytes with less intense staining of mature cells. Note, apoptotic sebocytes are not labelled with the anti-melanocortin-1 receptor antibody. Melanocortin-1 receptor immunoreactivity in sebaceous glands of normal skin is less intense (from [27])

levels may induce increased MC-1 receptor expression in sebaceous glands [27]. Initial functional studies on human sebocytes revealed that melanocortin peptides increase lipid droplet formation and [NIe (4), D-Phe (7)] α-MSH furthermore stimulates squalene synthesis in these cells [113]. In primary human sebocytes, the MC-1 receptor has significantly higher expression than the MC-5 receptor [113], while in SZ95 sebocytes MC-5 receptor expression is undetectable [13]. Interestingly, MC-5 receptor expression levels were found to be dependent on the differentiation level of normal and SZ95 sebocytes [114,118] and that the MC-5 receptor mediates the intracellular calcium increase induced by α-MSH [114]. On the other hand, the difference in MC-1 receptor immunoreactivity in the pilosebaceous unit between acne patients and healthy individuals is most accentuated in the ductus seboglandularis, which plays a crucial role in the pathogenesis of acne due to increased cellular turnover or keratinization [100]. α-MSH can, indeed, increase the metabolic activity in HaCaT keratinocytes [41] in contrast to sebocytes, in which this neuropeptide does not augment proliferation [13,119].

Based on these findings, melanocortins released upon psychoemotional stress may influence the course of acne vulgaris [66]. However, since α-MSH also has anti-inflammatory effects on sebocytes [13], ligands of melanocortin receptors may be interesting candidates for acne treatment [28,60].

20.1.7 Endogenous Opioids

β-Endorphin, another propiomelanocortin-derived peptide and member of the endogenous opioid family of neuropeptides, has sebotropic activity. The μ-opioid receptor, which binds β-endorphin with high affinity, is expressed by the human sebaceous gland in situ as well as by SZ95 sebocytes in vitro [14]. β-Endorphin inhibits proliferation and stimulates lipogenesis and specifically increases the amount of C16:0, C16:1, C18:0, C18:1 and C18:2 fatty acids to an extent similar to linoleic acid. These data highlight another prototypical stress-induced neuroendocrine mediator that acts on the human sebocyte and modulates proliferation, differentiation, and lipogenesis.

20.1.8 Other Neuropeptides

There is evidence that additional neuropeptides with hormonal and non-hormonal activity may control the development of clinical inflammation in acne. Nerve growth factor is only detected within the germinative sebocytes [103], but it does not influence sebaceous gland size or the number of lipid vacuoles in human sebocytes [100]. An increase in the number of mast cells and a strong expression of endothelial leukocyte adhesion molecule-1 on the post-capillary venules were observed in areas adjacent to the sebaceous glands. Among several molecules produced by mast cells, IL-6 selectively induced sebocytes to produce nerve growth factor [101] (Fig. 20.6).

VIP receptors were expressed in human sebocytes, with VIP receptor-2 exhibiting the most pronounced expression in cells surrounded by VIP immunoreactive nerve fibers [84]. Sebocytes also express CGRP and neuropeptide Y receptors. However, neither VIP, nor CGRP or neuropeptide Y influenced sebaceous gland size or the number of lipid vacuoles in human sebocytes [100].

20.1.9 Outlook

There is increasing evidence to suggest that the nervous system (for example emotional stress) can influence the course of acne (Fig. 20.7). Cutaneous neurogenic factors may contribute to the onset and/or exacerbation of acne inflammation. Central or topical stress may, indeed, influence the feedback regulation in the sebaceous gland, thus inducing the development of clinical inflammation in early acne lesions. The identification of the precise action of such neuromediators on the pilosebaceous unit can be expected to lead to novel treatment options with neuropeptide-like active substances and/or neuropeptide inhibitors in order to normalize the altered formation of sebaceous lipids and to reduce inflammation in acne.

20.2 Rosacea

Synonyms Box: Acne rosacea

20.2.1 Introduction

Rosacea is a chronic disorder of the interfollicular skin, affecting primarily the convexities of the central face (cheeks, nose, chin and central forehead) and progressing through stages over time [72]. Its onset usually occurs between the ages of 30 and 50 years, equally affecting both genders. Although rosacea occurs in all racial and ethnic groups, white-skin colour individuals of Celtic origin are thought to be particularly prone to the disorder [70], which is rather uncommon in persons with dark skin.

Although the precise aetiology of rosacea remains unknown, various factors have been suspected of

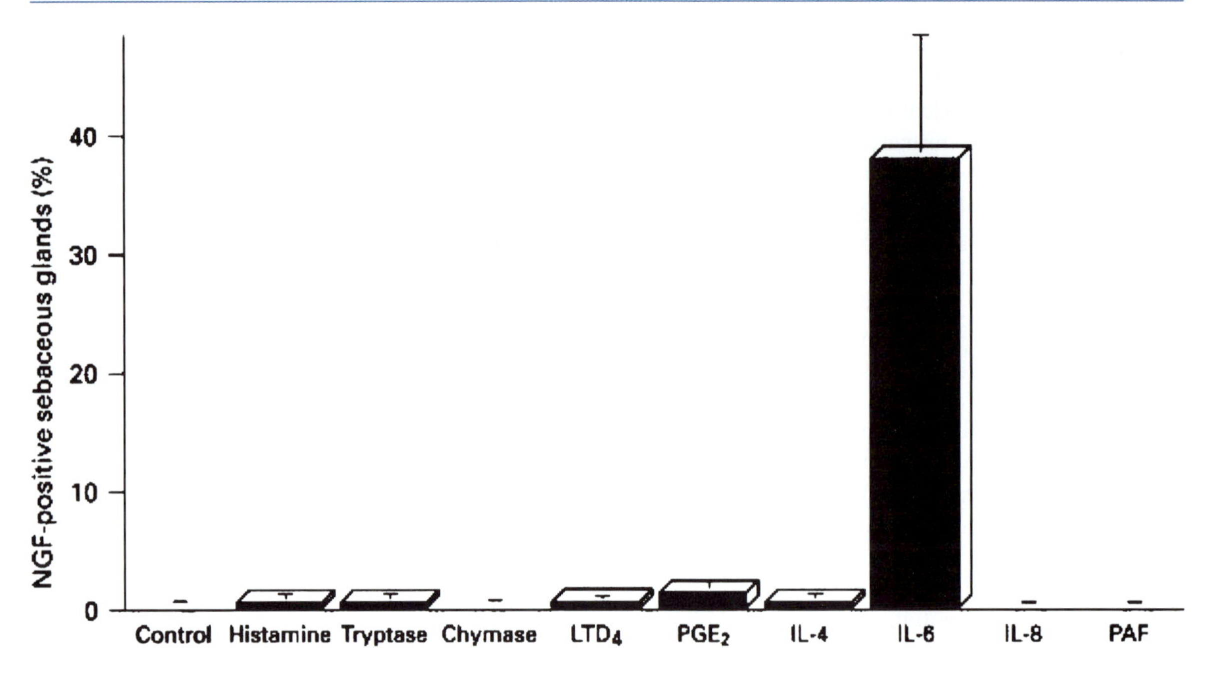

Fig. 20.6 Interleukin-6 (IL-6) specifically induces the expression of neural growth factor (NGF) in sebaceous glands maintained ex vivo. Further macrophages-derived mediators, such as histamine, tryptase, chymase, leukotriene D4 (LTD_4), prostaglandin E2 (PGE_2), interleukin-4 (IL-4), interleukin-6 (IL-6), interleukin-8 (IL-8), platelet-activating factor (PAF) and tumour necrosis factor-α, interferon-γ (the latter two not shown) did not affect NGF expression. Pre-incubation of explants with antibodies specific for the IL-6 receptor, followed by exposure to IL-6, abrogated the NGF induction (from [101])

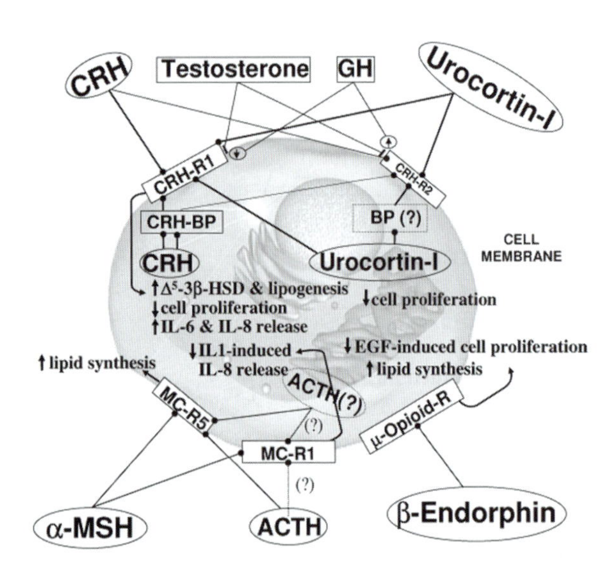

Fig. 20.7 Effects of the neuropeptide and sexual hormone network on human sebocytes in vivo and in vitro (modified from [121]). *CRH* corticotropin-releasing hormone, *GH* growth hormone, *ACTH* adrenocorticotropin, *α-MSH* α-melanocyte-stimulating hormone, *BP* binding protein, *EGF* epidermal growth factor, *MC-R* melanocortin receptor, *R* receptor, *IL* interleukin, *HSD* hydroxysteroid dehydrogenase

contributing to this condition with the most-cited pathogenic theory centred on inherent abnormalities in cutaneous vascular homeostasis [59,76,107]. Flushing or transient erythema is controlled by two vasodilatory mechanisms: humoral substances and neural stimuli [16,106]. In this context, cytokines, hormones and neuropeptides probably communicate within the network made of the endocrine, nervous and immune systems. The apparent inflammatory reaction in rosacea is likely the result of altered communication and/or reciprocal modulation between them [93,104,105]. The major neuropeptides probably involved in rosacea include substance P [49,73], VIP [109] and CRH [23] (Fig. 20.8). Apart from their proinflammatory properties, neuropeptides and neurohormones are also potent downregulators of immunity.

20.2.2 Ultraviolet Radiation as Initial Stress Factor for the Development of Inflammatory Lesions

Kligman's postulate, that rosacea should be viewed as an ultraviolet (UV)-induced dermatosis [43], meets a general consensus among clinicians, who consider rosacea to be at least a photoaggravated disorder. Pathophysiologic processes induced by UV radiation, which are processes similar to those seen in photoaging, contribute to the signs and symptoms of rosacea [62]. The pivotal role of sunlight is supported by the distribution of erythema and telangiectasias on the facial convexities. Sun-protected areas, such as the supraorbital and submental areas, are typically spared. However, several photoprovocation studies in rosacea patients have failed to show heightened skin sensitivity to the acute effects of UV radiation [29]. On the other hand, UV radiation stimulates angiogenesis, which promotes telangiectasia. UVB irradiation of human skin results in pronounced dermal angiogenesis accompanied by the up-regulation of the potent angiogenic factor vascular endothelial growth factor and the down-regulation of thrombospondin-1, an endogenous angiogenesis inhibitor [111]. These newly formed blood vessels facilitate the infiltration of inflammatory cells into the dermal tissue, resulting in damage to dermal matrix components.

At the cellular level, epidermal Langerhans cells are considered as the main target of UV radiation, since UV light inhibits their antigen-presenting activity and their capacity to stimulate allogeneic type 1 T cells [7]. CD11β+ macrophages and neutrophils infiltrate the epidermis after intense UV radiation. Although the mechanisms by which immune regulatory suppressor T cells act still remain unclear, there is increasing evidence that apoptosis of epidermal Langerhans cells or reactive T cells may play an important role. Even at suberythemal doses, UVB reduces Langerhans cell density, migration and maturation in the epidermis and regional lymphoid tissue [75]. It has been suggested that this immunosuppressive effect is in large part due to the effects of UVB in modulating cytokine gene expression in keratinocytes. Keratinocytes are also a target of UV light and they produce and release numerous soluble immunosuppressive mediators and pro-angiogenetic factors [36,46,85]. UVB induces an angiogenic switch on human keratinocytes by a gene expression mechanism [36].

20.2.3 Neuropeptides Mediate Effects of UV Radiation-Induced Immunosuppression

There is accumulating evidence that neurogenic mediators contribute to inflammation and immunosuppression following UV irradiation of the skin. The interaction between peripheral nerves and the immune system is

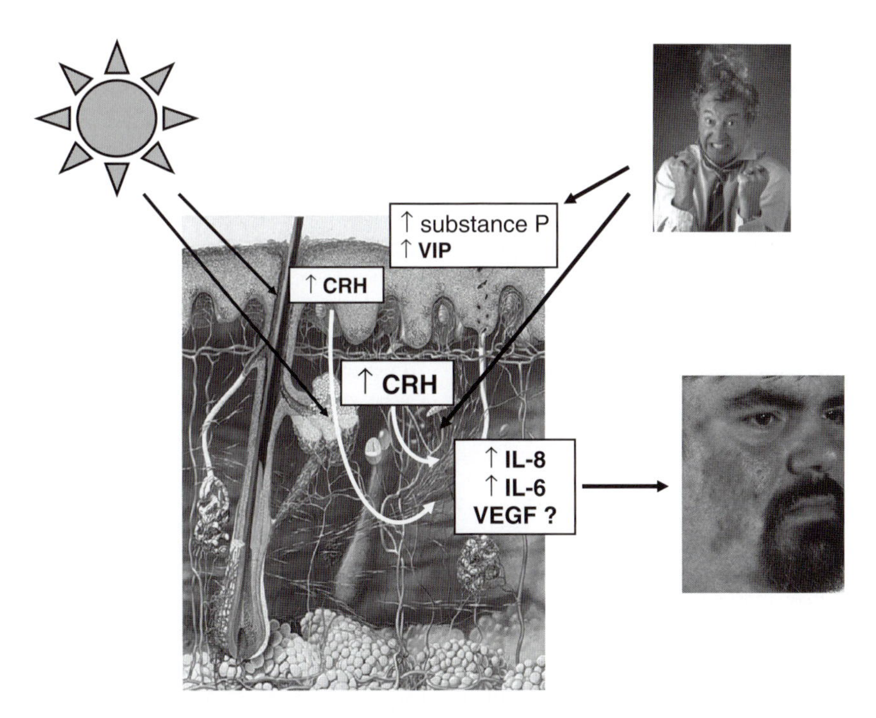

Fig. 20.8 Current concept of the induction of inflammation and modification of vascular homeostasis by ultraviolet light and neuropeptides in rosacea. *VIP* vasoactive intestinal peptide, *CRH* corticotropin-releasing hormone, *IL* interleukin

mediated by different types of cutaneous nerve fibres that release neuromediators and activate specific receptors on target cells in the skin such as keratinocytes, mast cells, Langerhans cells, micro-vascular endothelial cells, fibroblasts and infiltrating immune cells. Among the best-studied neuropeptides are CGRP and substance P; knowledge on CRH [89,91] and other neuropeptides is emerging. These neuropeptides are capable of mediating cutaneous neurogenic inflammation by induction of vasodilation, plasma extravasation and augmentation of cytokine, chemokine and cellular adhesion molecule expression [82,83,89]. CGRP may contribute to local immunosuppression in mice mediated by repeated sub-inflammatory UV irradiation [53]. The number of epidermal nerve fibers immunoreactive for CGRP in skin increases without altering their total number. Also in human skin samples CGRP is significantly enhanced after UVB exposure [88]. In Sprague–Dawley rats, repeated subinflammatory UVB irradiation locally augments the content of cutaneous substance P and CGRP by an increase of neuropeptide content of nerve fibres rather than by an accretion of overall nerve fibre density [52].

Substance P has been mentioned in previous rosacea studies [49,73]. Other proposed mediators include VIP [109], serotonin, histamine, prostaglandins [32] and somatostatin [68]. VIP is known to induce histamine release and to include a nitric oxide (NO)-dependent vasodilation in human skin [108]. A possible modulatory role for somatostatin in the outcome of rosacea has been reported [68]: Four patients presenting long-standing recalcitrant facial rosacea were treated with octreotide for diabetic retinopathy. Rosacea improved rapidly and even cleared without any recurrence in three of the patients. The beneficial effect may be attributed to inhibitory actions on the sebaceous gland, on the neovascularization and/or on the inflammatory process. In particular, somatostatin plays a prominent role in neuroendocrinological aspects of the skin and may help to clear some inflammatory dermatoses, including rosacea [90].

20.2.4 CRH and the Response to Peripheral Stress

A number of studies have indicated that CRH and its analogues can cause marked increases in vascular permeability in the skin microcirculation through the degranulation of mast cells, an action mediated via CRH-R1 receptors [20,87]. Both CRH and urocortin increase vascular permeability in rat skin via mast cell degranulation [87,94]. Mast cell degranulation involves the release and stimulation of numerous vasoactive molecules, including histamine and NO. CRH-induced vasodilation in human skin appears to be mediated, at least in part, by mast cell-derived histamine. Human mast cells synthesize and secrete both CRH and urocortin in response to immunoglobulin E receptor cross-linking. Mast cells also express CRH receptors, activation of which leads to the selective release of cytokines and other pro-inflammatory mediators [95]. Acute stress increases skin CRH that can trigger mast cell-dependent vascular permeability, effects inhibited by certain histamine-1 receptor antagonists, possibly acting to reduce intracellular Ca^{2+} ion levels [58]. CRH is upregulated in human sebocytes and keratinocytes in vitro by UVB [23].

20.2.5 Conclusions

An involvement of neurogenic components in the pathogenesis of vascular lesions of rosacea has been suggested and neuropeptides are thought to play a modulatory role in cutaneous inflammation. The role of neuropeptides and proinflammatory cytokines in producing the flush response is yet to be defined but the light sensitivity of rosacea skin and circulating neuropeptide vasoactive mediators are important factors to be considered.

Summary for the Clinician

Ongoing research is modifying the classical view of acne pathogenesis. A hereditary background, androgens, skin lipids, inflammatory signalling and regulatory neuropeptides seem to be involved in this multifactorial process. Emotional stress from external sources may have a significant influence on acne vulgaris. Cutaneous neurogenic factors (substance P, ectopeptidases, the CRH system, α-MSH and endogenous opioids) may contribute to the onset or exacerbation of acne inflammation. The identification of the precise action of such neuromediators on the pilosebaceous unit can lead to novel treatment options of inflammation in acne. On the other hand, flushing or transient erythema in rosacea is controlled by humoral substances and neural stimuli. The major neuropeptides probably involved in rosacea include substance P, VIP and CRH. Ultraviolet radiation is an initial stress factor for the development of inflammatory lesions in rosacea.

References

1. Aldana OL, Holland DB, Cunliffe WJ (1998) Variation in pilosebaceous duct keratinocyte proliferation in acne patients. Dermatology 196:98–99
2. Alesci S, Bornstein SR (2000) Neuroimmunoregulation of androgens in the adrenal gland and the skin. Horm Res 54:281–286
3. Alestas T, Ganceviciene R, Fimmel S, Müller-Decker K, Zouboulis CC (2006) Enzymes involved in the biosynthesis of leukotriene B_4 and prostaglandin E_2 are active in sebaceous glands. J Mol Med 84:75–87
4. Amann W (1975) Manifestation der Acne vulgaris nach mentaler Überförderung. Dermatol Monatsschr 161:499–501
5. Antczak C, De Meester I, Bauvois B (2001) Ectopeptidases in pathophysiology. Bioessays 23:251–260
6. Anttila HS, Reitamo S, Saurat JH (1992) Interleukin 1 immunoreactivity in sebaceous glands. Br J Dermatol 127:585–588
7. Aubin F (2003) Mechanisms involved in ultraviolet light-induced immunosuppression. Eur J Dermatol 13.515–523
8. Baldwin HE (2002) The interaction between acne vulgaris and the psyche. Cutis 70:133–139
9. Bhardwaj R, Becher K, Mahnke M, Hartmeyer T, Schwarz T, Sholzen T, Luger TA (1997) Evidence for the differential expression of the functional alpha-melanocyte-stimulating hormone receptor MC-1 on human monocytes. J Immunol 158:3378–3384
10. Boehm KD, Yun JK, Strohl KP, Elmets CA (1995) Messenger RNAs for the multifunctional cytokines interleukin-1 alpha, interleukin-1 beta and tumor necrosis factor-alpha are present in adnexal tissues and in dermis of normal human skin. Exp Dermatol 4:335–341
11. Böhm M, Luger TA (2000) The role of melanocortins in skin homeostasis. Horm Res 54:287–293
12. Böhm M, Metze D, Schulte U, Becher E, Luger T, Brzoska T (1999) Detection of melanocortin-1 immunoreactivity in human skin cells in culture and in situ. Exp Dermatol 8:453–461
13. Böhm M, Schiller M, Ständer S, Seltmann H, Li Z, Brzoska T, Metze D, Schiöth HB, Skottner A, Seiffert K, Zouboulis CC, Luger TA (2002) Evidence for expression of melanocortin-1 receptor in human sebocytes in vitro and in situ. J Invest Dermatol 118:533–539
14. Böhm M, Li Z, Ottaviani M, Picardo M, Zouboulis CC, Ständer S, Luger TA (2004) Beta-endorphin modulates lipogenesis in human sebocytes. J Invest Dermatol 123:A10
15. Bornstein SR, Böttner A, Chrousos GP (1999) Knocking out the stress response. Mol Psychiatr 4:403–407
16. Burnstock G (1971) Fine-structural identification of autonomic nerves and their relation to smooth muscle. Prog Brain Res 34:389–404
17. Chen W, Kelly MA, Opitz-Araya X, Thomas RE, Low MJ, Cone RD (1997) Exocrine gland dysfunction in MC5-R-deficient mice: evidence for coordinated regulation of exocrine gland function by melanocortin peptides. Cell 91:789–798
18. Chiu A, Chon SY, Kimball AB (2003) The response of skin disease to stress: changes in the severity of acne vulgaris as affected by examination stress. Arch Dermatol 139:897–900
19. Chrousos GP (1998) Stressors, stress, and neuroendocrine integration of the adaptive response. The 1997 Hans Selye Memorial Lecture. Ann N Y Acad Sci 30:311–335
20. Crompton R, Clifton VL, Bisits AT, Read MA, Smith R, Wright MR (2003) Corticotropin-releasing hormone causes vasodilation in human skin via mast cell-dependent pathways. J Clin Endocrinol Metab 88:5427–5432
21. De Meester I, Korom S, Van Damme J, Scharpe S (1999) CD26, let it cut or cut it down. Immunol Today 20:367–375
22. Ebling FJ (1977) Hormonal control and methods of measuring sebaceous gland activity. J Invest Dermatol 62:161–171
23. Fimmel S, Schnitger A, Glass E, Seiffert K, Granstein RD, Zouboulis CC (2004) Keratinocyte- and sebocyte-derived factors modify UVB activity on endothelial cells: a possible mechanism for the development of vascular changes in rosacea. J Invest Dermatol Res 122:A3
24. Fritsch M, Orfanos CE, Zouboulis CC (2001) Sebocytes are the key regulators of androgen homeostasis in human skin. J Invest Dermatol 116:793–800
25. Gabrilovac J, Cupic B, Breljak D, Zekusic M, Boranic M (2004) Expression of CD13/aminopeptidase N and CD10/neutral endopeptidase on cultured human keratinocytes. Immunol Lett 91:39–47
26. Ganceviciene R, Marciukaitiene I, Graziene V, Rimkevicius A, Zouboulis CC (2006) New accents in the pathogenesis of acne vulgaris. Acta Medica Lituanica 13:83–87
27. Ganceviciene R, Graziene V, Böhm M, Zouboulis CC (2007) Increased in situ expression of melanocortin-1 receptor in sebaceous glands of lesional skin of patients with acne vulgaris. Exp Dermatol 16:547–552
28. Getting SJ (2006) Targeting melanocortin receptors as potential novel therapeutics. Pharmacol Ther 111:1–15
29. Goetz H, Cronen J (1980) Die UV-Lichtempfindlichkeit der Haut bei der Rosacea. Z Hautkr 55:232–236
30. Goggin N, Enright F, Costigan C, Duff D, Oslizlok P, Wood AE, Watson R (1999) Striae and acne following cardiac surgery in a child. Br J Dermatol 140:734–736
31. Graham GM, Farrar MD, Cruse-Sawyer JE, Holland KT, Ingham E (2004) Proinflammatory cytokine production by human keratinocytes stimulated with Propionibacterium acnes and P. acnes GroEL. Br J Dermatol 150:421–428
32. Guerriera M, Parodi A, Cipriani C, Divano C, Rebora A (1982) Flushing in rosacea: a possible mechanism. Arch Dermatol Res 272:311–316
33. Guy R, Green MR, Kealey T (1996) Modeling acne in vitro. J Invest Dermatol 106:176–182
34. Hartmeyer M, Sholzen T, Becher E, Bhardwaj R, Schwarz T, Luger T (1997) Human dermal microvascular endothelial cells express the melanocortin receptor type 1 and produce increased levels of IL-8 upon stimulation with alpha-melanocyte-stimulating hormone. J Immunol 159:1930–1937
35. Hatta N, Dixon C, Ray AJ, Phillips SR, Cunliffe WJ, Dale M, Todd C, Meggit S, Birch-MacHin MA, Rees JL (2001) Expression, candidate gene, and population studies of the melanocortin 5 receptor. J Invest Dermatol 116:564–570
36. Howell BG, Wang B, Freed I, Mamelak AJ, Watanabe H, Sauder DN (2004) Microarray analysis of UVB-regulated genes in keratinocytes: downregulation of angiogenesis inhibitor thrombospondin-1. J Dermatol Sci 34:185–194
37. Hunyadi J, Simon M Jr, Kenderessy AS, Dobozy A (1993) Expression of monocyte/macrophage markers (CD13,

CD14, CD68) on human keratinocytes in healthy and diseased skin. J Dermatol 20:341–345
38. Ingham E, Eady EA, Goodwin CE, Cove JH, Cunliffe WJ (1992) Proinflammatory levels of interleukin-1 alpha-like bioactivity are present in the majority of open comedones in acne vulgaris. J Invest Dermatol 98:895–901
39. Ingham E, Walters CE, Eady EA, Cove JH, Kearney JN, Cunliffe WJ (1998) Inflammation in acne vulgaris: failure of skin micro-organisms to modulate keratinocyte interleukin 1 alpha production in vitro. Dermatology 196:86–88
40. Iwata S, Morimoto C (1999) CD26/dipeptidyl peptidase IV in context. The different roles of a multifunctional ectoenzyme in malignant transformation. J Exp Med 190:301–306
41. Jeong SK, Hwang SW, Choi SY, An JM, Seo JT, Zouboulis CC, Lee SH (2007) Intracellular calcium mobilization is mediated by melanocortin receptors in SZ95 sebocytes. J Invest Dermatol 127:S72
42. Jeremy AH, Holland DB, Roberts SG, Thomson KF, Cunliffe WJ (2003) Inflammatory events are involved in acne lesion initiation. J Invest Dermatol 121:20–27
43. Kligman AM (2004) A personal critique on the state of knowledge of rosacea. Dermatology 208:191–197
44. Kono M, Nagata H, Umemura S, Kawana S, Yoshiyuki R (2001) In situ expression of corticotropin-releasing hormone (CRH) and proopiomelanocortin (POMC) genes in human skin. FASEB J 15:2297–2299
45. Koo JY, Smith LL (1991) Psychologic aspects of acne. Pediatr Dermatol 8:185–188
46. Kosmadaki G, Yaar M, Arble BL, Gilchrest BA (2003) UV induces VEGF through a TNF-alpha independent pathway. FASEB J 17:446–448.
47. Kraus SJ (1970) Stress, acne and skin surface free fatty acids. Psychosom Med 32:503–588
48. Krause K, Schnitger A, Fimmel S, Glass E, Zouboulis CC (2007) Corticotropin-releasing hormone skin signalling is receptor-mediated and is predominant in the sebaceous glands. Horm Metab Res 39:166–170
49. Kurkcuoglu N, Alaybeyi F (1991) Substance P immunoreactivity in rosacea. J Am Acad Dermatol 25:725–726
50. Kurokawa I, Mayer-da-Silva A, Gollnick H, Orfanos CE (1988) Monoclonal antibody labeling for cytokeratins and filaggrin in the human pilosebaceous unit of normal, seborrhoeic and acne skin. J Invest Dermatol 91:566–571
51. Lee WJ, Jung HD, Lee HJ, Kim B-S, Lee S-J, Kim DW (2008) Influence of substance P on human sebocytes. Arch Dermatol Res 300:311–316
52. Legat FJ, Griesbacher T, Schicho R, Althuber P, Schuligoi R, Kerl H, Wolf P (2002) Repeated subinflammatory ultraviolet B irradiation increases substance P and calcitonin gene-related peptide content and augments mustard oil-induced neurogenic inflammation in the skin of rats. Neurosci Lett 329:309–313
53. Legat FJ, Jaiani LT, Wolf P, Wang MS, Lang R, Abraham T, Solomon AR, Glass JD, Ansel JC (2004) The role of calcitonin gene-related peptide in cutaneous immunosuppression induced by repeated subinflammatory ultraviolet irradiation exposure. Exp Dermatol 13:242–250
54. Lendeckel U, Arndt M, Frank K, Wex T, Ansorge S (1999) Role of alanyl aminopeptidase in growth and function of human T cells. Int J Mol Med 4:17–27
55. Li Z, Ständer S, Luger TA, Zouboulis CC, Böhm M (2003) Expression of opioid receptors in human sebocytes – another link between stress and acne? Exp Dermatol 12:332
56. Lorenz T, Graham DT, Wolf S (1953) The relation of life stress and emotions to human sebum secretion and to the mechanism of acne vulgaris. J Lab Clin Med 41:11–28
57. Luger TA (2002) Neuromediators – a crucial component of the skin immune system. J Dermatol Sci 30:87–93
58. Lytinas M, Kempuraj D, Huang M, Boucher W, Esposito P, Theoharides TC (2003) Acute stress results in skin corticotropin-releasing hormone secretion, mast cell activation and vascular permeability, an effect mimicked by intradermal corticotropin-releasing hormone and inhibited by histamine-1 receptor antagonists. Int Arch Allergy Immunol 130:224–231
59. Marks R, Beard RJ, Clark ML, Kwok M, Robertson WB (1967) Gastrointestinal observations in rosacea. 1:739–743
60. Millington GWM (2006) Proopiomelanocortin (POMC): the cutaneous roles of its melanocortin products and receptors. Clin Exp Dermatol 31:407–412
61. Mountjoy KG, Robbins LS, Mortrud MT, Cone RD (1992) The cloning of a family of genes that encode the melanocortin receptors. Science 257:1248–1251
62. Murphy GM (2004) Ultraviolet light and rosacea. Cutis 74:32–34
63. Nagy I, Pivarcsi A, Kis K, Koreck A, Bodai L, McDowell A, Seltmann H, Patrick S, Zouboulis CC, Kemény L (2006) *Propionibacterium acnes* and lipopolysaccharide induce the expression of antibacterial peptides and proinflammatory cytokines/chemokines in human sebocytes. Microbes Infect 8:2195–2205
64. Norris JF, Cunliffe WJ (1988) A histological and immunocytochemical study of early acne lesions. Br J Dermatol 118:651–659
65. Novelli M, Savoia P, Fierro MT, Verrone A, Quaglino P, Bernengo MG (1996) Keratinocytes express dipeptidyl-peptidase IV (CD26) in benign and malignant skin diseases. Br J Dermatol 134:1052–1056
66. Orel L, Simon M, Karlseder J, Bhardwaj R, Trautinger F, Schwarz T, Luger T (1997) Alpha-melanocyte stimulating hormone downregulates differentiation driven heat shock protein 70 expression in keratinocytes. J Invest Dermatol 108:401–405
67. Picardi A, Abeni D (2001) Stressful life events and skin diseases: disentangling evidence from myth. Psychother Psychosom 70:118–136
68. Pierard-Franchimont C, Quatresooz P, Pierard GE (2003) Incidental control of rosacea by somatostatin. Dermatology 206:249–251
69. Pivarcsi A, Nagy I, Koreck A, Kis K, Kenderessy-Szabo A, Szell M, Dobozy A, Kemeny L (2005) Microbial compounds induce the expression of pro-inflammatory cytokines, chemokines and human β-defensin-2 in vaginal epithelial cells, Microbes Infect 7:1117–1127
70. Plewig G, Kligman AM (2000) Acne and rosacea. Springer, Berlin, pp 433–475
71. Poli F, Dreno B, Verschoore M (2001) An epidemiological study of acne in female adults: results of a survey conducted in France. J Eur Acad Dermatol Venereol 15:541–545
72. Powell FC (2005) Clinical practice. Rosacea. N Engl J Med 352:793–803

73. Powell FC, Corbally N, Powell D (1993) Substance P and rosacea. J Am Acad Dermatol 28:132–133
74. Quevedo ME, Slominski A, Pinto W, Wei E, Wortsman J (2001) Pleiotropic effects of corticotropin releasing hormone on normal human skin keratinocytes. In Vitro Cell Dev Biol Anim 37:50–54
75. Rattis FM, Concha M, Dalbiez-Gauthier C, Courtellemont P, Schmitt D, Peguet-Navarro J (1998) Effects of ultraviolet B radiation on human Langerhans cells: functional alteration of CD86 upregulation and induction of apoptotic cell death. J Invest Dermatol 111:373–379
76. Rebora A (1993) The red face: rosacea. Clin Dermatol 11:225–234
77. Reinhold D, Bank U, Bühling F, Lendeckel U, Faust J, Neubert K, Ansorge S (1997) Inhibitors of dipeptidyl peptidase IV induce secretion of transforming growth factor-beta 1 in PWM-stimulated PBMC and T cells. Immunology 91:354–360
78. Reinhold D, Vetter RW, Mnich K, Bühling F, Lendeckel U, Born I, Faust J, Neubert K, Gollnick H, Ansorge S (1998) Dipeptidyl peptidase IV (DP IV, CD26) is involved in regulation of DNA synthesis in human keratinocytes. FEBS Lett 428:100–104
79. Reinhold D, Kähne T, Steinbrecher A, Wrenger S, Neubert K, Ansorge S, Bröcke S (2002) The role of dipeptidyl peptidase IV (DP IV) enzymatic activity in T cell activation and autoimmunity. Biol Chem 383:1133–1138
80. Riemann D, Kehlen A, Langner J (1999) CD13 – not just a marker in leukemia typing. Immunol Today 20:83–88
81. Scholzen T, Armstrong CA, Bunnett NW, Luger TA, Olerud JE, Ansel JC (1998) Neuropeptides in the skin: interactions between the neuroendocrine and the skin immune system. Exp Dermatol 7:81–96
82. Scholzen TE, Brzoska T, Kalden DH, O'Reilly F, Armstrong CA, Luger TA, Ansel JC (1999) Effects of ultraviolet light on the release of neuropeptides and neuroendocrine hormones in the skin: mediators of photodermatitis and cutaneous inflammation. J Invest Dermatol Symp Proc 4:55–60
83. Seiffert K, Granstein RD (2002) Neuropeptides and neuroendocrine hormones in ultraviolet radiation-induced immunosuppression. Methods 28:97–103
84. Seiffert K, Zouboulis CC, Seltmann H, Granstein RD (2000) Expression of neuropeptide receptors by human sebocytes and stimulatory effect of their agonists on cytokine production. Horm Res 53:102
85. Seiffert K, Fimmel S, Zouboulis CC, Granstein RD (2004) UV-B irradiation differentially affects VEGF production in human microvascular endothelial cells and keratinocytes. J Invest Dermatol 122:A147
86. Shuster S, Thody AJ (1974) The control and measurement of sebum secretion. J Invest Dermatol 61:172–190
87. Singh LK, Boucher W, Pang X, Letourneau R, Seretakis DI, Green M, Theoharis C, Theoharides TC (1999) Potent mast cell degranulation and vascular permeability triggered by urocortin through activation of corticotropin-releasing hormone receptors. J Pharmacol Exp Ther 288:1349–1356
88. Sleijffers A, Herreilers M, van Loveren H, Garssen J (2003) Ultraviolet B radiation induces upregulation of calcitonin gene-related peptide levels in human Finn chamber skin samples. J Photochem Photobiol B 69:149–152
89. Slominski A, Wortman J, Luger T, Paus R, Solomon S (2000) Corticotropin releasing hormone and proopiomelanocortin involvement in the cutaneous response to stress. Physiol Rev 80:979–1020
90. Slominski AT, Wortsman J (2000) Neuroendocrinology of the skin. Endocrin Rev 21:457–487
91. Slominski AT, Botchkarev V, Choudhry M, Fazal N, Fechner K, Furkert J, Krause E, Roloff B, Sayeed M, Wei E, Zbytek B, Zipper J, Wortsman J, Paus R (1999) Cutaneous expression of CRH and CRH-R. Is there a 'skin stress response system?' Ann NY Acad Sci 885:287–311
92. Stüttgen G (1964) Zentralnerren-system und Talgsekretion. Arch Klin Exp Dermatol 219:795–820
93. Ten Bokum AM, Hofland LJ, van Hagen PM (2000) Somatostatin and somatostatin receptors in the immune system. Eur Cytokine Network 1:161–176
94. Theoharides TC, Singh LK, Boucher W, Pang X, Letourneau R, Webster E, Chrousos G (1998) Corticotropin-releasing hormone induces skin mast cell degranulation and increased vascular permeability, a possible explanation for its proinflammatory effects. Endocrinology 139:403–413.
95. Theoharides TC, Donelan JM, Papadopoulou N, Cao J, Kempuraj D, Conti P (2004) Mast cells as targets of corticotropin-releasing factor and related peptides. Trends Pharmacol Sci 25:563–568
96. Thiboutot D, Sivarajah A, Gilliland K, Cong Z, Clawson G (2000) The melanocortin 5 receptor is expressed in human sebaceous glands and rat preputial cells. J Invest Dermatol 115:614–619
97. Thielitz A, Bukowska A, Wolke C, Vetter R, Lendeckel U, Wrenger S, Hashimoto Y, Ansorge S, Gollnick H, Reinhold D (2004) Identification of extra- and intracellular alanyl aminopeptidases as new targets to modulate keratinocyte growth and differentiation. Biochem Biophys Res Commun 321:795–801
98. Thielitz A, Reinhold D, Vetter R, Lendeckel U, Kähne T, Bank U, Helmuth M, Neubert K, Faust J, Hartig R, Wrenger S, Zouboulis CC, Ansorge S, Gollnick H (2007) Inhibitors of dipeptidyl peptidase IV (DP IV, CD26) and aminopeptidase N (APN, CD13) target major pathogenetic steps in acne initiation. J Invest Dermatol 127:1042–1051
99. Thody AJ, Shuster S (1989) Control and function of sebaceous glands. Physiol Rev 69:383–416
100. Toyoda M, Morohashi M (2001) Pathogenesis of acne. Med Electron Microsc 34:29–40
101. Toyoda M, Morohashi M (2003) New aspects in acne inflammation. Dermatology 206:17–23
102. Toyoda M, Nakamura M, Morohashi M (2002) Neuropeptides and sebaceous glands. Eur J Dermatol 12:422–427
103. Toyoda M, Nakamura M, Makino T, Kagoura M, Morohashi M (2002) Sebaceous glands in acne patients express high levels of neutral endopeptidase. Exp Dermatol 11:241–247
104. Weinstock JV, Elliott D (1998) The substance P and somatostatin interferon-gamma immunoregulatory circuit. Ann N Y Acad Sci 840:532–539
105. Weinstock JV, Elliott D (2000) The somatostatin immunoregulatory circuit present at sites of chronic inflammation. Eur J Endocrinol 143:S15–S19
106. Wilkin J (1988) Why is flushing limited to a mostly facial cutaneous distribution. J Am Acad Dermatol 19:309–313

107. Wilkin JK (1994) Rosacea, Pathophysiology and treatment. Arch Dermatol 130:359–362
108. Wilkins BW, Chung LH, Tublitz NJ, Wong BJ, Minson CT (2004) Mechanisms of vasoactive intestinal peptide-mediated vasodilation in human skin. J Appl Physiol 97:1291–1298
109. Wollina U (1996) Rhinophyma – unusual expression of simple-type keratins and S100A in sebocytes and abundance of VIP receptor-positive dermal cells. Histol Histopathol 11:111–115
110. Wollina U, Abdel-Naser MB, Ganceviciene R, Zouboulis CC (2007) Receptors of eccrine, apocrine and holocrine skin glands. Dermatol Clin 25:577–588
111. Yano K, Kadoya K, Kajiya K, Hong YK, Detmar M (2005) Ultraviolet B irradiation of human skin induces an angiogenic switch that is mediated by upregulation of vascular endothelial growth factor and by downregulation of thrombospondin-1. Br J Dermatol 152:115–121
112. Zbytek B, Mysliwski A, Slominski A, Wortsman J, Wei E T, Mysliwska J (2002) Corticotropin-releasing hormone affects cytokine production in human HaCaT keratinocytes. Life Sci 70:1013–1021
113. Zhang L, Anthonavage M, Huang Q, Li W, Eisinger M (2003) Proopiomelanocortin peptides and sebogenesis Ann NY Acad Sci 994:154–161
114. Zhang L, Li WH, Anthonavage M, Eisinger M (2006) Melanocortin-5 receptor: a marker of human sebocyte differentiation. Peptides 27:413–420
115. Zouboulis CC (2000) Human skin: an independent peripheral endocrine organ. Horm Res 54:230–242
116. Zouboulis CC (2001) Is acne vulgaris a genuine inflammatory disease? Dermatology 203:277–279
117. Zouboulis CC (2004) The human skin as a hormone target and an endocrine gland. Hormones 3:9–26
118. Zouboulis CC, Böhm M (2004) Neuroendocrine regulation of sebocytes – a pathogenetic link between stress and acne. Exp Dermatol 13(suppl 4):31–35
119. Zouboulis CC, Xia L, Akamatsu H, Seltmann H, Fritsch M, Hornemann S, Rühl R, Chen W, Nau H, Orfanos CE (1998) The human sebocyte culture model provides new insights into development and management of seborrhoea and acne. Dermatology 196:21–31
120. Zouboulis CC, Seltmann H, Neitzel H, Orfanos CE (1999) Establishment and characterization of an immortalized human sebaceous gland cell line (SZ95). J Invest Dermatol 113:1011–1020
121. Zouboulis CC, Seltmann H, Hiroi N, Chen W, Young M, Oeff M, Scherbaum WA, Orfanos CE, McCann SM, Bornstein SR (2002) Corticotropin releasing hormone: an autocrine hormone that promotes lipogenesis in human sebocytes. Proc Natl Acad Sci U S A 99:7148–7153
122. Zouboulis CC, Eady A, Philpott M, Goldsmith LA, Orfanos C, Cunliffe WJ, Rosenfield R (2005) What is the pathogenesis of acne? Exp Dermatol 14:143–152

Wound Healing and Stress

21

C.G. Engeland and P.T. Marucha

Contents

Synonyms Box: *ACTH* Adrenocorticotropic hormone, *CFU* Colony forming units, *CRH* Corticotropin-releasing hormone, *EPI* Epinephrine, *GC* Glucocorticoid, *HBO* Hyperbaric oxygen, *HPA axis* Hypothalamic–pituitary–adrenal axis, *KGF* Keratinocyte growth factor, *NE* Norepinephrine, *NK cells* Natural killer cells, *SNS* Sympathetic nervous system, *TEWL* Trans-epidermal water loss, *TNF* Tumor necrosis factor, *VEGF* Vascular endothelial growth factor

Key Features

- In both animals and humans it is well accepted that chronic stress delays dermal wound healing.
- This occurs through activation of both the HPA and SNS pathways. It is mediated by the release of cortisol and norepinephrine into the periphery.
- Through these pathways stress dysregulates the early wound repair process, prolonging the inflammatory phase of healing.
- A prolonged inflammatory phase results in higher bacterial counts and an increased incidence of infection, which in itself delays wound closure and worsens healing outcomes (e.g., scarring).
- Negative effects on wound healing caused by other factors (e.g., aging, diabetes) are exacerbated by stress.

21.1 Overview

The skin provides an essential, constantly renewing protective barrier to the external environment. If the barrier is broken by injury, infection, or immune destruction, rapid repair programs are activated (for reviews, see [34,107]). These programs are often described as three overlapping phases: an early inflammatory phase (minutes to days), a proliferative phase (days to weeks), and a remodeling phase

R.D. Granstein and T.A. Luger (eds.), *Neuroimmunology of the Skin*,

(weeks to months). These phases are interdependent and overlapping. Thus, inflammation during the early stages of healing induces growth factors required for the proliferative phase. Activity of the proliferating epithelium helps to resolve the inflammatory phase. Finally, the timing and activity of the remodeling phase depends upon the resolution of the proliferative phase. As a result, dysregulation of this cascade, especially early in the healing process, can impair healing on multiple levels.

At all stages of wound repair, psychological stress has been shown to dysregulate inflammatory, immune, and physiologic reactions important to both wound healing and protection from infection after injury. The negative outcomes of stress-impaired healing include infection, altered esthetics, poor tensile strength, and scarring. This chapter will briefly review how stress impacts upon dermal wound healing.

21.2 Stages of Wound Healing

21.2.1 The Inflammatory Phase

The first phase of healing, the inflammatory phase, is initiated as soon as the injury occurs and is important in minimizing damage, protecting from infection, removing debris, and initiating the downstream processes for tissue repair [34]. When plasma and formed elements of the blood, including platelets and white blood cells, come in contact with connective tissue, a number of important mediators are released or activated in response to the injury. These include complement products, kinins, fibrin, and prostaglandins. A provisional matrix, the clot, forms which temporizes the injury and later provides a scaffold for repair. The clot is not inert. Factors are released from the clot that is important in the activation and recruitment of phagocytic cells and tissue cells.

Endogenous cells are early participants in the development of the inflammatory phase of healing. Epithelial cells contain preformed IL-1α that is released in response to injury. A network of monocyte-derived Langerhans cells in the epidermis release inflammatory mediators, and mast cells deliver preformed histamine and tumor necrosis factor (TNF) in response to injury. These mediators increase blood flow to the injury site and help to recruit and activate inflammatory cells by inducing the expression of adhesion molecules on endothelial cells (e.g., E-selectin and ICAM) and chemokines (e.g., IL-8 for neutrophils; MCP-1 and MIP-1α for macrophages), which regulate the ordered migration of inflammatory cells. Within minutes, neutrophils begin to arrive and within hours monocytes follow. These recruited phagocytic cells debride damaged tissue and begin the clearance of microbial contaminants. They also help to recruit and produce growth factors for cells involved in the proliferative phase of healing, such as fibroblasts and endothelial cells.

The timeline and extent of phagocytic cell recruitment and inflammation is dependent on several local parameters, including the status of the tissue before wounding, wound site and dimensions, the degree of microbial contamination, wound coverings, and also patient-specific parameters. Under conditions of low bacterial load and rapid bacterial clearance, phagocytic recruitment is limited and resolved rapidly. When microbial load is sustained in wound tissue, continued recruitment occurs and inflammation is extended and/or exuberant [98]. It is well-documented that greater, more prolonged inflammation is detrimental to the outcome of healing [106]. Interestingly, the most idealized healing, fetal wound healing, is characterized by almost no inflammation and no scarring (for review, see [113]). Despite this, the inhibition of inflammation in adult skin, in the presence of microbial contamination, increases the risk of infection and associated complications. Thus, healing of cutaneous wounds in adults requires an appropriate level of inflammation, with appropriate kinetics, to proceed optimally without scarring.

21.2.2 The Proliferative Phase

The second phase of healing, the proliferative phase, involves the rebuilding of lost or damaged structures. The proliferative phase begins within hours of injury and involves the activity of fibroblasts, epithelial cells, and endothelial cells. There is a dramatic increase in the proliferation and recruitment of all of these cells and the resultant tissue is called granulation tissue. These cells accomplish the rebuilding processes required to restore the injured tissue's structure and function, including wound contraction, reformation of the epithelial barrier, redevelopment of blood supply, reformation of sensory connections, and regeneration of lost connective tissue. The control of these processes is through a myriad of growth factors that overlap in specificity for cell type and activity. Growth factors are produced by monocytes and tissue cells in a complex web of interactions among the cells, under the control of multiple factors, in an architecturally sensitive fashion. For example, vascular endothelial growth factor (VEGF) is produced by epithelial cells, monocytes, and even endothelial cells. VEGF is upregulated by inflammatory mediators (e.g., IL-1), other growth factors, and hypoxia. One of the important growth factors for epithelial cells is keratinocyte growth factor (KGF). It is produced by fibroblasts and, like VEGF, is upregulated by inflammation [111].

Early after injury endothelial cells migrate and proliferate to form tubular structures, which mature to form functional blood vessels. Revascularization of tissue is under the control of growth factors (e.g., VEGF), maturation factors (e.g., angiopoietins), inflammation, and oxygen balance (i.e., oxygen, oxidants, oxygen demand, etc.). Oxidative metabolism, which is needed to clear microbial contamination, further amplifies oxygen demands in tissue [58]. Since tissue repair has a high metabolic cost, the degree of vascularization exceeds what is required for normal tissue maintenance.

A central cell in the proliferative phase of healing is the fibroblast. Fibroblasts are responsible for manufacturing and remodeling the matrix of the connective tissue, as well as regulating many of the processes involved during the proliferative phase. Some fibroblasts migrate to the wound margin, form contractile elements, and become myofibroblasts [56]. Myofibroblasts and less well-defined contractile fibroblasts cause the wound to contract to minimize reconstructive costs by decreasing the area of the wound to be resurfaced. Many of the fibroblasts, under control of a cascade of growth factors, manufacture collagen, hyaluronic acid, peptidoglycans, elastin, and other components of the connective tissue matrix. The early collagen fibers formed by fibroblasts are relatively small in diameter, not well cross-linked, and somewhat disorganized. Over time collagen fibers become thicker, more cross-linked, and better organized. Under ideal conditions (i.e., limited inflammation) this reorganization is optimized, allowing healing tissues to most closely resemble that of normal mature connective tissues (i.e., scarring is minimal).

Epithelial cells are predominately involved in re-establishing the external barrier of the wound [93]. At the wound margins and at residual hair follicles, keratinocytes form a hyperproliferative advancing front of the epithelium. Epithelial cells migrate between the eschar and the connective tissue, remaining in contact with the advancing barrier, to cover the entire wound surface. As the wound surface is covered by the epithelium, signals to reduce inflammation are transmitted. To complete barrier function, the epithelium thickens and produces structural proteins such as involucrin and keratins [111].

21.2.3 The Remodeling Phase

The remodeling phase of healing occurs over a period of weeks to months, depending upon the parameters of the wound (e.g., wound site, size of injury, etc.). The relatively disorganized weak connective tissue matrix is remodeled into a more organized, stronger matrix [87]. It is the final outcome of remodeling that will determine how close original form and function is restored to a tissue (i.e., how much scarring occurs). Scarring is characterized by smaller, less cross-linked, less organized collagen fibrils with weaker tensile strength. Scarred dermal tissue may also have changes in the surface appearance, including pigment changes and loss of hair that compromise the esthetics of the tissue. In adult dermal tissue where wounding involves a significant amount of tissue, scarring nearly always occurs to some extent. Scarring is more likely to occur in wounds that require an extended period of time to close. Scarring will also occur when there is extended inflammation and/or exuberant granulation tissue, since the growth factors and other mediators that regulate matrix quantity and quality are produced in larger amounts [106].

As stated before, during the proliferative phase of healing the connective tissue is hypercellular as compared to what is required to maintain the tissue at homeostasis. Therefore, an important aspect of the remodeling phase is the ordered and timely removal of cellular machinery that is no longer required for rebuilding. Ideally, this occurs through a process of apoptosis where cells go through programmed death and removal by macrophages before cells lyse. If cells are allowed to lyse or go through necrosis, inflammation results with additional tissue damage and additional scarring. The process of cell removal is more likely to lead to scarring when a greater excess of cells are produced, again during extended inflammation or as a result of infection. Therefore, the avoidance of scarring is dependent upon the quick resolution of the earlier phases of healing.

21.2.4 Summary

Cutaneous healing after injury encompasses a series of overlapping, well-orchestrated processes to restore the form and function of the skin. If these processes are disrupted, wound healing is not optimal and the restoration of form and function is impaired. The proceeding sections of this chapter will provide evidence for a role of stress in dysregulating wound healing. Most of the studies presented focus on the early events of healing, particularly the inflammatory phase. This is because stress has been shown to have significant effects on inflammation and inflammatory mediated processes, including responses to infection and skin diseases. Furthermore, wound closure, an objective measure of healing, is well-correlated with inflammation. Since the final outcomes of healing, such as scarring and wound strength, are dependent upon the early phases, it is presumed that stress will impact these outcomes as well.

21.3 Stress

The degree of stress which one experiences from any given stressor is subjective, as multiple factors influence the nature and intensity of the stress response. In humans, especially, these factors are numerous and include genetics, past experiences, concurrent stress, health, social setting, social support, etc. That being said, this chapter will focus on chronic stress, defined for our purposes as being: (1) stress that is relatively unremitting for a period of longer than a few hours (e.g., Alzheimer's caregiving, presurgical stress); or (2) stress that is relatively intense and experienced repeatedly (e.g., restraint stress in animals). It is important to differentiate *chronic* stress from *acute* stress, which is shorter lasting (minutes to hours), often less intense and typically not repeated. As this chapter will illustrate, chronic stress has been shown in multiple models and species to be deleterious to wound healing. Conversely, acute stress may actually benefit some aspects of immunity including skin function, presumably by priming the immune system and allowing for quicker responses when challenged (for reviews, see [26,54]). While fascinating, a discussion of the effects of acute stress on immunity is beyond the scope of this chapter.

21.4 Stress Pathways

Most of the physiological effects of stress are mediated by two major neuroendocrine systems.

21.4.1 Hypothalamic–Pituitary–Adrenal (HPA) Axis

When stress is perceived by the integrative cortex, corticotropin-releasing hormone (CRH) is secreted by the paraventricular nucleus of the hypothalamus into the hypophyseal portal system. This induces the anterior pituitary to release adrenocorticotropic hormone (ACTH) into the systemic circulation which, in turn, causes glucocorticoids (GCs) and epinephrine (EPI) to be secreted from the adrenal cortex. Collectively, this neuroendocrine system is known as the hypothalamic–pituitary–adrenal (HPA) axis, and is one of the two major stress pathways. The primary GC in humans and non-human primates is cortisol, whereas in rodents it is corticosterone. GCs regulate HPA activity through a negative feedback mechanism, targeted at type II GC receptors located in various brain regions (e.g., hippocampus, hypothalamus) and on the pituitary. Activation of these receptors inhibits further release of both CRH and ACTH, hence the system is self-regulating.

Immune cells possess GC receptors and, thus, GCs have effects on virtually all aspects of immunity (for review, see [17]). It is the potent immunosuppressive effects of GCs, and particularly the anti-inflammatory and anti-mitotic properties, that account for their inhibitory effects on wound healing [6].

Multiple studies have shown that exogenous GC administration is detrimental to tissue repair. Such treatment reduces neutrophil [18] and monocyte recruitment [85], suppresses phagocytosis and bacterial killing [37], and may serve as an immunocompromised wound model in animals [53]. In humans, a single administration of corticosteroids reduces circulating lymphocytes and monocytes by 70% and 90%, respectively [8]. Inflammation at the wound site is decreased largely due to reduced production of pro-inflammatory cytokines (e.g., IL-1α, IL-1β, TNF-α) [56]. GCs also inhibit keratinocyte [31] and fibroblast proliferation, and the expression of collagen [6]. These effects result in poor bacterial elimination, delayed wound debridement, and a poorer quality of epithelialization. Wound strength is impaired and wound closure is delayed.

In mice, this delay in epithelialization has been associated with reduced mRNA expression for keratinocyte growth factor (KGF) following injury [12]. In vitro studies have further confirmed that GCs reduce KGF mRNA and protein levels in human dermal fibroblasts [16]. These effects on KGF levels may be indirect, stemming from the inhibition of factors which induce KGF-release (e.g., IL-1β, TNF-α), as the addition of IL-1β or TNF-α reverses the above inhibition of KGF in GC-treated fibroblasts.

Pro-inflammatory cytokines play an essential role in normal healing. For example, IL-6 knockout mice exhibit reduced inflammation and heal up to three times more slowly than wild-type controls, and the administration of recombinant IL-6 1 h before wounding reverses these effects [41,42]. In mice, GC treatment reduces the induction of pro-inflammatory cytokines (IL-1β and TNF-α) and impairs dermal wound closure [61]. Similarly, in rats, hydrocortisone treatment prior to wounding produces an immunocompromised state, which impairs healing by reducing cell proliferation, contraction, and wound tensile strength [53].

These effects of GCs on inflammation are of particular importance to healing. GCs inhibit the production of pro-inflammatory cytokines and also cause resistance of some cells to the inflammatory effects of these molecules [61,79]. Interestingly, the natural circadian rhythm of cortisol is inversely proportional to levels of circulating blood lymphocytes (i.e., T, B and natural killer (NK) cells) and pro-inflammatory cytokines [8]. Stress has been shown to disrupt the circadian rhythm of cortisol [33] which, in turn, can dysregulate inflammatory responses. Under severe stress, low grade systemic inflammation can result (for review, see [54]). As a result, stress has been shown to play a role in the onset, pathogenesis, and severity of numerous inflammatory diseases, including inflammatory bowel disease, hypertension, coronary artery disease, rheumatoid arthritis, psoriasis, and asthma [9]. Many of these effects stem directly from the actions of GCs on immune cells. However, stress also activates the sympathetic nervous system (SNS), which constitutes the second major stress pathway and regulates multiple immune parameters important for healing.

21.4.2 Sympathetic Nervous System (SNS)

The role of the SNS in "stress and wound healing" has often been overlooked, as research has focused more on the effects of GCs. Along with HPA activation, CRH stimulates the locus coeruleus, an autonomic region of the brainstem, to release norepinephrine (NE) from sympathetic nerve endings, which can alter both cell function and blood flow. The adrenal medulla is similarly stimulated to produce EPI. Recently, Gosain et al. [51] studied wound healing in mice that were NE-depleted (NED), via chemical axotomy using 6-hydroxydopamine, and compared them to NE-intact (NEI) mice. Early in the wound healing process, NEI mice had higher infiltrations of neutrophils and macrophages to the wound site, and exhibited quicker re-epithelialization, than NED mice. This suggests that the SNS plays a critical role in normal wound healing. Importantly, dysregulation of the SNS by stress can negatively impact upon important wound healing parameters.

For instance, these catecholamines (NE, EPI) contribute to local tissue edema, increase endothelial cell wall adhesion [75], and inhibit epidermal cell migration [27]. This suggests that increased levels of catecholamines at the wound site, either arising from the systemic circulation or released from local sympathetic nerve endings, can impact upon wound healing processes by increasing edema and retarding epithelialization. Moreover, norepinephrine binding to β-adrenergic receptors increases cAMP levels. This increase is associated with a reduction in pro-inflammatory cytokine levels [99], which can also impair wound healing.

Oxygen influences multiple healing processes, including collagen synthesis, angiogenesis, epithelialization, and metabolic reactions for leukocyte bactericidal action [112]. Moreover, oxygen tension significantly influences revascularization and epithelialization in wounds in a variety of animal models [88]. Overall, the disruption of blood supply and the accumulation of a large population of oxygen consuming cells in the wound make the availability of oxygen a vital requisite for physiologic healing [15,58]. Indeed, wounded tissue is typically hypoxic. Importantly, because catecholamines promote vasoconstriction via α-adrenergic receptor activation [1], stress drives these tissue oxygen levels lower again. This can have profound effects on wound healing.

Decreased oxygen in wounds predicts wound infection in both surgical patients [59] and in animal models [65]. Moreover, the infusion of epinephrine decreases wound oxygen by around 45% [64] and a discontinuation of epinephrine results in a rebound of wound oxygen levels. Finally, a local injection of epinephrine during vaginal hysterectomies has been associated with increased rates of infections in patients [36]. Therefore, stress-induced catecholamines appear to mediate vascular changes that play a key role in the dysregulation of oxygenation, and ultimately impact upon wound infection and healing.

Many other molecules for which immune cells have receptors are also released in association with stress such as arginine vasopressin, α-melanocyte-stimulating hormone, endorphins, enkephalins, substance P, dop amine, and serotonin [9]. However, GCs, NE, and EPI are the three primary stress hormones. The effects of these other mediators will not be discussed in detail in this chapter.

21.5 Human Studies

21.5.1 Effects of Psychological Stress

A number of well-controlled human studies have been conducted, which show that psychological stress impairs a variety of critical health parameters including dermal wound healing (for recent reviews, see [25,47]). Given that caregiving has been associated with immune dysregulation [46,69], Kiecolt-Glaser et al. [68] placed 3.5 mm punch biopsy wounds on the arms of 13 chronically stressed Alzheimer caregivers and 13 age-matched controls. Healing took 4–7 weeks. Alzheimer caregivers took an average of 9 days longer than controls (24% longer) to heal standardized wounds. Group differences in wound sizes were evident during week one and reached statistical significance during week two. Whole blood was obtained and stimulated with endotoxin, from which an ex vivo IL-1β gene expression assay was performed to test the subjects' abilities to mount an inflammatory response. Alzheimer caregivers produced significantly less IL-1β than controls. This concurs with other reports in humans that stress-induced increases in cortisol suppress IL-1β and TNF-α levels [22]. These findings suggest that chronic stress impairs the early inflammatory phase of wound repair, resulting in delayed wound closure.

To further examine this notion that stress affects inflammation, skin blisters were induced by suction on the forearms of 36 women. Women with higher scores on the Perceived Stress Scale had lower levels of IL-1α and IL-8 in the wound sites which, in turn, related to higher levels of salivary cortisol [45]. This is another instance in which stress appears to have inhibited normal inflammatory responses in the skin, this time demonstrated directly in the wound site. These effects were likely mediated by the anti-inflammatory effects of cortisol.

Ebrecht et al. [30] demonstrated that perceived stress and cortisol levels predicted healing rates in a cross sectional study of healthy young males. Subjects received a 4 mm dermal punch biopsy and healing was assessed using high resolution ultrasound scanning, which allows for the determination of width at the base of the wound. This is a more accurate marker of wound healing progress than surface photography [30]. Salivary (unbound) cortisol was measured two weeks before, directly after, and two weeks after wounding. Wound healing was negatively correlated with perceived stress and positively correlated with perceived optimism. Furthermore, when a median split was performed to subdivide slow healers from fast healers, slow healers had significantly higher stress levels, lower trait optimism, and higher cortisol levels to awakening. This study further confirms links among stress, anti-inflammatory stress hormones, and impaired healing.

It has been shown that higher preoperative stress relates to reductions in lymphocyte responses, lymphocyte blood counts [77], and NK cell activity [74,105] following surgery. To examine the effects of preoperative stress on early post-surgical wound repair, Broadbent et al. [13] determined Perceived Stress Scale scores in 47 patients prior to undergoing an open incisional hernia operation. Higher preoperative stress predicted lower IL-1 levels in wound fluid, and greater worry about the operation predicted lower metalloproteinase-9 levels, as well as a more painful, slower, and poorer recovery. Thus, preoperative stress appears to impair inflammatory and matrix degradation processes following surgery, resulting in poorer outcomes.

Multiple studies have examined the effects of stress on wound healing in nondermal tissues, most commonly in mucosal tissues of the mouth. Clinically, it is well accepted that stress affects inflammation of the mucosa and that stress is a risk factor for periodontal disease [11]. Using a within-subjects design, Marucha et al. [78] placed a 3.5 mm circular standardized wound on the hard palate of second year dental students at two different time points. Students were found to heal 40% slower when they were wounded three days prior to examinations (stress) than during summer vacation (nonstress). These data show that something as transient, predictable, and benign, as examination stress can significantly impair wound healing, even in young healthy adults.

21.5.2 Effects of Depression and Other Behavioral Constructs

In a study of marital interactions, 42 couples underwent on separate days a structured social support interaction and a discussion of a marital disagreement. Experimental dermal blister wounds were found to heal slower, and had lower levels of pro-inflammatory cytokines (IL-1β, IL-6, and TNF-α) in the wound fluid, following the conflictive interaction compared to the supportive interaction. In addition, high hostile couples were found to heal at only 60% of the rate of low hostile couples [70]. This further highlights the impact that everyday stressors and behavioral patterns can have on immune parameters such as inflammation and dermal healing.

Using the mucosal wound healing model described earlier, Bosch et al. [10] examined healing rates in individuals with higher depressive symptoms, as changes in many immune parameters occur in clinically depressed patients (for recent review, see [101]). It was found that individuals with higher depressive scores, as assessed by the Beck Depression Inventory, exhibited slower closure of mucosal wounds than subjects with lower scores. This is of particular interest as these subjects were not clinically depressed, but rather they placed further along the depressive spectrum than most individuals. Thus, sub-clinical depression can delay wound closure.

The above-mentioned studies have reported the effects of stress and depression on either acute experimental wounds or surgical recovery. Cole-King and Harding [19] studied the effects of anxiety and depression on the healing of chronic leg ulcers in 53 outpatients. Using the Hospital Anxiety and Depression Score, they found that slower healing was associated with higher anxiety and depression scores. Furthermore, patients who scored in the top 50% were four times more likely to be categorized as slow healers than patients who scored in the bottom 50%. Thus, psychological stress, such as anxiety or depression, can delay the healing of both acute and chronic wounds.

Post-operative pain can also act as a stressor. It was recently shown that acute and especially persistent pain after elective gastric bypass predicted slower closure of 2 mm dermal wounds. This was independent of depressive symptoms, pre-existing pain, or medical symptoms following discharge [81]. Although surgery dysregulates a variety of immune parameters (e.g., inflammation), these effects are lessened by analgesics [7]. Together, this suggests that pain is a unique form of stress. As a result, pain may cause or exacerbate stress-impaired healing.

To summarize, it is clear that psychological stress in many forms can negatively impact upon wound healing and surgical recovery in humans. Longer healing times, altered inflammatory responses, and higher circulating cortisol levels have been consistently related to chronic stress.

21.5.3 Health Behaviors

It is important to note that stress can affect immunity and healing both directly and indirectly. Its direct effects are primarily mediated by activation of the HPA and SNS pathways as previously discussed. Indirect stress effects stem from associated negative emotions and behavioral alterations such as changes in sleep or diet, increased consumption of alcohol or nicotine, reduced exercise, and self-neglect [11]. For instance, sleep deprivation can hinder immunity, disrupting macrophage/lymphocyte functions and pro-inflammatory cytokine levels (for review, see [63]). These behavioral changes may in turn promote stress, often in the form of depression, loneliness, or anxiety, which then impact further on immune parameters creating a downward spiral effect. Another behavioral change is smoking, which often increases during stress and has been shown to impair collagen deposition and alter extracellular matrix turnover in wound healing [66,73]. Interestingly, smoking intervention programs implemented 6–8 weeks prior to surgery have been found to reduce post-operative morbidity, including wound related complications [83]. Although a full discussion of these indirect behavioral effects is beyond the scope of this chapter, it is important to recognize their contribution to stress-impaired health.

Interventions that lower stress levels have shown some promise in reducing healing impairments. A meta-analysis of 191 studies by Devine [24] indicated that psychoeducational care can reduce anxiety and distress to patients, and ultimately lead to better recovery with reduced pain and hospital stays. Field et al. [39] demonstrated that massage therapy decreased levels of anxiety, depression, cortisol, and pain in burn patients undergoing debridement. In addition, Holden-Lund [57] showed that patients who underwent cholecystectomy and utilized Relaxation with Guided Imagery had lower state anxiety, lower urinary cortisol levels, and less wound erythema compared to randomly selected controls. Thus, interventions aimed at lowering stress such as the provision of patient education, massage therapy, or relaxation techniques appear to have some merit in reducing the negative effects of stress on surgical outcomes.

21.6 Animal Studies

One of the drawbacks to human research is that it is largely correlational in nature and, hence, it can be difficult to determine mechanism(s). Conversely, animal studies have less variability, and are better designed to examine both causation and mechanism(s) of action. Interestingly, the effects of stress on immunity and healing are quite consistent across both human and animal studies. Although mostly studied in mice [86,82], stress has been shown to affect dermal healing and/or healing parameters in other rodent species such as hamsters [71], voles [48], and rhesus monkeys [67], and typically does so to a degree similar to humans.

21.6.1 Stress and the HPA Axis

The previously described findings that GCs can alter the course of healing suggest that stress itself, through its effects on the HPA axis, might impair healing. In a seminal paper in 1998, Padgett et al. tested this hypothesis. Mice were restrained in 50 ml well-ventilated conical tubes for 12 h per day, 3 days prior to and 5 days after receiving two 3.5 mm dermal back wounds. Control mice were food- and water-deprived during the same time period. Mice that underwent restraint stress healed 27% more slowly (i.e., 3 days longer) than nonrestrained control mice. This negative effect of stress on healing was robust and has since been replicated many times using this same model [97,40,60,55].

Mice subjected to restraint stress had wounds, which exhibited lower levels of pro-inflammatory cytokines (e.g., IL-1β) and growth factors (e.g., KGF) [86,82,55], and higher counts of opportunistic bacteria [97] one day after wounding. Conversely, on days 3 and 5, postwounding neutrophil counts in the wounds were higher in stressed vs. nonstressed mice (unpublished observations). As the clearance of neutrophils from the wound site typically coincides with the end of the inflammatory phase of healing, this suggests that the wounds of stressed animals remained inflamed for a period of time beyond that of controls. In support of this, wounds of stressed mice had elevated gene expression for chemokines (KC, MCP-1, MIP-1α, and MIP-2) at days 3 and 5 (unpublished observations), and for cytokines (IL-1α and IL-1β) at day 5 [76], compared to nonstressed mice. These results are in line with evidence that neutrophils in the wound site can delay wound closure [28]. Fibroblast migration, differentiation and, in turn, wound contraction were also impaired in mice that were stressed [60]. Finally, granulation tissue in the wound site was reduced in the stressed vs. nonstressed condition when assessed 5 days post-wounding [82].

To summarize, stress appears to have effects throughout the healing process. It reduces inflammation, resulting in higher bacteria counts during the early healing process. By days 3–5, neutrophil counts and inflammatory infiltrate are higher in the wounds of stressed animals compared to controls, suggesting that prolonged inflammation occurs in the wound site due to stress. The outcome of these effects is slower wound closure in stressed vs. nonstressed animals.

Interestingly, treatment of stressed mice with the GC antagonist RU486 restored normal expression of IL-1β, but only partially ameliorated normal healing and to a lesser extent microbial clearance [97]. Because not all immune impairments were restored by GC receptor blockade, these findings indicate that a second stress-activated pathway impacts upon dermal healing. As previously discussed, this pathway is the SNS, which exerts its effects on wound healing largely through altering blood and oxygen supplies to tissues.

21.6.2 Stress and the SNS

Human studies have demonstrated the importance of tissue oxygen during healing, and healing rates are limited by oxygen availability [62]. Activation of the SNS can cause peripheral vasoconstriction, resulting in tissue hypoxia which hinders both tissue repair and microbial clearance. Not surprising, oxygen modulated genes are also dysregulated by stress. For example, inductible nitric oxide synthase (iNOS) gene expression is upregulated threefold in the wounds of restraint-stressed mice [40]. Interestingly, hyperbaric oxygen (HBO) treatment restores both healing and iNOS gene expression to normal levels in this same model [40], indicating a critical role for oxygen in this model of stress-impaired healing.

Studies in mice have shown that peripheral NE release modulates both the inflammatory and proliferative phases of wound healing. As a result, a dysregulation of NE often occurs during chronic stress and can be detrimental to wound healing. In restraint-stressed mice, nonspecific blockade of peripheral α-adrenergic receptors has been shown to attenuate impairments in wound closure and contraction, and normalize edema. This was not seen when peripheral β-adrenergic receptors were blocked [32]. Conversely, in vitro studies have shown that activation of the β2-adrenergic receptor subtype delays wound healing [90], and inhibition of this same receptor accelerates healing [91] via promigratory and pro-proliferative pathways [89].

To summarize, restraint stress delays dermal healing in mice through activation of the HPA axis. However, GC blockade attenuates, but does not fully block, these stress-induced healing impairments. Evidence shows that tissue oxygen levels mediate these effects as well, indicating a critical role for the SNS in this stress model. Both α- and β-adrenergic receptors appear to be involved in modulating these effects. Thus, inhibiting activation of *both* the HPA axis and the SNS simultaneously during wound healing may provide optimal healing responses in stressed individuals. Importantly, various stressors may differentially activate these two pathways. This may explain the inconsistencies of reported "stress effects" on immunity when different stress paradigms are used.

21.6.3 Alternative Stressors/Models

Aside from rodents and humans, stress has been seen to alter the molecular mechanisms of healing in non-human primates. Confinement stress decreased the expression of pro-inflammatory chemokines (IL-8, MIP-1α) associated with early healing in the dermal wounds of rhesus monkeys [67]. Interestingly, lesions of the central nucleus of the amygdala, which mediates anxiety and fear responses, increased the expression of these chemokines.

Other studies have shown that factors that are anxiolytic (reduce anxiety) can alleviate stress-induced wound healing deficits. For instance, immobilization stress in Siberian hamsters delays wound healing in isolated but not in socially housed animals [23]. Social housing, which is anxiolytic to these animals, appears to improve wound healing through a suppression of HPA axis reactivity to stress [23]. It has been similarly shown in some mouse species that social contact facilitates wound healing [48].

It is important to note that not all chronic stressors affect immunity and healing equally. Social disruption, in which an aggressive intruder mouse defeats all mice in a cage thereby changing the group's social order, has also been examined in the context of wound healing. This disruption occurs for 2 h per day, repeatedly for six days. Interestingly, although this stressor is severe enough to cause the formation of GC resistance in splenocytes [102], it has no effect on wound healing rates [103]. Similarly, restraint stress in mice for only a few hours per day does not delay wound healing (unpublished observations). Therefore, both the *type* and the *duration* of a stressor are critical components in determining the effects of stress on dermal healing.

21.7 Stress and the Skin Barrier

The epithelium provides a crucial barrier, providing resistance to pathogens and limiting the loss of water from the body. Tape stripping is a relatively noninvasive technique used to assess the repair of this barrier. Cellophane tape is repeatedly applied and removed from an area of dermis, commonly to the forearm, to cause skin barrier disruption. Barrier recovery is then assessed by using an evaporimeter to measure transepidermal water loss (TEWL) over time. Studies in hairless mice have shown that stress slows barrier recovery times, and anxiolytic drugs (i.e., tranquilizers) reduce this effect [21]. In humans, interview stress (Trier Social Stress Test) has been demonstrated to slow barrier recovery kinetics, and to increase circulating levels of pro-inflammatory cytokines [72,2]. Moreover, cytokine responses were inversely correlated to skin barrier recovery, suggesting that stress-induced changes in cytokine secretion may lead to changes in barrier function homeostasis [2]. Muizzuddin et al. [84] showed that the stress associated with marital dissolution related to slower barrier recovery times but not to initial skin barrier strength, defined as the number of tape strippings required to disrupt the skin barrier (TEWL $\geq 18\,g/m^2/h$). Other studies have demonstrated that slower barrier recovery times were related to the stress caused by university examinations [43] and by sleep deprivation [2]. These effects were generally the greatest in the individuals reporting the most stress. Thus, it is clear that stress affects skin barrier recovery, which is an important factor in many skin diseases (e.g., atopic dermatitis, psoriasis) and in dermal wound healing [43]. Interestingly, stress appears to impact these disease states and dermal healing, in part, by causing a generalized state of low grade inflammation.

21.8 Stress and Infection

Infection is the invasion and proliferation of microorganisms into body tissues. If the body's defenses are effective, these microorganisms will either be eliminated or remain localized with sub-clinical effects. However, under conditions of stress, microorganisms may continue to proliferate and produce acute infection with localized clinical effects. Chronic infection is the persistence of this state. Finally, systemic infection may result if the microorganisms gain entry into the lymphatic or vascular systems.

Infected wounds heal slower with an increased chance of scarring [95]. The ability to prevent wound infection is reduced by various stressors such as foot shock in rats [104] and restraint stress in mice [14,97]. For example, in mice, the bacterial count after restraint stress in either excisional [97] or incisional wounds (unpublished observations) is greater than four logs more than controls, and at a level considered an indication of wound infection (i.e., $>10^5$ colony forming units [CFU]/g tissue) [96]. Using this criterion, 85.4% of excisional wounds became infected in stressed mice vs. 27.4% in controls [97], resulting in a three-fold increase in infection. Clearly, stress can greatly increase the risk of infection in healing wounds.

In a separate experiment, 10^7 CFU of group G *Streptococcus* were applied to the wound surface immediately after wounding. Although sepsis has been shown to retard wound healing [94], this application of bacteria surprisingly improved healing rates in nonstressed control mice. This was presumably due to enhanced inflammation, which quickens wound contraction. Interestingly, in stressed mice healing rates were not affected by this application of bacteria, and bacterial numbers remained higher than controls. Thus, under conditions of high microbial contamination, nonstressed mice exhibited accelerated wound closure and retained the ability to clear microbial contaminants, whereas stressed mice were unable to do either.

As previously discussed, multiple studies have demonstrated that stress can slow wound closure. In as much as slower wound healing increases the chance of infection [95], persistence of bacteria in the wound site further delays wound closure by causing tissue damage and inhibiting re-epithelialization. Stress may also reduce neutrophil function and prolong neutrophil recruitment (unpublished observations), thereby increasing neutrophil numbers at the wound site. Neutrophils themselves cause tissue damage through the oxidative burst, which is an important mechanism for killing microbes. In addition, neutrophils release substances that accelerate keratinocyte differentiation, thereby slowing keratinocyte proliferation and re-epithelialization [29]. Thus, through a variety of mechanisms stress may slow wound closure and promote infection, both of which promote each other.

Oxygen is reduced in the wound sites of stressed mice, and these levels rebound once stress is terminated. Similarly, the number of microbes in the wounds of stressed mice drops dramatically once stress ends [40]. These findings suggest that oxygen is critical for microbial clearance. This seems likely, as neutrophils in the wound site have high oxygen requirements for killing bacteria. For this reason, keeping a wound open may help neutrophils function better, as oxygen levels are reduced upon wound closure. Interestingly, inhibiting neutrophils in wounds has been shown to accelerate wound closure without altering the quality of healing [28]. As discussed above, although neutrophils provide protection against infection they may simultaneously slow healing rates. Recently, their necessity in the healing of sterile (i.e., surgical) wounds has been questioned [28].

21.9 Aging and Wound Healing in Humans

It is well accepted that aging alters skin morphology [5]. Different studies have shown that older adult skin undergoes reductions in vascularization [110], collagen density [20,50,110], granulation tissue [110], elastin [4,109], and fibroblast numbers [110]. However, whether such morphological changes necessarily translate to delayed/impaired healing has been unclear. In addition, age-related differences in skin exist that are related to exposure to the sun, cell turnover, availability of stems cells, etc. [109]. Again, these differences do not necessarily translate to age-related deficits. To date, the mechanisms by which aging impairs wound healing are not fully understood [3].

Numerous studies have reported delayed wound healing in the elderly (65+ years) [38,44,49]. Such findings have been criticized, however, because confounding factors more common in the elderly, such as morbidity and medication use, have not been adequately controlled [3,5,109]. In addition, these have mostly been clinical reports so wounds were not standardized between individuals. Therefore, while it is generally accepted that aging slows wound healing in animal models [92,108], until recently it has been unclear that aging per se delays wound healing in humans.

Using the previously described mucosal wound model, Engeland et al. assessed the effects of aging on mucosal wound healing for 7 days post-wounding in humans. Older individuals (50–88 years; $n = 93$) healed mucosal wounds significantly slower than younger adults (18–35 years; $n = 119$), regardless of gender. At five days post-wounding, wounds were 56% larger in older vs. younger subjects, and younger individuals were 3.7 times more likely to be considered healed than older individuals [35]. Additional analyses were performed, which excluded individuals that were on any type of medication (excluding allergy medication, birth control, or nutritional supplements) and/or ever had a serious medical condition (e.g., diabetes, cancer, stroke, heart disease, hypertension, hypothyroidism, arthritis, other inflammatory disease, psychopathology). When these individuals were excluded from the analyses, older individuals were still found to heal slower than younger subjects. Surprisingly, the exclusion of individuals on medication strengthened the finding of an age-associated deficit in wound healing [35]. Thus, age-associated delays were not exaggerated by medication use. Moreover, these findings suggest the detrimental effects of aging on wound healing may be even stronger than previously suspected.

The effects of age appear to be interactive with other known risk factors for delayed healing. For instance, women were found to heal oral mucosal wounds slower than men, and older women (50+ years) healed the slowest of all subgroups (i.e., younger and older, men and women). Compared with younger men (18–35 years of age), who were the fastest healing adult group, wound closure in older women was delayed by up to 95% [35]. Elderly individuals also have a higher frequency of chronic wounds (e.g., diabetic ulcers) than young adults [109], and concomitant morbidities (e.g., diabetes) which are more prevalent in the aged can certainly impair wound healing.

Aging is known to influence the HPA axis. It is generally accepted that peak levels of GCs are not substantially different in older vs. younger adults, but that termination of the stress response becomes impaired with aging [80,100]. Thus, GCs are expressed for a longer period of time in the elderly following stress. Such an extended period of GC secretion can result in lower cell recruitment and an inhibition of cytokines and growth factors during healing. In addition, advanced age is associated with a hyper-inflammatory state, evidenced by higher circulating levels of pro-inflammatory cytokines and prostaglandins [76]. This is similar to what happens during chronic stress (for review, see [54]) and can hinder appropriate immune responses upon injury. Hence, the impact of stress and age on immunity is interactive, with stress exacerbating the effects of aging [52]. This interaction between stress and aging may partially explain the increased variability in healing among older adults, and the morphological differences in tissues between older and younger adults.

21.10 Summary

Optimized dermal healing requires carefully regulated inflammation, which is rapidly resolved once bacterial contamination has been eliminated. Recruitment of neutrophils is then discontinued, at which point remaining neutrophils undergo phagocytosis by macrophages. If these processes are delayed or impaired in any way, for instance through reduced early recruitment or functioning of neutrophils, a period of extended inflammation results, leading to higher risks of both infection and scarring [95].

Chronic stress, in many forms (e.g., depression), has been shown to negatively impact healing rates. Stress-induced activation of the HPA axis inhibits inflammatory cytokine expression during the healing process. This not only impacts on inflammatory cell function during early healing, but also reduces cytokine-induced growth factor expression. SNS activity causes vasoconstriction, thereby reducing oxygen levels in tissue, which can damage reparative cells. Oxygen is necessary for proper neutrophil function. Hence, tissue hypoxia impedes the clearance of microbes from the site of injury. As a result, stress prolongs the inflammatory phase of healing, inhibits re-epithelialization, and delays wound closure. Aside from infection, this delay can result in altered integrity of the injured tissue, compromised esthetics, and impaired function [95].

Many different factors act as stressors and/or interact with stress such as pain, age, and medical conditions such as diabetes. The average age of our society is on the incline. Thus, an increasing number of people are at risk for stress-impaired healing and the need to identify such individuals prior to elective surgery is growing. Fortunately, the quantity of research and resources being dedicated to this important area is on the rise. With the increasing stressful lifestyle of modern society and with an increase in elective surgery, including cosmetic surgery, the need for optimized healing is high. The simultaneous targeting of both the HPA axis and the SNS, along with behavioral therapies aimed at reducing stress, provides a good potential model for "normalizing" wound healing in chronically stressed individuals. However, a greater understanding of the mechanisms involved in stress-impaired healing is needed to address these problems therapeutically.

Summary for the Clinician

- The available data strongly suggest that individuals undergoing chronic stress are at higher risk for delayed healing following surgery. Delays in wound closure, in turn, relate to higher rates of infection and medical complications.
- The effects of chronic stress on dermal wound healing include reduced early inflammation and higher bacterial counts. This is followed by delayed clearance of cellular infiltrates and a period of prolonged inflammation. The end result is slower contraction and re-epithelialization of the wound, and a poorer healing outcome (e.g., increased scarring).
- HPA activation is potently immunosuppressive and inhibits inflammation. SNS activation causes vasoconstriction, which reduces oxygen supply to the wound site and, in turn, hinders bacterial clearance and re-epithelialization. Concurrent inhibition of the HPA and SNS pathways may be a good therapeutic target for ameliorating impaired healing in individuals who are chronically stressed.

- Attempts should be made to reduce or avoid stress prior to elective surgeries. Scheduling surgery immediately after a stressful period such as University examinations may be unwise. These considerations are especially important in the presence of other risk factors for delayed healing, such as diabetes or old age, as the negative effects of stress on healing are interactive with such factors.
- Techniques aimed at reducing pre-surgical stress have some benefit. Post-surgical stress and pain can similarly prolong healing rates and should be minimized to whatever extent possible.

References

1. Ahlquist RP (1976) Present state of alpha- and beta-adrenergic drugs I. The adrenergic receptor. Am Heart J 92:661–664
2. Altemus M, Rao B, Dhabhar FS, Ding W, Granstein RD (2001) Stress-induced changes in skin barrier function in healthy women. J Invest Dermatol 117:309–317
3. Ashcroft GS, Horan MA, Ferguson MW (1995) The effects of ageing on cutaneous wound healing in mammals. J Anat 187:1–26
4. Ashcroft GS, Kielty CM, Horan MA, Ferguson MW (1997) Age-related changes in the temporal and spatial distributions of fibrillin and elastin mRNAs and proteins in acute cutaneous wounds of healthy humans. J Pathol 183:80–89
5. Ashcroft GS, Mills SJ, Ashworth JJ (2002) Ageing and wound healing. Biogerontology 3:337–345
6. Beer HD, Fassler R, Werner S (2000) Glucocorticoid-regulated gene expression during cutaneous wound repair. Vitam Horm 59:217–239
7. Beilin B, Shavit Y, Trabekin E, Mordashev B, Mayburd E, Zeidel A, Bessler H (2003) The effects of postoperative pain management on immune response to surgery. Anesth Analg 97:822–827
8. Berczi I (1986) The influence of pituitary-adrenal axis on the immune system. In: Berczi I (ed) Pituitary Function and Immunity. CRC Press, Boca Raton, Florida, p 49
9. Black PH, Berman AS (1999) Stress and inflammation. In: Plotnikoff NP, Faith RE, Murgo AJ, Good RA (eds) Cytokines: Stress and Immunity. CRC Press, Boca Raton, Florida, pp 115–132
10. Bosch JA, Engeland CG, Cacioppo JT, Marucha PT (2007) Depressive symptoms predict mucosal wound healing. Psychosom Med 69:597–605
11. Boyapati L, Wang HL (2007) The role of stress in periodontal disease and wound healing. Periodontol 2000 44:195–210
12. Brauchle M, Fassler R, Werner S (1995) Suppression of keratinocyte growth factor expression by glucocorticoids in vitro and during wound healing. J Invest Dermatol 105:579–584
13. Broadbent E, Petrie KJ, Alley PG, Booth RJ (2003) Psychological stress impairs early wound repair following surgery. Psychosom Med 65:865–869
14. Brown DH, Sheridan J, Pearl D, Zwilling BS (1993) Regulation of mycobacterial growth by the hypothalamus-pituitary-adrenal axis: differential responses of Mycobacterium bovis BCG-resistant and -susceptible mice. Infect Immun 61:4793–4800
15. Chang N, Goodson WH III, Gottrup F, Hunt TK (1983) Direct measurement of wound and tissue oxygen tension in postoperative patients. Ann Surg 197:470–478
16. Chedid M, Hoyle JR, Csaky KG, Rubin JS (1996) Glucocorticoids inhibit keratinocyte growth factor production in primary dermal fibroblasts. Endocrinology 137:2232–2237
17. Chrousos GP (1998) Stressors, stress, and neuroendocrine integration of the adaptive response. The 1997 Hans Selye Memorial Lecture. Ann N Y Acad Sci 851:311–335
18. Clark RA, Gallin JI, Fauci AS (1979) Effects of in vivo prednisone on in vitro eosinophil and neutrophil adherence and chemotaxis. Blood 53:633–641
19. Cole-King A, Harding KG (2001) Psychological factors and delayed healing in chronic wounds. Psychosom Med 63:216–220
20. Cook JL, Dzubow LM (1997) Aging of the skin: implications for cutaneous surgery. Arch Dermatol 133:1273–1277
21. Denda M, Tsuchiya T, Elias PM, Feingold KR (2000) Stress alters cutaneous permeability barrier homeostasis. Am J Physiol Regul Integr Comp Physiol 278:R367–R372
22. DeRijk R, Michelson D, Karp B, Petrides J, Galliven E, Deuster P, Paciotti G, Gold PW, Sternberg EM (1997) Exercise and circadian rhythm-induced variations in plasma cortisol differentially regulate interleukin-1 beta (IL-1 beta), IL-6, and tumor necrosis factor-alpha (TNF alpha) production in humans: high sensitivity of TNF alpha and resistance of IL-6. J Clin Endocrinol Metab 82:2182–2191
23. Detillion CE, Craft TK, Glasper ER, Prendergast BJ, Devries AC (2004) Social facilitation of wound healing. Psychoneuroendocrinology 29:1004–1011
24. Devine EC (1992) Effects of psychoeducational care for adult surgical patients: a meta-analysis of 191 studies. Patient Educ Couns 19:129–142
25. Devries AC, Craft TK, Glasper ER, Neigh GN, Alexander JK (2007) 2006 Curt P. Richter award winner Social influences on stress responses and health. Psychoneuroendocrinology 32:587–603
26. Dhabhar FS (2003) Stress, leukocyte trafficking, and the augmentation of skin immune function. Ann N Y Acad Sci 992:205–217
27. Donaldson DJ, Mahan JT (1984) Influence of catecholamines on epidermal cell migration during wound closure in adult newts. Comp Biochem Physiol C 78:267–270
28. Dovi JV, He LK, DiPietro LA (2003) Accelerated wound closure in neutrophil-depleted mice. J Leukoc Biol 73:448–455
29. Dovi JV, Szpaderska AM, DiPietro LA (2004) Neutrophil function in the healing wound: adding insult to injury? Thromb Haemost 92:275–280
30. Ebrecht M, Hextall J, Kirtley LG, Taylor A, Dyson M, Weinman J (2004) Perceived stress and cortisol levels predict speed of wound healing in healthy male adults. Psychoneuroendocrinology 29:798–809
31. Edwards JC, Dunphy JE (1958) Wound healing. II. Injury and abnormal repair. N Eng J Med 259:275–285
32. Eijkelkamp N, Engeland CG, Gajendrareddy PK, Marucha PT (2007) Restraint stress impairs early wound healing in mice via alpha-adrenergic but not beta-adrenergic receptors. Brain Behav Immun 21:409–412

33. Elenkov IJ, Chrousos GP (2006) Stress system – organization, physiology and immunoregulation. Neuroimmunomodulation 13:257–267
34. Eming S, Kreig T, Davidson JM (2007) Inflammation in wound repair: molecular and cellular mechanisms. J Invest Dermato 127:514–525
35. Engeland CG, Bosch JA, Cacioppo JT, Marucha PT (2006) Mucosal wound healing: the roles of age and sex. Arch Surg 141:1193–1197
36. England GT, Randall HW, Graves WL (1983) Impairment of tissue defenses by vasoconstrictors in vaginal hysterectomies. Obstet Gynecol 61:271–274
37. Fauci AS, Dale DC, Balow JE (1976) Glucocorticosteroid therapy: mechanisms of action and clinical considerations. Ann Intern Med 84:304–315
38. Fenske NA, Lober CW (1986) Structural and functional changes of normal aging skin. J Am Acad Dermatol 15:571–585
39. Field T, Peck M, Krugman S, Tuchel T, Schanberg S, Kuhn C, Burman I (1998) Burn injuries benefit from massage therapy. J Burn Care Rehabil 19:241–244
40. Gajendrareddy PK, Sen CK, Horan MP, Marucha PT (2005) Hyperbaric oxygen therapy ameliorates stress-impaired dermal wound healing. Brain Behav Immun 19:217–222
41. Gallucci RM, Simeonova PP, Matheson JM, Kommineni C, Guriel JL, Sugawara T, Luster MI (2000) Impaired cutaneous wound healing in interleukin-6-deficient and immunosuppressed mice. FASEB J 14:2525–2531
42. Gallucci RM, Sugawara T, Yucesoy B, Berryann K, Simeonova PP, Matheson JM, Luster MI (2001) Interleukin-6 treatment augments cutaneous wound healing in immunosuppressed mice. J Interferon Cytokine Res 21:603–609
43. Garg A, Chren MM, Sands LP, Matsui MS, Marenus KD, Feingold KR, Elias PM (2001) Psychological stress perturbs epidermal permeability barrier homeostasis: implications for the pathogenesis of stress-associated skin disorders. Arch Dermatol 137:53–59
44. Gerstein AD, Phillips TJ, Rogers GS, Gilchrest BA (1993) Wound healing and aging. Dermatol Clin 11:749–757
45. Glaser R, Kiecolt-Glaser JK, Marucha PT, MacCallum RC, Laskowski BF, Malarkey WB (1999) Stress-related changes in proinflammatory cytokine production in wounds. Arch Gen Psychiatr 56:450–456
46. Glaser R, Sheridan J, Malarkey WB, MacCallum RC, Kiecolt-Glaser JK (2000) Chronic stress modulates the immune response to a pneumococcal pneumonia vaccine. Psychosom Med 62:804–807
47. Glaser R, Kiecolt-Glaser JK (2005) Stress-induced immune dysfunction: implications for health. Nat Rev Immunol 5:243–251
48. Glasper ER, Devries AC (2005) Social structure influences effects of pair-housing on wound healing. Brain Behav Immun 19:61–68
49. Goodson WH III, Hunt TK (1979) Wound healing and aging. J Invest Dermatol 73:88–91
50. Gosain A, DiPietro LA (2004) Aging and wound healing. World J Surg 28:321–326
51. Gosain A, Jones SB, Shankar R, Gamelli RL, DiPietro LA (2006) Norepinephrine modulates the inflammatory and proliferative phases of wound healing. J Trauma 60:736–744
52. Graham JE, Christian LM, Kiecolt-Glaser JK (2006) Stress, age, and immune function: toward a lifespan approach. J Behav Med 29:389–400
53. Gupta A, Jain GK, Raghubir R (1999) A time course study for the development of an immunocompromised wound model, using hydrocortisone. J Pharmacol Toxicol Methods 41:183–187
54. Hawkley LC, Bosch JA, Engeland CG, Cacioppo JT, Marucha PT (2007) Loneliness, dysphoria, stress, and immunity: a role for cytokines. In: Plotnikoff NP, Faith RE, Murgo AJ (eds) Cytokines: Stress and Immunity. CRC Press, Boca Raton, Florida, pp 67–85
55. Head CC, Farrow MJ, Sheridan JF, Padgett DA (2006) Androstenediol reduces the anti-inflammatory effects of restraint stress during wound healing. Brain Behav Immun 20:590–596
56. Hinz B (2007) Formation and function of the myofibroblast during tissue repair. J Invest Dermatol 127:526–537
57. Holden-Lund C (1988) Effects of relaxation with guided imagery on surgical stress and wound healing. Res Nurs Health 11:235–244
58. Hopf HW, Rollins MD (2007) Wounds: an overview of the role of oxygen. Antioix Redox Signal 8:1183–1192
59. Hopf HW, Hunt TK, West JM, Blomquist P, Goodson WH III, Jensen JA, Jonsson K, Paty PB, Rabkin JM, Upton RA, von Smitten K, Whitney JD (1997) Wound tissue oxygen tension predicts the risk of wound infection in surgical patients. Arch Surg 132:997–1004
60. Horan MP, Quan N, Subramanian SV, Strauch AR, Gajendrareddy PK, Marucha PT (2005) Impaired wound contraction and delayed myofibroblast differentiation in restraint-stressed mice. Brain Behav Immun 19:207–216
61. Hubner G, Brauchle M, Smola H, Madlener M, Fassler R, Werner S (1996) Differential regulation of pro-inflammatory cytokines during wound healing in normal and glucocorticoid-treated mice. Cytokine 8:548–556
62. Hunt TK, Pai MP (1972) The effect of varying ambient oxygen tensions on wound metabolism and collagen synthesis. Surg Gynecol Obstet 135:561–567
63. Irwin M (2002) Effects of sleep and sleep loss on immunity and cytokines. Brain Behav Immun 16:503–512
64. Jensen JA, Jonsson K, Goodson WH III, Hunt TK, Roizen MF (1985) Epinephrine lowers subcutaneous wound oxygen tension. Curr Surg 42:472–474
65. Jonsson K, Hunt TK, Mathes SJ (1988) Oxygen as an isolated variable influences resistance to infection. Ann Surg 208:783–787
66. Jorgensen LN, Kallehave F, Christensen E, Siana JE, Gottrup F (1998) Less collagen production in smokers. Surgery 123:450–455
67. Kalin NH, Shelton SE, Engeland CG, Haraldsson HM, Marucha PT (2006) Stress decreases, while central nucleus amygdala lesions increase, IL-8 and MIP-1alpha gene expression during tissue healing in non-human primates. Brain Behav Immun 6:564–568
68. Kiecolt-Glaser JK, Marucha PT, Malarkey WB, Mercado AM, and Glaser R (1995) Slowing of wound healing by psychological stress. Lancet 346:1194–1196
69. Kiecolt-Glaser JK, Glaser R, Gravenstein S, Malarkey WB, Sheridan J (1996) Chronic stress alters the immune response to influenza virus vaccine in older adults. Proc Natl Acad Sci U S A 93:3043–3047
70. Kiecolt-Glaser JK, Loving TJ, Stowell JR, Malarkey WB, Lemeshow S, Dickinson SL, Glaser R (2005) Hostile marital

interactions, proinflammatory cytokine production, and wound healing. Arch Gen Psychiatr 62:1377–1384
71. Kinsey SG, Prendergast BJ, Nelson RJ (2003) Photoperiod and stress affect wound healing in Siberian hamsters. Physiol Behav 78:205–211
72. Kirschbaum C, Pirke KM, Hellhammer DH (1993) The 'Trier Social Stress Test' – a tool for investigating psychobiological stress responses in a laboratory setting. Neuropsychobiology 28:76–81
73. Knuutinen A, Kokkonen N, Risteli J, Vahakangas K, Kallioinen M, Salo T, Sorsa T, Oikarinen A (2002) Smoking affects collagen synthesis and extracellular matrix turnover in human skin. Br J Dermatol 146:588–594
74. Koga C, Itoh K, Aoki M, Suefuji Y, Yoshida M, Asosina S, Esaki K, Kameyama T (2001) Anxiety and pain suppress the natural killer cell activity in oral surgery outpatients. Oral Surg Oral Med Oral Pathol Oral Radiol Endod 91:654–658
75. Koopman CF (1995) Cutaneous wound healing, an overview. Otolaryngol Clin North Am 28:835–845
76. Kovacs EJ (2005) Aging, traumatic injury, and estrogen treatment. Exp Gerontol 40:549–555
77. Linn BS, Linn MW, Klimas NG (1988) Effects of psychophysical stress on surgical outcome. Psychosom Med 50:230–244
78. Marucha PT, Kiecolt-Glaser JK, Favagehi M (1998) Mucosal wound healing is impaired by examination stress. Psychosom Med 60:362–365
79. Mastorakos G, Bamberger C, Chrousos GP (1999) Neuroendocrine regulation of the immune process. In: Plotnikoff NP, Faith RE, Murgo AJ, Good RA (eds) Cytokines: stress and immunity. CRC Press, Boca Raton, Florida, pp 17–37
80. McEwen BS, Stellar E (1993). Stress and the individual: mechanisms leading to disease. Arch Internal Med 153:2093–2101
81. McGuire L, Heffner K, Glaser R, Needleman B, Malarkey W, Dickinson S, Lemeshow S, Cook C, Muscarella P, Melvin WS, Ellison EC, Kiecolt-Glaser JK (2006) Pain and wound healing in surgical patients. Ann Behav Med 31:165–172
82. Mercado AM, Padgett DA, Sheridan JF, Marucha PT (2002) Altered kinetics of IL-1 alpha, IL-1 beta, and KGF-1 gene expression in early wounds of restrained mice. Brain Behav Immun 16:150–162
83. Moller AM, Villebro N, Pedersen T, Tonnesen H (2002) Effect of preoperative smoking intervention on postoperative complications: a randomised clinical trial. Lancet 359:114–117
84. Muizzuddin N, Matsui MS, Marenus KD, Maes DH (2003) Impact of stress of marital dissolution on skin barrier recovery: tape stripping and measurement of trans-epidermal water loss (TEWL). Skin Res Technol 9:34–38
85. Norris DA, Capin L, Weston WL (1982) The effect of epicutaneous glucocorticosteroids on human monocyte and neutrophil migration in vivo. J Invest Dermatol 78:386–390
86. Padgett DA, Marucha PT, Sheridan JF (1998) Restraint stress slows cutaneous wound healing in mice. Brain Behav Immun 12:64–73
87. Page-McCaw A, Ewald A, Werb Z (2007) Matrix metalloproteinases and the regulation of tissue remodeling. Nat Rev Mol Cell Biol 8:221–233
88. Pai MP, Hunt TK (1972) Effect of varying oxygen tensions on healing of open wounds. Surg Gynecol Obstet 135:756–758
89. Pullar CE, Isseroff RR (2006) The beta 2-adrenergic receptor activates pro-migratory and pro-proliferative pathways in dermal fibroblasts via divergent mechanisms. J Cell Sci 119:592–602
90. Pullar CE, Grahn JC, Liu W, Isseroff RR (2006) Beta2-adrenergic receptor activation delays wound healing. FASEB J 20:76–86
91. Pullar CE, Rizzo A, Isseroff RR (2006) beta-Adrenergic receptor antagonists accelerate skin wound healing: evidence for a catecholamine synthesis network in the epidermis. J Biol Chem 281:21225–21235
92. Quirinia A, Viidik A (1991) The influence of age on the healing of normal and ischemic incisional skin wounds. Mech Ageing Dev 58:221–232
93. Raja, Sivamani K, Garcia MS, Isseroff RR (2007) Wound re-epthelialization: modulating keratinocyte migration in wound healing. Front Biosci 12:2849–2868
94. Rico RM, Ripamonti R, Burns AL, Gamelli RL, DiPietro LA (2002) The effect of sepsis on wound healing. J Surg Res 102:193–197
95. Robson MC (1997) Wound infection. A failure of wound healing caused by an imbalance of bacteria. Surg Clin North Am 77:637–650
96. Robson MC, Heggers JP (1969) Bacterial quantification of open wounds. Mil Med 134:19–24
97. Rojas IG, Padgett DA, Sheridan JF, Marucha PT (2002) Stress-induced susceptibility to bacterial infection during cutaneous wound healing. Brain Behav Immun 16:74–84
98. Ryan TJ (2007) Infection following soft tissue injury: its role in wound healing. Curr Opin Infect Dis 20:124–128
99. Sanders VM, Straub RH (2002) Norepinephrine, the beta-adrenergic receptor, and immunity. Brain Behav Immun 16:290–332
100. Sapolsky RM, Krey LC, McEwen BS (1986). The neuroendocrinology of stress and aging: the glucocorticoid casade hypothesis. Endocrin Rev 7:284–301
101. Schiepers OJ, Wichers MC, Maes M (2005) Cytokines and major depression. Prog Neuropsychopharmacol Biol Psychiatry 29:201–217
102. Sheridan JF, Stark JL, Avitsur R, Padgett DA (2000) Social disruption, immunity, and susceptibility to viral infection. Role of glucocorticoid insensitivity and NGF. Ann N Y Acad Sci 917:894–905
103. Sheridan JF, Padgett DA, Avitsur R, Marucha PT (2004) Experimental models of stress and wound healing. World J Surg 28:327–330
104. Shurin MR, Kusnecov A, Hamill E, Kaplan S, Rabin BS (1994) Stress-induced alteration of polymorphonuclear leukocyte function in rats. Brain Behav Immun 8:163–169
105. Starkweather AR, Witek-Janusek L, Nockels RP, Peterson J, Mathews HL (2006) Immune function, pain, and psychological stress in patients undergoing spinal surgery. Spine 31:E641–E647
106. Stramer BM, Mori R, Martin P (2007) The Inflammation-fibrosis link? A Jekyll and Hyde role for blood cells during wound repair. J Invest Dermatol 127:1009–1017
107. Strecker-McGraw MK, Jones TR, Baer DG (2007) Soft tissue wounds and principles of healing. Emerg Med Clin N Am 25:1–22

108. Swift ME, Kleinman HK, DiPietro LA (1999) Impaired wound repair and delayed angiogenesis in aged mice. Lab Invest 79:1479–1487
109. Thomas DR (2001) Age-related changes in wound healing. Drugs Aging 18:607–620
110. van de Kerkhof PC, Van Bergen B, Spruijt K, Kuiper JP (1994) Age-related changes in wound healing. Clin Exp Dermatol 19:369–374
111. Werner S, Kreig T, Smola H (2007) Keratinocyte-fibroblast interactions in wound healing. J Invest Dermatol 127:998–1008
112. Whitney JD (1989) Physiologic effects of tissue oxygenation on wound healing. Heart Lung 18:466–474
113. Wilgus TA (2007) Regenerative healing in fetal skin: a review of the literature. Ostomy Wound Manage 53:16–31

Index